Lecture Notes in Applied and Computational Mechanics

Volume 27

Series Editors

Prof. Dr.-Ing. Friedrich Pfeiffer

Prof. Dr.-Ing. Peter Wriggers

Lecture Notes in Applied and Computational Mechanics

Edited by F. Pfeiffer and P. Wriggers

Vol. 27: Wriggers P., Nackenhorst U. (Eds.)
Analysis and Simulation of Contact Problems
410 p. 2006 [3-540-31760-0]

Vol. 26: Nowacki, J.P.
Static and Dynamic Coupled
Fields in Bodies with Piezoeffects
or Polarization Gradient
220 p. 2006 [3-540-31668-X]

Vol. 25: Chen C.-N.
Discrete Element Analysis Methods
of Generic Differential Quadratures
294 p. 2006 [3-540-28947-X]

Vol. 24: Schenk C., Schuëller G. (Eds.)
Uncertainty Assessment of Large Finite
Element Systems
400 p. 2005 [3-540-25343-2]

Vol. 23: Frémond M., Maceri F. (Eds.)
Mechanical Modelling and Computational Issues
in Civil Engineering
400 p. 2005 [3-540-25567-3]

Vol. 22: Chang C.H.
Mechanics of Elastic Structures
with Inclined Members:
Analysis of Vibration, Buckling
and Bending of X-Braced
Frames and Conical Shells
190 p. 2004 [3-540-24384-4]

Vol. 21: Hinkelmann R.
Efficient Numerical Methods
and Information-Processing
Techniques for Modeling Hydroand
Environmental Systems
305 p. 2005 [3-540-24146-9]

Vol. 20: Zohdi T.I., Wriggers P.
Introduction to Computational Micromechanics
196 p. 2005 [3-540-22820-9]

Vol. 19: McCallen R., Browand F., Ross J. (Eds.)
The Aerodynamics of Heavy Vehicles:
Trucks, Buses, and Trains
567 p. 2004 [3-540-22088-7]

Vol. 18: Leine, R.I., Nijmeijer, H.
Dynamics and Bifurcations
of Non-Smooth Mechanical Systems
236 p. 2004 [3-540-21987-0]

Vol. 17: Hurtado, J.E.
Structural Reliability: Statistical Learning Perspectives
257 p. 2004 [3-540-21963-3]

Vol. 16: Kienzler R., Altenbach H., Ott I. (Eds.)
Theories of Plates and Shells:
Critical Review and New Applications
238 p. 2004 [3-540-20997-2]

Vol. 15: Dyszlewicz, J.
Micropolar Theory of Elasticity
356 p. 2004 [3-540-41835-0]

Vol. 14: Frémond M., Maceri F. (Eds.)
Novel Approaches in Civil Engineering
400 p. 2003 [3-540-41836-9]

Vol. 13: Kolymbas D. (Eds.)
Advanced Mathematical and Computational
Geomechanics
315 p. 2003 [3-540-40547-X]

Vol. 12: Wendland W., Efendiev M. (Eds.)
Analysis and Simulation of Multifield Problems
381 p. 2003 [3-540-00696-6]

Vol. 11: Hutter K., Kirchner N. (Eds.)
Dynamic Response of Granular and Porous Materials
under Large and Catastrophic Deformations
426 p. 2003 [3-540-00849-7]

Vol. 10: Hutter K., Baaser H. (Eds.)
Deformation and Failure in Metallic Materials
409 p. 2003 [3-540-00848-9]

Vol. 9: Skrzypek J., Ganczarski A.W. (Eds.)
Anisotropic Behaviour of Damaged Materials
366 p. 2003 [3-540-00437-8]

Vol. 8: Kowalski, S.J.
Thermomechanics of Drying Processes
365 p. 2003 [3-540-00412-2]

Vol. 7: Shlyannikov, V.N.
Elastic-Plastic Mixed-Mode Fracture Criteria
and Parameters
246 p. 2002 [3-540-44316-9]

Vol. 6: Popp K., Schiehlen W. (Eds.)
System Dynamics and Long-Term Behaviour
of Railway Vehicles, Track and Subgrade
488 p. 2002 [3-540-43892-0]

Vol. 5: Duddeck, F.M.E.
Fourier BEM: Generalization
of Boundary Element Method by Fourier Transform
181 p. 2002 [3-540-43138-1]

Vol. 4: Yuan, H.
Numerical Assessments of Cracks
in Elastic-Plastic Materials
311 p. 2002 [3-540-43336-8]

Analysis and Simulation of Contact Problems

Peter Wriggers
Udo Nackenhorst
(Eds.)

Springer

Professor Dr. Peter Wriggers
Institute of Mechanics
and Computational Mechanics
University of Hannover
Appelstr. 9A
30167 Hannover
Germany

Professor Dr. Udo Nackenhorst
Institute of Mechanics
and Computational Mechanics
University of Hannover
Appelstr. 9A
30167 Hannover
Germany

With 328 Figures

ISSN print edition: 1613-7736
ISSN electronic edition: 1860-0816

ISBN 978-3-642-06861-4 ISBN 978-3-540-31761-6 (eBook)

Springer is a part of Springer Science+Business Media
springer.com

Softcover reprint of the hardcover 1st edition 2006

Cover design: *design & production* GmbH, Heidelberg

Karl Popp 1942-2005

During the preparation of the 4th CMIS one of the local organizers, Prof. Dr.-Ing. Prof. eh. Karl Popp passed away on April 24th 2005 after a serious illness. The scientific committee decided to dedicate this volume to him.

With the death of Karl Popp, the community of Mechanics has lost one of its most prominent researchers. Karl Popp, born in Regensburg in 1942, studied mechanical engineering initially at the Regensburg Polytechnic and later at the Technical University Munich, where he also obtained his doctorate in 1972 under Professor Magnus and received his habilitation in "Mechanics" in 1978. In 1981 he was first appointed Professor for the Mechanics of Systems in Hannover, and in 1985 he took over the vacant Professorship for Mechanics B.

With a large number of his own publications and publications completed under his supervision on highly different subjects concerning linear and nonlinear machine-, systems- and vehicle-dynamics, chaotic motion, friction phenomena and mechatronics he obtained an outstanding reputation well beyond the borders of Germany. His considerable experimental, and often constructional, contribution was always typical of his method of work.

In addition to his intensive and generous supervision and guidance of his co-workers he attached great importance to his work in national bodies such as GAMM and DEKOMECH, and also to maintaining close contacts with colleagues in the countries of Eastern Europe, as well as the Far East and several countries in America. In this connection he was awarded an Honorary Professorship of the University of Tongji (Shanghai, China) in 2003.

In the scientific community, all who met Prof. Popp were fascinated by his charming personality, his sense of humor, and, not in the least his scientific work. The research community will remember him as one of the outstanding researchers on linear and non-linear dynamics and contact problems.

We owe much to him and will always honor his memory.

Preface

Contact mechanics was and is an important branch in mechanics which covers a broad field of theoretical, numerical and experimental investigations. The study of contact problems in Mechanics is associated with the formulation and treatment of variational inequalities. This topic has led to developments of new and interesting areas in Mechanics, Applied Mathematics and the Engineering Sciences during the last three decades. New insight into contact problems has been gained, on the theoretical side, by the application of variational inequalities and linear complementarity problems in Mechanics. Furthermore the application range of contact to complex engineering was widened by new developments for computational schemes and algorithms.

With the goal to give leading researchers in the area of contact mechanics a platform for discussion and interchange, the 4th CMIS symposium was organized as one of the thematic ECCOMAS conferences in Loccum, Germany, following the symposia in Lausanne (1992), Carry Le Rouet (1994) and Praia da Consolacao (2001). Lectures were presented by scientists from Brazil, Croatia, France, Greece, Germany, Italy, Poland, Portugal, Russia, Slovenia, Spain, Sweden, Switzerland and the United States. The contributions collected in this volume summarize the lectures presented during the symposium. Many of the research papers were contributed by leading scientists in the area of contact mechanics. Hence the reader will obtain a state-of-the-art overview on formulation, mathematical analysis and numerical solution procedures of contact problems. In this respect the book should be of value to applied mathematicians and engineers who are concerned with Contact Mechanics.

The help of the staff of the Institute of Mechanics B and the Institute of Mechanics and Computational Mechanics at the University of Hannover is greatly appreciated. We thank especially Sven Reese for the thorough editing of this book which involved reformatting of some of the contributions.

Hannover,
December 2005

Peter Wriggers
Udo Nackenhorst

Contents

Part III Contact models, results and applications

FOURTH CONTACT MECHANICS INTERNATIONAL SYMPOSIUM

CMIS 2005 – Eccomas Thematic Conference
July 4-6, 2005
Loccum, University of Hannover, GERMANY

1 Organizing Committee

Chairmen

P. Wriggers, Dep. Civil Engg., Univ. of Hannover
K. Popp†, Dep. Mechnical Engg., Univ. of Hannover

Members

U. Nackenhorst, Dep. Civil Engg., Univ. of Hannover
E. Stephan, Dep. Appl. Math., Univ. of Hannover
M. Klüppel, DIK, Hannover
K. Wiechmann, Dep. Civil Engg., Univ. of Hannover
M. Kröger, Dep. Mech. Engg., Univ. of Hannover

2 Scientific Committee

P. Alart, Montpellier, France
J.R. Barber, Michigan, USA
S. Cescotto, Liege, Belgium
A. Curnier, Lausanne, Switzerland
G. Del Piero, Ferrara, Italy
M. Fremond, Champs sur Marne, France
M. Jean, Marseille, France
A. Klarbring, Linköping, Sweden
T. Laursen, Durham, USA
F. Maceri, Roma, Italy
J.A.C. Martins, Lisboa, Portugal
F. Pfeiffer, Munich, Germany
M. Raous, Marseille, France
K. Schweizerhof, Karlsruhe, Germany
W. Sextro, Graz, Austria
D. Stronge, Cambridge, U.K.
G. Zavarise, Torino, Italy

3 Sponsors and Acknowledgments

The Organizing Commitee of CMIS 2005 gratefully thanks the following institutions for their financial and material support:

- German Research Foundation (DFG)
- German Association for Computational Mechanics (GACM)
- International Center for Computational Engineering Sciences (ICCES)

Part I

Numerical methods

Mortar-based surface-to-surface contact algorithms in large deformation solid mechanics

T.A. Laursen

Computational Mechanics Laboratory, Department of Civil and Environmental Engineering, Duke University, Durham, North Carolina, USA,
laursen@duke.edu

Abstract. The mortar element method is extremely useful in a variety of settings in computational mechanics, often when the optimal connection or "tying" of dissimilarly meshed domains is desired. This paper describes the benefits of applying such methods to a more complex application: large deformation contact analysis. Although the enhanced accuracy of such contact formulations is to be expected given their sound theoretical grounding, it also turns out that the spatial smoothing provided by mortar contact operators lends considerably more robustness than more traditional, node-to-surface approaches. Issues associated with efficient searching in the surface-to-surface framework are discussed, and some examples are summarized which demonstrate the effectiveness of the approach.

1 Introduction

This paper reviews recent results pertaining to the effective implementation of mortar-based surface-to-surface contact algorithms in large deformation, large sliding finite element computations. The mortar element method (see, for example, [3, 1, 2, 8]) was originally conceived in the context of domain decomposition. The key idea behind the method is that continuity conditions across interdomain boundaries are imposed in integral form, rather than on a point to point basis. Optimal convergence rates (as expected from the underlying finite element method) are obtained, even in the context of non-conforming interfacial discretizations, provided suitable choices for the "mortar" spaces are made.

This paper focuses on deformable–deformable contact, where two bodies of non-negligible compliance mechanically interact, either with or without friction. Given that such contact analyses naturally give rise to dissimilar grids across interfaces, particularly where large relative sliding motions occur, the aforementioned properties of the mortar discretization technique are highly appealing for this more difficult class of problems. Furthermore, once we recognize that existing node-to-node strategies for contact interaction frequently give rise to violation of *inf-sup* conditions (see, for example, [4]), the stability properties afforded by these methods provide further motivation for their use. Using these ideas as motivation, a number of recent works (see [6, 7, 9]) have established the viability and remarkable degree of robustness associated with using mortar-based spatial discretization strategies for large deformation contact analysis. In this paper, we highlight key features of this approach, and then discuss some recent algorithmic developments related to implementation of these methods.

2 Mortar formulation of contact constraints

We consider in general a problem depicted in schematic form in Figure 1, where two deformable bodies $\Omega^{(1)}$ and $\Omega^{(2)}$ are expected to come into contact. For notational simplicity, one may assume that the surfaces $\Gamma_c^{(i)}$, $i = 1, 2$ are selected so that all potential points of contact are included over some time domain of interest. At any time t, the virtual work for the system can be expressed in terms of the deformation mapping φ and the admissible variations $\overset{*}{\varphi}$ as

$$
\begin{aligned}
G(\varphi, \overset{*}{\varphi}) &:= \sum_{i=1}^{2} G^{(i)}(\varphi^{(i)}, \overset{*}{\varphi}{}^{(i)}) \\
&= \sum_{i=1}^{2} \Bigg\{ \int_{\Omega^{(i)}} \left[\rho_0^{(i)} \overset{*}{\varphi}{}^{(i)} \cdot \boldsymbol{A}^{(i)} + Grad \overset{*}{\varphi} : \boldsymbol{P}^{(i)} \right] d\Omega \\
&\qquad - \int_{\Omega^{(i)}} \overset{*}{\varphi}{}^{(i)} \cdot \boldsymbol{F}^{(i)} d\Omega - \int_{\Gamma_\sigma^{(i)}} \overset{*}{\varphi}{}^{(i)} \cdot \bar{\boldsymbol{T}}^{(i)} d\Gamma \Bigg\} \\
&\qquad - \sum_{i=1}^{2} \int_{\Gamma_c^{(i)}} \overset{*}{\varphi}{}^{(i)} \cdot \boldsymbol{t}^{(i)} d\Gamma = 0 \\
&= G^{int,ext}(\varphi, \overset{*}{\varphi}) + G^c(\varphi, \overset{*}{\varphi}) = 0,
\end{aligned} \tag{1}
$$

where $G^{int,ext}(\varphi, \overset{*}{\varphi})$ is the sum of the virtual work arising from the internal and external forces, while the notation $G^c(\varphi, \overset{*}{\varphi})$ denotes the virtual work associated with the contact tractions. The notation $\boldsymbol{A}^{(i)}$ has been employed to denote the material acceleration field in body (i) (in the event that inertial effects are present), $\rho_0^{(i)}$ denotes the reference density, $\bar{\boldsymbol{T}}^{(i)}$ are the prescribed tractions, $\boldsymbol{P}^{(i)}$ is the first Piola-Kirchhoff stress tensor, and $\boldsymbol{F}^{(i)}$ denotes the body force. By transforming to the spatial configuration, the contact virtual work can be expressed as:

$$
G^c(\varphi, \overset{*}{\varphi}) := - \int_{\gamma_c^{(1)}} \boldsymbol{\lambda}^{(1)}(\boldsymbol{X}, t) \cdot \left(\overset{*}{\varphi}{}^{(1)}(\boldsymbol{X}, t) - \overset{*}{\varphi}{}^{(2)}(\bar{\boldsymbol{Y}}, t) \right) d\gamma, \tag{2}
$$

where $\varphi^{(2)}(\bar{\boldsymbol{Y}})$ is the current position of the contact point for $\boldsymbol{X}$.

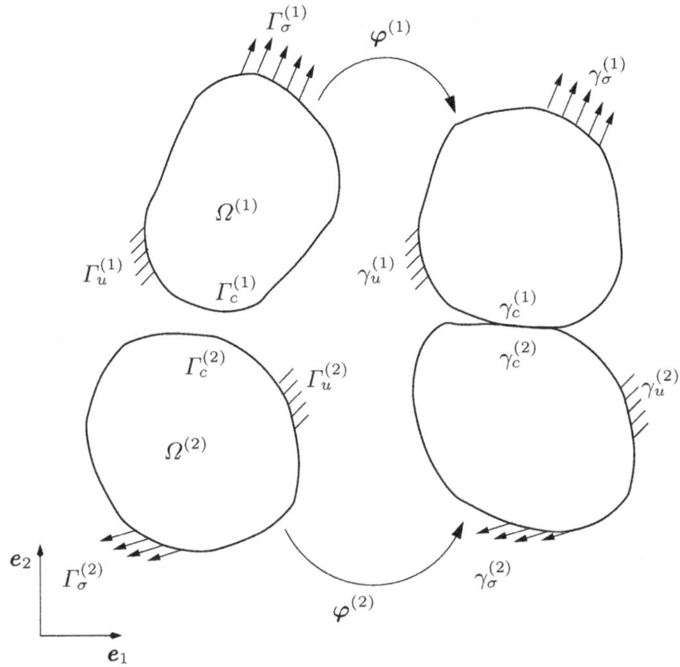

Fig. 1. Notation for the two body large deformation contact problem.

In the surface-to-surface, mortar-based approach to contact we explore here (see also [9, 7]), we discretize the contact virtual work by developing shape function expansions for the contact surface fields and substituting into (2). The discretized version of the contact virtual work then becomes

$$G^{cm}(\boldsymbol{\varphi}^h, \overset{*}{\boldsymbol{\varphi}}^h) = -\sum_A \sum_B \sum_C \boldsymbol{\lambda}_A \cdot \left[n_{AB}^{(1)} \overset{*}{\boldsymbol{\varphi}}{}_B^{(1)} - n_{AC}^{(2)} \overset{*}{\boldsymbol{\varphi}}{}_C^{(2)} \right] \tag{3}$$

where $n_{AB}^{(1)}$ and $n_{AC}^{(2)}$, refered as mortar integrals, are defined as:

$$n_{AB}^{(1)} = \int_{\gamma^{(1)h}} N_A^{(1)} \left(\boldsymbol{\xi}^{(1)}(\boldsymbol{X}) \right) N_B^{(1)} \left(\boldsymbol{\xi}^{(1)}(\boldsymbol{X}) \right) d\gamma \tag{4}$$

$$n_{AC}^{(2)} = \int_{\gamma^{(1)h}} N_A^{(1)} \left(\boldsymbol{\xi}^{(1)}(\boldsymbol{X}) \right) N_C^{(2)} \left(\boldsymbol{\xi}^{(2)} \left(\bar{\boldsymbol{Y}} (\boldsymbol{X}) \right) \right) d\gamma \tag{5}$$

where $N_A^{(1)} \left(\boldsymbol{\xi}^{(1)}(\boldsymbol{X}) \right)$ and $N_C^{(2)} \left(\boldsymbol{\xi}^{(2)} \left(\bar{\boldsymbol{Y}} (\boldsymbol{X}) \right) \right)$ are the shape functions defined on the discretized slave and master surfaces, respectively. A and B denote slave nodes (here assumed arbitrarily to be on body (1), and C denotes a master node (i.e., body (2)). Furthermore, implicit in our notation is the assumption that the mortar multipliers $\boldsymbol{\lambda}$ are interpolated over slave element facets on body (1), giving rise to nodal values $\boldsymbol{\lambda}_A$ to be found as part of the solution.

The normal and tangential portions of the contact operator are now exposed by splitting $\boldsymbol{\lambda}_A$ into normal and frictional parts:

$$\boldsymbol{\lambda}_A = \boldsymbol{\lambda}_{A_N} + \boldsymbol{\lambda}_{A_T}. \tag{6}$$

The normal contact constraints defining the normal traction may be represented as

$$\boldsymbol{\lambda}_{A_N} = -\lambda_{A_N} \boldsymbol{n}_A \text{ (no sum)} \tag{7}$$

where λ_{A_N} represents the contact pressure at node A. It is also noted that in our implementation, the normal $\boldsymbol{n}_A$ is a nodal normal on the slave surface, taken to be an average of neighboring slave facet normals. The contact pressure is subject to Kuhn-Tucker conditions via

$$\begin{aligned} \lambda_{A_N} &\geq 0 \\ g_A &\leq 0 \\ \lambda_{A_N} g_A &= 0 \end{aligned} \tag{8}$$

where the mortar projected gap g_A at slave node A is defined as

$$\begin{aligned} g_A &= \boldsymbol{n}_A \cdot \boldsymbol{g}_A, \\ \boldsymbol{g}_A &:= \sum_B \sum_C \left[n_{AB}^{(1)} \boldsymbol{\varphi}_B^{(1)} - n_{AC}^{(2)} \boldsymbol{\varphi}_C^{(2)} \right]. \end{aligned} \tag{9}$$

Although other choices are possible (Lagrange multiplier, augmented Lagrangian), we consider penalty regularization of these constraints here for simplicity.

In the case of the frictional constraints, a penalty regularized form of the frictional conditions is considered via

$$\begin{aligned} \mathcal{L}_v \boldsymbol{\lambda}_T &= \epsilon_T \left[\boldsymbol{v}_T - \dot{\gamma} \frac{\boldsymbol{\lambda}_T}{\left\| \boldsymbol{\lambda}_T \right\|} \right] \\ \Phi := \left\| \boldsymbol{\lambda}_T \right\| - \mu \left\| \boldsymbol{\lambda}_N \right\| &\leq 0 \\ \dot{\gamma} &\geq 0 \\ \Phi \dot{\gamma} &= 0 \end{aligned} \tag{10}$$

where ϵ_T is the frictional penalty parameter. $\mathcal{L}_v \boldsymbol{\lambda}_T$ is the *Lie derivative* of the frictional traction, and may be defined in the current context via

$$\mathcal{L}_v \boldsymbol{\lambda}_T = \dot{\lambda}_{T_\alpha} \boldsymbol{\tau}^\alpha, \tag{11}$$

where $\boldsymbol{\tau}^\alpha$ is covariant tangential base vector. The repeated indices α imply summation with respect to the repeated index α.

As detailed in [9, 7], a trial state-return map strategy may be readily employed to determine the Coulomb frictional traction at each node A, while utilizing a mortar projected version of equations (10). We first compute a trial state, assuming no slip during the increment:

$$\lambda^{trial}_{A_{T_{n+1}\alpha}} = \lambda_{A_{T_{n\alpha}}} - \epsilon_T \boldsymbol{\tau}_{A_\alpha} \cdot \left[\sum_C \left(n^{(2)}_{AC_{n+1}} - n^{(2)}_{AC_n} \right) \boldsymbol{\varphi}^{(2)}_C - \sum_B \left(n^{(1)}_{AB_{n+1}} - n^{(1)}_{AB_n} \right) \boldsymbol{\varphi}^{(1)}_B \right] \tag{12}$$

from which a trial value for the slip function can be computed via

$$\Phi^{trial}_{n+1} = \left\| \boldsymbol{\lambda}^{trial}_T \right\| - \mu \left\| \boldsymbol{\lambda}_N \right\|. \tag{13}$$

Return map strategies for the frictional stress updates are then conceived in the usual manner. Although not necessarily apparent from first glance, the expression in (12) is frame indifferent and represents a consistent time discrete approximation of $(10)_1$ (assuming zero incremental slip). In particular, it is to be noted that the proper notion of incremental tangential motion is written in terms of increments of $n^{(1)}_{AB}$ and $n^{(2)}_{AC}$, as indicated in (12). Effectively, one may show that use of this incremental motion measure corresponds to a frame indifferent version of a mortar-projected relative velocity measure; further details are given in [9].

3 Mortar contact searching and surface intersection detection

Algorithmically, the key to the algorithms we discuss in this paper is the computation of the mortar integrals, $n^{(1)}_{AB}$ and $n^{(2)}_{AC}$, defined in (4) and (5). In general, these are computed as a summation of contributions over *contact segments*, which in two dimensions are line or curve segments, and which in three dimensions amount to polygonal regions of intersection between opposing element facets. The key to these segments is that each involves only one element boundary from each surface, such that the mortar integrals $n^{(1)}_{AB}$ and $n^{(2)}_{AC}$ can be evaluated with sufficient quadrature to ensure stability (in a Babuska-Brezzi sense). Some simple schematics of these contact segments are given in Figure 2. Most of the complexity in implementing these algorithms in finite element analysis emanate from the necessity to linearize these mortar integral expressions; for much more detail on the operations involved, the interested reader should consult [9] or [7].

3.1 Bounding Volume Hierarchies

In general, dynamic detection of the aforementioned contact segments would be an $O(nm)$ operation, where n is the number of nodes on the non-mortar

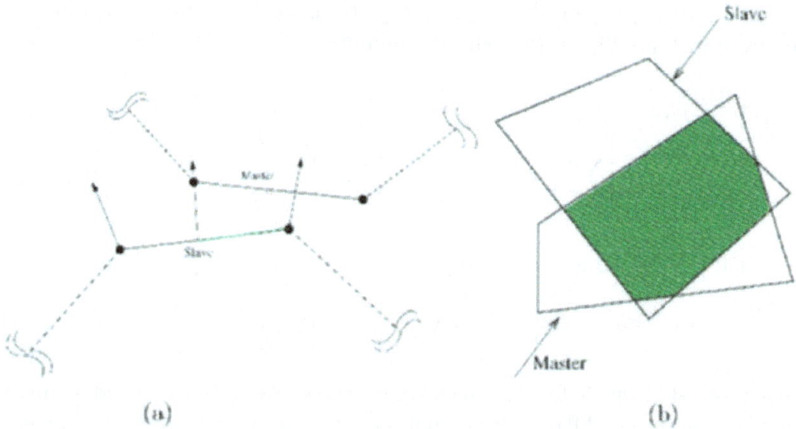

Fig. 2. Mortar segments: (a) in two dimensions; and (b) in three dimensions.

(slave) contact surface, and where m is the number of nodes on the opposing master surface. However, considerable savings can be achieved by use of a bounding volume based hierarchical data structure, as shown schematically in Figure 3. Use of this algorithm facilitates an $O(n)$ algorithm, in terms of binary tree updating, contact detection, and storage.

3.2 Surface to surface self contact algorithms

with a relatively minor adaptation to the above hierarchical data structure, self contact situations are readily and efficiently implemented. Since the algorithm is facet-based, and amounts to efficient location of facet-facet intersections, the extension of the algorithm to not only self contact, but also to automatic contact detection, is rather readily achieved. However, in the case of self contact, care must be taken to sort the facet definitions so that contiguous assemblies of "master" and "slave" contact facets result (so that multipliers, which are interpolated over slave contact segments, are defined over connected patches of contact facet assemblies). A schematic of the sort of situation which must be maintained when a surface folds on itself is shown in Figure 4.

4 Examples and conclusion

Figure 5 depicts two examples of the effectiveness of the mortar-based contact enforcement strategy, in the context of self contact. In the two dimensional example, the rings impact both with each other, and at various stages also experience self-contact as they collapse under dynamic impact loading. In the

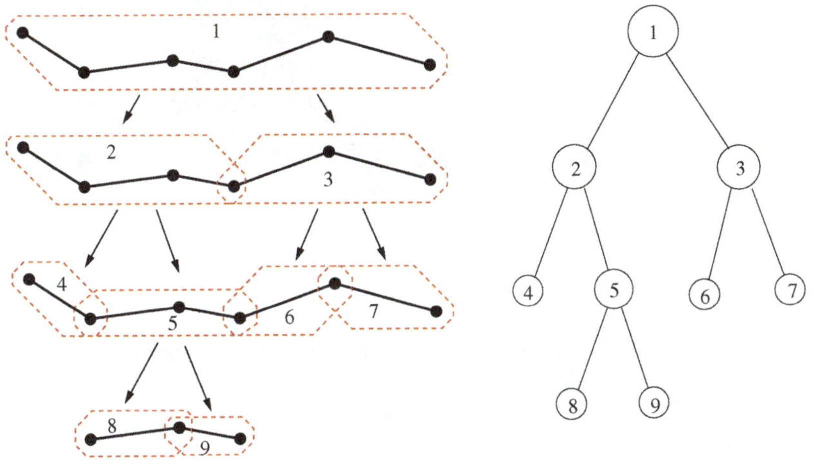

Fig. 3. Schematic of a bounding volume tree for a two dimensional surface.

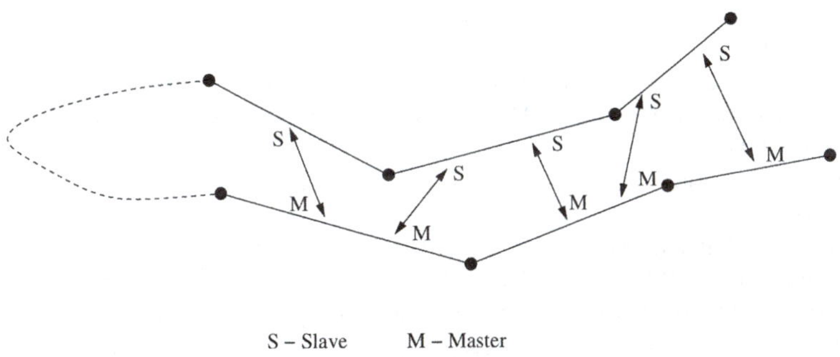

Fig. 4. Desirable array of contact element pairs after facet sorting in a self contact application.

three dimensional example, an idealized tire is simulated as it rolls with zero internal pressurization, giving rise to self contact and very large contact pressures (and frictional tractions) inside the tire. In both cases, the simulations execute much more slowly (or not at all) with traditional contact algorithms, as the faceted geometries typical of node to surface treatments cause a great deal of difficulty for the Newton-Raphson based iterative solvers. The robustness achieved by the the nonlocal character of the surface-to-surface contact operators is a somewhat unanticipated feature of these algorithms, but one which makes their application in such problems highly attractive.

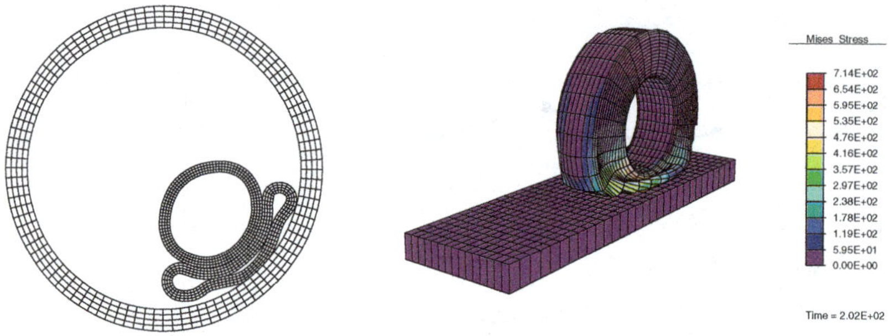

(a) (b)

Fig. 5. Deformed configurations for some representative self contact, large deformation frictional formulations: a) two dimensional ring impact problem; and b) idealized zero inflation pressure tire rolling problem.

References

1. F. Belgacem and Y. Maday. A spectral element methodology tuned to parallel implementations. *Computer Methods in Applied Mechanics and Engineering*, 116:59 67, 1994.
2. Z. Belhachmi and C. Bernardi. Resolution of fourth order problems by the mortar element method. *Computer Methods in Applied Mechanics and Engineering*, 116:53 58, 1994.
3. C. Bernardi, Y. Maday, and A. Patera. A new nonconforming approach to domain decomposition: The mortar element method. In H. Brezia and J. Lions, editors, *Nonlinear Partial Differential Equations and Their Applications*, pages 13 51. Pitman and Wiley, 1992.
4. N. El-Abbasi and K.-J. Bathe. Stability and patch test performance of contact discretizations and a new solution algorithm. *Computers and Structures*, 79:1473 1486, 2001.
5. T. Laursen. *Computational Contact and Impact Mechanics.* Springer, Berlin, 2002.
6. M. Puso and T. Laursen. A mortar segment-to-segment contact method for large deformation solid mechanics. *Computer Methods in Applied Mechanics and Engineering*, 193:601 629, 2004.
7. M. Puso and T. Laursen. A mortar segment-to-segment frictional contact method for large deformations. *Computer Methods in Applied Mechanics and Engineering*, 193:4891 4913, 2004.
8. B. Wohlmuth. *Discretization Methods and Iterative Solvers Based on Domain Decomposition.* Springer-Verlag, Heidelberg, 2001.
9. B. Yang, T. Laursen, and X. Meng. Two dimensional mortar contact methods for large deformation frictional sliding. *International Journal for Numerical Methods in Engineering*, 2004.

From inexact active set strategies to nonlinear multigrid methods

R. Krause

Institute for Numerical Simulation
krause@ins.uni-bonn.de

1 Introduction

Due to their efficiency and robustness, linear multigrid methods lend themselves to be a starting point for the development of nonlinear iterative strategies for the solution of nonlinear contact problems, see, e.g., [3, 1, 10, 5]. One nonlinear strategy is to reduce the contact problem to a sequence of linear problems and to solve each of these by a linear multigrid method. This approach is often connected to active set strategies or semismooth Newton methods [6]. To avoid solving the linear problems exactly, one can use inexact active set strategies, see [7, 8]. The convergence of this inexact strategy depends on the accuracy the inner problem is solved with, see [7], as well as on algorithmic parameters [8]. A second strategy is to deal directly with the nonlinearity within the multigrid method by using, e.g., nonlinear smoothers and nonlinear interpolation operators, see [10, 9, 1]. Using the convex energy for controlling the iteration process, globally convergent nonlinear multigrid methods can be constructed which allow for solving contact problems with the speed of a linear multigrid method [10]. A third possibility is to employ a saddle point approach [3] and to solve for the primal and dual variables simultaneously using an algebraic multigrid method.

For the construction of an efficient multigrid based strategy for contact problems it is of crucial importance to use a suitable multiscale representation of the constraints arising from non penetration condition. This is even the case, if active set strategies are used and only linear subproblems have to be solved, since the active constraints in general will lead to Dirichlet data which cannot be resolved on the coarser grids. This boundary data might influence the efficiency of the linear multigrid method if it is not properly dealt with: On the one hand, if the coarse grid spaces do not reflect the active constraints on the finest grid properly, the nonlinear scheme might not converge. On the other hand, the coarse grid spaces have to fulfill an approximation property in order to guarantee the efficiency of the multigrid method and therefore cannot be chosen too small. For the strongly related case of complicated boundaries not being resolved by the coarse grid, we refer the reader to [11, 4] and the references cited therein.

Here, we discuss different methods of adapting the multilevel basis to the actual zone of contact. Our discussion is carried out in the context of two multigrid based nonlinear iterative strategies, i.e., primal-dual active set strategies [6] and monotone multigrid methods [12].

2 Contact problem and discretization

We identify the body under consideration in its reference configuration with the polygonal domain $\Omega \subset \mathbb{R}^d, d = 2,3$. The boundary $\partial\Omega$ is decomposed into three disjoint parts, Γ_D, Γ_N, and Γ_C, the possible contact boundary. The actual zone of contact is assumed to be contained in Γ_C and we assume $\text{meas}_{d-1}(\Gamma_D) > 0$. Tensor and vector quantities are denoted by bold symbols, e.g., $\boldsymbol{v}$, and the components by v_i, $1 \leq i,j \leq d$. We define the usual Sobolev space of displacements with weak derivative in $\boldsymbol{L}^2$ by $\boldsymbol{H}^1(\Omega) := (H^1(\Omega))^d$ and set $\boldsymbol{H}_D := \{v \,|\, v \in \boldsymbol{H}^1(\Omega), v_{|\Gamma_D} = \boldsymbol{0}\}$. We consider linear elastic material where the stresses $\boldsymbol{\sigma} = (\sigma_{ij})_{i,j=1}^d$ are given by Hooke's law $\sigma_{ij}(\boldsymbol{u}) := E_{ijml}\, u_{l,m}$ with the symmetric, positive definite and sufficiently smooth tensor $\mathbf{E} = (E_{ijml})_{i,j,l,m=1}^d$. On $\partial\Omega$ the normal and tangential displacements are defined by $u_n = \boldsymbol{u} \cdot \boldsymbol{n}$ and $\boldsymbol{u}_T = \boldsymbol{u} - u_n \cdot \boldsymbol{n}$, $\boldsymbol{n}$ the outer normal vector. Similarly, $\sigma_n = n_i \sigma_{ij} n_j$ and $(\boldsymbol{\sigma}_T)_i = \sigma_{ij} n_j - \sigma_n \cdot \boldsymbol{n}$ are the normal and tangential stresses, respectively. The distance between the body and the rigid foundation, taken in normal direction with respect to the reference configuration is given by the continuous function $g : \mathbb{R}^d \supset \Gamma_C \to \mathbb{R}$ and for the data we assume $\boldsymbol{f} \in \boldsymbol{L}^2(\Omega)$ and $\boldsymbol{p} \in (\boldsymbol{H}^{1/2}(\Gamma_N))'$.

The boundary value problem constituting the elastic contact problem without friction consists of the equation of linear elasticity (1)-(3), and the contact conditions (4)-(6) on Γ_C.

$$-\sigma_{ij}(\boldsymbol{u})_{,j} = f_i \quad \text{in } \Omega \qquad (1)$$

$$\boldsymbol{u} = 0 \quad \text{on } \Gamma_D \qquad (2)$$

$$\sigma_{ij}(\boldsymbol{u}) \cdot n_j = p_i \quad \text{on } \Gamma_N \qquad (3)$$

$$\sigma_n \leq 0 \qquad (4)$$

$$\boldsymbol{u} \cdot \boldsymbol{n} \leq g \qquad (5)$$

$$(\boldsymbol{u} \cdot \boldsymbol{n} - g)\, \sigma_n = 0 \qquad (6)$$

$$\boldsymbol{\sigma}_T = 0 \qquad (7)$$

We define the closed convex set $\mathcal{K}$ of admissible displacements by

$$\mathcal{K} := \{\boldsymbol{v} \in \boldsymbol{H}_D \mid v_n \leq g \text{ on } \Gamma_C\}, \qquad (8)$$

the bilinear form $a(\boldsymbol{u}, \boldsymbol{v}) := \int_\Omega \sigma_{ij}(\boldsymbol{u}) v_{i,j}\, dx$ and set $f(\boldsymbol{v}) = \int_\Omega f_i v_i\, dx + \int_{\Gamma_N} p_i v_i\, ds$. Now, the weak solution $\boldsymbol{u}$ to the boundary value problem (1)-(7) can be characterized as the solution of the constrained minimization problem

$$J(\boldsymbol{u}) = \min_{\boldsymbol{v} \in \mathcal{K}}, \qquad (9)$$

where $J(\boldsymbol{u}) := \frac{1}{2}a(\boldsymbol{u},\boldsymbol{u}) - f(\boldsymbol{u})$ is the unconstrained quadratic energy. Since the functional J is convex and lower semi continuous, existence and uniqueness of a solution can be obtained by standard methods from convex analysis.

We approximate the solution $\boldsymbol{u}$ of (9) using finite elements. Let $(\mathcal{T}_\ell)_{\ell=1}^L$ denote a family of nested and shape regular meshes with meshsize parameter h_ℓ. We assume Γ_D to be resolved by $\mathcal{T}_1$ and use Lagrangian conforming finite elements $\boldsymbol{S}_\ell \subset \boldsymbol{H}_D$ of first order. The set of nodes of $\mathcal{T}_\ell$ is denoted by $\mathcal{N}_\ell$ and the nodal basis functions of S_ℓ are $\{\lambda_p^{(\ell)}\}_{p\in\mathcal{N}_\ell}$. The nodes on the possible contact boundary are $\mathcal{C}_\ell = \Gamma_C \cap \mathcal{N}_\ell$. As discretization of the convex set $\mathcal{K}$ we take, $\boldsymbol{n}(p)$ the outer normal at $p \in \mathcal{C}_\ell$,

$$\mathcal{K}_L := \{\boldsymbol{u} \in \boldsymbol{S}_L \mid \boldsymbol{u}(p)\cdot\boldsymbol{n}(p) \le g(p)\,,\, p \in \mathcal{C}_L\}\,.$$

We emphasize that in general we have $\mathcal{K}_{\ell-1} \not\subset \mathcal{K}_\ell$. Using this notations, the finite element solution $\boldsymbol{u}_\ell$ on Level ℓ of our multigrid hierarchy can be characterized as the solution of the discrete constrained minimization problem

$$J(\boldsymbol{u}_\ell) = \min_{\boldsymbol{v}\in\mathcal{K}_\ell}\,. \tag{10}$$

3 Multigrid based nonlinear strategies

In this section, we discuss different multigrid based nonlinear strategies for the solution of constrained minimization problems of the form (9). As first solution method, we consider the primal dual active set strategy given by Algorithm 1, see, e.g., [6] and as second method the fully nonlinear monotone multigrid method given by Algorithm 2, see [10].

In order to locally decouple the constraints $\boldsymbol{u}(p)\cdot\boldsymbol{n}(p) \le sg(p)$, we use different local coordinate systems. With each node $p \in \mathcal{C}_\ell$, $1 \le \ell \le L$, we associate an orthogonal transformation $\mathcal{O}_p \in \mathbb{R}^{d,d}$ with $\boldsymbol{n}(p) = \mathcal{O}_p\boldsymbol{e}_1$, $\{\boldsymbol{e}_i\}_{1\le i\le d}$ the standard basis of $\mathbb{R}^3$. We now represent $\boldsymbol{u}(p)$ w.r.t. the local coordinate system $\{\boldsymbol{E}_i(p)\}_{1\le i\le d}$, where $\boldsymbol{E}_i(p) = \mathcal{O}_p\boldsymbol{e}_i$. For all remaining nodes $p \in \mathcal{R}_\ell := \mathcal{N}_\ell \setminus \mathcal{C}_\ell$, we set $\boldsymbol{E}_i = \boldsymbol{e}_i$. By applying the following local basis transformation to the stiffness matrix $\boldsymbol{A}_\ell$ on Level ℓ

$$\hat{\boldsymbol{A}}_\ell = \begin{pmatrix} \mathbf{id}_{\mathcal{R}_\ell,\mathcal{R}_\ell} & 0 \\ \mathbf{0} & \mathcal{O}_{\mathcal{C}_\ell,\mathcal{C}_\ell} \end{pmatrix} \begin{pmatrix} \boldsymbol{A}_{\mathcal{R}_\ell,\mathcal{R}_\ell} & \boldsymbol{A}_{\mathcal{R}_\ell,\mathcal{C}_\ell} \\ \boldsymbol{A}_{\mathcal{C}_\ell,\mathcal{R}_\ell} & \boldsymbol{A}_{\mathcal{C}_\ell,\mathcal{C}_\ell} \end{pmatrix} \begin{pmatrix} \mathbf{id}_{\mathcal{R}_\ell,\mathcal{R}_\ell} & 0 \\ \mathbf{0} & \mathcal{O}^T_{\mathcal{C}_\ell,\mathcal{C}_\ell} \end{pmatrix}, \tag{11}$$

we can locally decouple the non-penetration constraints. Here, we have set $\mathcal{O}_{\mathcal{C}_\ell,\mathcal{C}_\ell} = \text{blockdiag}_{p\in\mathcal{C}_\ell}(\mathcal{O}_p)$ and **id** denotes the identity. Setting $\mathcal{O}_\ell = \text{blockdiag}(\mathbf{id}_{\mathcal{R}_\ell,\mathcal{R}_\ell}, \mathcal{O}_{\mathcal{C}_\ell,\mathcal{C}_\ell})$, we get with $\hat{\boldsymbol{u}} = \mathcal{O}_\ell\boldsymbol{u}_\ell$ and $\hat{\boldsymbol{f}} = \mathcal{O}_\ell\boldsymbol{f}$ the linear system $\hat{\boldsymbol{A}}_\ell\hat{\boldsymbol{u}}_\ell = \hat{\boldsymbol{f}}_\ell$. Here, $\boldsymbol{f}_\ell$ is the right hand side in the standard coordinate system. We note that the interpolation operator has to be transformed accordingly. More precisely, let $\boldsymbol{I}_{\ell-1}^\ell$ denote the algebraic representation of the natural injection $I_{\ell-1}^\ell\colon \boldsymbol{S}_{\ell-1} \longrightarrow \boldsymbol{S}_\ell$. Then, we set $\hat{\boldsymbol{I}}_{\ell-1}^\ell = \mathcal{O}_\ell\boldsymbol{I}_{\ell-1}^\ell\mathcal{O}_{\ell-1}^T$. Let us now give the two nonlinear strategies to be considered:

Algorithm 1 (Multigrid Based Primal Dual Active Set Strategy)

(1) Set $k = 0$ and choose $c > 0$. Initialize $\mathcal{A}_0 \subset \mathcal{C}_L$, $\mathcal{I}_0 = \mathcal{C}_L \setminus \mathcal{A}_0$ and u_L^0.

(2) Modify stiffness matrices $\boldsymbol{A}_\ell$ and interpolation operators $\boldsymbol{I}_{\ell-1}^{\ell}$ according to $\mathcal{A}^k$, $1 \le \ell \le L$

(3) Find $\boldsymbol{u}_L^k \in \mathcal{K}_L$ by applying N multigrid steps to the linear system:

$$\begin{aligned} \boldsymbol{A}_L \boldsymbol{u}_L^k + \boldsymbol{\eta}^k &= \boldsymbol{f}_L\,, \\ (\boldsymbol{u}_L^k(p))_n = g(p)\,, p \in \mathcal{A}_k; \quad \eta_n^k(p) &= 0\,, p \in \mathcal{I}_k; \quad \boldsymbol{\eta}_T^k = 0 \end{aligned} \tag{12}$$

(4) Set new active and inactive sets to

$$\mathcal{A}_{k+1} = \{h_p^{-1} s_n(p) + c(u_n^k(p) - g(p)) > 0)\} \tag{13}$$

$$\mathcal{I}_{k+1} = \{h_p^{-1} s_n(p) + c(u_n^k(p) - g(p)) \le 0)\} \tag{14}$$

(5) If $|\boldsymbol{u}_L^k - \boldsymbol{u}_L^{k-1}|/|\boldsymbol{u}_L^k| <$ TOL stop, else set $k \mapsto k+1$ and go to (2).

Here, we have set $h_p := \int_{\Gamma_C} \lambda_p \, ds$. Primal dual active set strategies are known to converge superlinearly, if the initial iterate $\boldsymbol{u}_L^0$ is sufficiently close to the solution. If, in addition the matrix $\boldsymbol{A}_L$ is a M-matrix and the linear systems (12) are solved exactly, also global convergence can be shown [13]. For efficiency reasons often inexact strategies are used, see [8, 6], which solve the problems (12) only up to some tolerance $\hat{\text{TOL}}$. For the inexact strategy, the choice of $\hat{\text{TOL}}$ and the choice of the parameter $c > 0$ influence the overall iteration processes, see [8] and the last Section..

The second algorithm to be considered is the monotone multigrid method for contact problems, see [10, 12]. Following [10], we seek the minimizer of J over $\mathcal{K}_L$ by successive minimization. To this end, we associate with each node $p \in \mathcal{N}_\ell$ a local d-dimensional subspace $\boldsymbol{V}_p$. On the finest level, we use Gauß-Seidel relaxation and set $\boldsymbol{V}_p = \text{span}\{\lambda_p^{(L)} \cdot \boldsymbol{e}_i\}_{1 \le i \le d}$. The search directions on the coarser levels are $\boldsymbol{V}_p = \text{span}\{\lambda_p^{(\ell)} \cdot \boldsymbol{E}_i\}_{1 \le i \le d}$ for nodes $p \in \mathcal{N}_\ell \setminus \mathcal{C}_\ell$. On the possible contact boundary we use so called *truncated* basis functions $\{\boldsymbol{\mu}_p^{(\ell)}\}$ and set $\boldsymbol{V}_p = \text{span}\{(\boldsymbol{\mu}_p^{(\ell)})_i \cdot \boldsymbol{e}_i\}_{1 \le i \le d}$, see [10]. We assume that all nodes on all levels are ordered "from fine to coarse".

Algorithm 2 (Nonlinear Monotone Multigrid Method)

(1) Initialize $\boldsymbol{u}_L^0$. Set $\boldsymbol{w}_0 = \boldsymbol{u}_L^0$

(2) Fine grid smoothing: Solve successively the local minimization problems

for $m = 1, \ldots, n_L = |\mathcal{N}_L|$ {

Find $\boldsymbol{v}_m \in \boldsymbol{V}_{p_m}$, such that $\boldsymbol{v}_m + \boldsymbol{w}_{m-1} \in \mathcal{K}_L$ and

$J(\boldsymbol{v}_m + \boldsymbol{w}_{m-1}) \le J(\boldsymbol{v} + \boldsymbol{w}_{m-1})\,, \quad \boldsymbol{v} \in \boldsymbol{V}_{p_m}$.

Update $\boldsymbol{w}_m = \boldsymbol{w}_{m-1} + \boldsymbol{v}_m$. } Set $\bar{\boldsymbol{u}}_L = \boldsymbol{w}_{n_L}$

(3) Compute the coarse grid search directions $\boldsymbol{\mu}_p^{(\ell)}$, $\ell < L$, w.r.t. $\bar{\boldsymbol{u}}_L$

(4) Coarse grid correction: Set $\boldsymbol{c}_0 = \bar{\boldsymbol{u}}_L$

for $m = 1, \ldots, M := n_1 + \cdots + n_{L-1}$ {

Find $\boldsymbol{v}_m \in \boldsymbol{V}_{p_m}$, such that $\boldsymbol{v}_m \in \mathcal{D}_{p_m}$ and

$J(\boldsymbol{v}_{p_m} + \boldsymbol{c}_{m-1}) \le J(\boldsymbol{v} + \boldsymbol{c}_{m-1})\,, \quad \boldsymbol{v} \in \boldsymbol{V}_{p_m}$.

Update $\boldsymbol{c}_m = \boldsymbol{c}_{m-1} + \boldsymbol{v}_m$. }

(5) Set $\boldsymbol{u}_L = \bar{\boldsymbol{u}}_L + \boldsymbol{c}_M$

Here, the convex sets $\mathcal{D}_{p_m}$ are constructed by *monotone restrictions*, see [10] for details, and are such that $\bar{\boldsymbol{u}}_L + c_m \in \mathcal{K}_L$ for all $m = n_L + 1, \dots, M$. As a consequence, the energy J can be shown to decrease over $\mathcal{K}_L$ in each iteration step of the multigrid method and global convergence can be shown.

We now present different multilevel splittings for the use in Algorithm 1, Step (2) and (3) or Algorithm 2, Step (3) and (4). We emphasize that the choice of the multilevel splitting influences convergence properties of the nonlinear methods Algorithm. Since in general the active set $\mathcal{A}^k$ cannot be represented on the coarser grids, i.e., $\mathcal{A}^k \not\subset \mathcal{C}_\ell$ for $\ell < L$, we modify the coarse grid spaces $\boldsymbol{S}_\ell$. We therefore introduce the index sets $\mathcal{A}_\ell^k \subset \mathcal{C}_\ell^k := \{(p,i) \,|\, p \in \mathcal{C}_\ell, 1 \le i \le d\}$ and $\mathcal{I}_\ell^k = \mathcal{C}_\ell^k \setminus \mathcal{A}_\ell^k$ for each Level $\ell \le L$. On Level L, we set $\mathcal{A}_L^k := \{(p,1) \,|\, p \in \mathcal{A}_k\}$. Moreover, for $q \in \mathcal{N}_\ell$ and $p \in \mathcal{N}_{\ell-1}$, $(\hat{\boldsymbol{I}}_{\ell-1}^{\ell})_{qp}$ denotes the $d \times d$-blockmatrix-entry of $\hat{\boldsymbol{I}}_{\ell-1}^{\ell}$. Let us note that for the case $\boldsymbol{n}(p) \ne \boldsymbol{n}(q)$ this block entry in general is not diagonal. We consider the multilevel splittings of $\boldsymbol{S}_L$ induced by the following index sets:

1. Set $\mathcal{A}_\ell^k = \{(p,1) \,|\, p \in \mathcal{A}_k \cap \mathcal{C}_\ell\}$.
2. Define recursively for $\ell < L$: $\mathcal{A}_{\ell-1}^k = \{(p,1) \,|\, \exists (q,1) \in \mathcal{A}_\ell^k : (\hat{\boldsymbol{I}}_{\ell-1}^{\ell})_{qp}^{11} \ne 0\}$.
3. Define recursively for $\ell < L$: $\mathcal{A}_{\ell-1}^k = \{(p,j) \,|\, \exists (q,i) \in \mathcal{A}_\ell^k : (\hat{\boldsymbol{I}}_{\ell-1}^{\ell})_{qp}^{ij} \ne 0\}$.
4. *Truncated* basis functions from [10].

For Splittings 1–3, the induced subspaces are $\boldsymbol{S}_\ell = \operatorname{span}\{\lambda_p^{(\ell)} \cdot \boldsymbol{E}_i(p) \,|\, p \in \mathcal{N}_\ell, 1 \le i \le d, (p,i) \notin \mathcal{A}_\ell^k\}$. Let us note, that also the interpolation operators have to be adapted to the chosen splitting.

Remark 1. In case of the outer normal being constant along the contact boundary, Splitting 2 and 3 coincide. Splittings (1)–(3) can easily be implemented by modifying the transfer operator and by introducing additional Dirichlet nodes. In contrast, the truncated basis functions (4) require reassembling of the stiffness matrices on the coarser levels at the contact boundary, if the discrete zone of contact changes, see [12]. For scalar problems, corresponding multilevel splittings have been used and analyzed in [4, 14]. In particular, for the scalar truncated basis functions and a hierarchical decomposition see [14]. Let us also note that by using algebraic multigrid methods no special care has to be taken in order to represent the active set on the coarser grids. However, this requires a new setup step for the creation of the coarse grid matrices each time the discrete contact zone changes.

4 Numerical results

In this section, we compare the different nonlinear strategies arising from the active set strategy Algorithm 1 and the nonlinear multigrid method Algorithm 2. The multilevel splittings for the respective methods are those given above.

Example 1 As our first example, we consider a Hertzian contact problem in 3*d*. A sphere of radius 1 with midpoint $(0.5, 0.5, 0.5)^T$ is pressed onto the rigid foundation $\{(x, y, z) \,|\, z \leq -0.5\}$ by prescribing the Dirichlet Boundary conditions $\boldsymbol{u}(x, y, z) = (0, 0, -0.5)^T$ on $\Gamma_D = \{(x, y, z) \in \partial\Omega \,|\, z > 1.1\}$. We have chosen this Dirichlet values in order to stress the nonlinearity at the contact boundary. We set $\Gamma_C = \{(x, y, z) \in \partial\Omega \,|\, z < 0.5\}$ and prescribe homogeneous Neumann conditions on the remaining boundary part. The material parameters are $E = 2 \cdot 10^5$ and $\nu = 0.3$. The initial mesh $\mathcal{T}_1$ consist of 60 pyramids and 35 nodes. We apply three uniform refinement steps and then solve the linear complementarity problem arising after discretization on Level $L = 4$ with Algorithm 1 with parameter $c = 20 \cdot 10^5$ and with Algorithm 2. The initial iterate $\boldsymbol{u}_L^0$ is given by random values in the interval $[-0.2, -0.1]$.In our first experiment, we take $\boldsymbol{n}(p)$ to be the outer normal to $\partial\Omega$. In the left Figure 1, for the Splittings 1-4 when applied for (12) the solution of the linear system within Algorithm 1 with $N = 1$ the reduction of the relative algebraic error $|\boldsymbol{u}_4^k - \boldsymbol{u}_4|_2 / |\boldsymbol{u}_4|_2$ in the Euclidian norm is given. For solving the linear problems (12) within Algorithm 1 and for the monotone multigrid Algorithm 2 method we use a $W(3, 3)$-cycle.

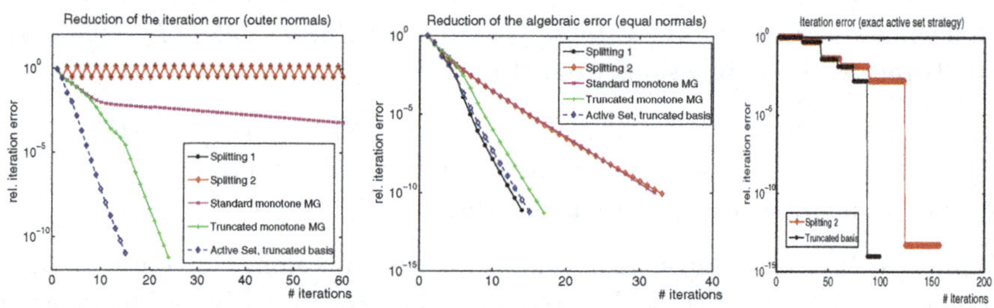

Fig. 1. Reduction of the algebraic error for the splittings 1–4 on Level $L = 4$

Truncated and standard Multigrid (MG) are the nonlinear multigrid method (Algorithm 2) with Splittings 4 and 3, respectively for the coarse grid search directions $\{\boldsymbol{\mu}_p^{(\ell)}\}$, $\ell < L$. For all methods, we employ $|\boldsymbol{u}_L^k - \boldsymbol{u}_L|_2 / |\boldsymbol{u}_L|_2 < 10^{-12}$ as stopping criterion. As can be seen, the inexact active set strategy does not converge for the Splitting 1 and 2. The standard monotone multigrid method shows convergence, which can be significantly improved by using the truncated basis functions. The inexact active set strategy converges, if the truncated basis functions are used, showing that the convergence of the nonlinear method depends significantly on the multilevel splitting used. The initial slowdown of the monotone multigrid method is due to the fact that during the iteration process all iterates are forced to stay in $\mathcal{K}_L$, which ensures convergence.

In our second experiment, see the middle of Figure 1, we take as normal direction at the contact boundary $\boldsymbol{n}(p) = (0, 0, -1)^T$ for all $p \in \mathcal{C}_\ell$, $1 \leq \ell \leq L$. For this choice, we get constraints on Γ_C only in z-Direction. Convergence of

Level	Splitting 1	Splitting 2	Splitting 3	Truncated basis	Truncated Monotone MG
1	6/4	6/3	4/3	4/3	2/2
2	8/15	10/30	9 /31	8/7	7/6
3	9/-	11/-	11/50	9/8	7/6
4	9/-	12/-	12/ 100	9/9	6/7
5	10/-	12/-	12/>100	10/9	6/7
6	10/-	12/-	12/>100	10/9	6/7

Table 1. Comparison of the inexact active set strategy Algo. 1 with $N = 5$ and the truncated monotone multigrid method Algo. 2 for the Hertzian contact problem.

the inexact active set strategy is now also observed using splitting 1. Again, the truncated basis performs best for either method. Finally, for this example in the right of figure the error reduction for the exact active set strategy can be seen. We note, that at least 100 instead of 18 multigrid cycles are necessary when compared to the monotone multigrid method.

We now investigate the mesh-dependence of our solution strategies. We apply 5 steps of adaptive refinement leading to a grid with 1.164.973 elements and set $N = 5$ in (12). In Table 1, the number of iteration steps for the case of equal normals (left number in each column) and outer normals (right number) are shown for Algorithm 1 with Splittings 1-4. For the truncated monotone MG also the W-cycles are given in units of five for comparison. A dash means no convergence. As can be seen, the behavior of the active set strategy depends highly on the representation of the constraints on the coarser grids. In case of varying normals, the multilevel splitting has to be chosen carefully in order to ensure convergence and efficiency of the method.

Example 2 Our second example is the mandible shown in the right of Figure 2 after two steps of adaptive refinement. The mandible is pressed against a rigid block above the teeth by Dirichlet boundary conditions of $u(x, y, z) = (0, 0, 0.14)$ prescribed at the condyles. In the left of Figure 2, the reduction of the algebraic error for Algorithm 1 with $N = 1$ and splitting 2 and the truncated basis 4 is shown for Levels $\ell = 1, 2, 3$. For the truncated basis, on Levels 2 and 3 we get the same convergence rate.

Concluding Remarks On the basis of multigrid methods, efficient nonlinear solution methods for contact problems can be constructed. However, the used multilevel splitting has to be chosen carefully in order to account for the maybe complicated zone of contact. For guaranteeing convergence it is necessary to control the nonlinear iteration process on all levels of the multigrid hierarchy. This is achieved within the monotone multigrid method by requiring monotonicity, thus giving rise to the most robust scheme. In all of our experiments the multilevel splitting best suited for the overall iteration process for both methods was that induced by the truncated basis functions.

All implementations have been done in the framework of the finite element toolbox [15].

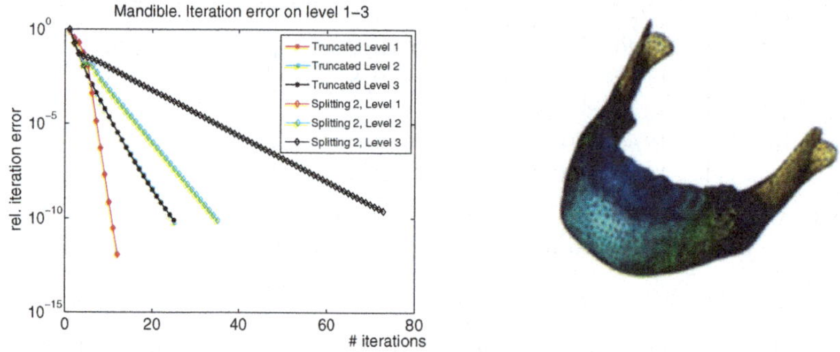

Fig. 2. Example 2: Reduction of the algebraic error and geometry (greyscales show u_2)

References

1. A. Brandt and C.W. Cryer, Multigrid algorithms for the solution of linear complementarity problems arising from free boundary problems *SIAM J. Sci. Stat. Comput.*, 4: 655-684, 1983.
2. P. Averyand G. Rebel and M. Lesoinne and C. Farhat, A numerically scalable dual-primal substructuring method for the solution of contact problems. I: The frictionless case, *Comput. Methods Appl. Mech. Eng.*, 193:2403-2426, 2004.
3. M. Adams, Algebraic methods for constrained linear systems with applications to contact problems in solid mechanics, *Num. Lin. Alg. w Appl.*, 1-6, 2000
4. R. Kornhuber and H. Yserentant, Multilevel Methods for Elliptic Problems on Domains not Resolved by the Coarse Grid, *Contemp. Math.*, 180:49-60, 1994
5. J. Schöberl, Solving the Signorini problem on the basis of domain decomposition techniques, *Computing*, 60:323-344, 1998
6. K. Kunisch and G. Stadler, Generalized Newton methods for the 2D-Signorini contact problem with friction in function space, *M2AN*, 4:827-854, 2005
7. W. Hackbusch and H.D. Mittelmann, On multi–grid methods for variational inequalities, *Numer. Math.*, 42:65-76, 1983
8. S. Hüeber and M. Mair and B. Wohlmuth, A priori error estimates and an inexact primal-dual active set strategy for linear and quadratic finite elements applied to multibody contact problems , *Appl. Num. Math.*, 54:555-570, 2005
9. Ana H. Iontcheva and Panayot S. Vassilevsky, Monotone multigrid methods based on element agglomeration coarsening away from the contact boundary for the Signorini's problem, *Num. Lin. Alg. w Appl.*, 11:189-204, 2004
10. R. Kornhuber and R. H. Krause, Adaptive multigrid methods for Signorini's problem in linear elasticity, *CVS*, 4:9-20, 2001, Springer
11. Tony F. Chan and Jinchao Xu and Ludmil Zikatanov, An Agglomeration Multigrid Method for Unstructured Grids, *Contemp. Math.*, 218:67-81, 1998
12. R. H. Krause, Monotone Multigrid Methods for Signorini's Problem with Friction,Freie Universität Berlin, 2001
13. M. Hintermüller and K. Ito and K. Kunisch, The Primal-Dual Active Set Strategy as a Semismooth Newton Method, *SIAM J. Optim.*, 13:865-888, 2003

14. R.H.W. Hoppe and R. Kornhuber, Adaptive multilevel–methods for obstacle problems, *SIAM J. Numer. Anal.*, 31:301-323, 1994
15. P. Bastian and K. Birken and K. Johannsen and S. Lang and N. Neuß and H. Rentz–Reichert and C. Wieners, UG – a flexible Software toolbox for solving partial differential equations, *CVS*, 1:27-40, 1997, Springer

On a geometrical approach in contact mechanics

A. Konyukhov and K. Schweizerhof

Institut fuer Mechanik,
Alexander.Konyukhov@ifm.uni-karlsruhe.de,
Karl.Schweizerhof@ifm.uni-karlsruhe.de

Abstract. The focus of the contribution is on a detailed discussion of the 2D formulation in contact which can be viewed either as a reduction of the general 3D case to a special cylindrical geometry, or as the contact of 2D bodies bounded by plane curves. In addition, typical frictional characteristics, such as the yield surface and the update of the sliding displacements allow a geometrical interpretation in the chosen coordinate system on the contact surface.

1 Geometry and kinematics of contact

In the literature various contact descriptions adapted for an effective finite element implementation are available, which can be characterized by the following: from 2D to 3D formulations, from non-frictional to frictional contact. One of the first contributions on finite element solutions of 2D frictional contact problems based on an elasto-plastic analogy has been made by Wriggers et. al. [1]. General overviews over contact conditions and contact algorithms which are nowadays used in practice, are covered by the books of Wriggers [2] and Laursen [3]. The covariant description, see Konyukhov and Schweizerhof [4] for the frictionless case and [5] for the frictional contact, has been found as a universal penalty based approach for contact with various approximation of the surfaces. In the current contribution, we will show the unity of 2D and 3D formulations, where the 2D case can be derived, from one hand, as a simplified case of the particular 3D geometry of contact surfaces and, from the other hand, can be constructed separately based on the differential geometry of 2D plane curves.

Considering a special contact case – contact between two cylindrical infinite bodies with plane strain deformations, see Fig. 1, leads to the definition of a 2D contact. In this case a generatrix **GH** of the first cylindrical body is a contact line and corresponds to a contact line **G'H'** which is also a generatrix but of the second cylindrical body. Thus, 3D contact which can be seen as an interaction between two surfaces is reduced to an interaction between two

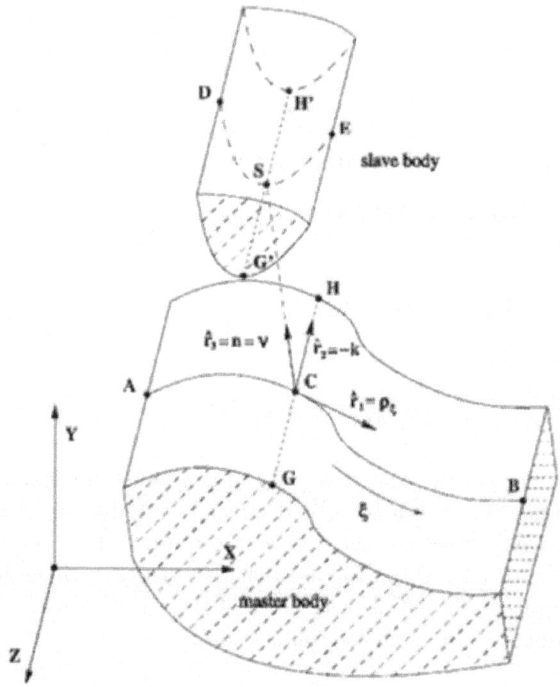

Fig. 1. Two dimensional contact as a special case of three dimensional contact

boundary curves in the 2D case. One of both boundary curves resp. surfaces is chosen as the master contact curve resp. surface. A coordinate system is considered on the boundary, either for a surface in 3D or for a curve in 2D.

On the plane we define a curvilinear coordinate system associated with the curve by introducing two principal vectors as a basis: the tangent vector $\boldsymbol{\rho}_\xi = \dfrac{\partial \boldsymbol{\rho}}{\partial \xi}$ and the unit normal vector $\boldsymbol{\nu}$. Then a slave contact point $\mathbf{S}$ is found as

$$\mathbf{r}_s(\xi, \zeta) = \boldsymbol{\rho}(\xi) + \zeta \boldsymbol{\nu}(\xi). \tag{1}$$

The normal unit vector $\boldsymbol{\nu}$ in the case of arbitrary Lagrangian parameterizations with ξ can be defined via a cross product in a Cartesian coordinate system as:

$$\boldsymbol{\rho}_\xi = \frac{\partial \boldsymbol{\rho}}{\partial \xi}; \quad \Longrightarrow \quad \boldsymbol{\nu} = \frac{[\mathbf{k} \times \boldsymbol{\rho}_\xi]}{\sqrt{\boldsymbol{\rho}_\xi \cdot \boldsymbol{\rho}_\xi}}, \tag{2}$$

where $\mathbf{k}$ is the third unit vector in this Cartesian coordinate system. According to this definition the traction vector $\mathbf{R}_s$ is defined via the contravariant basis vectors:

$$\mathbf{R}_s = T\boldsymbol{\rho}^1 + N\boldsymbol{\rho}^2 = T\frac{\boldsymbol{\rho}_\xi}{(\boldsymbol{\rho}_\xi \cdot \boldsymbol{\rho}_\xi)} + N\boldsymbol{\nu}. \tag{3}$$

The decomposition of the traction in eqn. (3) leads to the following contact integral

$$\delta W_c = \int_l (N\delta\zeta + T\delta\xi) dl. \tag{4}$$

A closer look reveals that the contact integral (4) contains the work of the contact tractions T and N defined on the master contact curve and is computed along the slave curve $l \equiv l_s$.

2 Regularization of the contact tractions

For the normal traction N, the following regularized equation in the closed form is taken

$$N = \epsilon_N \zeta, \quad if \quad \zeta \leq 0, \tag{5}$$

where ϵ_N is a penalty parameter for the normal interaction.

As a reasonable equation for the regularization of the tangent traction T we choose a proportional relation between the full time derivative $\frac{d\mathbf{T}}{dt}$ and the relative velocity vector expressed in covariant form on the tangent line $\zeta = 0$:

$$\frac{D_1 T}{dt} \boldsymbol{\rho}^1 = -\epsilon_T \dot{\xi} \boldsymbol{\rho}_\xi. \tag{6}$$

Expressing the covariant derivative in eqn. (6) the following equation is obtained

$$\frac{dT}{dt} = -\epsilon_T (\boldsymbol{\rho}_\xi \cdot \boldsymbol{\rho}_\xi) \dot{\xi} + \frac{\boldsymbol{\rho}_{\xi\xi} \cdot \boldsymbol{\rho}_\xi}{(\boldsymbol{\rho}_\xi \cdot \boldsymbol{\rho}_\xi)} \dot{\xi} - \frac{h_{11}}{a_{11}} \dot{\zeta}, \tag{7}$$

where ϵ_T is a penalty parameter for the tangential interaction, a_{11} and h_{11} are components of the metric tensor resp. of the curvature tensor for the cylindrical surface. If a path length $s = \xi$ is used for parameterization, then eqn. (7) is transformed as

$$\frac{dT}{dt} = -\epsilon_T \dot{s} - \kappa \dot{\zeta}, \tag{8}$$

where κ is a curvature of the master contact plane curve.

3 Linearization for a Newton type solution scheme

The derivation of the contact matrices can be performed either according to the cylindrical geometry, or according to the geometry of the plane curves. The latter provides a straightforward geometrical explanation of each part of the contact matrix and can be used for the judgment of their necessity within the solution process. Thus, the linearization of **the normal part** in eqn. (4) given by the following integral:

$$\delta W_c^N = \int_l N\delta\zeta dl, \tag{9}$$

leads to the following.

$$D(\delta W_c^N) = \int_l \left(\frac{dN}{dt}\delta\zeta + N\frac{d\delta\zeta}{dt} \right) dl$$

$$= \int_l \epsilon_N (\delta \mathbf{r}_s - \delta \boldsymbol{\rho}) \cdot (\boldsymbol{\nu} \otimes \boldsymbol{\nu})(\mathbf{v}_s - \mathbf{v}) dl \tag{10}$$

$$- \int_l \epsilon_N \zeta \left(\delta\boldsymbol{\tau} \cdot (\boldsymbol{\nu} \otimes \boldsymbol{\tau})(\mathbf{v}_s - \mathbf{v}) + (\delta \mathbf{r}_s - \delta \boldsymbol{\rho}) \cdot (\boldsymbol{\tau} \otimes \boldsymbol{\nu}) \frac{\partial \boldsymbol{\tau}}{\partial t} \right) dl \tag{10 a}$$

$$- \int_l \epsilon_N \zeta \kappa (\delta \mathbf{r}_s - \delta \boldsymbol{\rho}) \cdot (\boldsymbol{\tau} \otimes \boldsymbol{\tau})(\mathbf{v}_s - \mathbf{v}) dl. \tag{10 b}$$

The form written via the path length allows a simple geometrical interpretation of each part in eqn. (10) and even allows to establish situations where some of the parts are zero. The first part eqn. (10) is called main part and defines the constitutive relation for normal contact conditions. The second part eqn. (10 a) is called rotational part and defines the geometrical stiffness due to the rotation of the tangent vector of the master curve. It disappears when a master segment is moving in parallel, because only in this case the derivative of a unit vector $\boldsymbol{\tau}$ becomes zero. The third part eqn. (10 b) is called curvature part. This part disappears when the curvature κ of a master segment is zero, i.e. in the case of linear approximations of the master segment.

The structure of **the tangential part for the sticking case** is as follows:

$$D_v(\delta W_c^T) = \int_l \left(\frac{dT}{dt}\delta\xi + T\frac{d\delta\xi}{dt} \right) dl$$

$$= - \int_l \varepsilon_T (\delta \mathbf{r}_s - \delta \boldsymbol{\rho}) \cdot (\boldsymbol{\tau} \otimes \boldsymbol{\tau})(\mathbf{v}_s - \mathbf{v}) dl \tag{11}$$

$$- \int_l T \left[(\delta \mathbf{r}_s - \delta \boldsymbol{\rho}) \cdot (\boldsymbol{\tau} \otimes \boldsymbol{\tau}) \frac{\partial \boldsymbol{\tau}}{\partial l} + \delta\boldsymbol{\tau} \cdot (\boldsymbol{\tau} \otimes \boldsymbol{\tau}) \, (\mathbf{v}_s - \mathbf{v}) \right] dl \tag{11 a}$$

$$+ \int_l \kappa (\delta \mathbf{r}_s - \delta \boldsymbol{\rho}) \cdot \, (\boldsymbol{\tau} \otimes \boldsymbol{\nu} + \boldsymbol{\nu} \otimes \boldsymbol{\tau}) \, (\mathbf{v}_s - \mathbf{v}) dl. \tag{11 b}$$

Here T is a trial tangent traction computed from the discrete evolution equation (6) at each load step, e.g. under the assumption that the slave point obeys the elastic deformation law.

One can see that the symmetry is preserved for the full sticking case for any curvilinear geometry of the contact bodies.

If sliding is detected, i.e. if $\|\mathbf{T}\| > \mu|N|$, then the sliding force is computed according to Coulomb's friction law within the return-mapping scheme, see [2] and [3]. We also keep a covariant form:

$$T^{sl} = \mu|N| \, \frac{T_{tr}}{\|\mathbf{T}_{tr}\|} = \mu|N| \, (\boldsymbol{\rho}_\xi \cdot \boldsymbol{\rho}_\xi)^{1/2} sgn(T_{tr}). \tag{12}$$

The linearized contact integral **for the tangential part in the case of sliding** has the same geometrical structure, but contains non-symmetric parts due to the non-associativity of the Coulomb friction law.

Details for the implementation of the linear contact element as well as for the computation according to the return-mapping scheme can be found in Konyukhov and Schweizerhof [7].

4 Treatment of special cases

Some special cases can appear when a direct application of the return-mapping scheme can lead to improper results. The first problem is arising when the applied load is not simply modified proportionally. In this situation a trial load can not be computed only via the evolution equation, because the attraction point ξ^0 must be updated. Thus we have to extend the algorithm as is shown in the following. The second problem is arising when the projection point is crossing an element boundary during the incremental loading. In this case, the computation according to the incremental evolution equation will produce a jump, because the convective coordinate ξ belongs to different elements.

4.1 Update of the sliding displacements in the case of reversible loading

A geometrical interpretation of the trial step in the return-mapping scheme leads to the definition of the elastic region with an attraction point ξ^0

$$|T_{tr}^{(n)}| < \mu|N^{(n)}|\sqrt{a_{11}} \quad \Longrightarrow \quad \epsilon_T|\xi^{(n)} - \xi^0| < \mu|N^{(n)}| \tag{13}$$

If a point $\xi^{(n+1)}$ appears to be outside of the domain at load step $(n+1)$, then its only admissible position is on the boundary of the domain. A sliding force is applied then at the contact point. As long as we have a motion of the contact point only in one direction the sign function for the sliding force $sgn(T_{tr}^{(n+1)}) = sgn(\Delta\xi^{(n+1)})$ does not change and the computation will be correct. However, when a reversible load is applied and it forces the contact point to move forward or backward, the attraction point ξ^0 must be updated in order to define the sign function for the sliding force correctly. This update can be defined geometrically from Fig. 2:

$$|\Delta\xi^{(n+1)}| = |\Delta\xi^{(n+1)}| - \frac{\mu|N^{(n)}|}{\epsilon_T}. \tag{14}$$

Thus, computation of the trial force at the next load step $(n+1)$ has to be made in accordance to the update procedure, see more in Konyukhov and Schweizerhof [7].

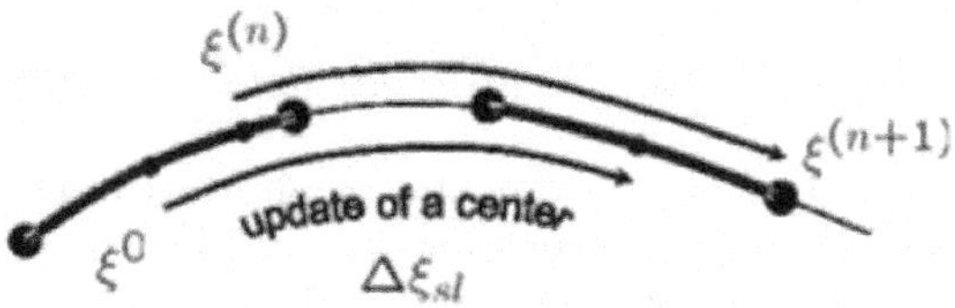

Fig. 2. Coulomb friction. Updating of sliding displacements in convective coordinates. Motion of friction cone and center of attraction

4.2 Crossing an element boundary – continuous integration scheme

Consider two adjacent elements **AB** and **BC**, see Fig. 3. If we follow the computation of the trial force according to the formula expressed in convective coordinates as

$$T_{tr}^{(n+1)} = -\epsilon_T a_{11}(\xi^{(n+1)} - \xi^{(n)}), \tag{15}$$

then this leads to a jump in the contact tractions. The maximum of the jump is computed as

$$T_{jump} = -\epsilon_T a_{11}\left(\lim_{\xi \to -1+0} (\xi)_{\xi \in \mathbf{BC}} - \lim_{\xi \to +1-0} (\xi)_{\xi \in \mathbf{AB}} \right) = 2\epsilon_T a_{11}. \tag{16}$$

This jump appears only due to the different approximation of the adjacent elements. In order to overcome this, we can compute the force in geometrical form. Following this procedure the incremental coordinate $\Delta\xi$ can be expressed as

$$\Delta\xi = \frac{\left(\boldsymbol{\rho}(\xi^{(n+1)})_{\xi \in \mathbf{BC}} - \left(\boldsymbol{\rho}(\xi^{(n)}) + \Delta\mathbf{u}(\xi^{(n)})\right)_{\xi \in \mathbf{AB}}\right) \cdot \boldsymbol{\rho}_\xi^{(n+1)}}{a_{11}^{(n+1)}}, \tag{17}$$

where $\Delta\mathbf{u}$ is an incremental displacement vector. The evolution equation becomes then

$$T_{tr}^{(n+1)} = T^{(n)} - \epsilon_T a_{11}^{(n+1)} \Delta\xi. \tag{18}$$

In the 2D case, the computation can be made via the length parameter s leading to the continuous scheme as well, see Wriggers [2].

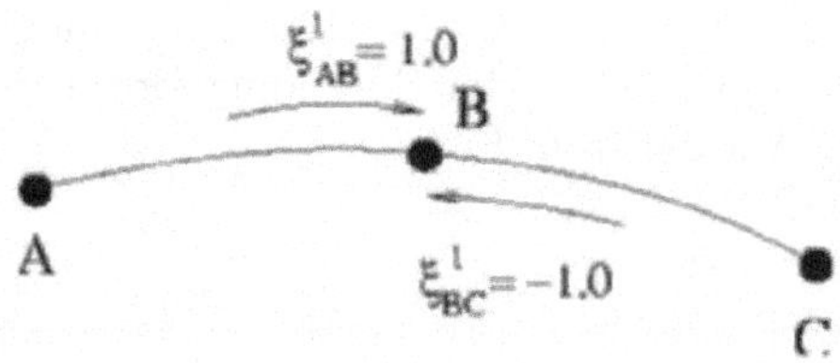

Fig. 3. Crossing an element boundary within a load increment. Typical case for the continuous integration scheme

5 Numerical examples

5.1 Sliding of a block. Linear approximation of the contact surfaces. Reversible loading process

We consider the sliding of an elastic block on the rigid base loaded with horizontally prescribed reversible displacements, see Fig. 4(a). The geometrical and mechanical parameters are elasticity modulus $E = 2.1 \cdot 10^4$, Poisson ratio $\nu = 0.3$, length $a = 20$, height $b = 5$, Coulomb friction coefficient $\mu = 0.3$. The penalty parameters are chosen as $\varepsilon_N = \varepsilon_T = 2.1 \cdot 10^6$. The main point is to show the update procedure for sliding displacements. The loading is applied as prescribed displacements at the top side of the deformable block. For contact the "node-to-segment" approach is taken. The hysteresis curve representing the computed horizontal displacement at point **D** vs. the applied displacement at point **C** is given in Fig. 4(b).

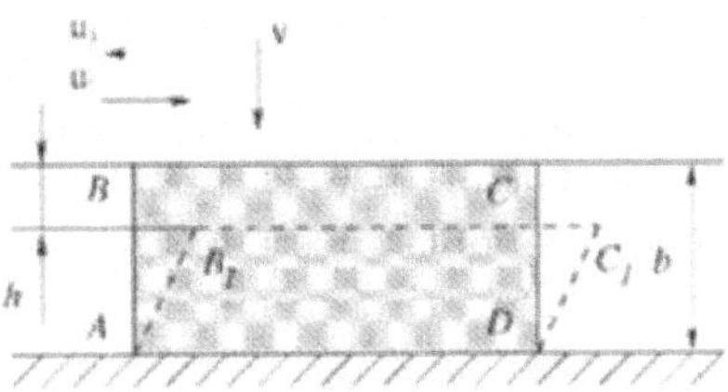

(a) Plane deformation of a block. Applied displacement loading at top of the block.

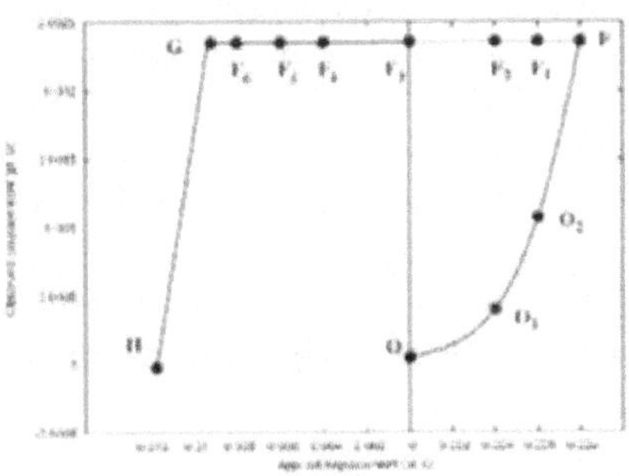

(b) Hysteresis curve. Observed horizontal displacement at point D vs. applied horizontal displacement at point C.

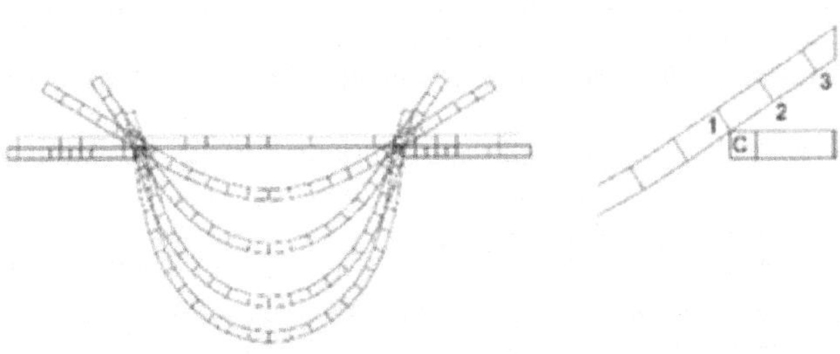

Fig. 4. Drawing of an elastic strip into a channel with sharp corners

5.2 Drawing of an elastic strip into a channel with sharp corners

A particular example in which application of the continuous integration is absolutely necessary is a deep drawing of an elastic strip into a channel with sharp corners, see Fig. 4. The crucial point during the analysis is the sliding of a sharp corner **C** over the element boundaries **1**, **2**, **3**. A load-displacement curve computed for the loading point is chosen as the representative parameter to compare various contact approaches. We obtain, see also [7], that the application of the "segment-to-segment" approach, as described in [6], without the continuous integration scheme allows to compute the force-displacement curve only for non-frictional cases. For frictional cases a more careful transport of the history variables is necessary as it is suggested here.

6 Conclusions

In this contribution a convective description is reconsidered for the 2D quasistatical frictional contact problem. Special attention is paid to the derivation of the necessary equations either as a reduction of the known 3D covariant formulation, or directly from the special 2D cylindrical geometry of the contact surfaces. The algorithmic linearization in the covariant form allows to obtain the tangent matrices before the linearization process. Thus, an implementation can be easily carried without providing any special attention to the approximation of the contact surfaces. Further it is shown that special algorithmic techniques have to be taken to provide robust answerers for load reversion and for the crossing of element boundaries.

References

1. Wriggers P., Vu Van, Stein E. Finite element formulation of large deformation impact-contact problem with friction. *Computers and Structures*, **37**, pp. 319–331, 1990.
2. Wriggers P. Computational Contact Mechanics. John Wiley & Sons, 2002.
3. Laursen T. A. Computational contact and impact mechanics. Fundamentals of modeling interfacial phenomena in nonlinear finite element analysis. Springer, 2002.
4. Konyukhov A., Schweizerhof K. Contact formulation via a velocity description allowing efficiency improvements in frictionless contact analysis. *Computational Mechanics*, **33**, pp. 165–173, 2004.
5. Konyukhov A., Schweizerhof K. Covariant description for frictional contact problems. *Computational mechanics*, **35**, 3, pp. 190–213, 2005.
6. Harnau M, Konyukhov A, Schweizerhof K. Algorithmic aspects in large deformation contact analysis using "Solid-Shell" elements. *Computers and Structures*, **83**, 1804–1823, 2005.
7. Konyukhov A., Schweizerhof K. A special focus on 2D formulations for contact problems using a covariant description. *submitted to International Journal for Numerical Methods in Engineering.*

On the discretization of contact problems in elastodynamics

H.B. Khenous[1], P. Laborde[2], and Y. Renard[1]

[1] INSA, Département de Génie Mathématique, 135 avenue de Rangueil, 31077 Toulouse cedex, France,
khenous@insa-toulouse.fr, renard@insa-toulouse.fr
[2] Laboratoire MIP, 118 route de Narbonne, 31062 Toulouse cedex, France,
laborde@mip.ups-tlse.fr

Abstract. In this work, we will presente a comparison of two formulation for the discretization of elastodynamic contact problems. The first approach consists on a midpoint scheme and a contact condition expressed in terms of velocity. This approach gives an energy conserving scheme. The second one we propose is a new distribution of the solid mass. The problem expressed with the new mass matrix is well posed, energy conserving and has a lipschitz solution. Finally, some numerical results are presented.

1 Introduction

The contact problem attracts considerable attention from the computational mechanics community, due in large part to its highly non-linear and discontinous nature. Indeed, engineering analysts charged with solving such problems will attest merely achieving convergence of non-linear solution schemes can be difficult under many circumstances. This difficulties stem primary from the fact that elastodynamic contact problems are not well posed [8].

For purely contact elastodynamic problems (hyperbolic problems), as far as we now, existence result has been proved in a scalar two dimensional case by Lebeau-Schatzman [4], Kim [3] and in the vector case with a modified contact law by Renard-Paumier [9]. It seems that no energy conserving result has been proved. In this work, we will present two numerical energy conserving strategies for elastodynamic contact problems.

The plan of the paper is as follows. Section 2 we propose a semi-discretization using a finite element method, give the equivalent contact condition and the time integration scheme used and discuss the energy conserving of the algorithm. Section 3 we consider an equivalent mass matrix so that the equivalent finite element problem is energy conserving. Some numerical tests are presented in Section 4 to demonstrate performance of the two approaches.

2 Method with the equivalent contact condition ECC

In elastodynamics, the classical contact condition is not sufficient to deal with the problem correctly. Of course, in the enrergy analysis of any time integration scheme for contact problems, we remark that a sort of contact condition in terms of velocity appears. The stake is how to define this contact condition which can replace the classical one. Then, many authors try to give a new formulation for the contact condition: Laursen-Love [6], Moreau [7] and obtain an interesting results. In this section, we deal with such condition.

We assume a vanishing initial gap between the structure and the rigid foundation and we denote u^0, u^1 and T the given initial displacement, initial velocity and time simulation respectively.

The space semi-discretization of the elastodynamic contact problem with nodal contact condition is defined for a Lagrange finite element method as follows. Find $u : [0,T] \longrightarrow \mathbb{R}^d$ such that

$$\begin{cases} M\ddot{u} + Ku = f + \sum_{i \in I_C} \lambda_N^i N_i, \quad \text{in } [0,T] \times \mathbb{R}^d, \\ \lambda_N^i \leq 0, \ u.N_i \leq 0, \ \lambda_N^i (u.N_i) = 0, \forall\, i \in I_C, \\ u(0) = u^0, \dot{u}(0) = u^1, \end{cases} \tag{1}$$

where d is the number of degrees of freedom for the displacement u, the notations M, K, f stend for the mass matrix, the stiffness matrix and the given force densities respectively, I_C is the set of the contact boundary indices. On each node $i \in I_C$, we denote λ_N^i and N_i the normal stress and the outward unit normal respectively.

Problem (1) is not well posed (see [7]). However, uniqueness can be recovered, for rigid bodies, by introducing an impact law with a restitution coefficient. This seems not to be completly satisfying for deformable bodies because whatever the restitution coefficient value, the system tends to a global restitution of energy when the mesh parameter goes to zero (more details will be presented in [1]).

We replace the classical Signorini condition in Problem (1) by the following condition

$$\begin{cases} u.N_i < 0 \Longrightarrow \lambda_N^i = 0, \\ u.N_i \geq 0 \Longrightarrow \dot{u}.N_i \leq 0, \ \lambda_N^i \leq 0, \ \lambda_N^i (\dot{u}.N_i) = 0. \end{cases} \tag{2}$$

The expression (2) in terms of velocity is very close to the one introduced in [7] and corresponds to the one introduced in [5]. The velocity is to be understood as a right derivative and the second condition implies in fact the non-interpenetration.

We discretize the elastodynamic part in Problem (1) by a midpoint scheme as follows:

$$\begin{cases} u^0 \text{ et } v^0 \text{ donnés}, \\ u^1 = u^0 + \Delta t\ v^0 + \Delta t\ z(\Delta t\ ) / \lim\limits_{\Delta t\ \longrightarrow 0} z(\Delta t\ ) = 0, \\ M\left(\dfrac{u^{n+1} - 2u^n + u^{n-1}}{\Delta t^2}\right) + K\left(\dfrac{u^{n+1} + 2u^n + u^{n-1}}{4}\right) \\ = f + \sum\limits_{i \in I_C} \lambda_N^{i,n} N_i, \forall\ n \geq 1, \end{cases} \tag{3}$$

where Δt is the time parameter. The contact condition (2) is approximated using a central difference scheme:

$$\begin{cases} u^n.N_i < 0 \Longrightarrow \lambda_N^{i,n} = 0, \\ u^n.N_i \geq 0 \Longrightarrow \begin{cases} \dfrac{(u^{n+1} - u^{n-1}).N_i}{2\Delta t} \leq 0, \quad \lambda_N^{i,n} \leq 0, \\ \lambda_N^{i,n} \left(\dfrac{(u^{n+1} - u^{n-1}).N_i}{2\Delta t}\right) = 0. \end{cases} \end{cases} \tag{4}$$

Theorem 1. *The stability of scheme (3)-(4) is ensured by the fact that the discrete energy*

$$J(u, v) = \frac{1}{2} < Mv, v > + \frac{1}{2} < Ku, u > - < f, u > \tag{5}$$

is conserved in the following sense

$$\Delta J = J(u^{n+\frac{1}{2}}, v^{n+\frac{1}{2}}) - J(u^{n-\frac{1}{2}}, v^{n-\frac{1}{2}}) = 0, \tag{6}$$

with $u^{n+\frac{1}{2}} = \dfrac{u^{n+1} + u^n}{2}, \qquad v^{n+\frac{1}{2}} = \dfrac{u^{n+1} - u^n}{\Delta t}.$

Proof. One has

$$\begin{aligned} \Delta J = & \frac{1}{2} < M\left(v^{n+\frac{1}{2}} + v^{n-\frac{1}{2}}\right), v^{n+\frac{1}{2}} - v^{n-\frac{1}{2}} > \\ & + \frac{1}{2} < K\left(u^{n+\frac{1}{2}} + u^{n-\frac{1}{2}}\right), u^{n+\frac{1}{2}} - u^{n-\frac{1}{2}} > - < f, u^{n+\frac{1}{2}} - u^{n-\frac{1}{2}} > . \end{aligned}$$

Then, using the definition of $u^{n+\frac{1}{2}}$ and $v^{n+\frac{1}{2}}$, we obtain:

$$\begin{aligned} \Delta J = & \frac{1}{2\Delta t^{\ 2}} < M\left(u^{n+1} - 2u^n + u^{n-1}\right), u^{n+1} - u^{n-1} > \\ & + \frac{1}{8} < K\left(u^{n+1} + 2u^n + u^{n-1}\right), u^{n+1} - u^{n-1} > - \frac{1}{2} < f, u^{n+1} - u^{n-1} > . \end{aligned}$$

Hence, from (3):

$$\Delta J = \frac{1}{2} < \sum_{i \in I_C} \lambda_N^{i,n} N_i, u^{n+1} - u^{n-1} > = \Delta t \sum_{i \in I_C} \lambda_N^{i,n} \left(\frac{u^{n+1}.N_i - u^{n-1}.N_i}{2\Delta t}\right).$$

Finally, (4) leads to $\Delta J = 0$. Then, the first approach is energy conserving.

Remark.

- The major difficulty with scheme proposed in [5] is that the contact condition (2) is discretized using a central diffence scheme.
- In this section, we proved conservation energy of the discretized contact elastodynamic problem using an appropriate scheme and a choice of contact condition in terms of velocity. But this condition, as it was approximated, allows some interpenetration. Then, we opt for an other approach which can take in acounte our no-iterpenetration property and conserve energy. This is what we are going to do in the following section.

3 Equivalent mass matrix method (EMM)

The non-well-posedness of Problem (1) comes from the fact that the nodes on the contact boundary have their own inertia. This leads to instabilities especially for energy conserving schemes. We propose here to introduce a new distribution of the mass with the same total mass, center of gravity and inertia momenta. This distribution of the mass is done so that there is no inertia for the contact nodes (similarly to what happens in the continuous case). We refer the reader to [1] for further details. We assume that the modified mass matrix is still denoted M such that ${N_i}^T M N_j = 0, \forall\, i,j \in I_C$. Then, the two following results hold.

Theorem 2. *Problem (1) with the equivalent mass matrix is well posed and has a Lipschitz continuous solution.*

Proof. If we order the degrees of freedom such that the last ones are the nodes on the contact boundary, we can split each matrix and vector in interior part and contact boundary part as follows:

$$M = \begin{pmatrix} \bar{M} & 0 \\ 0 & 0 \end{pmatrix}, K = \begin{pmatrix} \bar{K} & C^T \\ C & D \end{pmatrix}, N_i = \begin{pmatrix} 0 \\ \tilde{N}_i \end{pmatrix} \text{ and } u = \begin{pmatrix} \bar{u} \\ \tilde{u} \end{pmatrix}.$$

Then, Problem (1) becomes:

$$\begin{cases} \bar{M}\ddot{\bar{u}} + \bar{K}\bar{u} = f + C^T\tilde{u}, \\ D\tilde{u} + C\bar{u} = \sum\limits_{i \in I_C} \lambda_N^i \tilde{N}_i, \\ \lambda_N^i \le 0,\ u.N_i \le 0,\ \lambda_N^i (u.N_i) = 0, \forall\, i \in I_C, \\ u(0) = u^0, \dot{u}(0) = u^1. \end{cases} \tag{7}$$

The second equation together with the contact condition uniquely define $\tilde{u}$ as soon as $\bar{u}$ is given. Furthermore, $\tilde{u}$ depends Lipschitz continuously on $\bar{u}$. The first equation is a lipschitz ordinary differential equation.

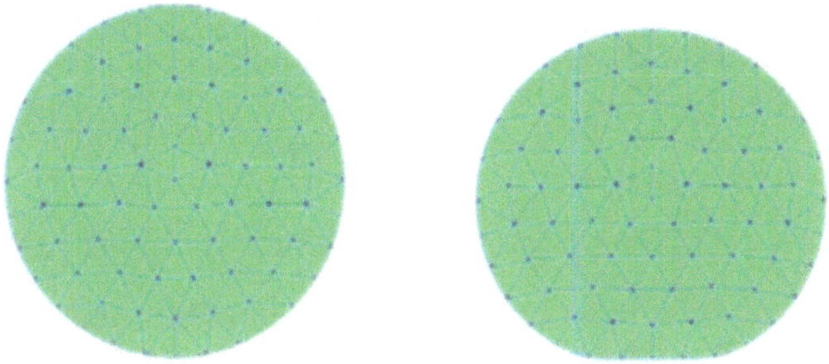

Fig. 1. *A disc before and during the first contact.*

Theorem 3. *Problem (1) with the equivalent mass matrix is energy conserving.*

We refer to [1] for a completely proof. This result comes from the fact that $\bar{u}_i, \tilde{u}_i$, and λ^i_N satisfy $\bar{u}_i \in \mathcal{C}^2([0,T],\mathbb{R})$, $\tilde{u}_i \in W^{1,\infty}([0,T],\mathbb{R})$ and $\lambda^i_N \in W^{1,\infty}([0,T],\mathbb{R})$.

4 Numerical results

In this section, we study the dynamic contact of an elastic disc (see Fig. 1) the properties of which are summerized in Tab. 1. We denote A the lowest point of the disc (the first point which will be in contact).

Table 1. *Characteristics of the elastic disc and the resolution method*

Disc property	Value	Property of the resolution method	Value
ρ, diameter	$6\ 10^{-6} kg/cm^3$, $20\ cm$	Time step	$10^{-4} s$
Lamé coefficients	$\lambda = 10\ GP$, $\mu = 5\ GP$	Simulation time	$0.3\ s$
u^0, v^0	$1\ cm$, $-100\ cm/s$	Mesh parameter	$\simeq 2\ cm$

The results of simulations for the midpoint scheme with the equivalent contact condition ECC are presented in Fig. 2. The energy is indeed constant, however the normal stress in point A is very noisy and inexploitable. Concerning Problem (1) with a modified mass matrix, the simulations are done using a Newmark scheme with $\beta = \gamma = 0.5$. Fig. 3 shows that the normal stress in A for the second approach is more regular than for the first one. Moreover, there is very small fluctuations in the energy evolution which is quasi-conserved.

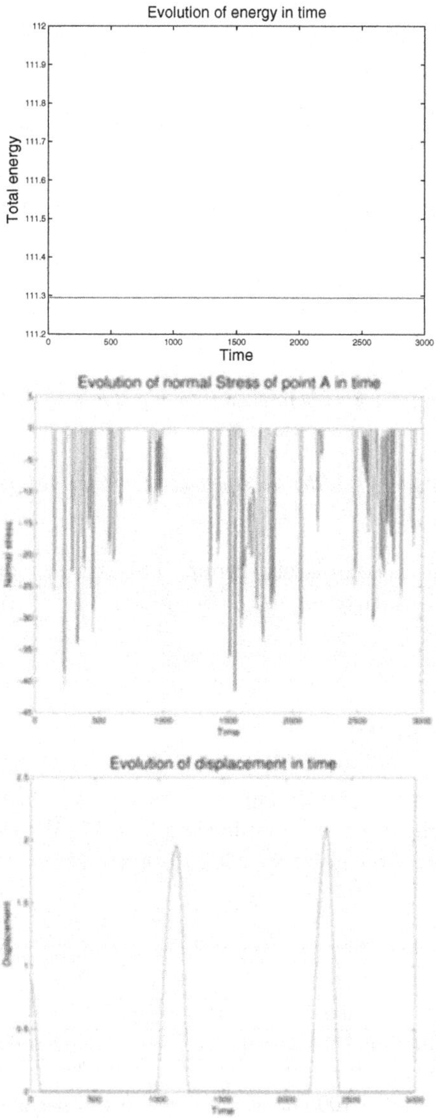

Fig. 2. *Energy, normal stress and displacement evolution for the midpoint scheme with ECC condition.*

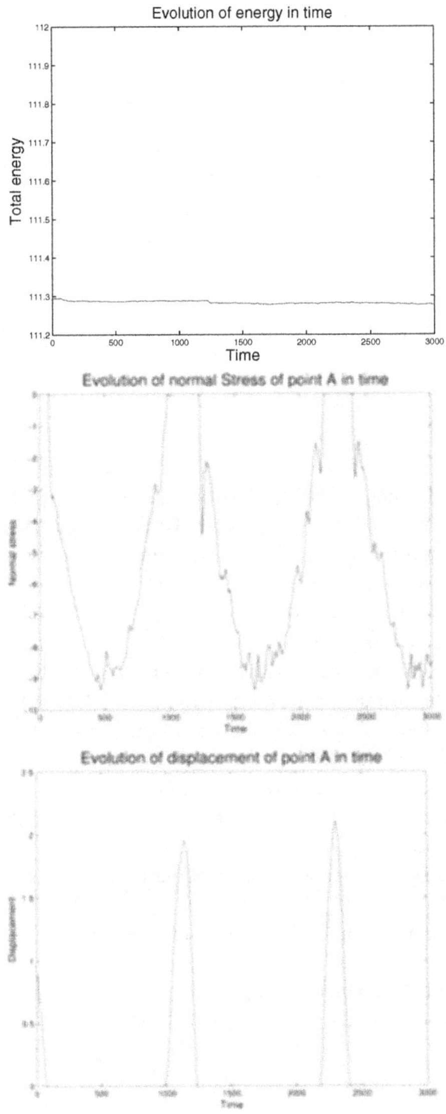

Fig. 3. *Energy, normal stress and displacement evolution for the Newmark scheme with EMM method.*

5 Concluding remarks

In this work, we proved the stability of elastodynamic contact problem using an appropriate time integration scheme. The first approach we presented ensure conservation of energy but allows a small interpenetration and the computed normal stress is badely approximated. It could be interesting to see that this scheme is well adapted for rigid bodies. The second approach is very simple to implement and gives a good approximation of normal stress. For both approaches, adding a Coulomb friction condition is not a difficulty from the stability view point. However, this condition could be badly approximated for the first approach because it depends on the normal stress.

References

1. H.B. Khenous, PhD thesis, INSA of Toulouse, in preparation.
2. H.B. Khenous, Y. Renard & J. Pommier, Hybrid discretization of the Signorini problem with Coulomb friction. Theoretical aspects and comparison of some numerical solvers, to appear.
3. J.U. Kim, A boundary thin obstacle problem for a wave equation, Com. part. diff. eqs., 1989, **14 (8&9)**, 1011-1026.
4. G. Lebeau & M. Schatzman, A wave problem in a half-space with a unilateral constraint at the boundary, J. diff. eqs., 1984, **55**, 309-361.
5. T.A. Laursen & V. Chawla, Design of energy conserving algorithms for frictionless dynamic contact problems, Int. J. Num. Meth. Engrg, Vol. 40, 1997, 863-886.
6. T.A. Laursen & G.R. Love, Improved implicit integrators for transient impact problems-geometric admissibility within the conserving framework, Int. J. Num. Meth. Eng, 2002, **53**, 245-274.
7. J.J. Moreau, Numerical aspects of the sweepig process, Comp. Meth. Appl. Mech. Engrg., 1999, **177**, 329-349.
8. L. Paoli, Time discretization of vibro-impact, Phil. Trans. R. Soc. Lond. A., 2001, **359**, 2405-2428.
9. Paumier & Y. Renard, Surface perturbation of an elastodynamic contact problem wih friction, European Journal of Applied Mathematics, vol. 14, 2003, 465-483.

Mortar methods for contact problems

S. Hüeber and B.I. Wohlmuth

Institut für Angewandte Analysis und Numerische Simulation (IANS), Universität Stuttgart, Pfaffenwaldring 57, D-70569 Stuttgart, Germany,
hueeber@ians.uni-stuttgart.de, wohlmuth@ians.uni-stuttgart.de

Abstract. For the numerical approximation of nonlinear contact problems, mortar methods provide a powerful and efficient tool. To detect the correct contact zone and to decompose it into the sliding and sticking part for Coulomb friction, we use primal-dual active set strategies. Combining these strategies with optimal multigrid methods, we get an efficient inexact approach. The extension of the mortar approach to thermal contact problems, to the case of large deformations and nearly incompressible materials is shown. Numerical results in 2D and 3D are given to illustrate the flexibility of the algorithm.

1 Introduction

We denote by $\Omega \subset \mathbb{R}^n$, $n = 2, 3$, a given domain representing the union of the two bodies Ω_s and Ω_m which we expect to come in contact. The boundaries are divided in the Dirichlet part Γ_D, the Neumann part Γ_N and the possible contact zone Γ_c, where we assume the bodies to come in contact under the applied loads and displacements. As material we use either a standard linearized St. Venant–Kirchhoff material or a nonlinear Neo–Hooke material. Let us denote by $\mathbf{n}$ the outward normal vector on Γ_c with respect to the body Ω_s. Since we use a mortar approach, we introduce the Lagrange multiplier $\boldsymbol{\lambda}$ on Γ_c with respect to the domain Ω_s, playing the role of the slave side, modeling the contact stress. The well known contact conditions using Coulomb friction on Γ_c are given by, see, e.g., [10, 14, 13],

$$\mathbf{u}\cdot\mathbf{n} \le d, \quad (\boldsymbol{\lambda}\cdot\mathbf{n})([\mathbf{u}\cdot\mathbf{n}] - d) = 0, \quad \boldsymbol{\lambda}\cdot\mathbf{n} \ge 0, \tag{1}$$

$$\|\boldsymbol{\lambda}_\tau\| \le \mathfrak{F}|\boldsymbol{\lambda}\cdot\mathbf{n}|, \quad \mathbf{u}_\tau = \alpha\boldsymbol{\lambda}_\tau, \text{ where } \begin{cases} \alpha = 0, & \text{if } \|\boldsymbol{\lambda}_\tau\| < \mathfrak{F}|\boldsymbol{\lambda}\cdot\mathbf{n}|, \\ \alpha \ge 0, & \text{if } \|\boldsymbol{\lambda}_\tau\| = \mathfrak{F}|\boldsymbol{\lambda}\cdot\mathbf{n}|. \end{cases} \tag{2}$$

Here $[\star] := \star^s - \star^m$ denotes the jump of the corresponding value between the two bodies and $\star_\tau := \star - (\star\cdot\mathbf{n})\mathbf{n}$ the tangential part of $\star$. d is the distance in normal direction between the two bodies and $\mathfrak{F}$ denotes the friction coefficient. We enforce the constraints (1) and (2) in a weak integral sense. Due to the use of dual Lagrange multipliers our problem results in a saddle point formulation with strong point-wise constraints. Let $\mathcal{S}$ be the set of all nodes of the finite element mesh on the possible contact boundary Γ_c of the slave side Ω_s. We apply a local basis transformation of the finite element basis such that the constraints for the displacements are only associated with the coefficients $\hat{\mathbf{u}}_p$, $p \in \mathcal{S}$, with respect to the new basis. The coefficient vector is denoted by $\hat{u}_h$. The vector of the displacements with respect to the original nodal basis is denoted by u_h. Furthermore we define the local diagonal matrix $D[p,p]$ describing the coupling between the finite element basis and the Lagrange multiplier at the node $p \in \mathcal{S}$. The discrete form of the weak non-penetration (1) condition reads as $(\hat{\mathbf{u}}_n)_{p,s} := \mathbf{n}_p^\top D[p,p]\hat{\mathbf{u}}_p \leq d_p$. For the discrete form of the upper bound for the Lagrange multiplier (2), we get $\|(\boldsymbol{\lambda}_\tau)_{p,s}\| \leq g_p$, where $(\boldsymbol{\lambda}_\tau)_{p,s} := D[p,p](\boldsymbol{\lambda}_\tau)_p$, where $(\star_\tau)_p := \star_p - (\star_p \cdot \mathbf{n}_p)\mathbf{n}_p$. For details of the formulation and the discretization, we refer to [8, 7]. The application of mortar methods for contact problems was also considered in [2, 6, 9, 11, 4]. Our implementation is based on the finite element toolbox UG, see [1].

2 Inexact primal-dual active set algorithm

In this chapter, we briefly present the inexact primal-dual active set algorithm. Primal-dual active set strategies can be interpreted as semi-smooth Newton methods, see, e.g., [5]. To handle the nonlinearity of the contact conditions in normal direction, the set $\mathcal{S}$ is decomposed into two disjoint subsets $\mathcal{A}^n$ and $\mathcal{I}^n$ such that $(\hat{\mathbf{u}}_n)_{p,s} = d_p$ for $p \in \mathcal{A}^n$ and $(\boldsymbol{\lambda}_n)_{p,sq} = 0$ for $p \in \mathcal{I}^n$. The scaled value $(\boldsymbol{\lambda}_n)_{p,sq}$ is given by $(\boldsymbol{\lambda}_n)_{p,sq} := \mathbf{n}_p^\top (D[p,p])^2 \boldsymbol{\lambda}_p$. We call $\mathcal{A}^n$ active set and $\mathcal{I}^n$ inactive set. To find the correct sets $\mathcal{A}^n$ and $\mathcal{I}^n$ an iterative scheme is used. Let $\hat{u}_h^k$ and λ_h^k be the solution for the sets $\mathcal{A}_k^n$ and $\mathcal{I}_k^n$. The next sets $\mathcal{A}_{k+1}^n$ and $\mathcal{I}_{k+1}^n$ of the iterative scheme are defined by

$$\mathcal{A}_{k+1}^n := \left\{p \in \mathcal{S} : (\boldsymbol{\lambda}_n)_{p,sq} + c_n\big((\hat{\mathbf{u}}_n)_{p,s} - d_p\big) > 0\right\},$$
$$\mathcal{I}_{k+1}^n := \left\{p \in \mathcal{S} : (\boldsymbol{\lambda}_n)_{p,sq} + c_n\big((\hat{\mathbf{u}}_n)_{p,s} - d_p\big) \leq 0\right\},$$

where $c_n > 0$. Thus in each step a linear system has to be solved. For all nodes $p \in \mathcal{A}_k^n$, Dirichlet boundary conditions in normal direction are given and for all nodes $p \in \mathcal{I}_k^n$, homogeneous Neumann data are specified. The constant c_n has no influence on the choice of the sets if we solve the linear system exactly. Due to the use of an optimal multigrid solver for the linear system, we update the sets $\mathcal{A}_k^n$ and $\mathcal{I}_k^n$ after each multigrid step. For details of the algorithm we refer to [8].

To show the performance of the algorithm, we consider a frictionless contact problem for a linearized St. Venant–Kirchhoff material in the 3D case. In the left picture of Figure 1, a cross section of the problem definition is shown. The lower domain Ω_m models a half-bowl which is fixed at the outer boundary as shown in the left picture of Figure 1. Against this bowl, we press

Fig. 1. Left: problem definition; middle: cut through the distorted domains with the effective von Mises stress on level 3; right: contact stress $\boldsymbol{\lambda}_h \cdot \mathbf{n}$ on level 3.

the body modeled by the domain Ω_s. At the top of Ω_s, we apply the Dirichlet data $(0, 0, -0.2)^\top$. The geometry is given by $r_i = 0.7$, $r_a = 1.0$, $r = 0.6$, $h = 0.5$ and $d = 0.3$. For Young's modulus and Poisson's ratio, we use on Ω_s the parameters $E_s = 300$ and $\nu_s = 0.3$ and on Ω_m the parameters $E_m = 400$ and $\nu_m = 0.3$. The picture in the middle of Figure 1 shows a cut through the distorted domains with the effective von Mises stress. In the right picture the contact stress $\boldsymbol{\lambda}_h \cdot \mathbf{n}$ is presented.

Table 1. Comparison between exact and inexact active set strategy for the example given in Figure 1.

level	NON	exact strategy		inexact strategy	
		K_l	$\lvert\mathcal{A}_k^n\rvert$	M_l	$\lvert\mathcal{A}_k^n\rvert$
0	312	3	0 9 6	3	0 9 6
1	1623	4	14 26 21 21	4	14 26 22 21
2	10062	3	66 88 85	3	66 91 85
3	71082	4	306 347 336 337	5	306 341 336 336 337

To determine the active set we follow an multilevel approach. The initial active set $\mathcal{A}_1^n$ on level $l+1$ is inherited from the final active set on level l. On the coarsest level, we start with the initial active set $\mathcal{A}_1^n = \emptyset$. We get the mesh on level $l+1$ by a uniform refinement step of the mesh on level l. The comparison of the exact and inexact strategy is shown in Table 1. The second column shows the number of nodes (NON) on level l. For the exact strategy, we denote by K_l the step in which the correct active set $\mathcal{A}^n$ on level l is found for the first time. For the inexact strategy, M_l denotes the multigrid step in which the exact active set $\mathcal{A}^n$ is found for the first time. By $|\mathcal{A}_k^n|$ we denote the number of active nodes in the k-th. iteration step for the exact strategy, and for the inexact strategy $|\mathcal{A}_k^n|$ denotes the number of active nodes in the k-th. multigrid step. These numbers are shown in column 4 for the exact strategy

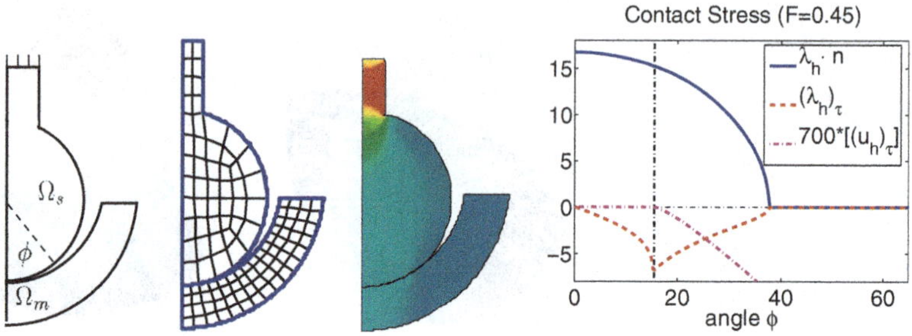

Fig. 2. Left: problem setting and grid on level 0; middle: distorted grid with effective von Mises stress; right: normal and tangential component of the Lagrange multiplier and the amplified tangential component of the jump $[(\mathbf{u}_h)_\tau]$.

and in column 6 for the inexact strategy. If we compare the exact with the inexact strategy, we observe that K_l and M_l are almost the same. Thus there is no need for solving the resulting linear problems exactly, and the additional cost for solving the nonlinear problem can be neglected.

3 Application to Coulomb friction

In this section, we generalize the presented algorithm to frictional problems. Now we have to handle the tangential constraints (2). Primal-dual active set strategies for problems with friction were also considered in [12]. In contrast to [12], we do not use a penalty parameter and use a different nonlinear complementarity function. Again we decompose the set $\mathcal{S}$ into two disjoint subsets $\mathcal{A}^\tau$ and $\mathcal{I}^\tau$ such that $\|(\boldsymbol{\lambda}_\tau)_{p,s}\| = g_p$ for $p \in \mathcal{A}^\tau$ and $(\hat{\mathbf{u}}_\tau)_p = 0$ for $p \in \mathcal{I}^\tau$, where g_p is a given friction bound. For the iterative scheme the choice of the next sets $\mathcal{A}^\tau_{k+1}$ and $\mathcal{I}^\tau_{k+1}$ in 2D is done by

$$\mathcal{A}^\tau_{k+1} := \left\{ p \in \mathcal{S} \, : \, (\hat{\mathbf{u}}_\tau)_p (\boldsymbol{\lambda}_\tau)_p + c_\tau \big(|(\boldsymbol{\lambda}_\tau)_{p,s}| - g_p \big) > 0 \right\},$$
$$\mathcal{I}^\tau_{k+1} := \left\{ p \in \mathcal{S} \, : \, (\hat{\mathbf{u}}_\tau)_p (\boldsymbol{\lambda}_\tau)_p + c_\tau \big(|(\boldsymbol{\lambda}_\tau)_{p,s}| - g_p \big) \le 0 \right\},$$

$c_\tau > 0$. In the case of Coulomb friction, we use an inexact approach, i.e. the friction bound g_p is updated in each step. For details, we refer to [7].

As examples we consider two axisymmetric problems in the 3D case with a linearized St. Venant–Kirchhoff material. Due to the symmetry of the problem setting, we can use cylinder coordinates and end up with a 2D formulation. To construct dual Lagrange multipliers for this case, we consider an arbitrary edge of the triangulation of Ω_s on Γ_c. We assume $s \in [0,1]$ to be the local coordinate of the parameterization of the edge. For $s = 0$, we get the first point $p_a = (r_a, z_a)^\top \in \mathcal{S}$ in the rz-plane and for $s = 1$, we have the second point $p_e = (r_e, z_e)^\top \in \mathcal{S}$. Due to the use of cylinder coordinates, we have to work with weighted integrals and the dual Lagrange multiplier is defined by

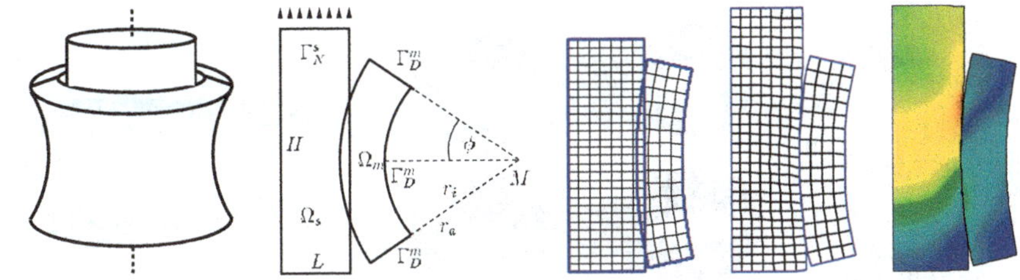

Fig. 3. Left: problem definition in 3D and 2D; middle: initial grid and distorted grid on level 0; right: effective von Mises stress on level 2.

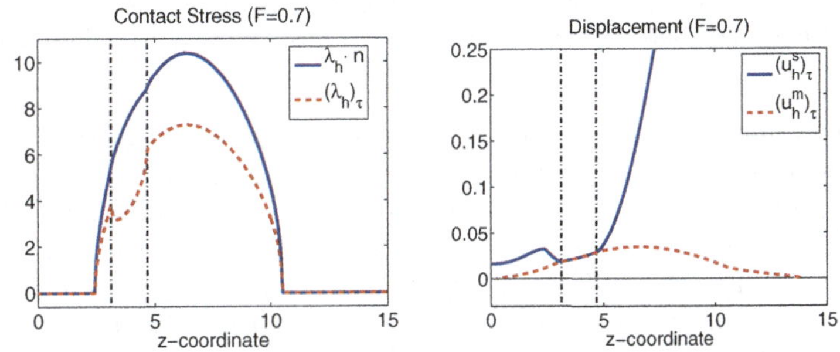

Fig. 4. Left: normal part $\boldsymbol{\lambda}_h \cdot \mathbf{n}$ and tangential part $(\boldsymbol{\lambda}_h)_\tau$ of the Lagrange multiplier $\boldsymbol{\lambda}_h$; right: tangential part $(\mathbf{u}_h^s)_\tau$ of the displacement on the slave side and tangential part $(\mathbf{u}_h^m)_\tau$ of the displacement on the master side.

$$\int_0^1 \psi_p(s)\phi_q(s)r(s)\,ds = \delta_{pq}\int_0^1 \phi_p(s)r(s)\,ds, \qquad p,q \in \mathcal{S}, \tag{3}$$

where $r(s) = r_a + s(r_e - r_a)$. If we use linear elements $\psi_{p_a} = \alpha_a + \beta_a s$ and $\psi_{p_e} = \alpha_e + \beta_e s$, the coefficients α_a, β_a, α_e and β_e can be easily computed. The weighted biorthogonality (3) yields

$$\alpha_a = \frac{(r_e + 2r_a)(3r_e + r_a)}{r_e^2 + 4r_e r_a + r_a^2}, \qquad \beta_a = -\frac{2(2r_a^2 + 2r_e^2 + 5r_e r_a)}{r_e^2 + 4r_e r_a + r_a^2},$$

$$\alpha_e = -\frac{(2r_e + r_a)(r_e + r_a)}{r_e^2 + 4r_e r_a + r_a^2}, \qquad \beta_e = \frac{2(2r_a^2 + 2r_e^2 + 5r_e r_a)}{r_e^2 + 4r_e r_a + r_a^2}.$$

Our first numerical example will be the axisymmetric example from Section 2 restricted to the 2D case. Now we consider the case with Coulomb friction for the friction coefficient $\mathfrak{F} = 0.45$. As material parameters, we set $\lambda_s = 173.077$, $\mu_s = 115.385$, $\lambda_m = 230.769$ and $\mu_m = 153.846$. At the top we apply Dirichlet data given by $(0.0, -0.2)^\top$. The grid on level 0, the distorted grid with the effective von Mises stress, the normal and tangential component of the Lagrange multiplier and the tangential component of the jump $[(\mathbf{u}_h)_\tau]$ are presented in Figure 2.

Fig. 5. Left: problem definition; middle: effective von Mises stress for stress-displacement formulation for Neo–Hooke material (top) and for linearized St. Venant Kirchhoff material (bottom); right: effective von Mises stress for Q1 formulation.

The geometry of our second example is shown in the left picture of Figure 3. On the top of the inner cylinder, we apply Neumann boundary condition. The resulting geometry can be seen in the second picture of Figure 3. For the geometry, we choose $L = 5$, $H = 15$, $r_i = 27$, $r_a = 30$, $M = (24.5, 7.0)^\top$ and $\phi = 13$. For the softer inner body Ω_s, we use for $\lambda_s = 57.6923$ and $\mu_s = 38.4615$. For the outer body Ω_m, we set $\lambda_m = 576.923$ and $\mu_m = 384.615$. At the top of the inner cylinder, we apply Neumann data given by $(0, 17)^\top$. Due to the axisymmetry, we have on the left side of Ω_s, where $r = 0$ holds, homogeneous Dirichlet data in radial direction. In z-direction and on the bottom, we have homogeneous Neumann boundary conditions. We fix Ω_m at the three segments Γ_D^m, as shown in the second picture of Figure 3. The friction coefficient is set to be $\mathfrak{F} = 0.7$. Figure 4 shows the normal and the tangential parts of the Lagrange multiplier and the displacements on Γ_c.

4 Nearly incompressible materials and large deformation

In this section, we give a numerical example for self contact with a nearly incompressible material. We consider a torus with inner radius 8 and outer radius 10. It is subjected to a vertical load $(0, 0, \pm 2.75)^\top$ along the diagonally opposite external edges in the plane $z = 0$ with a range of $d = 8.25$ as shown in the left picture of Figure 5. In addition to this, the torus is fixed in x-direction in the plane $x = 0$, in y-direction in the plane $y = 0$ and in z-direction in the plane $z = 0$. A nonlinear Neo–Hooke material law is considered with $E = 200$ and $\nu = 0.499$ in the nearly incompressible range. We compare the standard Q1 formulation with a stress-displacement formulation, see [3]. It can be seen in Figure 5, that the Q1 formulation shows volume locking, and it does not yield a reliable result in the incompressible limit. Furthermore in the middle of this figure we compare the Neo–Hooke material with the linearized St. Venant–Kirchhoff material. Due to the large deformations, we apply the Neumann data in 110 incremental steps. In each step, we update the values for the gap d_p and the coupling matrices between the Lagrange multiplier and the finite element basis with respect to the grid of the actual configuration.

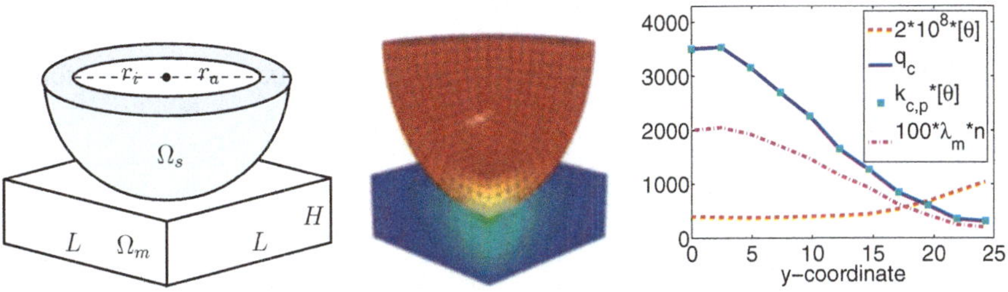

Fig. 6. Left: problem definition; middle: temperature θ on distorted domains; right: heat flux q_c and temperature jump $[\theta]$ at $x = 0$.

5 Extension to thermo-mechanical contact

In this section, we want to give the mortar-formulation for a thermo-mechanical problem. For the stress-strain relation of the material, we consider $\boldsymbol{\sigma} = (\lambda + 2/3\mu)\big(\mathrm{tr}(\boldsymbol{\varepsilon}) - 3\alpha\theta\big) + 2\mu\mathrm{dev}(\boldsymbol{\varepsilon})$. α denotes the thermal expansion coefficient. The balance conditions are given by $-\mathrm{div}(\boldsymbol{\sigma}) = \mathbf{0}$ and $\mathrm{div}(\mathbf{q}) = \varrho\mathbf{r}$, where $\mathbf{q} = -\mathbf{K}\mathrm{grad}(\theta)$ denotes the heat flux, ϱ the mass density and $\mathbf{r}$ the heat source. On the contact zone, we use the heat law $q_c = k_c[\theta]$ with the contact heat flux $q_c = \mathbf{q}\cdot\mathbf{n}$. For the heat conductivity at the contact zone, we use the nonlinear law $k_c = 0.19245 \times 10^8 |\boldsymbol{\lambda}_m \cdot \mathbf{n}|^{1.5}$, see, e.g., [14], where $\boldsymbol{\lambda}_m$ denotes the mechanical Lagrange multiplier. Let $\lambda_\theta = \sum_{p\in\mathcal{S}} q_p\psi_p$ be the thermal part of the Lagrange multiplier representing the heat flux q_c. Furthermore, we define the approximation of k_c at $p \in \mathcal{S}$ by $k_{c,p} := 0.19245 \times 10^8 |(\boldsymbol{\lambda}_m)_p \cdot \mathbf{n}_p|^{1.5}$. We enforce the coupling between $[\theta]$ and λ_θ at the contact zone in a weak sense by

$$\int_{\Gamma_c} [\theta]\,\psi_p\,ds = \int_{\Gamma_c} \frac{1}{k_{c,p}} \lambda_\theta \phi_p\,ds, \qquad p \in \mathcal{A}^n. \tag{4}$$

The algebraic representation of (4) is $D\hat{\theta}_{\mathcal{S}} = DD_{\lambda_m}\lambda_\theta$. We remark that due to the use of dual Lagrange multipliers the matrix D is diagonal. The scaling matrix D_{λ_m} depends on the contact pressure. We note that the nonlinear heat flux represents a Robin type interface condition. Due to the different regularity of the temperature and the heat flux, we use different test functions on the left and the right side of (4). The temperature part of the algebraic saddle point system has the form

$$\begin{pmatrix} \ddots & \vdots & 0 \\ \dots & \hat{A}_{\theta\theta}^{\mathcal{SS}} & D \\ 0 & \mathrm{Id} & -D_{\lambda_m} \end{pmatrix} \begin{pmatrix} \vdots \\ \hat{\theta}_{\mathcal{S}} \\ \lambda_\theta \end{pmatrix} = \begin{pmatrix} \vdots \\ r_{\mathcal{S}} \\ 0 \end{pmatrix}, \tag{5}$$

where the nonlinearity in the heat law enters in terms of D_{λ_m}. Static condensation of the heat flux in (5) yields for the second line of (5)

$$\ldots + (\hat{A}_{\theta\theta}^{SS} + DD_{\lambda_m}^{-1})\hat{\theta}_S = r_S.$$

To handle the nonlinearity in D_{λ_m} we use an inexact fixed point iteration.

We consider the problem depicted in the left picture of Figure 6. The body Ω_m is fixed at the bottom, where we set $\theta = 0$. On the inner surface of the cup, we set $\theta = 100$ and apply a radial pressure in normal direction of value 1.5. For the material parameter, we use $\lambda = 144.231$, $\mu = 96.154$, $\alpha = 0.001$ and $\mathbf{K} = 10^3\,\mathbf{Id}$. To define the geometry, we set $L = 130$, $H = 50$, $r_i = 75$ and $r_a = 100$. For the heat source, we set $\mathbf{r} = 0$. The temperature distribution is shown in the middle picture of Figure 6. The right picture shows the heat flux q_c, the normal part $\boldsymbol{\lambda}_m \cdot \mathbf{n}$ of the mechanical Lagrange multiplier and the temperature jump $[\theta]$ along a radial line on the contact zone.

Acknowledgements

This work was supported in part by the Deutsche Forschungsgemeinschaft, SFB 404, B8.

References

1. P. Bastian, K. Birken, K. Johannsen, S. Lang, N. Neuss, H. Rentz Reichert, and C. Wieners. UG - a flexible software toolbox for solving partial differential equations. *Comput. Vis. Sci.*, 1:27-40, 1997.
2. F. Ben Belgacem. Numerical simulation of some variational inequalities arisen from unilateral contact problems by the finite element methods. *SIAM J. Numer. Anal.*, 37(4):1198-1216, 2000.
3. K. Chavan, B. Lamichhane, and B. Wohlmuth. Locking-free finite element methods based on the Hu-Washizu formulation for linear and geometrically nonlinear elasticity. In preparation, 2005.
4. K. Fischer and P. Wriggers. Frictionless 2d Contact formulations for finite deformations based on the mortar method. *Comput. Mech.*, 36:226-244, 2005.
5. M. Hintermüller, K. Ito, and K. Kunisch. The primal-dual active set strategy as a semi-smooth Newton method. *SIAM J. Optim.*, 13(3):865-888, 2003.
6. P. Hild and P. Laborde. Quadratic finite element methods for unilateral contact problems. *Appl. Numer. Math.*, 41:410-421, 2002.
7. S. Hüeber, A. Matei, and B. Wohlmuth. Effcient algorithms for problems with friction. Technical Report 007, Universität Stuttgart SFB 404, 2005.
8. S. Hüeber and B. Wohlmuth. A primal-dual active set strategy for non-linear multibody contact problems. *Comput. Methods Appl. Mech. Engrg.*, 194:3147-3166, 2005.
9. R. Kornhuber and R. Krause. Adaptive multigrid methods for Signorini's problem in linear elasticity. *Comput. Vis. Sci.*, 4(1):9-20, 2001.
10. T. Laursen. *Computational Contact and Impact Mechanics.* Springer Berlin Heildelberg, 2002.
11. M. Puso and T. Laursen. A mortar segment-to-segment contact method for large deformation solid mechanics. *Comput. Methods Appl. Mech.* Engrg., 193(6-8):601-629, 2004.
12. G. Stadler. Semismooth Newton and augmented Lagrangian methods for a simplified friction problem. *SIAM J. Optim.*, 15(1):39-62, 2004.
13. K. Willner. Kontinuums- und Kontaktmechanik. Springer, 2003.
14. P. Wriggers. *Computational Contact Mechanics.* J. Wiley and Sons Ltd, 2002.

Contact mechanics for analysis of fracturing and fragmenting solids in the combined finite-discrete element method

A. Munjiza

Engineering Department, Queen Mary University of London, Mile End Road, E1 4NS, London, United Kingdom

Abstract. The combined finite-discrete element method is a new computational method for the simulation of fracturing and fragmenting solids or particulate media where individual particles are deformable. The method combines the power of the finite element method in capturing deformability of solid particles with the power of the discrete element method to accurately represent interaction between individual particles for systems comprising millions of particles. In addition, the method is suitable for the simulation of extensive fracture and/or fragmentation processes. In recent years the method has captured the attention of researchers in a wide field of potential applications and the interest in the method has grown significantly. In this paper key features of the method are explained with special attention being paid to the processing of contact interaction. Also, the new generation of linear time complexity search algorithms has been presented including the NBS and the MR algorithm. In the computational mechanics context, these algorithms are also relevant for the discrete element method, the finite element method and grid generation.

1 The combined finite-discrete element method

The most widely used computational method in the computational solid mechanics field is the finite element method. In recent decades a set of computational methods have been developed to deal with particulates, jointed rock, granular flows and problems where the so called emergent properties of a system are a result of interaction between large numbers of individual solid particles. The most widely used method for a large class of these problems is the discrete element method. In the early 1990s the two methods have been combined and the resulting method was termed the combined finite-discrete element method. The combined finite-discrete element method is in essence a discrete element method with individual discrete elements discretised into finite elements. Alternatively, one can say that the combined finite-discrete element method is a finite element method comprising of millions even billions

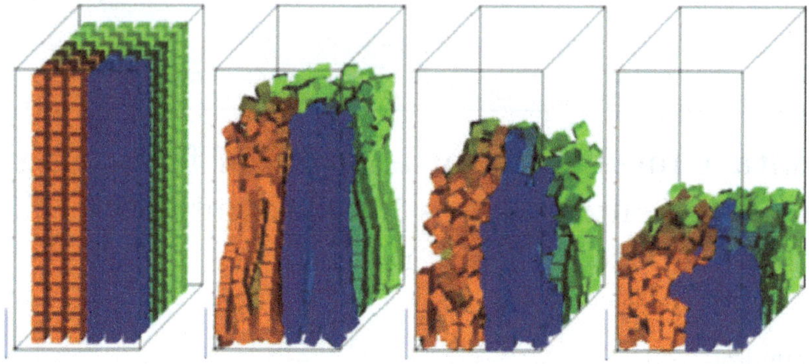

Fig. 1. A typical combined finite-discrete element simulation showing a number of discrete elements moving, deforming and interacting with each other.

of finite element meshes where interaction between these meshes is enforced through contact, Fig. 1.

1.1 Discretisation of discrete elements into finite elements

In the combined finite-discrete element method finite element discretisation of discrete elements is used to process both contact interaction and deformation of discrete elements. However, unlike the finite element method, the discrete elements are always subject to large displacements, including large rotations. Thus, in the combined finite-discrete element method a finite-strain elasticity element formulation is adopted. This is best demonstrated on a constant strain triangle. The matrix of the deformation gradient tensor for this element is obtained as follows

$$\mathbf{F} = \begin{bmatrix} \dfrac{\partial x_c}{\partial x_i} & \dfrac{\partial x_c}{\partial y_i} \\ \dfrac{\partial y_c}{\partial x_i} & \dfrac{\partial y_c}{\partial y_i} \end{bmatrix} = \begin{bmatrix} x_{1c} - x_{0c} & x_{2c} - x_{0c} \\ y_{1c} - y_{0c} & y_{2c} - y_{0c} \end{bmatrix} \begin{bmatrix} \hat{i}_x & \hat{j}_x \\ \hat{i}_y & \hat{j}_y \end{bmatrix}^{-1} \tag{1}$$

where the transformation matrix is given by

$$\begin{bmatrix} \hat{i}_x & \hat{j}_x \\ \hat{i}_y & \hat{j}_y \end{bmatrix} = \begin{bmatrix} x_{1i} - x_{0i} & x_{2i} - x_{0i} \\ y_{1i} - y_{0i} & y_{2i} - y_{0i} \end{bmatrix} \tag{2}$$

and x_{0i} ,$y_{0i}x_{1i}$,y_{1i}, x_{2i} , y_{2i} , x_{0c}, y_{0c} ,x_{1c} ,y_{1c} ,x_{2c} , y_{2c} are initial and current coordinates of node 0, 1 and 2 respectively. From the deformation gradient, the matrix of the left Cauchy-Green strain tensor $\mathbf{B}$ and the matrix of the Green-St.Venant strain tensor are calculated

$$\mathbf{B} = \mathbf{F}\mathbf{F}^T = \begin{bmatrix} \frac{\partial x_c}{\partial x_i} & \frac{\partial x_c}{\partial y_i} \\ \frac{\partial y_c}{\partial x_i} & \frac{\partial y_c}{\partial y_i} \end{bmatrix} \begin{bmatrix} \frac{\partial x_c}{\partial x_i} & \frac{\partial y_c}{\partial x_i} \\ \frac{\partial x_c}{\partial y_i} & \frac{\partial y_c}{\partial y_i} \end{bmatrix} \quad ;\ldots. \quad \check{\mathbf{E}} = \frac{1}{2}(\mathbf{B} - \mathbf{I}) \tag{3}$$

From the matrix of the strain tensor the matrix of the Cauchy stress tensor is obtained using the constitutive law.

1.2 Fracture and fragmentation of discrete elements

Several approaches are available to handle fracture and fragmentation, and these include global approaches, local approaches, smeared crack models and single crack models. Global approaches to fracture are based on the representation of the singularity of the stress field at the crack tip. Local approaches to crack analysis usually employ a smeared crack approach, with a single crack being replaced by a blunt crack band. Alternatively, the local approaches to crack analysis employ a single crack concept where the zone of strained material near crack tip is replaced by a zone of weakened bonds between the crack walls.

Recently a so called combined smeared-local approach has been successfully employed for mode I cracks in rock. It is based on the approximation of stress-strain curves for rock in direct tension. The strain-hardening part of the stress-strain curve is implemented in a standard way through the constitutive law. The strain-softening part of the stress-strain curve is expressed by means of stress and displacements.

1.3 Contact kinematics

The combined finite-discrete element method is aimed at simulation of deforming solids that can fracture and fragment. Fracture and fragmentation processes usually produce very complicated geometries. This is best demonstrated by a 2D block of concrete fragmenting under explosive charge shown in Fig. 2. The block is 2 m in length, while the depth (thickness) of the block is 0.1 m. A continuous cut from the centre of the block to the top edge containing the charge is assumed. The ignition of the explosive charge takes place at the bottom of the cut and spreads toward the top of the cut. Due to the detonation gas pressure, a stress wave is produced in the concrete. It is followed by radial cracks. The wave eventually reaches the free surface and reflects from it causing cracks on the boundary. The final fracture pattern is a result of inner cracks propagating outwards and outer cracks propagating inwards together with a system of secondary cracks appearing.

From the above example it is evident that the combined finite-discrete element method is capable of producing complicated geometry with contact forces being produced at boundaries of discrete elements. Robust accurate and

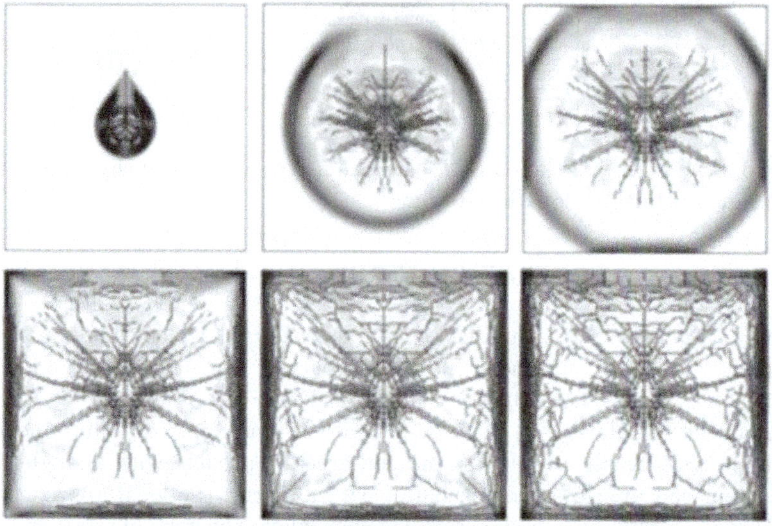

Fig. 2. Fracture sequence for a 2 m square block.

stable solutions for contact resolution between boundaries of discrete elements are therefore of major importance.

In the combined finite-discrete element method contact enforcement involves: a) contact interaction, i.e. integration of contact forces; b) contact detection, also called neighbour search.

There are two aspects to contact interaction, namely contact kinematics and physics of contact. Physics of contact is aimed at producing realistic contact forces between interacting boundaries. A good example of a model dealing with physics of interaction is a rock joint model. Contact kinematics is concerned with geometry of the interacting boundary, velocity fields, displacements fields and stress fields and integration of contact forces in such a way that for instance energy balance is preserved.

Contact detection (also called neighbour search) is a necessary part of contact processing especially when large systems are handled. It enables elimination of couples of discrete elements that are not in contact, thus leaving contact interaction to process interaction between the discrete elements that are actually in contact or are very close to each other.

In the rest of the paper an overview of contact interaction and contact detection algorithms employed in the combined finite-discrete element method is given together with some numerical results demonstrating the essential properties of the algorithms. Detailed description of these can be found elsewhere [1].

2 Contact interaction in the combined finite-discrete element method

As Fig. 2 demonstrates, the geometry of contact in the combined finite-discrete element method can be quite complicated. In addition, a typical combined finite-discrete element system comprises a few thousand to a few million separate interacting solids each associated with separate finite element meshes. In this context one of the key issues is the treatment of contact and especially the handling of the kinematics of contact. Contact kinematics in the combine finite-discrete element method must preserve energy and momentum balance. At the same time it must be robust enough to enable incorporation of various interaction laws [2-4] including for instance complex interaction in rock joints. In recent years it has been generally accepted that these can be achieved through the so-called potential contact force. The idea of the potential contact force is relatively simple: it is assumed that the penetration of any elemental area dA of the contactor into the target results in an infinitesimal contact force, given by

$$d\,\mathbf{f} = [grad\varphi_c - grad\varphi_t]\; dA \tag{4}$$

where the potentials φ_c and φ_t are defined over both the target and the contactor discrete element. For each of the discrete elements in contact, the contact force is calculated as a gradient of the corresponding potential function. The field of contact forces is therefore a conservative field.

Nodal forces are obtained by integrating the distributed contact force over the overlapping area. As individual discrete elements are discretised into finite elements, the potentials φ_c and φ_t are written as a sum of potentials associated with individual finite elements. By providing different potentials theoretically any interaction law can be obtained.

Potentials are concerned with normal force. Tangential force in the interaction is obtained as a function of normal force and so called state variables. The simplest interaction law is frictionless contact. Coulomb friction is also an example of a relatively simple interaction law. More complicated interaction laws such as rock joints are implemented through the concept of a so called yield surface that is used as a criterion for two surfaces in contact sliding over each other. Yield surface is usually a function of both normal stress and state variables that for instance define the wear of a joint.

3 Contact detection algorithms in the combined finite-discrete element method

Processing interaction between discrete elements in the combined finite-discrete element method requires a contact detection algorithm, otherwise the CPU time to process the interaction among the finite elements would be

Fig. 3. Example of a balanced binary tree.

proportional to N^2, where N is the total number of finite elements present in the system.

In the early days of the combined finite-discrete element method, binary tree based searches were mostly employed [5-8]. An example of a binary tree is given in Fig. 3. Given a tree in which the items are randomly received, the total time required to perform a search is proportional to the total number of nodes in the tree, i.e:

$$t \propto \ln(N) \tag{5}$$

However, Eq. (5) is valid only when the tree is balanced - a nearly balanced tree is easily built using a balanced tree storage and retrieval algorithm [9]. As the CPU time for building a balanced tree is proportional to $N \ln N$, the total contact detection CPU time is proportional to $N \ln N$ where N is the total number of discrete elements. For moderate size systems this can be fine. However, with very large systems CPU constraints can become very important.

Thus, in recent years the first linear contact detection algorithms (with total detection time T proportional to N) such as C-GRID and NBS [10,11], MR and MMR have been developed [12]. These algorithms have the total detection time proportional to the total number of entities, i.e. discrete elements and can therefore be several times to several hundred times faster than the equivalent binary searches. The NBS algorithm was the first linear search algorithm. It was developed by Munjiza in 1995. Prof. Williams at MIT generalised it for discrete elements of general size. The result was the so called C-GRID algorithm.

4 NBS contact detection algorithm

The NBS (No Binary Search) algorithm is based on space decomposition into identical cubical (3D) or square (2D) cells. The mapping of the discrete elements onto cells is done in such a way that each discrete element is assigned to one and only one cell according to the coordinates of its centre. In addition to the mapping of discrete elements onto cells, mapping of discrete elements onto columns and rows of cells is performed. All these mappings are numerically represented as singly connected lists of discrete elements. In its original form, the NBS algorithm comprises of the following steps:

Step 1: Mapping of discrete elements onto rows of cells is performed and a single connected list of discrete elements for each row is built.

Step 2: Mapping of discrete elements to individual cells. For each row of cells, discrete elements of that row and of the neighbouring row are mapped onto the individual cells.

Step 3: Detection of contact. Detection of contact is accomplished by checking all the discrete elements mapped to a particular cell against all the discrete elements mapped to neighbouring cells. In order to do this, a contact mask is introduced as shown in Fig. 4.

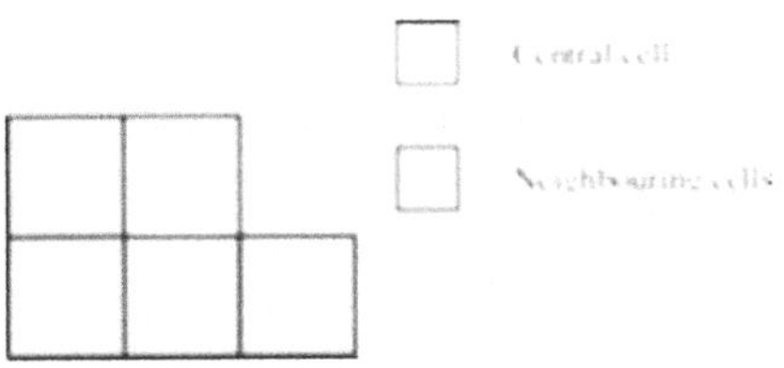

Fig. 4. 2D Contact mask.

CPU time and RAM requirements of the NBS contact detection algorithm are given by

$$T \propto cN \tag{6}$$

$$M = n_y + 2n_x + 2N \text{ integer numbers} \tag{7}$$

where c is a constant independent of either packing density or number of discrete elements.

The CPU requirements for the NBS algorithm are demonstrated by numerical examples shown in Fig. 5. Contact detection is solved 10 times for the problem, each time all contacting couples are detected. The cumulative CPU times as a function of the total number of discrete elements comprising the problem for packing A is shown in Fig. 6. The results accurately fit a linear relation, which confirms that the total detection time is indeed proportional to the total number of discrete elements present in the system. In

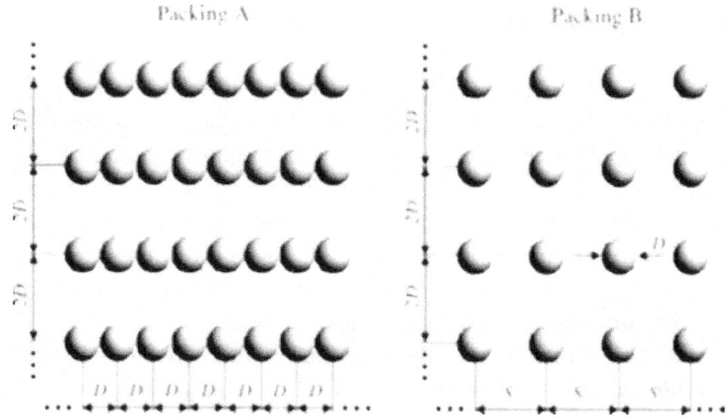

Fig. 5. Example 1: packing A (left) and packing B (right)

a similar way the fact that CPU time is constant regardless of the packing density is demonstrated using "packing B". The graph shows the CPU time being constant with packing density changing from 1 to $1/(200 \cdot 200)$, i.e. 1 to 1/40,000.

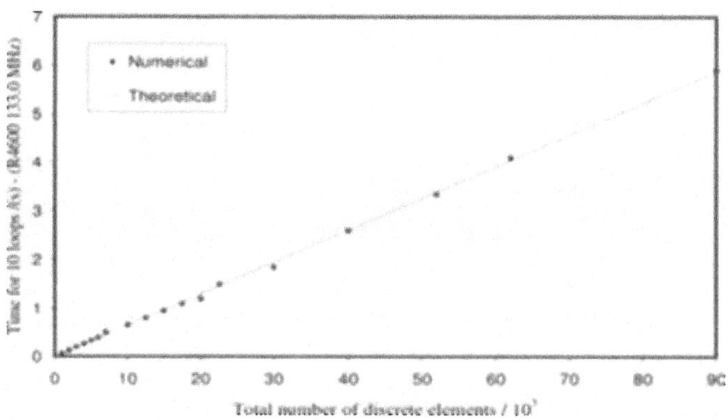

Fig. 6. CPU time as function of number of discrete elements.

5 MR contact detection algorithm

The MR contact detection algorithm comprises of three parts: a) Mapping of discrete elements onto cells. b) Sorting of discrete elements according to the cell to which they are mapped. For this operation Quick-Sort is readily available. However, the so called MR sort is used instead because it has a total sorting time proportional to the total number of discrete elements, which

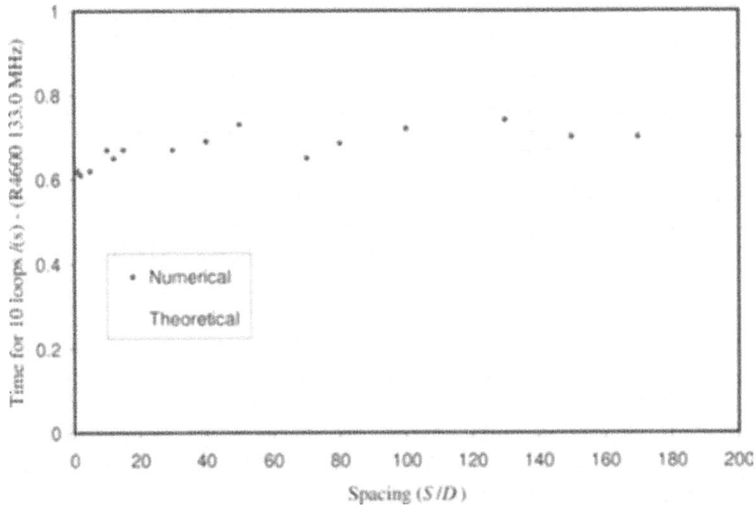

Fig. 7. CPU time as function of the packing density.

is much better performance than the often used Quick-Sort algorithm. c) Searching the sorted list for contacts. Again a binary search could be used. However, a novel MR search is used instead because it has a total search time proportional to the total number of discrete elements and is therefore much faster than the binary search. The total contact detection time comprises of the sort time and search time and is given by

$$T \propto N \tag{8}$$

The mapping of discrete elements onto the cells is represented by a list. In order for the list to represent the spatial distribution of the bounding boxes, it is necessary to sort the list according to a criterion that has spatial meaning as shown in Fig. 8.

This is achieved by using the MR-linear sort. The MR-Linear sort algorithm is based on the assumptions that no discrete body can move more than the size of a single cell in a given time interval. The list of discrete elements is parsed starting from the list head and discrete elements are moved within the list as required. This is facilitated by a 3 by 3 matrix of pointers to the discrete elements immediately before each of the neighbouring cells of the current discrete elements. This way the list is parsed only once and all the pointers are advanced only forward resulting in the theoretical CPU time for the MR-linear sort being given by.

$$T \propto N \tag{9}$$

Detection of contact is done by checking all the discrete elements mapped to a particular cell against all the discrete elements in neighbouring cells

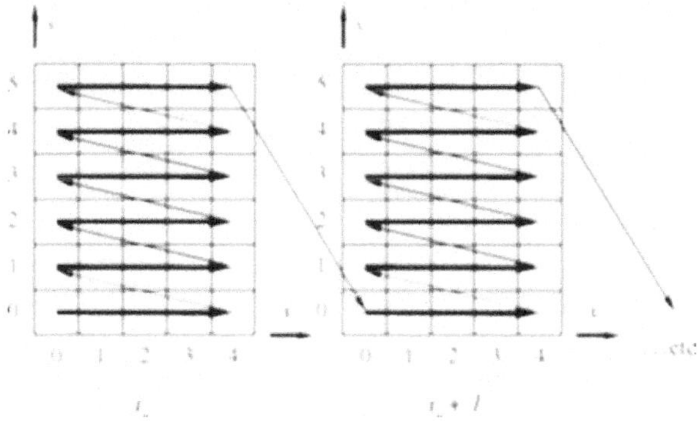

Fig. 8. Ordering of cells.

according to the contact mask. This is done by parsing the list of discrete elements starting with the list head and in the process remembering a set of pointers according to the contact mask. As the parsing advances the pointers are updated. Since the list is parsed only once and all the pointers are advanced only forward, it follows that the theoretical CPU time for the MR-Linear complexity search algorithm is given by

$$T \propto N \tag{10}$$

The memory requirements of the MR-linear contact detection algorithm are given by

$$\mathrm{M} = 4N \ (2\mathrm{D}); \ \ \mathrm{M} = 5N \ (3\mathrm{D}); \ \ \mathrm{M} = (2+n)\,N \ (\mathrm{nD\ -\ space}) \tag{11}$$

integer numbers. Neither memory nor CPU requirements are a function of the spatial distribution of the discrete elements.

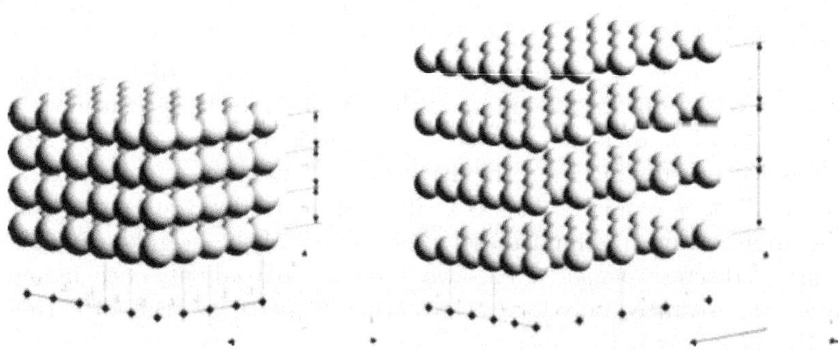

Fig. 9. Packing A (left) and packing B (right).

These properties are demonstrated using numerical examples shown in Fig. 9. For packing A contact detection is solved 10 times and the total CPU

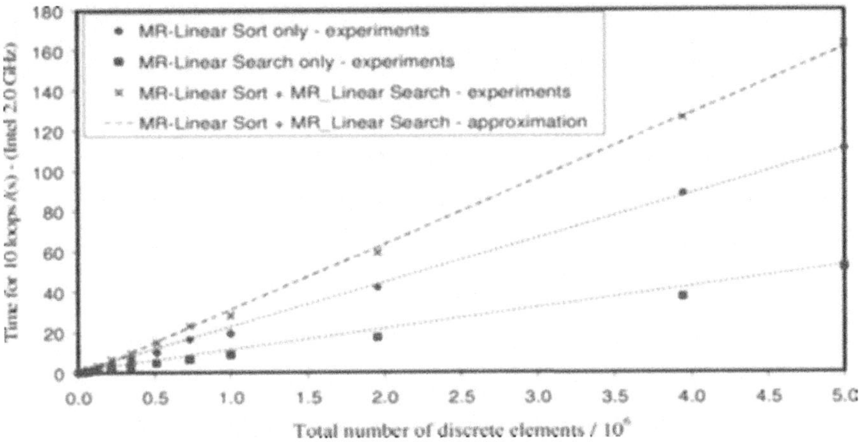

Fig. 10. MR search: CPU time as a function of the number of discrete elements - (Dell Dimension 4400 PC - Intel 2.0 GHz with 768 Mb RAM space).

time is recorded as shown in Fig. 10. The results clearly demonstrate linear complexity of MR sort, MR search and MR contact detection algorithms. Should one use Quick sort instead of MR-sort and binary search instead of MR search, the total CPU time to detect all contacts would be much longer, Fig. 11.

It is evident that MR-sort combined with MR-search improves CPU performance by 500%. The independence of performance of packing density is demonstrated in Fig. 12, where CPU time is recorded for packing B as a function of spacing distance s. The results obtained clearly show that a total detection time is not dependent on packing density, relative change of which goes from 1 to $1/(200 \cdot 200 \cdot 200) = 1/8{,}000{,}000$.

6 Concluding remarks

The essential features of the combined finite-discrete element method have been summarised together with contact resolution algorithms. Recently developed state of the art contact detection algorithms have been presented in more detail. The performance of these algorithms has been illustrated using numerical examples.

It is worth mentioning that the combined finite-discrete element method is a fast growing research area with a broad field of potential applications. It is therefore beyond the scope of this paper to go into the details of the method. These can be found elsewhere.

However, the author hopes that enough detail has been given for the interested reader to get the appreciation of the numerical composition of the combined finite-discrete element method and the classes of problems that can

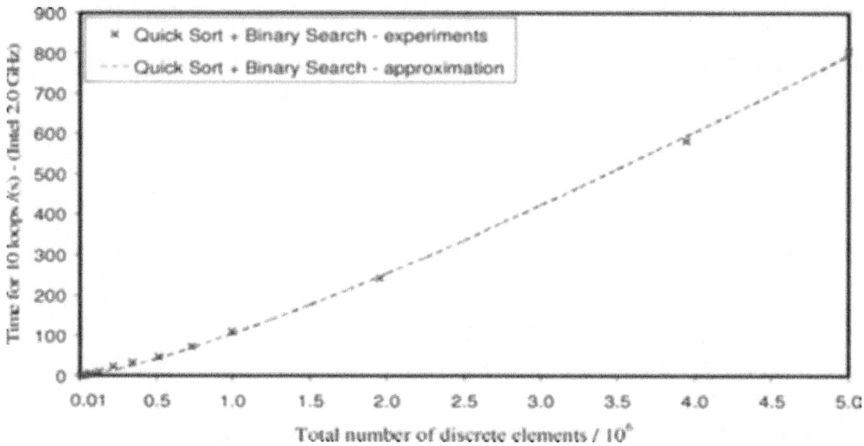

Fig. 11. Binary search: CPU time as a function of the number of discrete elements.

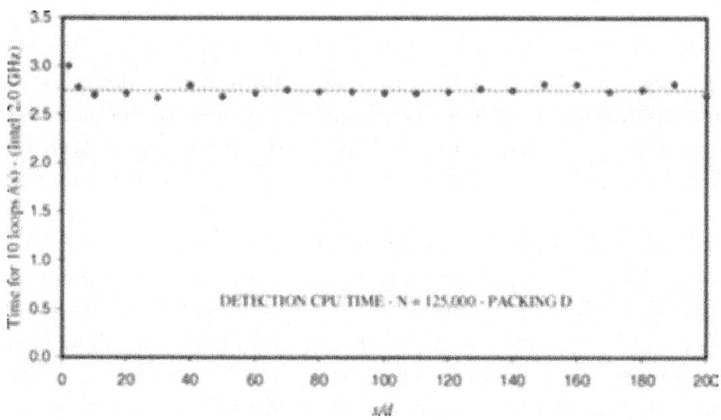

Fig. 12. CPU time as a function of packing density.

be addressed using the method. The author especially hopes that the neighbour search algorithms presented may find some applications in other areas of computational mechanics apart from the combined finite-discrete element method.

The method will be demonstrated further using the so called "movies from virtual worlds", which were made using the combined finite element simulations. The movies include some classic combined finite-discrete element applications, but also some atomic scale simulations that clearly demonstrate the ability of the method to produce the so called emergent properties.

Acknowledgements

The author wishes to express thanks to Prof. Wriggers for giving him an opportunity to present the method to this very special audience.

References

1. A. Munjiza (2004) 'The combined finite-discrete element method' John Wiley & Sons Inc, Chichester, UK.
2. P. Wriggers (2002) 'Computational contact mechanics' John Wiley & Sons Inc, Chichester, UK.
3. S.H. Chung and G.G.W. Mustoe (2002) 'Effects of particle shape and size distribution on stemming performance in blasting' Proc. 3rd International Conf. on Discrete Element Methods, Santa Fe, New Mexico, USA.
4. D.S. Preece and S.H. Chung, (2002) 'An algorithm for improving 2-D and 3-D spherical element behavior during formation of muck piles resulting from rock blasting', Proc. 3rd International Conf. on Discrete Element Methods, Santa Fe, New Mexico, USA.
5. M. Oldenburg and L. Nilsson (1994) 'The position code algorithm for contact searching', Int. J. Numer. Meth. Engng., 37, 359-386.
6. D.R.J. Owen, Y.T. Feng, M.G. Cottrel and J. Yu, (2002) 'Discrete/finite element modelling of industrial applications with multi-fracturing and particulate phenomena', Proc. 3rd International Conf. on Discrete Element Methods, Santa Fe, New Mexico, USA.
7. J.R. Williams and G.G.W. Mustoe (1993) Proc. 2nd U.S. Conf. on Discrete Element Methods, MIT, MA.
8. J. Bonet and J. Peraire (1991) 'An alternating digital tree (ADT) algorithm for 3D geometric searching and intersection problems', Int. J. Numer. Meth. Engng., 31, 1-17.
9. G.D.Knott (1971) 'A balanced tree storage and retrieval algorithm', Annual ACM Conference on Research and Development in Information Retrieval, p175-196.
10. A. Munjiza and K.R.F. Andrews (1998) 'NBS contact detection algorithm for bodies of similar size', Int. J. Num. Methods Eng., 43.
11. E. Perkins and J.R. Williams (2002) 'Generalized spatial binning of bodies of different sizes', Proc. 3rd International Conf. on Discrete Element Methods, Santa Fe, New Mexico, USA.
12. A. Munjiza, E. Rougier, N.W.M. John(2004) 'MR-linear contact detection algorithm for fragmenting solids', Advances in Boundary Element Techniques V, Lisbon Portugal, 279-286.

Numerical analysis of a dynamic viscoelastic contact problem with damage

M. Campo[1], J.R. Fernández[1], W. Han[2], and M. Sofonea[3]

[1] Departamento de Matemática Aplicada, Facultade de Matemáticas, Campus Univ. Sur s/n, Universidade de Santiago de Compostela, 15782 Santiago de Compostela, Spain,
macampo@usc.es, jramon@usc.es

[2] Department of Mathematics, University of Iowa, Iowa city, IA 52242-1410, USA,
whan@math.uiowa.edu

[3] Laboratoire de Mathématiques et Physique pour les Systèmes, Université de Perpignan, France,
sofonea@univ-perp.fr

Abstract. In this work, a dynamic frictionless viscoelastic contact problem is considered. The contact with the foundation is modelled by a normal compliance contact condition. The mechanical damage of the material, caused by excessive stress or strain, is included into the model through a differential inclusion. The weak formulation leads to a nonlinear system including a parabolic variational inequality for the damage field coupled with a variational equation for the displacement field. The existence of a unique weak solution is stated. Then, a fully discrete scheme is introduced using the finite element method to approximate the spatial variable and a finite difference to discretize the time derivatives. Error estimates are obtained, from which the linear convergence of the scheme, under suitable regularity conditions, can be derived. Finally, some numerical results on a two-dimensional problem are presented to show the performance of the scheme.

1 Introduction

Contact problems involving deformable bodies abound in industrial processes and everyday life, and the engineering literature relative to this topic is extensive (see, e.g., [6, 8] and references therein). As a result of the tensile or compressive stresses in the body, materials may suffer a dramatic decrease in their load bearing capacity, because of development of internal microcracks. This subject is extremely important in design engineering, since it directly affects the useful life-span of the designed structure and components.

In this work we deal with a model of dynamic contact between a viscoelastic body and a reactive foundation. Following [3], the evolution of the microscopic cracks causing the damage is taking into account and it is modelled with a function, the damage field, which satisfies a differential inclusion. The material is supposed to behave following a viscoelastic constitutive law in which the elasticity operator depends on the damage field. The contact is frictionless and it is modelled with normal compliance (see, e.g., [5]).

The paper is structured as follows. In Section 2 the mechanical model is presented together with its variational formulation, and an existence and uniqueness result is stated. It can be proved by using results of evolutionary variational inequalities and the Banach fixed point theorem. Then, a fully discrete scheme is introduced in Section 3, using the finite element method to approximate the spatial variable and the forward Euler scheme to discretize the time derivatives. An error bound is presented for the approximate solutions, from which linear convergence is deduced under suitable regularity conditions. Finally, in Section 4 some numerical results are shown in solving a two-dimensional test problem.

2 Mechanical problem and variational formulation

We follow [1] and we refer the reader there for details. We consider a viscoelastic body which occupies a domain $\Omega \subset \mathbb{R}^d$ for $d = 1, 2, 3$. The boundary Γ of Ω is assumed to be Lipschitz and is divided into three disjoint measurable parts Γ_D, Γ_F and Γ_C, such that meas $(\Gamma_D) > 0$. The body is clamped on $\Gamma_D \times (0, T)$ and so the displacement field vanishes there; a volume force of density $\boldsymbol{f}_0$ acts in $\Omega \times (0, T)$, and surface tractions of density $\boldsymbol{f}_2$ act on $\Gamma_F \times (0, T)$. Here, $T > 0$ and $[0, T]$ is the time interval of interest. The body may come in contact with a reactive foundation over the part Γ_C. A gap g exists between the potential contact surface Γ_C and the foundation, and is measured along the outward normal vector $\boldsymbol{\nu}$.

We denote by $\boldsymbol{u}$ the displacement field, $\boldsymbol{\sigma}$ the stress tensor and $\boldsymbol{\varepsilon}(\boldsymbol{u})$ the linearized strain tensor. Moreover, we denote by β, taking values in $[0, 1]$, the damage field which measures the density of the microcracks in the material. The material is assumed viscoelastic with the constitutive law ([7]),

$$\boldsymbol{\sigma} = \mathcal{A}\boldsymbol{\varepsilon}(\dot{\boldsymbol{u}}) + \mathcal{G}(\boldsymbol{\varepsilon}(\boldsymbol{u}), \beta),$$

where $\mathcal{A}$ and $\mathcal{G}$ are prescribed nonlinear constitutive functions. To simplify the notation, we do not indicate explicitly the dependence of various functions on the independent variables $\boldsymbol{x} \in \Omega \cup \Gamma$ and $t \in [0, T]$. Also, dots above a variable represent time derivatives.

We assume that the normal stress on Γ_C, denoted σ_ν, satisfies the *normal compliance* contact condition (see [5]),

$$-\sigma_\nu = p_\nu(u_\nu - g).$$

Here, $u_\nu = \boldsymbol{u} \cdot \boldsymbol{\nu}$ denotes the normal displacement, and $\sigma_\nu = (\boldsymbol{\sigma}\boldsymbol{\nu}) \cdot \boldsymbol{\nu}$ is the normal component of $\boldsymbol{\sigma}$. When $u_\nu > g$, the difference $u_\nu - g$ represents the interpenetration of the body's asperities into those of the foundation. The normal compliance function p_ν is prescribed and satisfies $p_\nu(r) = 0$ for $r \leq 0$, since the reaction of the foundation vanishes when there is no contact. In the numerical example presented in Section 4, we consider

$$p_\nu(r) = \frac{1}{\mu} r_+, \tag{1}$$

where μ is a positive constant and $r_+ = \max\{0, r\}$. We assume that the contact is frictionless, i.e. the tangential component of the stress, denoted $\boldsymbol{\sigma}_\tau$, vanishes on the contact surface.

We turn to describe the damage process. As a result of the tensile or compressive stresses in the body, microcracks open and grow and this, in turn, causes the load bearing of the material to decrease. This reduction in the strength of the material is modelled by introducing the damage field $\beta = \beta(\boldsymbol{x}, t)$. Following the derivation in [3], the evolution of the microscopic cracks responsible for the damage is described by the differential inclusion

$$\dot{\beta} - \kappa \triangle \beta + \partial \psi_K(\beta) \ni \phi(\varepsilon(\boldsymbol{u}), \beta).$$

Here, $\kappa > 0$ is a constant, $\partial\psi_K$ denotes the subdifferential of the indicator function ψ_K of the set $K = [\beta_*, 1]$, and ϕ is a given constitutive function which describes damage sources in the system. We restrict the damage to $\beta_* \leq \beta$, for some $\beta_* > 0$ small, since when the damage field takes values close to zero, the material is full of microcracks and it no longer makes sense to model it by a viscoelastic material law. Moreover, a homogeneous Neumann condition is assumed on Γ.

As an example of damage source function, we can take

$$\phi(\varepsilon(\mathbf{u}), \beta) = \lambda_1 \left(\frac{1 - \eta(\beta)}{\eta(\beta)} \right) - \frac{1}{2} \lambda_2 \Psi_{q^*}(\boldsymbol{u}) + \lambda_3, \tag{2}$$

where λ_1, λ_2 and λ_3 are process parameters, $\Psi_{q^*}(\boldsymbol{u}) = \min\{\varepsilon(\boldsymbol{u}) \cdot \varepsilon(\boldsymbol{u}), q^*\}$ for some constant $q^* > 0$ and $\eta : \mathbb{R} \mapsto \mathbb{R}$ is a function defined by

$$\eta(\beta) = \begin{cases} \beta & \text{if } \beta > \beta_*, \\ \beta_* & \text{if } \beta \leq \beta_*. \end{cases}$$

Let $\boldsymbol{u}_0$, $\boldsymbol{v}_0$ and β_0 represent the initial displacement, the initial velocity and the initial damage field, respectively.

With these assumptions, the variational formulation of the mechanical problem is as follows.

Problem P_V**.** *Find a displacement field* $\boldsymbol{u} : [0, T] \to V$*, a stress field* $\boldsymbol{\sigma} : [0, T] \to Q$*, and a damage field* $\beta : [0, T] \to \mathcal{K}$ *such that*

$$\begin{aligned}
&\boldsymbol{\sigma}(t) = \mathcal{A}\varepsilon(\dot{\boldsymbol{u}}(t)) + \mathcal{G}(\varepsilon(\boldsymbol{u}(t)), \beta(t)), \\
&(\rho\ddot{\boldsymbol{u}}(t), \boldsymbol{w})_H + (\boldsymbol{\sigma}(t), \varepsilon(\boldsymbol{w}))_Q + j(\boldsymbol{u}(t), \boldsymbol{w}) = \langle \boldsymbol{f}(t), \boldsymbol{w} \rangle_{V' \times V} \quad \forall \boldsymbol{w} \in V, \\
&(\dot{\beta}(t), \xi - \beta(t))_{L^2(\Omega)} + a(\beta(t), \xi - \beta(t)) \\
&\qquad \geq (\phi(\varepsilon(\boldsymbol{u}(t)), \beta(t)),\ \xi - \beta(t))_{L^2(\Omega)} \quad \forall \xi \in \mathcal{K},
\end{aligned}$$

for almost any $t \in (0,T)$, and

$$\boldsymbol{u}(0) = \boldsymbol{u}_0, \quad \dot{\boldsymbol{u}}(0) = \boldsymbol{v}_0, \quad \beta(0) = \beta_0.$$

Here, $H = [L^2(\Omega)]^d$, $Q = [L^2(\Omega)]^{d\times d}$ and $(\cdot\,,\,\cdot)_H$, $(\cdot\,,\,\cdot)_Q$ denote the usual inner products on these spaces. Also, V is the closed subspace of the space $[H^1(\Omega)]^d$ given by

$$V = \{\boldsymbol{v} \in [H^1(\Omega)]^d \ ; \ \boldsymbol{v} = \boldsymbol{0} \text{ on } \Gamma_D\}.$$

Moreover, V' denotes the dual space of the space V and $\langle\cdot\,,\,\cdot\rangle_{V'\times V}$ represents the duality pairing mapping between V' and V. The following notations are also used: ρ is the mass density, $\mathcal{K}$ is the set of admissible damage functions,

$$\mathcal{K} = \{\xi \in H^1(\Omega) \ ; \ \beta_* \le \xi \le 1 \quad \text{a.e. in } \Omega\},$$

and $\boldsymbol{f} : [0,T] \to V'$ is the function

$$\langle \boldsymbol{f}(t), \boldsymbol{v}\rangle_{V'\times V} = (\boldsymbol{f}_0(t), \boldsymbol{v})_H + (\boldsymbol{f}_2(t), \boldsymbol{v})_{[L^2(\Gamma_F)]^d} \qquad \forall \boldsymbol{v} \in V, \ t \in [0,T].$$

Also, $j : V \times V \to \mathbb{R}$ and $a : H^1(\Omega) \times H^1(\Omega) \to \mathbb{R}$ are the forms given by

$$j(\boldsymbol{u}, \boldsymbol{v}) = \int_{\Gamma_C} p_\nu(u_\nu - g)\, v_\nu \, da \qquad \forall \boldsymbol{u}, \boldsymbol{v} \in V,$$
$$a(\xi, \psi) = \kappa \int_\Omega \nabla\xi \cdot \nabla\psi \, dx \qquad \forall\, \xi, \, \psi \in H^1(\Omega),$$

where $v_\nu = \boldsymbol{v} \cdot \boldsymbol{\nu}$ for all $\boldsymbol{v} \in V$.

Under appropriate assumptions on the data, it was proved in [1] that Problem P_V has a unique solution which satisfies

$$\begin{aligned} &\boldsymbol{u} \in H^1(0,T;V) \cap C^1([0,T];H) \cap L^2(0,T;V'),\\ &\boldsymbol{\sigma} \in L^2(0,T;Q), \quad \text{Div}\,\boldsymbol{\sigma} \in L^2(0,T;V'),\\ &\beta \in H^1(0,T;L^2(\Omega)) \cap L^2(0,T;H^1(\Omega)). \end{aligned}$$

Here and below, we use standard notation for functions defined in the interval $[0,T]$ with values on a real normed space.

3 A fully discrete scheme

In this section, we consider a numerical scheme for Problem P_V. It is done in two steps. First, we use two finite dimensional spaces $V^h \subset V$ and $B^h \subset H^1(\Omega)$ to approximate the spaces V and $H^1(\Omega)$, respectively, and we denote by $\mathcal{K}^h \subset B^h$ a convex set which approximates the convex set $\mathcal{K}$. Here and below $h > 0$ denotes the discretization parameter. Secondly, to discretize the time derivatives, consider a uniform partition of the time interval $[0,T]$, denoted by $0 = t_0 < t_1 < \ldots < t_N = T$, and let $k = T/N$ be the time step size. For a continuous function $f(t)$, we use the notation $f_n = f(t_n)$. In this section, c denotes a positive constant independent of the discretization parameters h and k.

Let $\boldsymbol{u}_0^h, \boldsymbol{v}_0^h \in V^h$ and $\beta_0^h \in \mathcal{K}^h$ be appropriate approximations of the initial conditions $\boldsymbol{u}_0$, $\boldsymbol{v}_0$ and β_0, respectively. A fully discrete approximation of Problem P_V is the following.

Problem P_V^{hk}**.** *Find a discrete velocity field* $\boldsymbol{v}^{hk} = \{\boldsymbol{v}_n^{hk}\}_{n=0}^N \subset V^h$ *and a discrete damage field* $\beta^{hk} = \{\beta_n^{hk}\}_{n=0}^N \subset \mathcal{K}^h$ *such that* $\boldsymbol{v}_0^{hk} = \boldsymbol{v}_0^h$, $\beta_0^{hk} = \beta_0^h$ *and for* $n = 1, 2, \ldots, N$,

$$\begin{aligned}&(\rho\delta\boldsymbol{v}_n^{hk}, \boldsymbol{w}^h)_H + (\mathcal{A}\boldsymbol{\varepsilon}(\boldsymbol{v}_n^{hk}), \boldsymbol{\varepsilon}(\boldsymbol{w}^h))_Q + (\mathcal{G}(\boldsymbol{\varepsilon}(\boldsymbol{u}_{n-1}^{hk}), \beta_{n-1}^{hk}), \boldsymbol{\varepsilon}(\boldsymbol{w}^h))_Q\\ &\qquad + j(\boldsymbol{u}_{n-1}^{hk}, \boldsymbol{w}^h) = \langle \boldsymbol{f}_n, \boldsymbol{w}^h\rangle_{V'\times V} \quad \forall \boldsymbol{w}^h \in V^h,\end{aligned}$$

$$\begin{aligned}&(\delta\beta_n^{hk}, \xi^h - \beta_n^{hk})_{L^2(\Omega)} + a(\beta_n^{hk}, \xi^h - \beta_n^{hk})\\ &\qquad \geq (\phi(\boldsymbol{\varepsilon}(\boldsymbol{u}_{n-1}^{hk}), \beta_{n-1}^{hk}), \xi^h - \beta_n^{hk})_{L^2(\Omega)} \quad \forall \xi^h \in \mathcal{K}^h,\end{aligned}$$

$$\boldsymbol{u}_n^{hk} = \sum_{j=1}^n k\boldsymbol{v}_j^{hk} + \boldsymbol{u}_0^h.$$

Here, $\delta\boldsymbol{v}_n^{hk} = (\boldsymbol{v}_n^{hk} - \boldsymbol{v}_{n-1}^{hk})/k$, $\delta\beta_n^{hk} = (\beta_n^{hk} - \beta_{n-1}^{hk})/k$.

Using classical arguments on variational inequalities (see [4]), we can deduce the existence of a unique solution to Problem P_V^{hk}.

It was proved in [1] that, under the additional regularity condition

$$\boldsymbol{u} \in C^1([0,T];V) \cap C^2([0,T];V'), \quad \beta \in H^2(0,T;L^2(\Omega)) \cap C([0,T];H^2(\Omega)),$$

the following error relation holds for all $\{\boldsymbol{w}_j^h\}_{j=1}^N \subset V^h$ and $\{\xi_j^h\}_{j=1}^N \subset \mathcal{K}^h$:

$$\begin{aligned}&\max_{0\leq n\leq N}\Big\{\|\boldsymbol{v}_n - \boldsymbol{v}_n^{hk}\|_H^2 + \|\beta_n - \beta_n^{hk}\|_{L^2(\Omega)}^2\Big\}\\ &\qquad + k\sum_{j=1}^N (\|\boldsymbol{v}_j - \boldsymbol{v}_j^{hk}\|_V^2 + \|\nabla(\beta_j - \beta_j^{hk})\|_H^2)\\ &\leq c\Big\{\sum_{j=1}^N k\Big(\|\dot{\boldsymbol{v}}_j - \delta\boldsymbol{v}_j\|_{V'}^2 + I_j^2 + \|\boldsymbol{u}_j - \boldsymbol{u}_{j-1}\|_V^2 + \|\boldsymbol{v}_j - \boldsymbol{w}_j^h\|_V^2\Big)\\ &\quad + \|\boldsymbol{v}_0 - \boldsymbol{v}_0^h\|_H^2 + \|\boldsymbol{u}_0 - \boldsymbol{u}_0^h\|_V^2 + \|\boldsymbol{v}_1 - \boldsymbol{w}_1^h\|_H^2 + \max_{0\leq n\leq N}\|\boldsymbol{v}_n - \boldsymbol{w}_n^h\|_H^2\\ &\quad + \|\beta_0 - \beta_0^h\|_{L^2(\Omega)}^2 + \|\beta_1 - \xi_1^h\|_{L^2(\Omega)}^2 + \max_{0\leq n\leq N}\|\beta_n - \xi_n^h\|_{L^2(\Omega)}^2 + k^2\\ &\quad + \frac{1}{k}\sum_{j=1}^{N-1}\|(\beta_{j+1} - \xi_{j+1}^h) - (\beta_j - \xi_j^h)\|_{L^2(\Omega)}^2 + k\sum_{j=1}^N\|\delta\beta_j - \dot{\beta}_j\|_{L^2(\Omega)}^2\\ &\quad + k\sum_{j=1}^N\|\phi(\boldsymbol{\varepsilon}(\boldsymbol{u}_j), \beta_j) - \delta\beta_j + \kappa\Delta\beta_j\|_{L^2(\Omega)}\cdot\|\beta_j - \xi_j^h\|_{L^2(\Omega)}\end{aligned}$$

$$+k\sum_{j=1}^{N}\|\beta_j-\xi_j^h\|^2_{H^1(\Omega)}+\frac{1}{k}\sum_{j=1}^{N-1}\|\boldsymbol{v}_j-\boldsymbol{w}_j^h-(\boldsymbol{v}_{j+1}-\boldsymbol{w}_{j+1}^h)\|^2_H\Big\}, \quad (3)$$

where

$$I_j=\Big\|\int_0^{t_j}\boldsymbol{v}(s)ds-\sum_{l=1}^{j}k\boldsymbol{v}_l\Big\|_V.$$

and $\boldsymbol{v}$ denotes the velocity field, i.e. $\boldsymbol{v}=\dot{\boldsymbol{u}}$.

The error relation (3) is a basis for error estimation. For instance, assume Ω is a polyhedral domain and denote by $\{\mathcal{T}^h\}$ a regular family of triangulations of $\overline{\Omega}$ compatible with the partition of the boundary $\Gamma=\partial\Omega$ into Γ_D, Γ_F and Γ_C. Let the spaces V^h and B^h consist of continuous and piecewise affine functions,

$$V^h=\{\boldsymbol{v}^h\in[C(\overline{\Omega})]^d\ ;\ \boldsymbol{v}^h|_T\in[P_1(T)]^d,\quad T\in\mathcal{T}^h\}, \quad (4)$$
$$B^h=\{\xi^h\in C(\overline{\Omega})\ ;\ \xi^h|_T\in P_1(T),\quad T\in\mathcal{T}^h\}, \quad (5)$$

and define the following convex subset of B^h,

$$\mathcal{K}^h=\{\xi^h\in B^h\ ;\ \beta_*\le\xi^h\le 1\}.$$

Assume now the additional solution regularity

$$\boldsymbol{u}\in H^2(0,T;V)\cap C^1([0,T];[H^2(\Omega)]^d),\quad \dddot{\boldsymbol{u}}\in L^2(0,T;V'),$$
$$\beta\in C([0,T];H^2(\Omega))\cap H^2(0,T;L^2(\Omega))\cap H^1(0,T;H^1(\Omega)),$$

and let the discrete initial conditions $\boldsymbol{u}_0^h$, $\boldsymbol{v}_0^h$ and β_0^h be defined by interpolation:

$$\boldsymbol{u}_0^h=\Pi^h\boldsymbol{u}_0,\quad \boldsymbol{v}_0^h=\Pi^h\boldsymbol{v}_0,\quad \beta_0^h=\Pi^h\beta_0,$$

where Π^h is the standard finite element interpolation operator (see [2]). Then, it can be proved that there exists $c>0$, independent of h and k, such that,

$$\max_{0\le n\le N}\Big\{\|\boldsymbol{v}_n-\boldsymbol{v}_n^{hk}\|_H+\|\beta_n-\beta_n^{hk}\|_{L^2(\Omega)}\Big\}\le c(h+k),$$

i.e., the fully discrete scheme is linearly convergent.

4 Numerical results

As a test example we consider an L-shaped body which is subject to the action of traction forces on its upper horizontal boundary, see Fig. 1 for details. The body is clamped on Γ_D and is in contact with a foundation on Γ_C.

The elasticity part $\mathcal{G}(\varepsilon(\boldsymbol{u}),\beta)$ has the form $\mathcal{G}(\varepsilon(\boldsymbol{u}),\beta)=\beta\Phi(\mathcal{B}\varepsilon(\boldsymbol{u}))$, where $\mathcal{B}$ is the two-dimensional elasticity tensor under the plane stress hypothesis (E and ν are Young's modulus and Poisson's ratio of the material which occupies

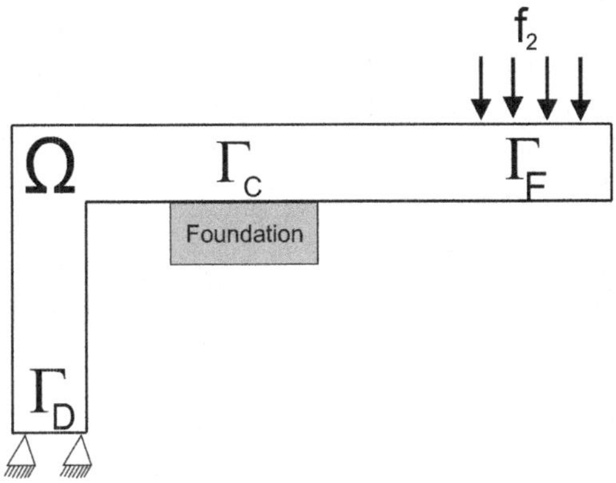

Fig. 1. Contact problem of an L-shaped domain.

Ω, respectively). Moreover, $\Phi : Q \to Q$ is a truncation operator defined by $(\Phi(\boldsymbol{\tau}))_{ij} = L$ if $\tau_{ij} > L$, τ_{ij} if $\tau_{ij} \in [-L, L]$, and $-L$ if $\tau_{ij} < -L$. Here, $L > 0$ is a given constant. We notice that the existence of L is justified by taking into account that the small displacement theory is used. The value $L = 1000$ has been used in this example.

The fourth-order viscoelasticity tensor $\mathcal{A}$ has the following form,

$$(\mathcal{A}\boldsymbol{\tau})_{\alpha\beta} = \eta_1(\tau_{11} + \tau_{22})\delta_{\alpha\beta} + \eta_2\tau_{\alpha\beta}, \quad 1 \le \alpha, \beta \le 2, \ \boldsymbol{\tau} \in Q,$$

where η_1 and η_2 are viscosity constants. Recall that the von Mises norm for a plane stress field $\boldsymbol{\tau} = (\tau_{\alpha\beta})$ is given by

$$\|\boldsymbol{\tau}\|_{VM} = \left(\tau_{11}^2 + \tau_{22}^2 - \tau_{11}\tau_{22} + 3\tau_{12}^2\right)^{\frac{1}{2}}.$$

We employ (1) and (2) as the normal compliance function and the damage source function, respectively. Moreover, we use the finite element spaces V^h and B^h defined in (4) and (5).

The rest of the data are as follows:

$$\begin{aligned}
&T = 10.5\,s, \quad \boldsymbol{f}_0 = \boldsymbol{0}\,N/m^3, \quad \boldsymbol{f}_2(x_1, x_2, t) = (0, -7)|\sin \pi t|\,N/m^2,\\
&\mu = 0.5 \times 10^{-5}\,N/m^2, \quad \kappa = 1, \quad E = 10000\,N/m^2, \quad \nu = 0.3, \quad \beta^* = 0.01,\\
&\rho = 5 \times 10^{-2}\,N \cdot s^2/m^3, \quad \lambda_1 = 5 \times 10^{-4}, \quad \lambda_2 = 10^3, \quad \lambda_3 = 0,\\
&\eta_1 = 57.69\,N \cdot s/m^2, \quad \eta_2 = 38.46\,N \cdot s/m^2, \quad q^* = 1, \quad g = 0\,m,\\
&\boldsymbol{u}_0 = \boldsymbol{0}\,m, \quad \dot{\boldsymbol{u}}(0) = \boldsymbol{0}\,m/s, \quad \beta_0 = 1.
\end{aligned}$$

In order to show the influence of the damage in the evolution of the process, two simulations have been done: the first one assuming that the material keeps undamaged (that is, $\beta(\boldsymbol{x}, t) = 1$ for all $\boldsymbol{x} \in \Omega$ and $t \in [0, T]$), and the second one when the damage takes place. The value $k = 0.01$ was employed for the time step.

The deformed meshes (amplified by 100) at final time and the initial configurations are plotted in Fig. 2 (the undamaged case on the left-hand side and the damaged one on the right-hand side). It results from these figures that inclusion of the damage effect leads to an increase of the magnitude of the deformation field.

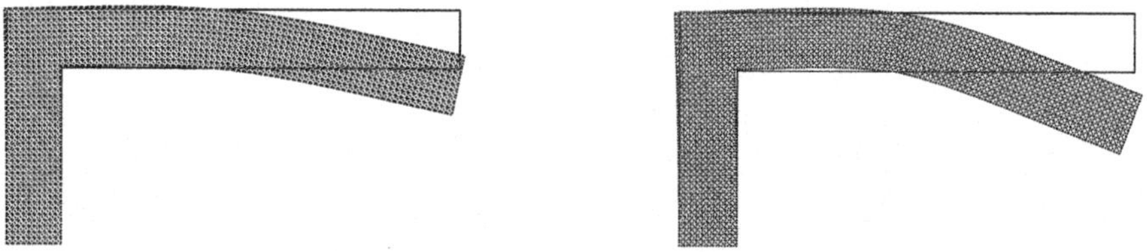

Fig. 2. Deformed meshes (amplified by 100) and initial configuration.

References

1. Campo M, Fernández JR, Han W, Sofonea M (200x) Finite Elem. Anal. Des. (to appear).
2. Ciarlet PG (1991), The Finite element method for elliptic problems. In: Ciarlet PG, Lions JL (eds) Handbook of Numerical Analysis, Volume II, Part 1. North Holland, 17–352.
3. Frémond M, Nedjar B (1996) Internat. J. Solids Structures 33(8):1083–1103.
4. Glowinski R (1984) Numerical Methods for Nonlinear Variational Problems. Springer-Verlag, New York.
5. Klarbring A, Mikelic A, Shillor M (1988) Internat. J. Engrg. Sci. 26:811–832.
6. Laursen TA (2002) Computational contact and impact mechanics: fundamentals of modeling interfacial phenomena in nonlinear finite element analysis. Springer, New York.
7. Shillor M, Sofonea M, Telega JJ (2004) Models and analysis of quasistatic contact. Lecture Notes in Physics, Vol. 655. Springer, Berlin.
8. Wriggers P (2002) Computational contact mechanics. John Wiley and Sons Ltd.

An energy-conserving algorithm for nonlinear elastodynamic contact problems – Extension to a frictional dissipation phenomenon

M. Barboteu

University of Perpignan, 52 avenue Paul-Alduy, 66860 Perpignan, France, barboteu@univ-perp.fr

1 Introduction

During the last years, the construction of energy conserving time integration methods to solve nonlinear elastodynamic problems has attracted the interest of many researchers ([12, 5, 3] ...). Furthermore, many works have been devoted to extend these conservative formulations to frictionless impact; more precisely, Laursen and Chawla [9] and Amero and Petocz [2] have shown the interest of the persistency condition to conserve the energy in the discrete framework. But these contributions concede a contact interpenetration which vanishing as the time step tends towards zero. Recently, this drawback is resolved by Laursen and Love [10] by introducing a discrete jump in velocity and by Hauret [6] by considering a specific penalized enforcement of the contact conditions. In this work, we present an energy-conserving algorithm for hyperelastodynamic contact problems which differs from the approaches mentionned above ([10] and [6]); this approach permits to ensure both the Kuhn-Tucker and persistency conditions at the end of each time step. These two laws are enforced during each time increment by using an extended Newton method. In section 2, we recall some general aspects of nonlinear elastodynamic problems with contact and friction. The section 3 permits also to recall the usual energy conserving frameworks used to solve nonlinear elastodynamic problems. In section 4, we present an energy-conserving algorithm to treat impact problems with an extension to frictional dissipation phenomenon. In the last section 5, representative numerical simulations are presented to assess the performance and also to underscore the conservative or dissipative behaviour of the proposed method.

2 Nonlinear elastodynamic problems

Dynamic deformable body systems in large deformations are governed by nonlinear time dependent equations. In this work, we consider nonlinear elastic behaviours which are characterized by hyperelastic constitutive laws. The first Piola-Kirchoff tensor $\mathbf{\Pi}$ is given by the relation $\mathbf{\Pi} = \partial_{\mathbf{F}} W(\mathbf{F})$ where $W(\mathbf{F})$ is the internal hyperelastic energy and $\mathbf{F}$ is the deformation gradient defined by $\mathbf{F} = \mathbf{I} + \nabla \mathbf{u}$ ($\mathbf{u}$ represents the displacement vector). Moreover, the frictional contact phenomenon is modelled by combining the normal unilateral contact law and the tangential frictional Coulomb law with *variable* pressure [11]. These multivalued relations depend on the normal gap distance $d_{\boldsymbol{\nu}}$, the tangential contact velocity $\overset{o}{\mathbf{d}}_{\boldsymbol{\tau}}$ and the frictional contact stress $\boldsymbol{\Gamma}$ (split into a normal contact pressure $\Gamma_{\boldsymbol{\nu}}$ and a tangential part $\boldsymbol{\Gamma}_{\boldsymbol{\tau}}$); they can be written as follows:

$$\text{contact law:} \begin{cases} d_{\boldsymbol{\nu}} \geq 0, \\ \Gamma_{\boldsymbol{\nu}} \leq 0, \\ d_{\boldsymbol{\nu}} \Gamma_{\boldsymbol{\nu}} = 0. \end{cases} \qquad \text{friction law:} \begin{cases} \overset{o}{\mathbf{d}}_{\boldsymbol{\tau}} = \| \overset{o}{\mathbf{d}}_{\boldsymbol{\tau}} \| \frac{\boldsymbol{\Gamma}_{\boldsymbol{\tau}}}{\|\boldsymbol{\Gamma}_{\boldsymbol{\tau}}\|}, \\ \|\boldsymbol{\Gamma}_{\boldsymbol{\tau}}\| + \mu \Gamma_{\boldsymbol{\nu}} \leq 0, \\ \| \overset{o}{\mathbf{d}}_{\boldsymbol{\tau}} \| (\|\boldsymbol{\Gamma}_{\boldsymbol{\tau}}\| + \mu \Gamma_{\boldsymbol{\nu}}) = 0. \end{cases} \tag{1}$$

where $\boldsymbol{\nu}$ and $\boldsymbol{\tau}$ are the normal and tangent contact unit vectors on $\partial_c \Omega$; μ is the friction coefficient. For a definition of the quantities $d_{\boldsymbol{\nu}}$ and $\overset{o}{\mathbf{d}}_{\boldsymbol{\tau}}$ in the framework of large deformations, we can refer to [4]. These tribological laws can be written in the form of subdifferential inclusions which derive from conjugate non differentiable convex potentials (in the sense of the subgradient [11]),

$$\Gamma_{\boldsymbol{\nu}} \in \partial \Psi_{\mathbb{R}^+}(d_{\boldsymbol{\nu}}) \quad \text{and} \quad \boldsymbol{\Gamma}_{\boldsymbol{\tau}} \in \partial \Psi^*_{C[\Gamma_{\boldsymbol{\nu}}]}(\overset{o}{\mathbf{d}}_{\boldsymbol{\tau}}), \tag{2}$$

where $\partial \Psi_{\mathbb{R}^+}$ and $\partial \Psi^*_{C[\Gamma_{\boldsymbol{\nu}}]}$ denote, respectively, the subdifferential of the indicator function $\Psi_{\mathbb{R}^+}$ of the positive half-line $\mathbb{R}$ and the subdifferential of Fenchel conjuguated of the indicator function $\Psi^*_{C[\Gamma_{\boldsymbol{\nu}}]}$ of $C[\Gamma_{\boldsymbol{\nu}}]$. $C[\Gamma_{\boldsymbol{\nu}}]$ denotes the convex disk of radius $-\mu \Gamma_{\boldsymbol{\nu}}$. We can note that this convex disk depends on the unknown normal contact stress $\Gamma_{\boldsymbol{\nu}}$. So, a typical nonlinear elastodynamic problem defined in a reference configuration can be governed by the following variational form,

$$\begin{cases} \text{Find } \mathbf{u} \in L^2(]0;T[;U) \quad \text{such that for each } t \in]0;T[, \\ \int_\Omega \rho \ddot{\mathbf{u}}.\mathbf{v} + \int_\Omega \mathbf{\Pi} : \nabla \mathbf{v} - \int_\Omega \mathbf{f}.\mathbf{v} - \int_{\partial_g \Omega} \mathbf{g}.\mathbf{v} + \mathcal{P}_{cont-frot} = 0, \quad \forall \mathbf{v} \in U \\ \frac{\partial^2 \mathbf{u}}{\partial t^2} = \ddot{\mathbf{u}} \quad \text{in } U \end{cases} \tag{3}$$

where ρ denotes the mass density; $\mathbf{f}$ and $\mathbf{g}$ are the external force densities. A dot superscript indicates the time derivative. The set $U = \{\mathbf{v} \in H^1(\Omega)^{dim}; \mathbf{v} = \mathbf{0} \text{ on } \partial_0 \Omega\}$ represents the space of kinematically admissible displacement fields. The frictional contact term is given by

$$\mathcal{P}_{cont-frot}(\mathbf{u}, \mathbf{v}) = \int_{\partial_c \Omega} \left[\Gamma_{\boldsymbol{\nu}} \delta d_{\boldsymbol{\nu}} + \boldsymbol{\Gamma}_{\boldsymbol{\tau}}.\delta \mathbf{d}_{\boldsymbol{\tau}} \right], \tag{4}$$

where $\partial_c \Omega$ denotes the surface contact. $\delta d_{\boldsymbol{\nu}}$ and $\delta \mathbf{d}_{\boldsymbol{\tau}}$ represents respectively the directional derivatives of $d_{\boldsymbol{\nu}}$ and $\mathbf{d}_{\boldsymbol{\tau}}$ belong the direction $\mathbf{v}$ (see [8]). The previous weak form (3) is verified for all admissible virtuel displacements such that the frictional contact conditions (1) are satisfied. If the material variation $\mathbf{v}$ is replaced by the material velocity $\mathbf{v}_t$, the frictional contact term (4) becomes the following frictional contact work

$$\mathcal{W}_{cont-frot} = \mathcal{P}_{cont-frot}(\mathbf{u}, \mathbf{v}_t) = \int_{\partial_c \Omega} \left[\Gamma_\nu \dot{d}_\nu + \boldsymbol{\Gamma}_\tau . \overset{o}{\mathbf{d}}_\tau \right],$$

where $\dot{d}_\nu$ and $\overset{o}{\mathbf{d}}_\tau$ represents respectively the material time derivatives of d_ν and $\mathbf{d}_\tau$ (see [8]).

In addition, in order to obtain energy conservation properties and in the absence of external forces, the work of normal contact reactions at time t ($\mathcal{W}_{cont} = \int_{\partial_c \Omega} \Gamma_\nu \dot{d}_\nu$) must vanished [9]; so for energy conservation purpose, the following persistency condition [9, 2] has to be added,

$$\text{persistency condition: } \dot{d}_\nu \Gamma_\nu = 0. \tag{5}$$

This condition means that normal contact reactions can only appear during persistent contact. One can easily prove [8] that the addition of the persistency condition (5) to the unilateral contact law gives the following law:

$$\text{persistent contact law: } \begin{cases} \text{if } d_\nu > 0 & \Gamma_\nu = 0 \\ \text{if } d_\nu = 0 & \Gamma_\nu \in \partial \Psi_{\mathbb{R}^+}(\dot{d}_\nu) \end{cases} \tag{6}$$

3 Usual energy conserving frameworks

In order to solve the problem (3), we have to use a time integration scheme. When one considers nonlinear dynamic problems, the standard implicit schemes (θ-method, Newmark schemes, midpoint or HHT methods, see for example [7, 8] ...) lose their unconditional stability. So we have to use implicit energy conservative schemes [12, 5, 3, 8, 6] which are appropriate due to their long term time integration accuracy and stability. These methods are based on the satisfaction of discrete mechanical conservation properties. In the following, we consider a collection of discrete times $(t_p)_{p=1..P}$ which define a partition of the time interval $[0;T] = \bigcup_{p=1}^{P} [t_p; t_{p+1}]$ with $t_{p+1} = t_p + \Delta t$ and $\Delta t = \frac{T}{P}$. By using a second order time integration scheme (midpoint scheme) with energy conservation properties [12, 5, 6], the weak form (3) integrated between the times t_p and t_{p+1} gives the following system,

$$\begin{cases} \text{Find } \mathbf{u}_{p+1} \in U_0 \quad \text{such that} \\ \frac{1}{\Delta t} \int_\Omega \rho(\dot{\mathbf{u}}_{p+1} - \dot{\mathbf{u}}_p).\mathbf{v} + \int_\Omega \mathbf{\Pi}_{algo} : \nabla \mathbf{v} - \int_\Omega \mathbf{f}_{p+\frac{1}{2}}.\mathbf{v} - \int_{\partial_g \Omega} \mathbf{g}_{p+\frac{1}{2}}.\mathbf{v} \\ \qquad + \frac{1}{\Delta t} \int_{t_p}^{t_p+1} \mathcal{P}_{cont-frot} dt = 0 \end{cases} \tag{7}$$

In the sytem (7), $\square_{p+\frac{1}{2}} = \frac{1}{2}(\square_p + \square_{p+1})$ and $\square_p$ denotes the approximation of $\square(t_p)$. The midpoint scheme gives the following relation: $\dot{\mathbf{u}}_{p+1} = -\dot{\mathbf{u}}_p +$

$\frac{2}{\Delta t}(\mathbf{u}_{p+1} - \mathbf{u}_p)$. Moreover, the time integration scheme (7) used in this work, is characterized by the tensor $\mathbf{\Pi}_{algo}$ proposed by Gonzalez [5] and defined by

$$\begin{cases} \mathbf{\Pi}_{algo} = \mathbf{F}_{p+\frac{1}{2}} \mathbf{\Sigma}_{algo}, \\ \mathbf{\Sigma}_{algo} = 2\frac{\partial W}{\partial \mathbf{C}}(\mathbf{C}_{p+\frac{1}{2}}) + 2[W(\mathbf{C}_{p+1}) - W(\mathbf{C}_p) \\ \qquad\qquad -\frac{\partial W}{\partial \mathbf{C}}(\mathbf{C}_{p+\frac{1}{2}}) : \Delta \mathbf{C}_p] \frac{\Delta \mathbf{C}_p}{\Delta \mathbf{C}_p : \Delta \mathbf{C}_p}, \end{cases} \tag{8}$$

with $\Delta \mathbf{C}_p = \mathbf{C}_{p+1} - \mathbf{C}_p$. The previous relations (8) were introduced in order to ensure exact energy conservation characterized by the following condition,

$$\mathbf{\Pi}_{algo} : (\nabla \mathbf{u}_{p+1} - \nabla \mathbf{u}_p) = W(\mathbf{C}_{p+1}) - W(\mathbf{C}_p) \quad \text{with} \quad \mathbf{C}_{p+1} = \mathbf{F}^t_{p+1} \mathbf{F}_{p+1}.$$

Furthermore, many works have been devoted to extend these conservative formulations to frictionless impact; more precisely, Laursen and Chawla [9] and Amero and Petocz [2] have shown the interest of the persistency condition to conserve the energy in the discrete framework. But these works are characterized by a contact interpenetration that only can vanished when the time step tends towards zero. Recently, in order to overcome this drawback Laursen and Love [10] have developed an efficient method by introducing a discrete jump in velocity; but this method requires the solution of an auxilary system in order to compute the velocity update results. Furthermore, Hauret [6] has considered a specific penalized enforcement of the contact conditions which permits implicitly to ensure the persistency condition.

4 Adaptation to impact problems

In this section, we present an energy-conserving algorithm for hyperelastodynamic contact problems which differs from the approaches mentionned above ([10] and [6]). Indeed, as we said in introduction, this method permits to ensure both the Kuhn-Tucker and persistency conditions at the end of each time step without any contact penetrations and with lower cost.

Furthermore, we choose to enforce the frictional contact conditions at the time t_{p+1}, i.e. the dicrete versions of the unilateral contact law ($\Gamma_{\boldsymbol{\nu}_{p+1}} \in \partial \Psi_{\mathbb{R}^+}(d_{\boldsymbol{\nu}_{p+1}})$), and of the Coulomb's friction law ($\boldsymbol{\Gamma}_{\boldsymbol{\tau}_{p+1}} \in \partial \Psi^*_{C[\Gamma_{\boldsymbol{\nu}_{p+1}}]}(\delta_p \overset{o}{\mathbf{d}}_{\boldsymbol{\tau}_{p+1}})$) but also of the persistent contact law ($\Gamma_{\boldsymbol{\nu}_{p+1}} \delta_p \dot{d}_{\boldsymbol{\nu}_{p+1}} = 0$). $\delta_p \square$ represents the incremental discretization of $\square$. Indeed, the fact of adding the unilateral contact and the persistency condition permits to ensure the persistent contact law (6) at the time t_{p+1}. To do that, we developed an extended Newton method which can be decomposed in two steps, a preliminary step (a) in which we solve the system (7) with the frictional contact law (1) by using a Newton method and a final step (b) in which we continue the newton iterations by replacing the laws (1) by the law (6) and the associated Coulomb's friction law. This extended strategy can be written as follows:

step (a): Newton scheme to solve the system (7) with

$$\Gamma_{\boldsymbol{\nu}_{p+1}} \in \partial \Psi_{\mathbb{R}^+}(d_{\boldsymbol{\nu}_{p+1}}) \text{ and } \boldsymbol{\Gamma}_{\boldsymbol{\tau}_{p+1}} \in \partial \Psi^*_{C[\Gamma_{\boldsymbol{\nu}_{p+1}}]}(\delta \overset{o}{\mathbf{d}}_{\boldsymbol{\tau}_{p+1}})$$

step (b): Continuation of the Newton scheme to solve the system (7) with

$$\begin{cases} \text{if } d_{\boldsymbol{\nu}_{p+1}} > 0 & \Gamma_{\boldsymbol{\nu}_{p+1}} = 0 \\ \text{if } d_{\boldsymbol{\nu}_{p+1}} = 0 & \Gamma_{\boldsymbol{\nu}_{p+1}} \in \partial \Psi_{\mathbb{R}^+}(\delta \dot{d}_{\boldsymbol{\nu}_{p+1}}) \text{ and } \boldsymbol{\Gamma}_{\boldsymbol{\tau}_{p+1}} \in \partial \Psi^*_{C[\Gamma_{\boldsymbol{\nu}_{p+1}}]}(\delta \overset{o}{\mathbf{d}}_{\boldsymbol{\tau}_{p+1}}) \end{cases}$$

This approach allows to avoid some penetrations during impacts which arrive when one solves only the step (b). Once the status of contact was found (step (a)), we apply then the persistency conditions (6) which will restore the conservation of energy without disturbing too much the state of the contact. The drawback of this approach is that it requires some additional Newton iterations in order to solve the problem of the step (b). In a practical point of view, one notes an overcost from 3 to 4 iterations for the convergence of the step (b).

After a fully discretization step (time and space), we deduce the nonlinear systems defined by

$$\begin{aligned} \mathcal{R}(\mathbf{u}_{p+1}, \boldsymbol{\lambda}_{p+1}) \;=\; & \frac{1}{\Delta t}\mathcal{M}(\dot{\mathbf{u}}_{p+1} - \dot{\mathbf{u}}_p) \;+\; \mathcal{G}_{algo}(\mathbf{u}_{p+1}, \mathbf{u}_p) \\ & + \mathcal{F}(\mathbf{u}_{p+1}, \boldsymbol{\lambda}_{p+1}) \;-\; \mathbf{q}_{p+\frac{1}{2}} = \mathbf{0} \end{aligned} \qquad (9)$$

where $\mathcal{M}(\dot{\mathbf{u}}_{p+1} - \dot{\mathbf{u}}_p)$ comes from the discretization of the inertia term $\frac{1}{\Delta t}\int_\Omega \rho(\dot{\mathbf{u}}_{p+1} - \dot{\mathbf{u}}_p).\mathbf{v}$ and $\mathcal{G}_{algo}(\mathbf{u}_{p+1}, \mathbf{u}_p)$ is due to the discretization of the hyperelastic part $\int_\Omega \mathbf{\Pi}_{algo}(\mathbf{u}_{p+1}, \mathbf{u}_p) : \nabla \mathbf{v}$ and $\mathbf{q}_{p+\frac{1}{2}}$ comes from the discretization of the external forces $\int_\Omega \mathbf{f}_{p+\frac{1}{2}}.\mathbf{v} + \int_{\partial_g \Omega} \mathbf{g}_{p+\frac{1}{2}}.\mathbf{v}$. We note $\mathcal{F}(\mathbf{u}_{p+1}, \boldsymbol{\lambda}_{p+1})$ the discretization of the frictional contact operator obtained by using a quasi-Lagrangean formulation which permits to treat in an exact way the frictional contact laws (step (a)) but also the frictional persistency contact conditions (step (b)). This formulation is characterized by the lagrangean multiplier $\boldsymbol{\lambda}$ which permits to enforce respectively the frictional contact conditions in steps (a) and (b). During the step (a) the multiplier $\boldsymbol{\lambda}$ takes the value $\hat{\boldsymbol{\lambda}} = (\hat{\lambda}_{\boldsymbol{\nu}}\boldsymbol{\nu}, \hat{\boldsymbol{\lambda}}_{\boldsymbol{\tau}})$ which is in relation with the frictional contact law (1). In addition, during the step (b) $\boldsymbol{\lambda}$ takes the value $\tilde{\boldsymbol{\lambda}} = (\tilde{\lambda}_{\boldsymbol{\nu}}\boldsymbol{\nu}, \tilde{\boldsymbol{\lambda}}_{\boldsymbol{\tau}})$ which is dedicated to the frictional persistency contact conditions. The operators $\mathcal{R}$ and $\mathcal{F}$ can take respectively the form $\mathcal{R}^a$ or $\mathcal{R}^b$ and $\mathcal{F}^a$ or $\mathcal{F}^b$ according the step (a) or (b).The forms of the operator $\mathcal{F}$ are the following (we remove the indices $p+1$ with an aim of reducing the writing):

$$\text{step } (a) \quad \mathcal{F}^a(\mathbf{u}, \hat{\boldsymbol{\lambda}}) = \begin{pmatrix} \nabla_{\mathbf{u}} \hat{l}^r_{\boldsymbol{\nu}}(\mathbf{u}, \hat{\lambda}_{\boldsymbol{\nu}}) + \nabla_{\mathbf{u}} \hat{l}^r_{\boldsymbol{\tau}}(\mathbf{u}, \hat{\boldsymbol{\lambda}}_{\boldsymbol{\tau}}; \Gamma_{\boldsymbol{\nu}}) \\ \nabla_{\hat{\boldsymbol{\lambda}}} \hat{l}^r_{\boldsymbol{\nu}}(\mathbf{u}, \hat{\lambda}_{\boldsymbol{\nu}}) + \nabla_{\hat{\boldsymbol{\lambda}}} \hat{l}^r_{\boldsymbol{\tau}}(\mathbf{u}, \hat{\boldsymbol{\lambda}}_{\boldsymbol{\tau}}; \Gamma_{\boldsymbol{\nu}}) \end{pmatrix},$$

$$\text{step } (b) \quad \text{if } d_{\boldsymbol{\nu}} = 0 \quad \mathcal{F}^b(\mathbf{u}, \tilde{\boldsymbol{\lambda}}) = \begin{pmatrix} \nabla_{\mathbf{u}} \tilde{l}^r_{\boldsymbol{\nu}}(\mathbf{u}, \tilde{\lambda}_{\boldsymbol{\nu}}) + \nabla_{\mathbf{u}} \tilde{l}^r_{\boldsymbol{\tau}}(\mathbf{u}, \tilde{\boldsymbol{\lambda}}_{\boldsymbol{\tau}}; \Gamma_{\boldsymbol{\nu}}) \\ \nabla_{\tilde{\boldsymbol{\lambda}}} \tilde{l}^r_{\boldsymbol{\nu}}(\mathbf{u}, \tilde{\lambda}_{\boldsymbol{\nu}}) + \nabla_{\tilde{\boldsymbol{\lambda}}} \tilde{l}^r_{\boldsymbol{\tau}}(\mathbf{u}, \tilde{\boldsymbol{\lambda}}_{\boldsymbol{\tau}}; \Gamma_{\boldsymbol{\nu}}) \end{pmatrix},$$

$$\text{if } d_{\boldsymbol{\nu}} > 0 \quad \mathcal{F}^b(\mathbf{u}, \tilde{\boldsymbol{\lambda}}) = 0.$$

The terms $\hat{l}^r_{\boldsymbol{\nu}}$, $\tilde{l}^r_{\boldsymbol{\nu}}$, $\hat{l}^r_{\boldsymbol{\tau}}$ and $\tilde{l}^r_{\boldsymbol{\tau}}$ represent respectively the regularization of the functions $\Psi_{\mathbb{R}^+}(d_{\boldsymbol{\nu}})$, $\Psi_{\mathbb{R}^+}(\delta_p \dot{d}_{\boldsymbol{\nu}})$, $\Psi^*_{\hat{C}}(\delta_p \overset{o}{\mathbf{d}}_{\boldsymbol{\tau}})$ and $\Psi^*_{\tilde{C}}(\delta_p \overset{o}{\mathbf{d}}_{\boldsymbol{\tau}})$ and take the following forms,

$$\begin{cases} \hat{l}^r_{\boldsymbol{\nu}}(\mathbf{v}, \hat{\lambda}_{\boldsymbol{\nu}}) = \left(d_{\boldsymbol{\nu}}(\mathbf{v}), \hat{\lambda}_{\boldsymbol{\nu}}\right) + \frac{r}{2}\|d_{\boldsymbol{\nu}}(\mathbf{v})\|^2 - \frac{1}{2r}\mathrm{dist}^2_{\mathbb{R}^-}\left\{\hat{\lambda}_{\boldsymbol{\nu}} + r d_{\boldsymbol{\nu}}(\mathbf{v})\right\} \\ \tilde{l}^r_{\boldsymbol{\nu}}(\mathbf{v}, \tilde{\lambda}_{\boldsymbol{\nu}}) = \left(\delta_p \dot{d}_{\boldsymbol{\nu}}(\mathbf{v}), \tilde{\lambda}_{\boldsymbol{\nu}}\right) + \frac{r}{2}\|\delta_p \dot{d}_{\boldsymbol{\nu}}(\mathbf{v})\|^2 - \frac{1}{2r}\mathrm{dist}^2_{\mathbb{R}^-}\left\{\tilde{\lambda}_{\boldsymbol{\nu}} + r\delta_p \dot{d}_{\boldsymbol{\nu}}(\mathbf{v})\right\} \\ \hat{l}^r_{\boldsymbol{\tau}}(\mathbf{v}, \hat{\boldsymbol{\lambda}}_{\boldsymbol{\tau}}) = \left(\delta_p \overset{o}{\mathbf{d}}_{\boldsymbol{\tau}}(\mathbf{v}), \hat{\boldsymbol{\lambda}}_{\boldsymbol{\tau}}\right) + \frac{r}{2}\|\delta_p \overset{o}{\mathbf{d}}_{\boldsymbol{\tau}}(\mathbf{v})\|^2 - \frac{1}{2r}\mathrm{dist}^2_{\hat{c}}\left\{\hat{\boldsymbol{\lambda}}_{\boldsymbol{\tau}} + r\delta_p \overset{o}{\mathbf{d}}_{\boldsymbol{\tau}}\right\} \\ \tilde{l}^r_{\boldsymbol{\tau}}(\mathbf{v}, \tilde{\boldsymbol{\lambda}}_{\boldsymbol{\tau}}) = \left(\delta_p \overset{o}{\mathbf{d}}_{\boldsymbol{\tau}}(\mathbf{v}), \tilde{\boldsymbol{\lambda}}_{\boldsymbol{\tau}}\right) + \frac{r}{2}\|\delta_p \overset{o}{\mathbf{d}}_{\boldsymbol{\tau}}(\mathbf{v})\|^2 - \frac{1}{2r}\mathrm{dist}^2_{\tilde{c}}\left\{\tilde{\boldsymbol{\lambda}}_{\boldsymbol{\tau}} + r\delta_p \overset{o}{\mathbf{d}}_{\boldsymbol{\tau}}\right\}, \end{cases} \tag{10}$$

where r is a positive penalty factor; The sets $\hat{c}$ and $\tilde{c}$ represent the augmented convex sets (see [1]) corresponding respectively to frictional contact conditions of the steps (a) and (b). Moreover, this approach of "quasi" augmented Lagrangean permits to satisfy exactly the contact constraints and friction criteria contrary to penalty techniques [3]. For more details about the quasi-Lagrangean formulation see [1]. The nonlinear system (9) can be solved by a generalized Newton method developed in [1]. This method leads to the following iterative scheme (indexed by i): $\mathbf{x}_{i+1,p+1} = \mathbf{x}_{i,p+1} - \left(\partial_{\mathbf{x}_{p+1}} \mathcal{R}(\mathbf{x}_{i,p+1})\right)^{-1} \mathcal{R}(\mathbf{x}_{i,p+1})$, where the variable $\mathbf{x}_{i+1,p+1}$ denotes the pair $(\mathbf{u}_{i+1,p+1}, \boldsymbol{\lambda}_{i+1,p+1})$. Through this Newton method, we choose to treat both variables $\mathbf{u}_{p+1}$ and $\boldsymbol{\lambda}_{p+1}$ simultaneously. So this method leads to the solution of a linear system:

$$\begin{aligned} \mathcal{K}_{i,p+1} \Delta \mathbf{x}_{i,p+1} &= -\frac{1}{\Delta t}\mathcal{M}(\dot{\mathbf{u}}_{i,p+1} - \dot{\mathbf{u}}_p) - \mathcal{G}_{algo}(\mathbf{u}_{i,p+1}, \mathbf{u}_p) \\ &\qquad - \mathcal{F}(\mathbf{u}_{i,p+1}, \boldsymbol{\lambda}_{i,p+1}) + \mathbf{q}_{p+\frac{1}{2}} \end{aligned} \tag{11}$$

$$\text{with} \quad \mathcal{K}_{i,p+1} = \frac{2}{\Delta t^2}\mathbf{M} + \mathbf{K}^e_{i,p+1} + \mathbf{K}^c_{i,p+1}$$

$$\text{and} \quad \Delta \mathbf{x}_{i,p+1} = (\mathbf{u}_{i+1,p+1} - \mathbf{u}_{i,p+1}, \boldsymbol{\lambda}_{i+1,p+1} - \boldsymbol{\lambda}_{i,p+1})$$

where $\mathbf{M} = \partial_{\dot{\mathbf{u}}_{p+1}} \mathcal{M}$ represents the mass matrix and $\mathbf{K}^e_{i,p+1} = \partial_{\mathbf{u}_{p+1}} \mathcal{G}_{algo}$ the hyperelastic tangent matrix and $\mathbf{K}^c_{i,p+1} = \partial_{\mathbf{x}_{p+1}} \mathcal{F}(\mathbf{x}_{i,p+1})$ denotes the frictional contact tangent matrix.

5 A representative simulation: bouncing of a ring

This representative impact problem [8] describes the bouncing of an hyperelastic ring against a rigid surface. The elastic ring is thrown with an initial velocity at 45° angle to a flat rigid surface as depicted in figure 1. The material properties and dimensions of the ring are as follows: Ogden material constants $c_1 = 0.5MPa, c_2 = 0.5 * 10^{-2}MPa, a = 0.35MPa$, outer radius of the ring $r_0 = 1m$, inner radius $= 0.9m$, density $\rho = 1000kg/m^3$. Moreover,

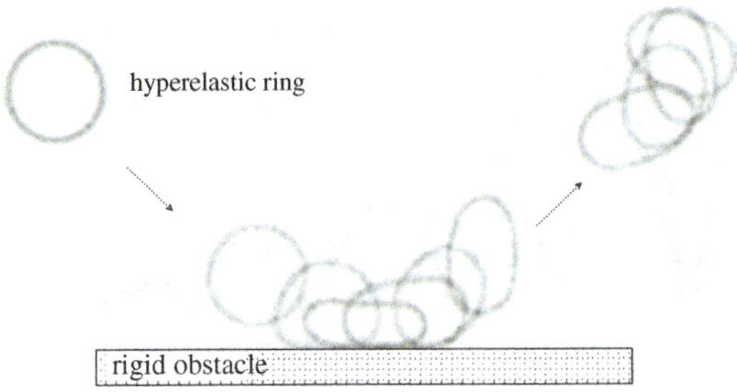

Fig. 1. Deformed sequence of the hyperelastic ring during and after the first impact.

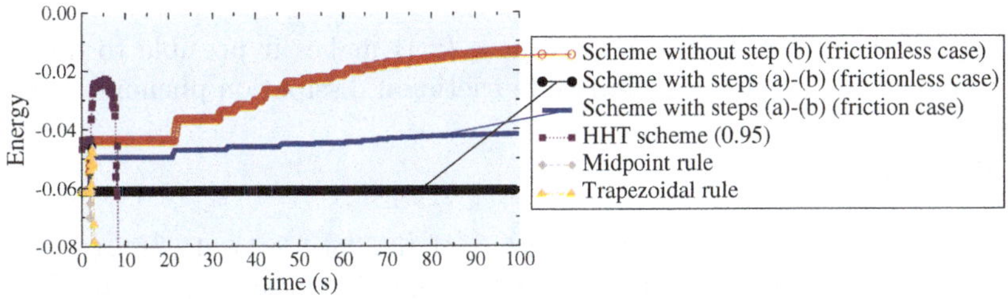

Fig. 2. Discrete energy of the ring bouncing problem.

we consider the final time $T = 100s$ and a time step $\Delta t = 0.01s$. The figure (2) represents the evolution of the discrete energy of the dynamic system in respect to various time integration schemes. Furthermore, we focus our attention on the method presented in section (4) and we present in the figure (3) the displacement of the ring center during the bounces according three alternative simulations. To analyse the behaviour of our method, we consider the following three simulations:

- the curve (—) represents our method to solve the considered problem with contact and friction ($\mu = 0.2$),

- the curve (●) denotes our method to solve the same problem but without friction,

- a variant of the method in which we use only the step (a) (without the step (b)) to solve the problem without friction (curve (⊖)).

Firstly, we can note that after the first impact (time $t = 2$), the methods (curve ◀, ■ and ♦) not using the Gonzalez form of Π_{algo} (8), are characterized by a energy blow up. In an other hand, the proposed scheme enables a long term time integration with a perfect energy conservation in the case we consider the problem without friction (curve (●)). But if we consider only the step (a) to solve this problem (curve (⊖)), a dissipation appears and that represents a physically nonrealistic phenomenon because the unilateral condition induces a dissipation. Moreover, in the case we solve the problem with

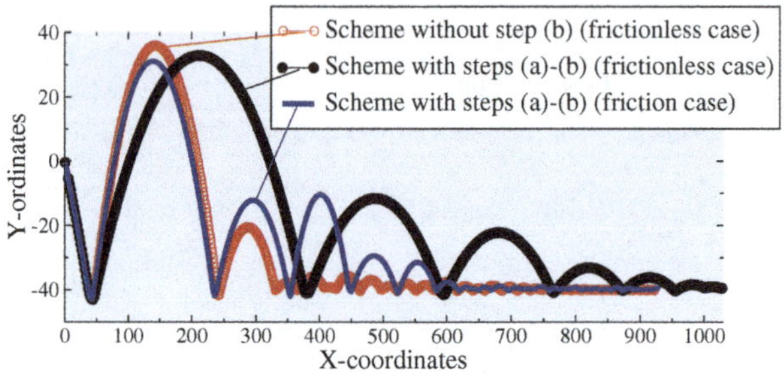

Fig. 3. Displacement of the ring center during the bounces.

contact and friction, our method (curve (—)) makes it possible to dissipate energy reasonably with an admissible frictional dissipation phenomenon.

References

1. Alart P, Curnier A (1991) A mixed formulation for frictional contact problems prone to Newton like solution methods. Comput Meth Appl Mech Engrg 92:353–375
2. Armero F, Petocz E (1998) Formulation and analysis of conserving algorithms for frictionless dynamic contact/impact problems. Comput Meth Appl Mech Engrg 158:269–300
3. Armero F, Romero I (2001) On the formulation of high-frequency dissipative time-stepping algorithms for nonlinear dynamics. Part II: second-order methods. Comput Meth Appl Mech Engrg 190:6783–6824
4. Curnier A, He Q-C, Klarbring A (1995) Continuum mechanics modelling of large deformation contact with friction. Contact Mechanics, ed. plenum Press
5. Gonzalez O (2000) Exact energy and momentum conserving algorithms for general models in non linear elasticity. Comput Meth Appl Mech Engrg 190:1763–1783
6. Hauret P (2004) Numerical methods for the dynamic analysis of two-scale incompressible nonlinear structures. PhD Thesis, Ecole Polytechnique, France
7. Hilber H, Hughes T, Taylor R (1977) Improved numerical disspation for time integration algorithms in structural dynamics. Earth Engrg Struc Dyna 5:283 292
8. Laursen T (2002) Computational contact and impact mechanics. Springer
9. Laursen T, Chawla V (1997) Design of energy conserving algorithms for frictionless dynamic contact problems. Int J Num Meth Engrg 40:863–886
10. Laursen T, Love G (2003) Improved implicit integrators for transient impact problems; dynamic frictional dissipation within an admissible conserving framework. Comput Meth Appl Mech Engrg 192:2223–2248
11. Moreau J-J (1973) On unilateral contraints, friction and plasticity. New Variational Techn in Math Phys, G. Capriz and G. Stampacchia. CIME II:175-322
12. Simo J, Tarnow N (1992) The discrete energy-momentum method. part I: Conserving algorithms for nonlinear elastodynamics. ZAMP 43:757–793
13. Wriggers P (2002) Computational Contact Mechanics. Wiley-Verlag

DDM-based sensitivity analysis and optimization for smooth contact formulations

J. Lengiewicz[1], S. Stupkiewicz[1], J. Korelc[2], and T. Rodic[3]

[1] Institute of Fundamental Technological Research (IPPT), Warsaw, Poland, `jleng@ippt.gov.pl`, `sstupkie@ippt.gov.pl`
[2] Faculty of Civil and Geodetic Engineering, University of Ljubljana, Slovenia, `jkorelc@fgg.uni-lj.si`
[3] Centre for Computational Continuum Mechanics (C3M), Ljubljana, Slovenia, `tomaz.rodic@c3m.si`

Abstract. Sensitivity analysis (SA) is developed for three-dimensional multi-body frictional contact problems. The direct differentiation method (DDM) is applied to obtain response sensitivities with respect to arbitrary design parameters (parameter and shape SA). The FE formulation of contact employs smoothing of the master surface, and the augmented Lagrangian technique is used to enforce the contact and friction conditions. Numerical examples, including application for optimization, illustrate the approach.

1 Introduction

Sensitivity analysis (SA) is a technique that allows efficient computation of derivatives of the solution to the problem at hand, the *direct problem*, with respect to parameters defining the problem, the *design parameters*. These derivatives are implicit, as the dependence of the solution on the design parameters is defined through the governing equation. The general framework of SA for path-dependent problems, such as those of incremental elasto-plasticity, frictional contact, etc., can be found in [1, 2].

Several applications for frictional contact problems have recently been published [3, 4, 5, 6], however, restricted classes of problems have only been considered. As the SA framework is readily available, the main difficulty in applying it for frictional contact problems is the complexity of the expressions (e.g. the sensitivity pseudo-load vector) that have to be derived and subsequently implemented in the finite element environment. This is particularly true in the case of 3D smooth contact formulations.

In this work, sensitivity analysis is developed for a general class of 2D and 3D multi-body contact problems. The smooth contact formulation of [7, 8] is adopted, along with the augmented Lagrangian treatment of contact and friction conditions. The direct differentiation method (DDM) is used, and both parameter and shape sensitivity analysis is efficiently developed using *AceGen* [9], a symbolic code generation system.

Below, the contact formulation is outlined briefly, for more details refer to [7, 8], and also to [10, 3]. The present framework of sensitivity analysis is based on that presented in [1], and due to space limitations the details are omitted in this work.

2 Direct problem: smooth contact formulation

The problem of frictional contact, after the usual finite element discretization, can be written in the following residual form

$$\mathbf{R}(\mathbf{a},\mathbf{b}) = \mathbf{R}_1(\mathbf{a},\mathbf{b}) + \mathbf{R}_2(\mathbf{a},\mathbf{b}) + \mathbf{R}_c(\mathbf{a},\mathbf{b}) = \mathbf{0}, \tag{1}$$

where $\mathbf{R}_1$ and $\mathbf{R}_2$ denote the contributions of the two contacting bodies, excluding the contact interaction, $\mathbf{R}_c$ is the contact contribution, $\mathbf{a}$ is the vector of all global unknowns, i.e. the nodal displacements and Lagrange multipliers, and $\mathbf{b}$ is the vector of dependent variables (state variables). The dependent variables $\mathbf{b}$ are defined by additional equations, namely

$$\mathbf{H}(\mathbf{a},\mathbf{b}) = \mathbf{0}. \tag{2}$$

The residual equation (1) and the equation of the dependent problem (2) correspond to the *current* time increment $t = t_{n+1}$ and are solved for the current values of $\mathbf{a}$ and $\mathbf{b}$, while ${}^n\mathbf{a}$ and ${}^n\mathbf{b}$, the values at the *previous* time increment $t = t_n$, are known. Below, we focus on the contact-specific issues, the part related to the contacting bodies ($\mathbf{R}_1$ and $\mathbf{R}_2$ with the corresponding dependent problems) is standard, and it is left unspecified.

Within the node-to-segment strategy, a smooth contact formulation is obtained by defining a smooth interpolation of the master surface, cf. [7, 12, 13]. The kinematic contact variables are defined by the *closest point projection* of the slave node onto the master surface, cf. Box 1. This *local contact search procedure* constitutes the dependent problem, defined locally at the contact element level, and written in the symbolic form as

$$\mathbf{H}_e(\mathbf{a}_e,\mathbf{b}_e) = \mathbf{0}, \qquad \mathbf{b}_e = \{\xi^1, \xi^2, g_N\}, \tag{3}$$

where $\mathbf{a}_e$ are the global unknowns involved in the description of the contact element (displacement of the slave node, displacements of the nodes defining the patch on the master surface, and Lagrange multipliers associated with the slave node).

The dependent problem (7) is solved at fixed $\mathbf{a}_e^i$ by the iterative Newton method (inner loop),

$$\frac{\partial \mathbf{H}_e}{\partial \mathbf{b}_e}\,\delta\mathbf{b}_e^j = -\mathbf{H}(\mathbf{a}_e^i,\mathbf{b}_e^j), \qquad \mathbf{b}_e^{j+1} = \mathbf{b}_e^j + \delta\mathbf{b}_e^j. \tag{4}$$

1. parametrization of the master surface, $\mathbf{x}_m = \mathbf{x}_m(\xi^1, \xi^2)$,
2. local tangent basis, $\boldsymbol{\tau}_\alpha = \partial \mathbf{x}_m / \partial \xi^\alpha$,
3. unit normal vector, $\mathbf{n} = \boldsymbol{\tau}_1 \times \boldsymbol{\tau}_2 / |\boldsymbol{\tau}_1 \times \boldsymbol{\tau}_2|$,
4. closest point projection: condition

$$\mathbf{H}_e = \mathbf{x}_m(\xi^1, \xi^2) + g_N \mathbf{n} - \mathbf{x}_s = \mathbf{0}$$

to be solved for $\{\xi^1, \xi^2, g_N\}$, where $\mathbf{x}_s$ is the position of the slave node.

Box 1. Dependent problem: local contact search procedure.

Clearly, the solution depends on $\mathbf{a}_e^i$, thus $\mathbf{b}_e = \mathbf{b}_e(\mathbf{a}_e^i)$, and this dependence is implicitly defined by $\mathbf{H}_e(\mathbf{a}_e^i, \mathbf{b}_e(\mathbf{a}_e^i)) = \mathbf{0}$.

In the augmented Lagrangian formulation, the contact element residual $\mathbf{R}_e$ is obtained by constructing a contact Lagrangian L_e and by taking its total derivative with respect to $\mathbf{a}_e$, cf. Box 2. The contact contribution to the global residual $\mathbf{R}_c$ is then obtained by assembling the element residuals $\mathbf{R}_e$.

1. tangential slip increment,

$$\Delta \mathbf{g}_T = \boldsymbol{\tau}_\alpha \Delta \xi^\alpha, \qquad \Delta \xi^\alpha = \xi^\alpha - {}^n\xi^\alpha,$$

2. Lagrange multipliers: normal λ_N and tangential $\lambda_{T\alpha}$ components,
3. augmented Lagrange multipliers ($\varrho > 0$, $m_{\alpha\beta} = \boldsymbol{\tau}_\alpha \cdot \boldsymbol{\tau}_\beta$),

$$\hat{\lambda}_N = \lambda_N + \varrho g_N, \qquad \hat{\lambda}_{T\alpha} = \lambda_{T\alpha} + \varrho m_{\alpha\beta} \Delta \xi^\beta$$

4. radius of the Coulomb friction disk (μ – friction coefficient)

$$\hat{k} = \begin{cases} -\mu \hat{\lambda}_N, & \hat{\lambda}_N \le 0 \\ 0, & \hat{\lambda}_N > 0 \end{cases}$$

5. contact Lagrangian (A – tributary area),

$$L_e(\mathbf{a}_e, \mathbf{b}_e) = A(l_N + l_T),$$

where ($\boldsymbol{\tau}_\alpha \cdot \boldsymbol{\tau}^\beta = \delta_\alpha^\beta$)

$$l_N = \lambda_N g_N + (\varrho/2)\, g_N^2 - (1/2\varrho) \max(0; \hat{\lambda}_N)^2,$$
$$l_T = \lambda_{T\alpha} \Delta \xi^\alpha + (\varrho/2)\, |\boldsymbol{\tau}_\alpha \Delta \xi^\alpha|^2 - (1/2\varrho) \max(0; |\boldsymbol{\tau}^\alpha \hat{\lambda}_{T\alpha}| - \hat{k})^2,$$

6. element residual

$$\mathbf{R}_e = \left.\frac{\partial L_e}{\partial \mathbf{a}_e}\right|_{\hat{k}=\text{const}} + \left(\left.\frac{\partial L_e}{\partial \mathbf{b}_e}\right|_{\hat{k}=\text{const}} \right) \frac{\partial \mathbf{b}_e}{\partial \mathbf{a}_e}, \qquad \frac{\partial \mathbf{b}_e}{\partial \mathbf{a}_e} = -\left(\frac{\partial \mathbf{H}_e}{\partial \mathbf{b}_e} \right)^{-1} \frac{\partial \mathbf{H}_e}{\partial \mathbf{a}_e}$$

Box 2. Contact contribution to global residual.

The global residual (1) is solved by the iterative Newton method (outer loop),

$$\left[\frac{\partial \mathbf{R}}{\partial \mathbf{a}} - \frac{\partial \mathbf{R}}{\partial \mathbf{b}}\left(\frac{\partial \mathbf{H}}{\partial \mathbf{b}}\right)^{-1}\frac{\partial \mathbf{H}}{\partial \mathbf{a}}\right]\delta \mathbf{a}^i = -\mathbf{R}(\mathbf{a}^i, \mathbf{b}(\mathbf{a}^i)), \quad \mathbf{a}^{i+1} = \mathbf{a}^i + \delta \mathbf{a}^i, \tag{5}$$

where the dependent variables $\mathbf{b}(\mathbf{a}^i)$ are defined by the solution of the dependent problem in the inner loop (3), and the term in square brackets is the *independent tangent operator.*

Note that, in the present augmented Lagrangian formulation [7, 8], the global problem (1) is solved simultaneously for displacements and Lagrange multipliers. Note also that the problem of frictional contact is a *path-dependent problem* because the solution $(\mathbf{a}, \mathbf{b})$ at the current time increment depends on the solution $({}^n\mathbf{a}, {}^n\mathbf{b})$ at the previous time increment (through ${}^n\xi^\alpha$ in the definition of the slip increment in Box 2.

3 Sensitivity analysis

Assume now that our path-dependent problem depends on a parameter ϕ, representing any material or shape parameter. As a result the solution $(\mathbf{a}, \mathbf{b})$ also depends on this parameter, and this dependence is implicitly defined by the equations of the problem (1) and (2). Sensitivity analysis deals with the evaluation of the derivative of this implicit dependence.

Rewrite equations (1) and (2), so that the dependencies on ϕ are visible

$$\mathbf{R}(\mathbf{a}(\phi), {}^n\mathbf{a}(\phi), \mathbf{b}(\phi), {}^n\mathbf{b}(\phi), \phi) = \mathbf{0}, \tag{6}$$

$$\mathbf{H}(\mathbf{a}(\phi), {}^n\mathbf{a}(\phi), \mathbf{b}(\phi), {}^n\mathbf{b}(\phi), \phi) = \mathbf{0}. \tag{7}$$

Within the direct differentiation method, the sensitivity problem is derived by taking the total derivative of (5) and (6) with respect to ϕ, namely

$$\frac{\partial \mathbf{R}}{\partial \mathbf{a}}\frac{\partial \mathbf{a}}{\partial \phi} + \frac{\partial \mathbf{R}}{\partial {}^n\mathbf{a}}\frac{\partial {}^n\mathbf{a}}{\partial \phi} + \frac{\partial \mathbf{R}}{\partial \mathbf{b}}\frac{\partial \mathbf{b}}{\partial \phi} + \frac{\partial \mathbf{R}}{\partial {}^n\mathbf{b}}\frac{\partial {}^n\mathbf{b}}{\partial \phi} + \frac{\partial \mathbf{R}}{\partial \phi} = \mathbf{0}, \tag{8}$$

$$\frac{\partial \mathbf{H}}{\partial \mathbf{a}}\frac{\partial \mathbf{a}}{\partial \phi} + \frac{\partial \mathbf{H}}{\partial {}^n\mathbf{a}}\frac{\partial {}^n\mathbf{a}}{\partial \phi} + \frac{\partial \mathbf{H}}{\partial \mathbf{b}}\frac{\partial \mathbf{b}}{\partial \phi} + \frac{\partial \mathbf{H}}{\partial {}^n\mathbf{b}}\frac{\partial {}^n\mathbf{b}}{\partial \phi} + \frac{\partial \mathbf{H}}{\partial \phi} = \mathbf{0} \tag{9}$$

The dependent sensitivity $\partial \mathbf{b}/\partial \phi$ is then obtained from (21), namely

$$\frac{\partial \mathbf{b}}{\partial \phi} = -\left(\frac{\partial \mathbf{H}}{\partial \mathbf{b}}\right)^{-1}\left(\frac{\partial \mathbf{H}}{\partial \mathbf{a}}\frac{\partial \mathbf{a}}{\partial \phi} + \frac{\partial \mathbf{H}}{\partial {}^n\mathbf{a}}\frac{\partial {}^n\mathbf{a}}{\partial \phi} + \frac{\partial \mathbf{H}}{\partial {}^n\mathbf{b}}\frac{\partial {}^n\mathbf{b}}{\partial \phi} + \frac{\partial \mathbf{H}}{\partial \phi}\right), \tag{10}$$

once the sensitivity $\partial \mathbf{a}/\partial \phi$ is known. The latter is obtained by solving (2)

$$\left[\frac{\partial \mathbf{R}}{\partial \mathbf{a}} - \frac{\partial \mathbf{R}}{\partial \mathbf{b}}\left(\frac{\partial \mathbf{H}}{\partial \mathbf{b}}\right)^{-1}\frac{\partial \mathbf{H}}{\partial \mathbf{a}}\right]\frac{\partial \mathbf{a}}{\partial \phi} = -\left[\frac{\partial \mathbf{R}}{\partial {}^n\mathbf{a}}\frac{\partial {}^n\mathbf{a}}{\partial \phi} + \frac{\partial \mathbf{R}}{\partial {}^n\mathbf{b}}\frac{\partial {}^n\mathbf{b}}{\partial \phi} + \frac{\partial \mathbf{R}}{\partial \phi}\right.$$

$$\left. - \frac{\partial \mathbf{R}}{\partial \mathbf{b}}\left(\frac{\partial \mathbf{H}}{\partial \mathbf{b}}\right)^{-1}\left(\frac{\partial \mathbf{H}}{\partial {}^n\mathbf{a}}\frac{\partial {}^n\mathbf{a}}{\partial \phi} + \frac{\partial \mathbf{H}}{\partial {}^n\mathbf{b}}\frac{\partial {}^n\mathbf{b}}{\partial \phi} + \frac{\partial \mathbf{H}}{\partial \phi}\right)\right], \tag{11}$$

where (10) has been used to replace $\partial \mathbf{b}/\partial \phi$ in (2). We note that $\partial \mathbf{H}/\partial \mathbf{b}$ in (10) is the dependent tangent operator and the bracketed operator at left-hand side of (11) is the independent tangent operator, both are the respective operators of the iteration-subiteration procedure, cf. Eqs. (3) and (5).

Concluding, the sensitivity analysis is performed at each time increment after the direct problem is solved. Firstly, the independent pseudo-load vector, i.e. the right-hand side in (11), is computed with the response sensitivities at the previous time increment, $\partial^n \mathbf{a}/\partial \phi$ and $\partial^n \mathbf{b}/\partial \phi$, known from the solution of the sensitivity problem at the previous time increment. Secondly, the linear sensitivity problem (11) is solved for independent sensitivity $\partial \mathbf{a}/\partial \phi$. Finally, the dependent sensitivity $\partial \mathbf{b}/\partial \phi$ is evaluated from (10) and the solution proceeds at the next time increment.

As mentioned above, the augmented Lagrangian treatment [7, 8] of frictional contact fully fits into the general framework of sensitivity analysis [1]. As the direct problem is solved simultaneously for the displacements and Lagrange multipliers, the independent tangent operator in (5) and (11) is the exact tangent operator of the problem, so that sensitivity analysis is a single linear problem to be solved at each time increment. This is in contrast to the so-called Uzawa algorithm, in which the exact sensitivity analysis is not a single linear problem, but requires iterations corresponding to the iterative update scheme for Lagrange multipliers, see [6] for a more detailed discussion.

4 Illustrative examples

Finite element implementation of the present DDM-based sensitivity analysis has been performed within the *Computational Templates* environment [14], and a symbolic code generation system *AceGen* [9], extending the symbolic capabilities of *Mathematica* [15], has been used to derive and automatically generate the numerical codes.

A 3D problem of two hyperelastic pipes in frictional contact has been used to check the accuracy of the sensitivity analysis, cf. Fig. 1. The analytical DDM-based sensitivities have been compared to the numerical finite difference

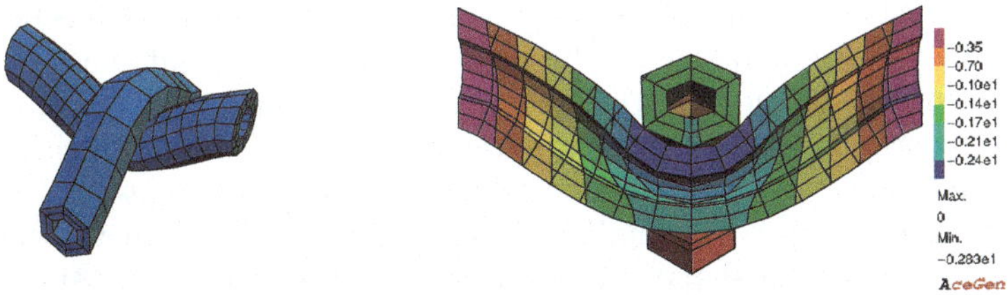

Fig. 1. Hyperelastic pipes: deformed mesh (left) and shape sensitivity of the vertical displacement u_z with respect to the outer radius of the upper pipe, $\mathrm{D}u_z/\mathrm{D}R_o$ (right).

Table 1. Comparison of DDM and FD sensitivities.

Node	DDM	FD	Error	DDM	FD	Error
	$\partial u_z/\partial \mu$			$\partial u_z/\partial R_o$		
32	-0.01126618	-0.01126612	5.3E-6	-0.64451519	-0.64450526	1.5E-5
64	-0.03031119	-0.03031102	5.6E-6	-1.49473608	-1.49470542	2.1E-5
393	-0.00535480	-0.00535474	1.1E-5	-0.27672344	-0.27672016	1.2E-5
667	0.00811398	0.00811385	1.6E-5	-0.34567623	-0.34565159	7.1E-5

(FD) approximation (with the FD perturbation step of $\Delta\mu/\mu = \Delta R_o/R_o = 10^{-5}$), and an excellent agreement has been obtained with the relative error below 10^{-4}, cf. Table 1.

The second example is a 3D-generalization of the optimization problem studied in [16]. Consider an elasto-plastic bar with a uniform, square cross-section compressed between two flat, elastic dies, cf. Fig. 3. As a result of friction, barrelling is observed after upsetting. An optimization problem is thus posed to find the initial shape of the workpiece, such that the final shape, after 33 per cent height reduction, is a regular cube.

undeformed mesh

deformed mesh

Fig. 2. Optimization of free boundary shape in forging: initial design.

The initial shape of the free boundary of the workpiece has been described using 4 shape parameters: the fourfold symmetry about the vertical axis and the mirror symmetry with respect to the horizontal mid-plane have been used to reduce the number of shape parameters. Thanks to availability of analytical response sensitivities, a gradient-based optimization algorithm (BFGS) could efficiently be used to minimize the discrepancy between the computed and the desired final shape. Figure 3 presents the solution obtained in 8 BFGS iterations, and the convergence history of the minimization algorithm is shown in Fig. 4.

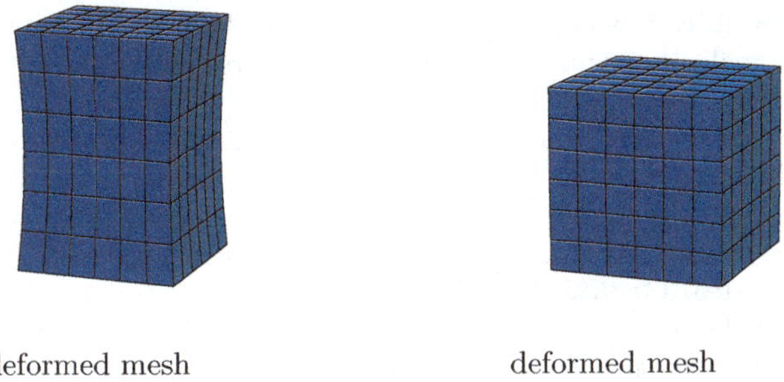

Fig. 3. Optimization of free boundary shape in forging: optimized design.

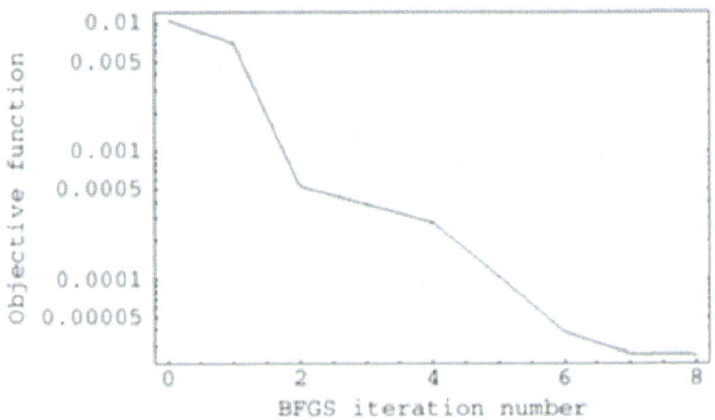

Fig. 4. Optimization of free boundary shape in forging: convergence history of the BFGS algorithm.

Acknowledgements

This work has been partially supported by the European Commission through FP5-GROWTH projects IMPRESS (G1RD-CT-2000-02002) and ENLUB (G1RD-CT-2002-00740).

References

1. P. Michaleris, D.A. Tortorelli, and C.A. Vidal. Tangent operators and design sensitivity formulations for transient non-linear coupled problems with applications to elastoplasticity. *Int. J. Num. Meth. Engng.*, 37:2471–2499, 1994.
2. M. Kleiber, H. Antunez, T.D. Hien, and P. Kowalczyk. *Parameter Sensitivity in Nonlinear Mechanics.* Wiley, Chichester, 1997.

3. N.H. Kim, K.K. Choi, J.S. Chen, and Y.H. Park. Meshless shape design sensitivity analysis and optimization for contact problem with friction. *Comp. Mech.*, 25(2–3):157–168, 2000.
4. A. Srikanth and N. Zabaras. Shape optimization and preform design in metal forming processes. *Comp. Meth. Appl. Mech. Engng.*, 190:1859–1901, 2000.
5. S. Stupkiewicz, J. Korelc, M. Dutko, and T. Rodic. Shape sensitivity analysis of large deformation frictional contact problems. *Comp. Meth. Appl. Mech. Engng.*, 191(33):3555–3581, 2002.
6. S. Stupkiewicz. Augmented Lagrangian formulation and sensitivity analysis of contact problems. In E. Onate and D.R.J. Owen, editors, *COMPLAS 2003, VII International Conference on Computational Plasticity*, CIMNE, Barcelona, 2003. proceedings on CD.
7. G. Pietrzak. *Continuum mechanics modelling and augmented Lagrangian formulation of large deformation frictional contact problems.* PhD thesis, Lausanne, EPFL, 1997.
8. G. Pietrzak and A. Curnier. Large deformation frictional contact mechanics: continuum formulation and augmented Lagrangian treatment. *Comp. Meth. Appl. Mech. Engng.*, 177(3–4):351–381, 1999.
9. J. Korelc. Multi-language and multi-environment generation of nonlinear finite element codes. *Engineering with Computers*, 18:312–327, 2002.
10. T.A. Laursen. *Computational Contact and Impact Mechanics.* Springer-Verlag, Berlin, 2002.
11. P. Wriggers. *Computational Contact Mechanics.* Wiley, Chichester, 2002.
12. L. Krstulović-Opara, P. Wriggers, and J. Korelc. A C^1-continuous formulation for 3D finite deformation frictional contact. *Comp. Mech.*, 29(1):27–42, 2002.
13. M.A. Puso and T.A. Laursen. A 3D contact smoothing method using Gregory patches. *Int. J. Num. Meth. Engng.*, 54:1161–1194, 2002.
14. J. Korelc. Computational Templates. User manual. Available at http://www.-fgg.unilj.si/Symech/, 2000.
15. S. Wolfram. *The Mathematica Book, 4th ed.* Wolfram Media/Cambridge University Press, 1999.
16. L. Fourment, T. Balan, and J.L. Chenot. Optimal design for non-steady-state metal forming processes — II. Application of shape optimization in forging. *Int. J. Num. Meth. Engng.*, 39:51–65, 1996.

On the modeling of contact/impact problems between rubber materials

Z.-Q. Feng[1], Q.-C. He[2], B. Magnain[1], and J.-M. Cros[1]

[1] Laboratoire de Mécanique d'Evry, Université d'Evry, 40 rue du Pelvoux 91020 Evry, France,
feng@iup.univ-evry.fr

[2] Laboratoire de Mécanique, Université de Marne-la-Vallée, 19 rue A. Nobel 77420 Champs sur Marne, France,
he@univ-mlv.fr

Abstract. This work is concerned with the finite element modeling of contact/impact problems between rubber materials. The developed algorithm, namely here Bi-First, combines the bi-potential method for solution of contact problems and the first order algorithm for integration of the time-discretized equation of motion. Numerical examples are given in two cases: multi-contact problem between Blatz-Ko hyperelastic bodies and Love-Laursen's test with a novel hyperelastic model.

1 Introduction

Problems involving contact and friction are among the most difficult ones in mechanics and at the same time of crucial practical importance in many engineering branches. A large number of algorithms for the modeling of contact problems by the finite element method have been presented in the literature. See for example the monographs by Kikuchi and Oden [1], Zhong [2], Wriggers [3], Laursen [4] and the references therein. De Saxcé and Feng [5] have proposed a bi-potential method, in which an augmented Lagrangian formulation was developed. Feng *et al.* [6, 7] have successfully applied this method for the modeling of static contact problems between elastic and Blatz-Ko hyperelastic bodies.

For dynamic implicit analysis in structural mechanics, the most commonly used time integration algorithm is the second order algorithm such as Newmark, Wilson, HHT [8]. The first order algorithm has also been proposed by Jean [9] for time stepping in rigid-body dynamic contact problems. Recently, Feng *et al.* [10] have applied this algorithm for the modeling of impact problems between elastic bodies.

In nonlinear elasticity, there exist many constitutive models to describe the hyperelastic behavior of foam-like or rubber-like materials, such as Blatz-Ko

[11], Ogden [12], Gent [13], etc. These models are available in many modern commercial finite element codes. In 1999, Lainé *et al.* [14] proposed a new third order hyperelastic model, namely here the LVF model. The aim of the present paper is to apply the Bi-First algorithm for contact modeling in dynamic cases between rubber materials described by Blatz-Ko and the LVF model. Two numerical examples are performed in this study to show the validity and efficiency of the algorithm developed.

2 Hyperelastic models

In the case of hyperelastic laws, there exists an elastic potential function W (or strain energy density function) which is a scale function of one of the strain tensors, whose derivative with respect to a strain component determines the corresponding stress component. This can be expressed by

$$\mathbf{S} = \frac{\partial W}{\partial \mathbf{E}} = 2\frac{\partial W}{\partial \mathbf{C}} \tag{1}$$

where $\mathbf{S}$ is the second Piola-Kirchoff stress tensor, $\mathbf{C}$ the right Cauchy-Green deformation tensor and $\mathbf{E}$ the Green-Lagrangian strain tensor. The Blatz-Ko constitutive law is used to model compressible foam-type polyurethane rubbers [11]. The strain energy density function is given as follows

$$W = \frac{G}{2}\left(\frac{I_2}{I_3} + 2\sqrt{I_3} - 5\right) \tag{2}$$

where I_2 and I_3 are respectively the second and third invariant of $\mathbf{C}$ and G is the shear modulus. Reporting (2) in (1) gives the constitutive relation as follows

$$\mathbf{S} = G\Big[\sqrt{I_3}(2\mathbf{E} + \mathbf{I})^{-1} - (2\mathbf{E} + \mathbf{I})^{-2}\Big] \tag{3}$$

The LVF constitutive law is proposed by Lainé *et al.* [14] to describe the isotropic compressible or incompressible rubber-like material. New invariants of $\mathbf{E}$: $(x,\ y,\ z)$ are introduced as follows

$$x = \sqrt{tr(\mathbf{E}_d)^2}\cos\vartheta, \quad y = \sqrt{tr(\mathbf{E}_d)^2}\sin\vartheta, \quad z = \frac{tr(\mathbf{E})}{\sqrt{3}} \tag{4}$$

where $\mathbf{E}_d$ is the distortional part of $\mathbf{E}$ and ϑ the Lode's angle. The strain energy density function of fourth order expressed in terms of the new invariants of the strain tensor is given as follows

$$\begin{aligned} W(x,y,z) &= \left(G + \frac{a_3}{4}z^2\right)\left(x^2+y^2\right) + \frac{3K}{2}z^2 \\ &+ \frac{a_1}{3}\left(x^3 - 3xy^2\right) + \frac{a_2}{3}z^3 + \frac{a_4}{4}z^4 + \frac{a_5}{4}\left(x^2+y^2\right)^2 \end{aligned} \tag{5}$$

where G is the shear modulus, K the bulk modulus and a_i $(i = 1, \ldots, 5)$ are parameters of the model. By deriving the energy density (13) with respect to the strain tensor, we obtain

$$\mathbf{S} = z\left(3K + a_2 z + a_4 z^2\right)\frac{\mathbf{I}}{\sqrt{3}} + \left[2G + \frac{a_5}{2}(x^2 + y^2)\right]\mathbf{E}_d$$
$$+ \sqrt{6}a_1\left[(\mathbf{E}_d)^2 - \frac{x^2 + y^2}{3}\mathbf{I}\right] \tag{6}$$

3 Local contact modeling

For notational convenience, we assume that the contact with friction may occur between some points of two bodies A and B. The contact and friction laws are written in terms of relative velocity $\dot{\mathbf{u}} = \dot{\mathbf{u}}_A - \dot{\mathbf{u}}_B$ and of contact reactions $\mathbf{r}$. The following contact bi-potential is introduced by de Saxcé and Feng [5]:

$$b_c(-\dot{\mathbf{u}}, \mathbf{r}) = \bigcup_{\mathbb{R}_-}(-\dot{u}_n) + \bigcup_{K_\mu}(\mathbf{r}) + \mu\, r_n\| - \dot{\mathbf{u}}_t\| \tag{7}$$

where $\mathbb{R}_- =]-\infty, 0]$, K_μ is the Coulomb's cone and $\bigcup$ stands for the indicator function. In order to avoid nondifferentiable potentials that occur in contact problems, it is convenient to use the Augmented Lagrangian Method [5, 15]. For the contact bi-potential b_c, we have:

$$\forall\, \mathbf{r}' \in K_\mu, \quad \varrho\mu(r_n' - r_n)\|\dot{\mathbf{u}}_t\| + \left(\mathbf{r}' - (\mathbf{r} - \varrho\dot{\mathbf{u}})\right)(\mathbf{r}' - \mathbf{r}) \geq 0 \tag{8}$$

where ϱ is a solution parameter which is not user-defined. The inequality (8) means that $\mathbf{r}$ is the projection of $\boldsymbol{\tau}$ onto the closed convex Coulomb's cone:

$$\mathbf{r} = \mathrm{proj}(\boldsymbol{\tau}, K_\mu) \tag{9}$$

For the numerical solution of the implicit equation (9), Uzawa's algorithm can be used, which leads to an iterative process involving one predictor-corrector step:

$$\begin{array}{l} \text{Predictor } \boldsymbol{\tau}^{i+1} = \mathbf{r}^i - \varrho^i\left(\dot{\mathbf{u}}_t^i + (\dot{u}_n^i + \mu\|\dot{\mathbf{u}}_t^i\|)\,\mathbf{n}\right) \\ \text{Corrector } \mathbf{r}^{i+1} = \mathrm{proj}(\boldsymbol{\tau}^{i+1}, K_\mu) \end{array} \tag{10}$$

4 Global time stepping

Generally, mechanical behaviors of solids under contact/impact conditions are governed by a set of nonlinear equations

$$\mathbf{M}\,\ddot{\mathbf{u}} = \mathbf{F} + \mathbf{R}, \qquad \text{where } \mathbf{F} = \mathbf{F}_{ext} - \mathbf{F}_{int} - \mathbf{A}\dot{\mathbf{u}} \tag{11}$$

where $\mathbf{M}$ is the mass matrix, $\mathbf{F}_{ext}$ the applied forces vector, $\mathbf{F}_{int}$ the internal forces vector and $\mathbf{R}$ the reaction forces vector. Taking the derivative of $\mathbf{F}_{int}$ with respect to the nodal displacements $\mathbf{u}$ gives the tangent stiffness matrix $\mathbf{K}$. The most common method to integrate Eq. (11) is the Newmark method which is based on a second order algorithm. However, in impact problems, higher order approximation does not necessarily mean better accuracy, and may even be superfluous. At the moment of a sudden change of contact conditions (impact, release of contact), the velocity and acceleration are not continuous, and excessive regularity constraints may lead to serious errors. For this reason, Jean [16] has proposed a first order algorithm which is used in this work. This algorithm is based on the following approximations:

$$\int_t^{t+\Delta t} \mathbf{M}\, d\dot{\mathbf{u}} = \mathbf{M}\left(\dot{\mathbf{u}}^{t+\Delta t} - \dot{\mathbf{u}}^t\right) \tag{12}$$

$$\int_t^{t+\Delta t} \mathbf{F}\, dt = \Delta t\left((1-\xi)\,\mathbf{F}^t + \xi\,\mathbf{F}^{t+\Delta t}\right) \tag{13}$$

$$\int_t^{t+\Delta t} \mathbf{R}\, dt = \Delta t\,\mathbf{R}^{t+\Delta t} \tag{14}$$

$$\mathbf{u}^{t+\Delta t} - \mathbf{u}^t = \Delta t\left((1-\theta)\,\dot{\mathbf{u}}^t + \theta\,\dot{\mathbf{u}}^{t+\Delta t}\right) \tag{15}$$

where $0 \le \xi \le 1$; $0 \le \theta \le 1$. In the iterative solution procedure, all the values at time $t + \Delta t$ are replaced by the values of the current iteration $i + 1$. Without going into details, we obtain the recursive form of (11) in terms of displacements:

$$\begin{aligned} \bar{\mathbf{K}}^i\,\Delta\mathbf{u} &= \bar{\mathbf{F}}^i + \bar{\mathbf{F}}^i_{acc} + \mathbf{R}^{i+1} \\ \mathbf{u}^{i+1} &= \mathbf{u}^i + \Delta\mathbf{u} \end{aligned} \tag{16}$$

where the so-called effective terms are given by

$$\bar{\mathbf{K}}^i = \xi\,\mathbf{K}^i + \frac{1}{\theta\,\Delta t^2}\mathbf{M}^i \tag{17}$$

$$\bar{\mathbf{F}}^i_{acc} = -\frac{1}{\theta\Delta t^2}\,\mathbf{M}^i\left[\mathbf{u}^i - \mathbf{u}^t - \Delta t\,\dot{\mathbf{u}}^t\right] \tag{18}$$

$$\bar{\mathbf{F}}^i = (1-\xi)\left(\mathbf{F}^t_{int} + \mathbf{F}^t_{ext}\right) + \xi\left(\mathbf{F}^i_{int} + \mathbf{F}^{t+\Delta t}_{ext}\right) \tag{19}$$

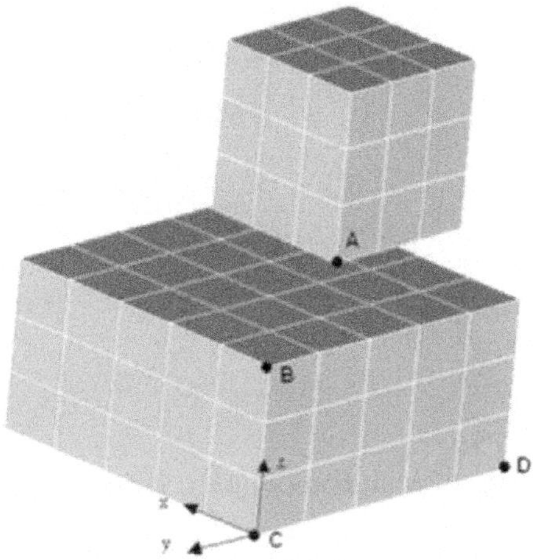

Fig. 1. Initial configurations and meshes.

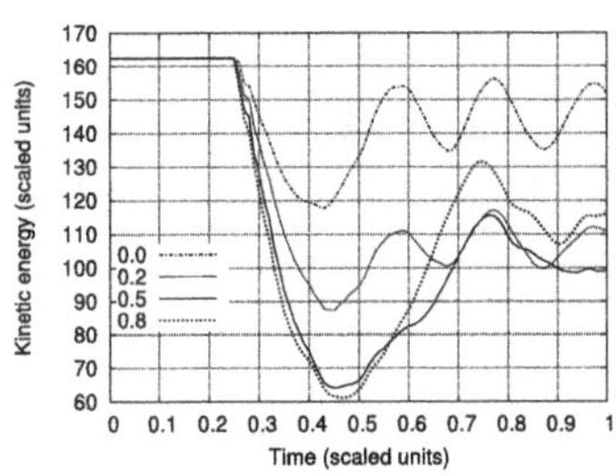

Fig. 2. Kinetic energy with different μ.

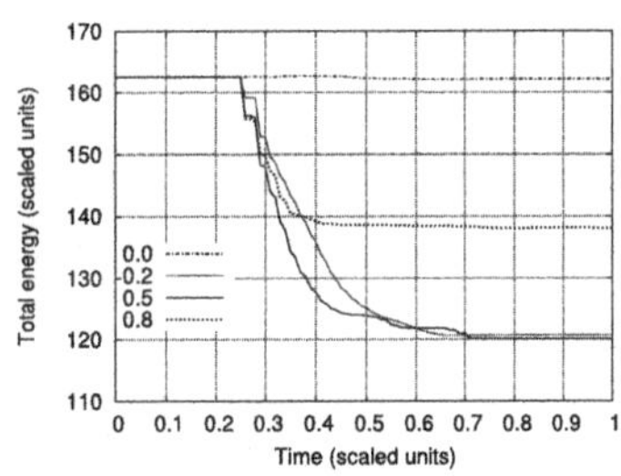

Fig. 3. Total energy with different μ.

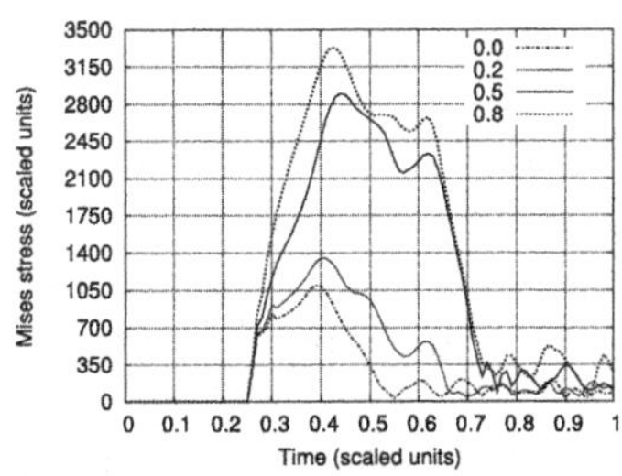

Fig. 4. von Mises stress at point A.

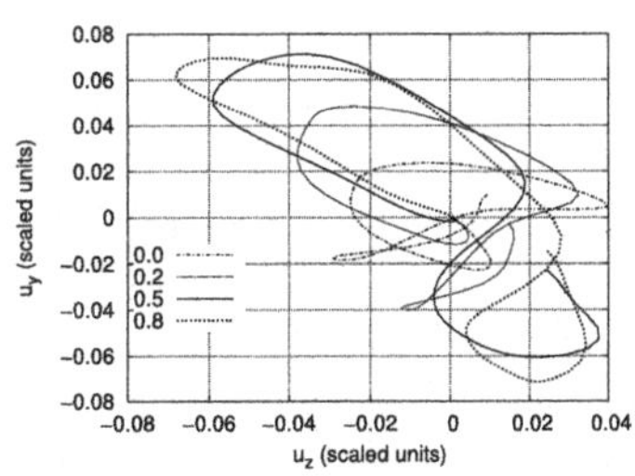

Fig. 5. Trace of point B.

Fig. 6. Multibody contact: initial mesh.

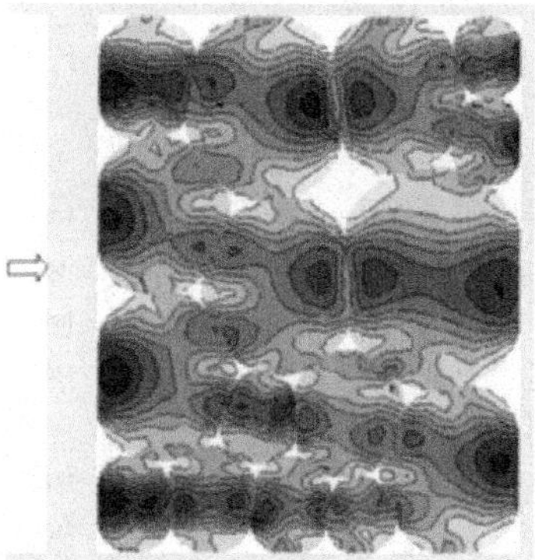

Fig. 7. Multibody contact: von Mises stress after loading.

5 Numerical results

The Bi-First algorithm presented above has been implemented and tested in the finite element code FER/Impact [17]. Due to the limitation of pages, we briefly present two examples of application.

The first example, proposed initially by Love and Laursen [18] who consider only linearly elastic materials, accounts for hyperelastic large deformations in the present work. The simulation consists of two three-dimensional blocks (Figure 1) that impact with relative tangential motion. The LVF model is considered here with the initial shear modulus G and bulk modulus K same as in [18] (scaled units): $G = 5000, K = 3333$. Other parameters are: $a_1 = 50$, $a_2 = 50$, $a_3 = 0$, $a_4 = 2000$ and $a_5 = 100$. The total simulation time is 1 scaled time unit and the solution parameters are: $\Delta t = 10^{-2}$, $\xi = \theta = 0.5$. In order to investigate the frictional effects on the energy dissipation, different coefficients of Coulomb friction are used: $\mu = 0.0, 0.2, 0.5, 0.8$.

Figures 2,3 show the plots of the kinetic energy E_k and the total energy E_t. We observe that the total energy is quite well conserved in the case of frictionless contact. However, in the case of frictional contact, the total energy decreases. So the total energy is dissipated by frictional effects as expected. It is worth noting that the dissipated energy is quantitatively calculated.

It is also interesting to examine another question: is the dissipated energy proportional to the friction coefficient? The answer is negative according to numerical results. The proof is illustrated by Figure 3 in which we observe almost the same dissipated energy even with two different friction coefficients ($\mu = 0.2, 0.5$). In addition, the dissipated energy is less in the case $\mu = 0.8$ than in the case $\mu = 0.2$ or $\mu = 0.5$. In fact, when the friction coefficient increases, the friction forces increase. However, the tangential slips will decrease. We know that the dissipated energy depends not only on the friction forces but also on the tangential slips on the contact surface.

Figure 4 shows the evolution of the von Mises stress at point A (see Figure 1). It can be seen that when the friction coefficient increases, the stress level becomes more important. The trajectory of point B in the plane BCD (see Figure 1 is depicted in Figure 5. We observe that the amplitude of the displacements increases with friction coefficient as expected.

The second example simulates the deformable multibody contact between Blatz-Ko hyperelastic bodies. In doing so, we wish to further explore the performance of the present method and the developed code FER/Impact in a large strain context and with complicated contact surfaces. In addition, this example would illustrate the possibility to investigate the heterogeneous behavior of granular materials involving both deformations of grains and the interaction of grains with friction. The problem is displayed in Figure 6. Several grains meshed with triangular elements are locked up in a rigid box. The left side of the box is given an horizontal motion so as to compress the grains. The contours of von Mises stress are depicted in Figure 7 from which we observe the effect of friction on the top and bottom surfaces. We observe also the stress concentrated zones as expected.

6 Conclusion

The main purpose of this paper is to briefly present the recent development of the bi-potential method applied to dynamic analysis of contact problems with Coulomb friction between hyperelastic bodies. Numerical results demonstrate that the Bi-First algorithm for local analysis of frictional contact problems and for global time integration of dynamics equations is suitable for a wide range of engineering applications.

References

1. N. Kikuchi and Oden, J.T. (1988) Contact problems in elasticity: A study of variational inequalities and finite elements, Philadelphia: SIAM
2. Z.H. Zhong (1993) Finite element procedures in contact-impact problems, Oxford University Press
3. P. Wriggers (2002) Computational contact mechanics, John Wiley & Sons
4. T.A. Laursen (2002) Computational contact and impact mechanics: Fundamentals of Modeling Interfacial Phenomena in Nonlinear Finite Element Analysis, Springer Verlag
5. G. De Saxcé and Z.-Q. Feng (1998) The bi-potential method: a constructive approach to design the complete contact law with friction and improved numerical algorithms, Mathematical and Computer Modeling 28(4-8): 225-245, Special issue: Recent Advances in Contact Mechanics
6. Z.-Q. Feng (1998) Some test examples of 2D and 3D contact problems involving Coulomb friction and large slip, Mathematical and Computer Modeling 28(4-8): 469-477, Special issue: Recent Advances in Contact Mechanics
7. Z.-Q. Feng, F. Peyraut and N. Labed (2003) Solution of large deformation contact problems with friction between Blatz-Ko hyperelastic bodies, Int. J. Engng. Science 41: 2213-2225
8. H.M. Hilber, T.J.R. Hughes and R.L. Taylor (1977) Improved numerical dissipation for the time integration algorithms in structural dynamics, Earthquake Engng. Struc. Dyn. 5: 283-292
9. M. Jean (1999) The non-smooth contact dynamics method, Comp. Meth. Appl. Mech. Engng. 177: 235-257
10. Z.-Q. Feng and B. Magnain and J.-M. Cros and P. Joli (2004) Energy dissipation by friction in dynamic multibody contact problems.In Z.-H. Yao, M.-W. Yuan and W.-X. Zhong, editors, Computational Mechanics, Beijing, China, Sept.2004. WCCM VI in conjunction with APCOM04, Springer
11. P.J. Blatz and W.L. Ko (1962) Application of Finite Elastic Theory to the Deformation of Rubbery Materials, Transactions of the Society of Rheology 6: 223-251
12. R.W. Ogden (1984) Non-linear elastic deformations, Ellis Horwood
13. A.N. Gent (1996) A new constititive relation for rubber, Rubber Chem. Technol. 69: 59-61
14. E. Lainé and C. Vallée and D. Fortuné (1999) Nonlinear isotropic constitutive laws: choice of the three invariants, convex potentials and constitutive inequalities, Int. J. Engng. Science 37: 1927-1941
15. J.C. Simo and T.A. Laursen (1992) An augmented lagrangian treatment of contact problems involving friction. Computers & Structures, 42: 97-116
16. M. Jean (1989) Dynamics with partially elastic shocks and dry friction: double scale method and numerical approach, 4th Meeting on unilateral problems in structural analysis, Capri
17. Z.-Q. Feng. http://gmfe16.cemif.univ-evry.fr:8080/~feng/FerImpact.html
18. G.R. Love and T.A. Laursen (2003) Improved implicit integrators for transient impact problems: dynamic frictional dissipation within an admissible conserving framework, Comp. Meth. Appl. Mech. Engng. 192: 2223-2248

The quadrilateral parametric contact element based on the moving friction cone formulation

L. Krstulović-Opara[1] and P. Wriggers[2]

[1] Faculty of Electrical Engineering, Mechanical Engineering and Naval Architecture, University of Split, R. Boškovića bb, HR-21000 Split, Croatia, Lovre.Krstulovic-Opara@fesb.hr

[2] Institut für Baumechanik und Numerische Mechanik, Universität Hannver, Appelstrasse 9a, D-30167 Hannover, Germany, wriggers@ibnm.uni-hannover.de

Abstract. Contact between bodies is most commonly analyzed using quadrilateral contact elements that are based on 8-node brick (hexahedral) continuum finite elements. As a quadrilateral contact surface, in comparison to a triangular contact surface (tetrahedral continuum elements), is not necessarily flat, or it deforms as deformable body deforms, contact formulation turns to be a complex problem. Recent developments in contact routines based on the Moving Friction Cone (MFC) approach for flat triangular contact elements enabled significant simplifications in the element formulation, what is used herein. The MFC formulation of contact is based on the single gap vector, instead of two vectors (slip and stick one). The curved contact surface is defined in a parametric form, thus enabling finite deformations and a Lagrangian definition of contact.

1 Introduction

Contact between three-dimensional (3D) solids is most commonly analyzed using quadrilateral contact elements that are based on 8-node brick (hexahedral) continuum finite elements. However, when these quadrilateral contact elements are used to describe the contact surface, due to the contact between bodes, corresponding contact loads deform the continuum element resulting in uneven quadrilateral contact surface. Since the closest point on uneven surface cannot be obtained by simply projecting orthogonally slave node to the master surface, an iterative procedure must be used. This iterative procedure must be taken into account when linearizing the residual vector in order to obtain the tangent matrix [1, 2].

To overcome these obstacles several approaches were developed to approximate the curved contact surface when using the quadrilateral contact elements. E.g., in [3] the curved contact surface is approximated by a flat contact surface, achieved by averaging normal. In [4, 5, 6] the curved contact surface is approximated by four flat triangles. In these approaches the explicit projection of the slave node is defining the closest point.

The MFC approach, used for flat triangular contact elements enables the explicit definition of the closest point [7, 8], is used herein for the definition of contact on curved quadrilateral contact surfaces.

2 The moving friction cone description of contact

When describing a contact between two 3D bodies, the master-node to slave-surface approach based on the contact constraint in the form of normal and tangential gap (Fig. 1) is often used, e.g. [9, 10]. In the frictional penalty formulation presented herein, the contact constraint is defined using the single gap vector (Fig. 2).

Supposing that there is a contact, in the case of stick, the spring-back effect of the elastic-like behavior is characterizing the case when the slave node is still within the Coulomb frictional cone. For the slip case, arising from the fact that the slave node is on the Coulomb frictional cone surface, the unknown position of the slave node on the contact surface can be determined. The position of the slave node on the surface, characterized by parameters ξ, η, is evaluated from the fact that the gap vector is perpendicular to the normal on the conical Coulomb surface. Therefore, for the slip case, the cone has moved from the initial stick position to the one depicted in Fig. 2. For this position the spring-back effect is again pushing the penetrated slave node back to the master surface. The whole formulation is defined in the parametric form, thus enabling the finite deformations and the Lagrangian definition of contact.

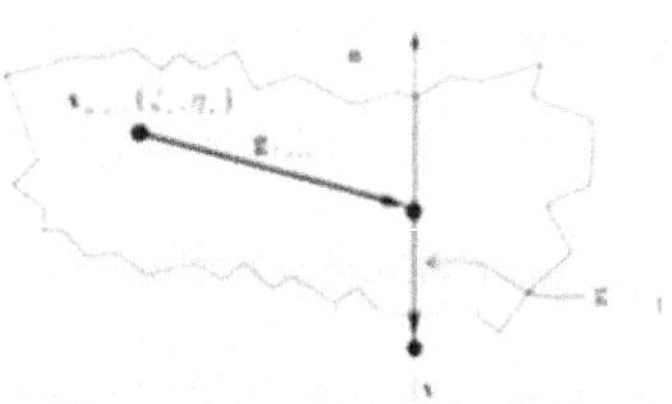

Fig. 1. Normal and tangential gap.

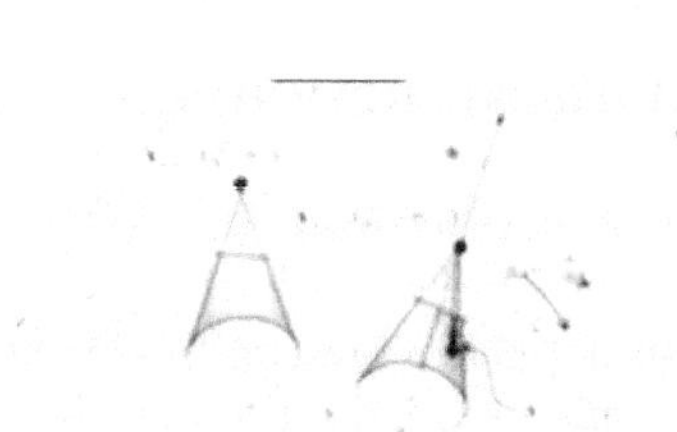

Fig. 2. Single gap vector.

A contact of two discretized bodies of interest is defined within the time interval $t \in [0, T]$, while a position of a continuum point at the time t_{n+1}, is defined with the current coordinate $\mathbf{x}_{n+1}=\mathbf{X}+\mathbf{u}_{n+1}$, where $\mathbf{X}=\mathbf{x}(t=0)$ represents an undeformed reference configuration of a point, while $\mathbf{u}_{n+1}$ is a displacement vector for the current time step.

In the case of sliding, parameters ξ_{n+1}, η_{n+1} defining the position of the slave node on the master surface, are evaluated from the orthogonality condition within the parameter search. Parameters ξ_{n+1}, η_{n+1} are saved as the history variables needed for the next loading step. In the case of sliding between several contact elements, nodal coordinates of the previous master surface are used, together with the parameters ξ_n, η_n, to construct the last solution coordinate $\mathbf{x}_{m\ n+1}(\xi_n, \eta_n)$. This is enabling large steps and sliding over several master surfaces. For the sliding case, the friction cone is moved from the point $\mathbf{x}_{m\ n+1}(\xi_n, \eta_n)$ to the point $\mathbf{x}_{m\ n+1}(\xi_{n+1}, \eta_{n+1})$ (Fig. 2). The point $\mathbf{x}_{m\ n+1}(\xi_{n+1}, \eta_{n+1})$ is undetermined, as well as the local normal $\mathbf{n}_{n+1}$ and tangents $\mathbf{t}_{\xi\ n+1}$gv , $\mathbf{t}_{\eta\ n+1}$ (Fig. 3). The tangent $\mathbf{t}_{S\ n+1}$ is defined as the projection of the vector $\mathbf{a}_{S\ n+1}$ to the tangential plane defined by $\mathbf{t}_{\xi\ n+1}$, $\mathbf{t}_{\eta\ n+1}$. The tangent $\mathbf{t}_{S\ n+1}$, together with the normal, is used for the definition of the orthogonality condition when finding $\mathbf{x}_{m\ n+1}(\xi_n, \eta_n)$ (Fig. 3 and 4).

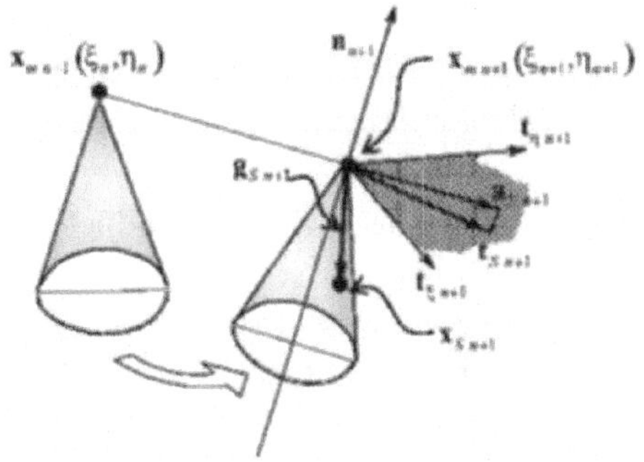

Fig. 3. Tangential plane.

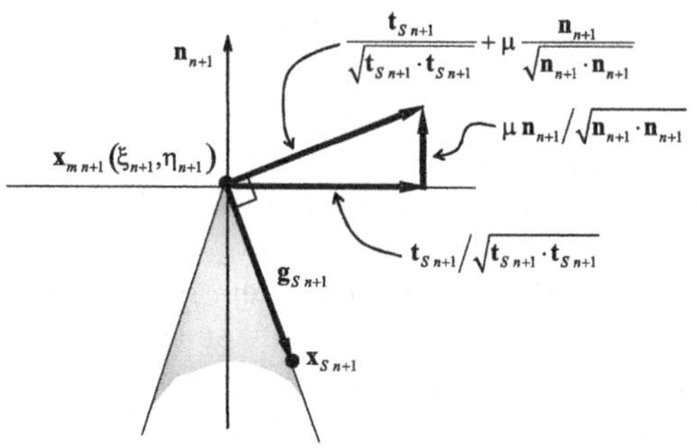

Fig. 4. Orthogonality condition.

3 The parametric contact surface description, the residual vector and the tangent matrix

Based on the 8-node brick continuum element, a curved quadrilateral contact surface is defined within the current configuration in the parametric form as a function of parameters ξ, η,

$$\begin{aligned}\mathbf{x}_{n+1}(\xi_{n+1},\eta_{n+1}) = \mathbf{x}_{1\ n+1}\frac{(1-\xi_{n+1})(1-\eta_{n+1})}{4} + \mathbf{x}_{2\ n+1}\frac{(1+\xi_{n+1})(1-\eta_{n+1})}{4}\\ +\mathbf{x}_{3\ n+1}\frac{(1+\xi_{n+1})(1+\eta_{n+1})}{4} + \mathbf{x}_{4\ n+1}\frac{(1-\xi_{n+1})(1+\eta_{n+1})}{4}\end{aligned} \tag{1}$$

The surface normal is defined as a vector product of the two tangents in direction of parametric axis, i.e.,

$$\mathbf{n}_{n+1} = \mathbf{t}_{\xi\ n+1} \times \mathbf{t}_{\eta\ n+1}, \quad \mathbf{t}_{\xi\ n+1} = \frac{\partial\,\mathbf{x}_{n+1}(\xi_{n+1},\eta_{n+1})}{\partial\,\xi_{n+1}}, \tag{2}$$

$$\mathbf{t}_{\eta\ n+1} = \frac{\partial\,\mathbf{x}_{n+1}(\xi_{n+1},\eta_{n+1})}{\partial\,\eta_{n+1}}.$$

The solution point from the last converged configuration ($t = t_n$) is mapped to the current configuration, i.e. $\mathbf{x}_{m\quad n+1}(\xi_n,\eta_n)$=$\mathbf{x}_{n+1}(\xi_n,\eta_n)$. The elastic gap vector is defined between this point and the slave node as:

$$\mathbf{g}_{S\ n+1}(\xi_n,\eta_n) = \mathbf{x}_{S\ n+1} - \mathbf{x}_{m\ n+1}(\xi_n,\eta_n)\,. \tag{3}$$

In the case of contact, i.e.

$$\mathbf{g}_{S\ n+1}(\xi_n,\eta_n)\cdot\mathbf{n}_{n+1} \leqslant 0, \tag{4}$$

the stick case is supposed to occur, and the trial traction vector is defined using constant penalty parameter ε as:

$$\mathbf{t}^{tr}_{n+1}(\xi_n,\eta_n)=\varepsilon\,\mathbf{g}_{S\;n+1}(\xi_n,\eta_n)\,. \tag{6}$$

Projection of the trial traction vector on normal vector and tangential plane defines normal pressure and trial traction vector as:

$$p_{N\;n+1}(\xi_n,\eta_n)=\mathbf{t}^{tr}_{n+1}(\xi_n,\eta_n)\cdot\frac{\mathbf{n}_{n+1}}{\sqrt{\mathbf{n}_{n+1}\cdot\mathbf{n}_{n+1}}}, \tag{7}$$

$$\mathbf{t}^{tr}_{T\;n+1}(\xi_n,\eta_n)=\mathbf{t}^{tr}_{n+1}(\xi_n,\eta_n)-p_{N\;n+1}(\xi_n,\eta_n)\cdot\frac{\mathbf{n}_{n+1}}{\sqrt{\mathbf{n}_{n+1}\cdot\mathbf{n}_{n+1}}}$$

The Coulomb criterion for the trial state is

$$f^{tr}_S=\left\|\mathbf{t}^{tr}_{T\;n+1}(\xi_n,\eta_n)\right\|-\mu\,Abs\left[p_{N\;n+1}(\xi_n,\eta_n)\right]\leqslant 0, \tag{8}$$

were μ is the constant Coulomb friction coefficient. If relation (8) is satisfied, the supposed stick case is the real case that has occurred. The slave node is positioned within the friction cone, and the friction cone is not moving from the initial position, i.e., $\mathbf{x}_{m\;n+1}(\xi_{n+1},\,\eta_{n+1})$=$\mathbf{x}_{m\;n+1}(\xi_n,\,\eta_n)$ and ξ_{n+1}=ξ_n, η_{n+1}=η_n. If not, sliding occurs and the frictional cone's tip is moved to the $\mathbf{x}_{m\;n+1}(\xi_{n+1},\,\eta_{n+1})$, what is evaluated from the orthogonality condition (Fig. 4)

$$\mathbf{g}_{S\;n+1}(\xi_{n+1},\eta_{n+1})\cdot\left(\frac{\mathbf{t}_{S\;n+1}}{\sqrt{\mathbf{t}_{S\;n+1}\cdot\mathbf{t}_{S\;n+1}}}+\mu\frac{\mathbf{n}_{n+1}}{\sqrt{\mathbf{n}_{n+1}\cdot\mathbf{n}_{n+1}}}\right)=0 \tag{9}$$

$$\mathbf{g}_{S\;n+1}(\xi_{n+1},\eta_{n+1})=\mathbf{x}_{S\;n+1}-\mathbf{x}_{m\;n+1}(\xi_{n+1},\eta_{n+1})\,.$$

The second member of the scalar product is the normal to the Coulomb cone. The tangent in relation (9) is defined as:

$$\mathbf{t}_{S\;n+1}(\xi_{n+1},\eta_{n+1})=\left[\mathbf{n}_{n+1}(\xi_{n+1},\eta_{n+1})\times\mathbf{a}_{S\;n+1}(\xi_{n+1},\eta_{n+1})\right] \tag{10}$$
$$\mathbf{n}_{n+1}(\xi_{n+1},\eta_{n+1}),$$

$$\mathbf{a}_{S\;n+1}(\xi_{n+1},\eta_{n+1})=\mathbf{x}_{m\;n+1}(\xi_{n+1},\eta_{n+1})-\mathbf{x}_{m\;n+1}(\xi_n,\eta_n)\,.$$

To solve two unknowns ξ_{n+1}, η_{n+1} within the Newton-Raphson procedure, an additional condition is introduced requiring orthogonality of normal vector and plane defined by $\mathbf{g}_{S_{n+1}}$ $(\xi_{n+1},\,\eta_{n+1})$ and $\mathbf{t}_{Sn+1}$, i.e.

$$\left[\mathbf{g}_{S\;n+1}(\xi_{n+1},\eta_{n+1})\times\mathbf{t}_{S\;n+1}\right]\cdot\mathbf{n}_{n+1}=0. \tag{11}$$

Once the new position of the vector $\mathbf{g}_{S_{n+1}}(\xi_{n+1},\,\eta_{n+1})$ is determined, as the contact surface is curved, the contact condition (4) has to be verified.
In the case where no sliding occurred (stick), as parameters ξ_n, η_n are evaluated in the last loading step. The residual vector and the tangent matrix

are obtained explicitly, what represents significant simplification of presented formulation:

$$\Psi^c_{stick_i} = \varepsilon \; \mathbf{g}_{S_{n+1}} \frac{\partial \, \mathbf{g}_{S \; n+1}}{\partial \; u_{i \; n+1}} \qquad i = 1, ...15 \tag{12}$$

$$K^c_{stick_{ij}} = \frac{\partial \, \Psi^c_{stick_i}}{\partial \, u_j} \qquad i = 1, ...15, \;\; j = 1, ...15$$

For the case when sliding occurred, new solution parameters ξ_{n+1}, η_{n+1} are evaluated using the Newton-Raphson procedure, taking into account relations (9, 11), what enables the consistent linearisation. The residual vector and the tangent matrix for this case are:

$$\Psi^c_{slip_i} = \varepsilon \; \mathbf{g}_{S \; n+1} \left(\xi_{n+1}, \eta_{n+1}\right) \cdot \left(\frac{\partial \, \mathbf{x}_{S \; n+1}}{\partial \, u_{i \; n+1}} - \frac{\partial \, \mathbf{x}_{m \; n+1} \left(\xi_{n+1}, \eta_{n+1}\right)}{\partial \, u_{i \; n+1}} \right) \qquad i = 1, ...15$$

$$K^c_{slip_{ij}} = \frac{\partial \, \Psi^c_{slip_i}}{\partial \, u_j} + \frac{\partial \, \Psi^c_{slip_i}}{\partial \, \xi_{n+1}} \frac{\partial \, \xi_{n+1}}{\partial \, u_j} + \frac{\partial \, \Psi^c_{slip_i}}{\partial \, \eta_{n+1}} \frac{\partial \, \eta_{n+1}}{\partial \, u_j} \qquad i = 1, ...15, \;\; j = 1, ...15 \tag{13}$$

$$\frac{\partial \, \xi_{n+1}}{\partial \, u_j} = - \frac{\dfrac{\partial \, F_{NR}}{\partial \, u_j}}{\dfrac{\partial \, F_{NR}}{\partial \, \xi_{n+1}}}, \qquad \frac{\partial \, \eta_{n+1}}{\partial \, u_j} = - \frac{\dfrac{\partial \, F_{NR}}{\partial \, u_j}}{\dfrac{\partial \, F_{NR}}{\partial \, \eta_{n+1}}}$$

$$F_{NR} = \begin{bmatrix} \mathbf{g}_{S \; n+1} \left(\xi_{n+1}, \eta_{n+1}\right) \cdot \left(\dfrac{\mathbf{t}_{S \; n+1}}{\sqrt{\mathbf{t}_{S \; n+1} \cdot \mathbf{t}_{S \; n+1}}} + \mu \dfrac{\mathbf{n}_{n+1}}{\sqrt{\mathbf{n}_{n+1} \cdot \mathbf{n}_{n+1}}} \right) \\ \left[\mathbf{g}_{S \; n+1} \left(\xi_{n+1}, \eta_{n+1}\right) \times \mathbf{t}_{S \; n+1} \right] \cdot \mathbf{n}_{n+1} \end{bmatrix}$$

4 Numerical example

Two blocks are brought in contact, bigger block (steel, bulk modulus K=*175*$\cdot$ *10*^{9}, shear modulus $G = 80.77 \cdot 10^9$, dimension 0.08×0.08×0.028, contact surface is master one) is clamped and smaller block (rubber, $K = 5.56 \cdot 10^9$, $G = 0.34 \cdot 10^9$, 0.02×0.02×0.01, contact surface is slave one) is slid as depicted in Fig. 5. Blocs are discretized using a 8-node brick element mesh, Coulomb friction coefficient μ=*0.3* and hyperelastic, isotropic, neo-hookean material. Sliding is performed in 100 steps. Total tangential reaction, for the nodes where displacement is applied, is depicted in Fig. 7. For the same nodes, total normal reaction multiplied with the friction coefficient is depicted, proving the correct description of the contact and Coulomb's law. Comparison of tangential reactions with the cases when loading is applied in 10 steps only (Fig. 7) is showing the ability of describing the sliding over several contact surfaces.

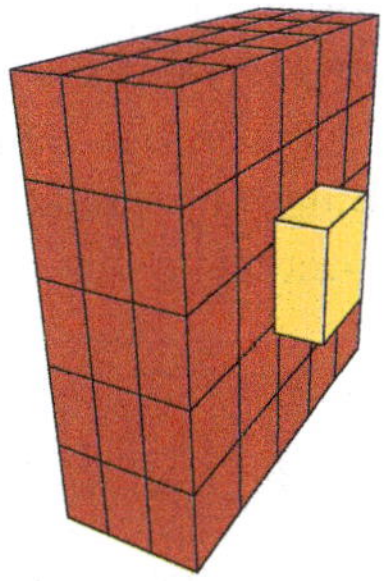

Fig. 5. Initial configuration.

Fig. 6. Final configuration.

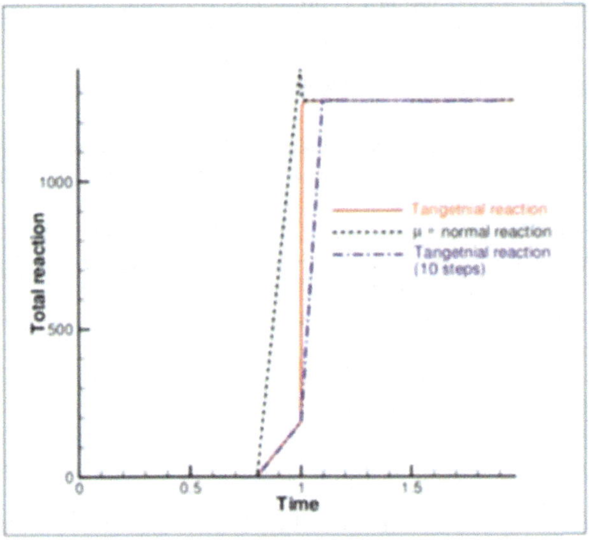

Fig. 7. Total tangential reaction.

5 Concluding remarks

The significant simplification of this single gap approach, in comparison with a standard one based on the gap splitting (normal and tangential component), is reducing the code size and the complexity of contact element. This results into consistent linearisation, faster contact element routine and better overall performance of the numerical simulation. The use of one penalty parameter instead of two, as it is the case in a standard approach, reduces the effort of finding proper penalty parameter such that penetration is minimal and ill-conditioning is avoided. The possibility of describing large steps represents significant advantage in comparison with common contact approaches.

References

1. T. A. Laursen and J. C. Simo: A Continuum-Based Finite Element Formulation for the Implicit Solution of Multibody, Large Deformation Frictional Contact Problems, *International Journal for Numerical Methods in Engineering,* **36**, 3451-3485, 1993.
2. H. Parisch and Ch. Lübbing: A Formulation of Arbitrarily Speed Surface Elements for Three-Dimensional Large Deformation Contact with Friction, *International Journal for Numerical Methods in Engineering,* **40**, 3359-3383, 1997.
3. E. Bittencourt, G. J. Creus: Finite element analysis of tree-dimensional contact and impact in large deformation problems, *Computers & Structures,* **69**, 219-234, 1998.
4. A. B. Chaudary and K. J. Bathe: A Solution Method for Static and Dynamic Analysis of Three-Dimensional Contact Problems with Friction, *Computers & Structures,* **24**, **6**, 855-873, 1986.
5. P. Papadopulos and R. L. Taylor: A simple algorithm for three-dimensional finite element analysis of contact problems, *Computers & Structures,* **46**, 1107-1118, 1993.
6. Sheng Ping Wang and Eiiji Nakamachi: The Inside-Outside Contact Search Algorithm for Finite Element Analysis, *International Journal for Numerical Methods in Engineering,* **40**, 3665-3685, 1997.
7. L. Krstulović-Opara, P. Wriggers: A three-dimensional contact element based on the Moving Friction Cone approach, eds.: D.R.J. Owen, E. Onate and B. Suarez, *Proceedings of the Computational Plasticity VII, Fundaments and Applications,* April 7-10, Barcelona, Spain, 2003.
8. P. Wriggers, L. Krstulović-Opara-Opara: The moving friction cone approach for three-dimensional contact simulations, *International Journal of Computational Methods,* **1**, 105-119, 2004.
9. P. Wriggers (1987), On Consistent Tangent Matrices for Frictional Problems, *Proceedings of NUMETA 87 Conference,* eds. J. Middleton, G. N. Pande, Nijhoff, Drobrecht.
10. Giannokopoulos, A.E. (1989), The Return Mapping Method for the Integration of Friction Constitutive Relations, *Computer & Structures,* **32**, 157-168.

for an abbreviated version of

3D beam-to-beam contact within coupled electromechanical fields: a finite element model

D.P. Boso[1], P. Litewka[2], B.A. Schrefler[1], and P. Wriggers[3]

[1] Department of Structural and Transportation Engineering, University of Padua, Via Marzolo 9, 35131 Padua, Italy,
boso@caronte.dic.unipd.it, bas@caronte.dic.unipd.it

[2] Institute of Structural Engineering, Poznan University of Technology, ul. Piotrowo 5, PL 60-965 Poznań, Poland,
przemyslaw.litewka@ikb.poznan.pl

[3] Institute of Mechanics and Computational Mechanics, University of Hannover, Appelstr. 9A, D 30167 Hanover, Germany,
wriggers@ibnm.uni-hannover.de

Abstract. In this paper a 3D beam-to-beam contact element is presented, to deal with contact problems in the coupled electric - mechanical fields. The beams are supposed to get in contact in a pointwise manner, the detection of the contact points and the computation of all contributions are carried out using a fully symmetric treatment of the two beams. Concerning the mechanical field, Hertz theory of contact for elastic bodies is considered. The contact area is varying according to the beam-to-beam angle, being circular only in the case of perpendicular beams. This variation of the shape is taken into account too. The problem is semi-coupled: the mechanical field influences the electric one because of the dependence of the voltage distribution on the contact area.

Within the finite element discretization, the mechanical and the electric treatment of the beam element is formulated in the usual way, considering nodal displacements and voltages as main unknowns. The electromechanical contact constraints are enforced with the penalty method. Starting from the virtual work equation the consistent linearization of all contributions is computed to achieve the quadratic convergence within the Newton-Raphson iterative scheme. The complete set of equations - arranged in a matrix form suitable for the finite element implementation - is solved with a monolithic approach. Finally some numerical examples are discussed to show the effectiveness of the model.

1 Introduction

The beam-to-beam contact in the coupled electric - mechanical fields has been recently investigated [1]. This paper is a further development of that approach. For sake of brevity, we describe here only the improvement, interested readers can find all the details in the cited paper [1] and reference therein.

We consider beams having circular cross sections which preserve shape, size and remain plane. Homogeneous, isotropic, linear elastic material is assumed; large displacements and small strains are allowed.

The beams are assumed to get in contact in a pointwise manner, so that this work does not apply to the case of parallel beams. The detection of the contact points and the computation of all contributions are carried out using a fully symmetric treatment of the two beams. Hertz contact theory for elastic bodies is considered. The contact area is varying according to the beam-to-beam angle, being circular only in the case of perpendicular beams. The problem is semi-coupled, the voltage distribution being affected by the mechanical contact areas.

The electric – mechanical contact constraints are enforced with the penalty method within the finite element technique. The virtual work equations for the mechanical and electric fields are consistently linearized to achieve a quadratic convergence within Newton Raphson iterative scheme. The equations set is solved with a monolithic approach.

2 Electric – mechanical contact

In the following just two contacting beams with radius r and $\bar{r}$, elastic modulus E and $\bar{E}$, Poisson's ratio ν and $\bar{\nu}$ are considered (Fig. 1). The extension to several beams is straightforward.

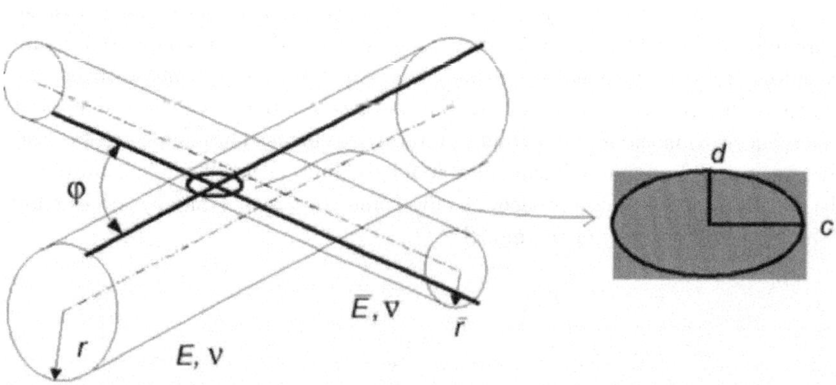

Fig. 1. Two contacting beams and detail of their contact area.

At first the contacting beams have ideally only one point in contact, then – increasing the external loads - the point becomes a contact area which can generally be considered elliptic. The semi-axes c and d of the ellipse can be written as a function of the contact normal force F_N

$$c = \alpha \cdot \left(\frac{3}{2} \frac{F_N \cdot r_{av}}{E_{av}} \right)^{\frac{1}{3}}, \quad d = \beta \cdot \left(\frac{3}{2} \frac{F_N \cdot r_{av}}{E_{av}} \right)^{\frac{1}{3}} \tag{1}$$

where $r_{av} = r \cdot \bar{r}/(r + \bar{r})$ and $E_{av} = E \cdot \bar{E}/\left[\bar{E}\left(1-\nu^2\right) + E\left(1-\bar{\nu}^2\right)\right]$ are the equivalent properties of the two beams.

The parameters α and β depend upon the beams radii and their axes angle φ. By interpolating some values found in [2] we can write

$$\alpha = 1 - \ln(1-x) - 0.32x + 3.35x^{40}, \; \beta = 1 - 0.57x - 0.19x^{20}$$

$$x = r_{av} \cdot \sqrt{\frac{1}{r^2} + \frac{1}{\bar{r}^2} + \frac{2}{r \cdot \bar{r}} \cdot \cos 2\varphi} \tag{2}$$

With these properties at hand we can define the radius of an equivalent circular area

$$a = \sqrt{cd} = \sqrt{\alpha\beta} \cdot \left(\frac{3}{2} \frac{F_N \cdot r_{av}}{E_{av}} \right)^{\frac{1}{3}} \tag{3}$$

Concerning the electric field, considering the model explained in [1], the conductance of a flux tube having a flat circular narrowing is

$$h_V = 2 \frac{K \cdot \bar{K}}{K + \bar{K}} a = 2 K_{av} \sqrt{\alpha\beta} \cdot \left(\frac{3}{2} \frac{F_N \cdot r_{av}}{E_{av}} \right)^{\frac{1}{3}} \tag{4}$$

The electric current I can be expressed as a function of the normal force F_N and the voltage difference ΔV

$$I = h_V \Delta V = 2 K_{av} \sqrt{\alpha\beta} \left(\frac{3}{2} \frac{r_{av}}{E_{av}} \right)^{1/3} F_N^{1/3} \Delta V \tag{5}$$

3 Global set of equations and contact contributions

From a mathematical point of view the solution of a contact problem leads to the minimization of a functional subjected to the contact inequality constraints. In our case this constraint is the non-penetration condition. To solve this mathematical problem the active set strategy is used. With the help of the contact search routine the active constraints are found and treated as equality ones. The formulation is similar to that described in [1], [3] and [4] where the contact element definition and the explanation of all vectors used are detailed.

As described above we assume a pointwise contact, hence without lack of generality we can suppose that the global set of equations is obtained by adding to the global virtual work of the two beams the virtual work given by the contact force and the analogous quantities in the electric field

$$
\begin{aligned}
\delta_u \left(\Pi_m\right)_{tot} &= \delta_u \left[\bar{\Pi}_m \left(\bar{u}, \bar{V}\right)\right] + \delta_u \left[\Pi_m \left(u, V\right)\right] + \bigcup_{act} (F_N \cdot \delta_u g_N) = 0 \\
\delta_V \left(\Pi_e\right)_{tot} &= \delta_V \left[\bar{\Pi}_e \left(\bar{u}, \bar{V}\right)\right] + \delta_V \left[\Pi_e \left(u, V\right)\right] + \bigcup_{act} (I \cdot \delta_V g_V) = 0
\end{aligned} \tag{6}
$$

In general the weak form (6) represents a set of non linear equations which can be solved iteratively by Newton-Raphson method. In the following we write explicitly the contact element contributions

$$
\begin{aligned}
\Delta_u \delta_u \left(\Pi_m\right)_{tot} &= \Delta_u \delta_u \left(\bar{\Pi}_m\right) + \Delta_u \delta_u \left(\Pi_m\right) + \bigcup_{act} \Delta_u \left(F_N \cdot \delta_u g_N\right) = 0 \\
\Delta_V \delta_u \left(\Pi_m\right)_{tot} &= \Delta_V \delta_u \left(\bar{\Pi}_m\right) + \Delta_V \delta_u \left(\Pi_m\right) + \bigcup_{act} \Delta_V \left(F_N \cdot \delta_u g_N\right) = 0 \\
\Delta_u \delta_V \left(\Pi_e\right)_{tot} &= \Delta_u \delta_V \left(\bar{\Pi}_e\right) + \Delta_u \delta_V \left(\Pi_e\right) + \bigcup_{act} \Delta_u \left(I \cdot \delta_V g_V\right) = 0 \\
\Delta_V \delta_V \left(\Pi_e\right)_{tot} &= \Delta_V \delta_V \left(\bar{\Pi}_e\right) + \Delta_V \delta_V \left(\Pi_e\right) + \bigcup_{act} \Delta_V \left(I \cdot \delta_V g_V\right) = 0
\end{aligned} \tag{7}
$$

Considering two-node contacting beam elements, the nodal degrees of freedom of the contact element are ordered as

$$
\mathbf{q}_{(28\times 1)} = \left\{ \bar{\mathbf{u}}_M^T, \mathbf{u}_M^T, \bar{\mathbf{V}}_M^T, \mathbf{V}_M^T \right\} \tag{8}
$$

where $\bar{\mathbf{u}}_M = \left\{\bar{\mathbf{u}}_1^T, \bar{\mathbf{u}}_2^T\right\}^T$, $\mathbf{u}_M = \left\{\mathbf{u}_1^T, \mathbf{u}_2^T\right\}^T$ collect the nodal displacement values and $\bar{\mathbf{V}}_M = \{\bar{V}_1, \bar{V}_2\}^T$, $\mathbf{V}_M = \{V_1, V_2\}^T$ collect the nodal voltage values for the two beam elements where the contact points lie.

By expressing all contributions in matrix notation [1], we arrive at the following form for the tangent stiffness matrix and the residual vectors

$$
\mathbf{K}_{[28\times 28]} = \begin{bmatrix} \mathbf{K}_{mm} & \mathbf{0} \\ (\mathbf{K}_{me1} + \mathbf{K}_{me2} + \mathbf{K}_{me3}) & \mathbf{K}_{ee} \end{bmatrix}, \quad \mathbf{R}_{[28\times 1]} = \begin{bmatrix} \mathbf{R}_m \\ \mathbf{R}_e \end{bmatrix} \tag{9}
$$

where $\mathbf{K}_{mm}$ and $\mathbf{R}_m$are the mechanical parts [3], [5] and $\mathbf{K}_{ee}$, $\mathbf{R}_m$ are the electric ones [1]

$$
\mathbf{K}_{ee} = 2K_{av}\sqrt{\alpha\beta}\left(\frac{3\varepsilon_N g_N r_{av}}{2E_{av}}\right)^{\frac{1}{3}} \begin{bmatrix} \bar{\mathbf{N}}_V \\ -\mathbf{N}_V \end{bmatrix} \otimes \begin{bmatrix} \bar{\mathbf{N}}_V \\ -\mathbf{N}_V \end{bmatrix}
$$

$$\mathbf{R}_e = 2K_{av}g_V\sqrt{\alpha\beta}\left(\frac{3\varepsilon_N g_N r_{av}}{2E_{av}}\right)^{\frac{1}{3}}\begin{bmatrix}\bar{\mathbf{N}}_V^T\\ -\mathbf{N}_V^T\end{bmatrix} \tag{10}$$

while $\mathbf{K}_{me1}$, $\mathbf{K}_{me2}$, $\mathbf{K}_{me3}$ are the coupling terms [1]

$$\mathbf{K}_{me1} = 2K_{av}\sqrt{\alpha\beta}\left(\frac{3\varepsilon_N g_N r_{av}}{2E_{av}}\right)^{\frac{1}{3}}\begin{bmatrix}\bar{\mathbf{N}}_V^T\\ -\mathbf{N}_V^T\end{bmatrix}\left[\bar{V}_{,\xi}\bar{\mathbf{F}} - V_{,\xi}\mathbf{F}\right]$$

$$\mathbf{K}_{me2} = \frac{2g_V K_{av}\sqrt{\alpha\beta}}{3}\left(\frac{3\varepsilon_N r_{av}}{2E_{av}}\right)^{\frac{1}{3}}(g_N)^{-\frac{2}{3}}\begin{bmatrix}\bar{\mathbf{N}}_V^T\\ -\mathbf{N}_V^T\end{bmatrix}\mathbf{n}^T\left[\bar{\mathbf{N}}_u, -\mathbf{N}_u\right]$$

$$\mathbf{K}_{me3} = 2K_{av}\sqrt{\alpha\beta}\left(\frac{3\varepsilon_N g_N r_{av}}{2E_{av}}\right)^{\frac{1}{3}} g_V\begin{bmatrix}\mathbf{D}_V^T\bar{\mathbf{F}}\\ -\mathbf{D}_V^T\mathbf{F}\end{bmatrix} \tag{11}$$

4 Numerical examples

4.1 Example 1

In this example the influence of the beam-to-beam angle and of the normal force upon the voltage gap is tested. Two beams loaded with four equal tip displacements $\Delta = 0.2$ are analysed (Fig. 2). The following data [N, cm, Ω] are used for both beams: Young's modulus $E = 205 \cdot 10^5$, Poisson's ratio $\nu = 0.3$, electric resistivity $\rho = 8 \cdot 10^{-6}$, beam radius $r = 0.1$, beam length $l = 4.0$, initial normal gap $g_N = 0.06$. Each beam is modelled using 10 finite elements. The axes angle is varied from 10 to 90 degrees. At the tips of one beam a voltage of 10mV is applied and at the tips of the other beam the applied voltage is –20mV. In Fig. 3 the deformed shape of the beams axes for one configuration is presented. The results of the analysis are illustrated in the graph of Fig. 3. It shows an almost linear dependence of the voltage gap upon the normal force. The axes angle vs. voltage gap relation is not linear and it can be seen from the isolines that it is more pronounced for acute angles, smaller than 40-50 degrees.

4.2 Example 2

Three twisted helix cables clamped at one end and loaded with three equal tip rotation angles as shown in Fig. 4 are analysed.

The following data [N, cm, Ω] are used for the three beams: Young's modulus $E = 205 \cdot 10^5$, Poisson's ratio $\nu = 0.3$, electric resistivity $\rho = 8 \cdot 10^{-6}$, cable radius $r = 0.8$, cable axes clearance 2.0, length of cables in plane 10.0. Each beam is modelled using 36 straight beam finite elements. Two cases of electric boundary conditions are considered. In Case 1 the three cables have

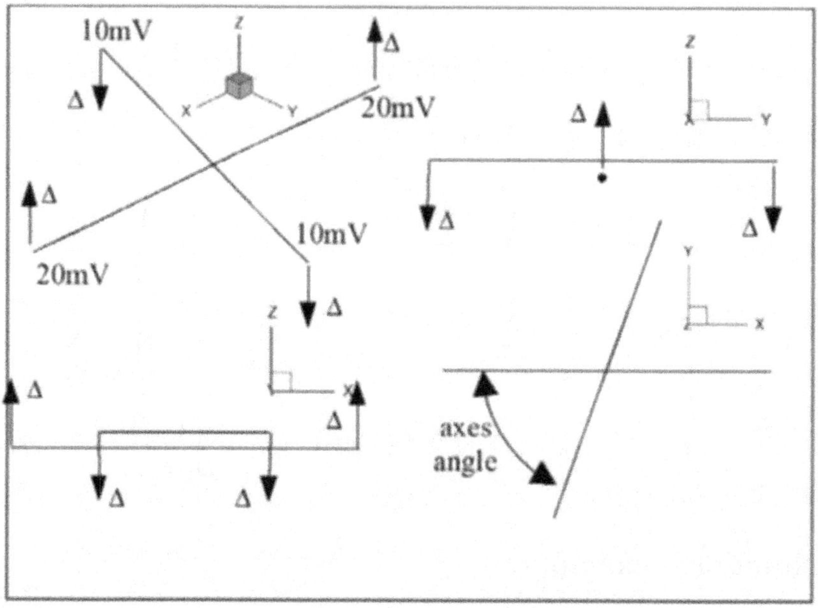

Fig. 2. Beam axes layout and boundary conditions for Example 1.

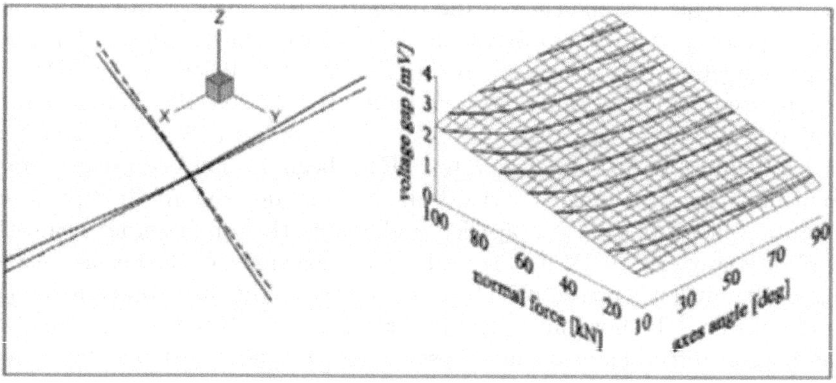

Fig. 3. Results of Example 1.

a voltage of – 20mV applied at the clamped ends and 10mV at the free ends. In Case 2 every cable has the same voltage applied at its ends: 10mV, 20mV and 30mV, respectively for each of the three cables.

The deformed configuration of the cables is shown in Fig. 5 together with the voltage distribution along the cables and a contact zone for two considered cases. As a matter of comparison, the dashed lines represent the solution if the contact is ignored. For Case 1 it can be noticed that the contact solution features only a small deviation from the linear voltage distribution of the

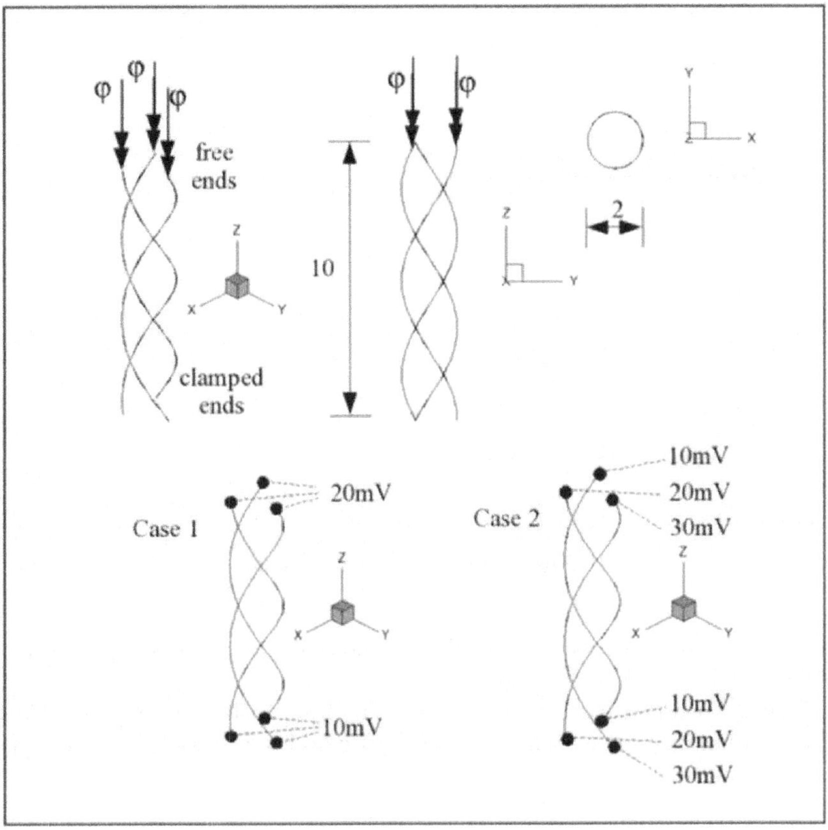

Fig. 4. Cables axes arrangement for Example 2.

non-contact case. In Case 2 there is a relatively long section of all three cables where the voltage is the same and it significantly differs from the non-contact solution.

5 Concluding remarks

A 3D beam-to-beam finite element is presented, suitable to be used to solve problems in the coupled electric-mechanical fields. The fields are semi-coupled, with the mechanical part influencing the electric one. The model described in [1] has been further improved by refining the description of the contact area. The tangent stiffness matrix and the residual vector for the contact finite element are presented.

The dependence of the contact area upon the beam-to-beam contact angle is taken into consideration. As shown in the second example, the axes angle

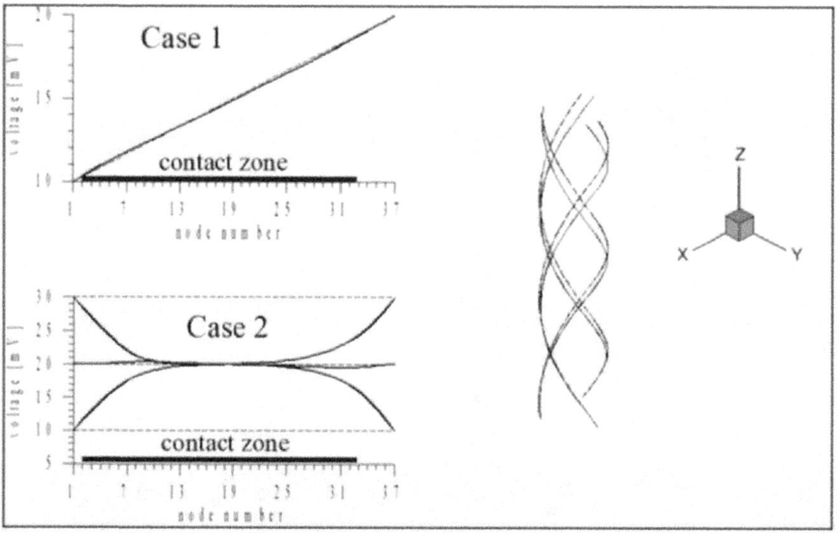

Fig. 5. Results for Example 2.

vs. voltage gap dependence is not linear, and it is more pronounced for acute angles, smaller than 40-50 degrees.

The governing equations are solved simultaneously with the monolithic scheme, to have a model ready for the future addition of the thermal field.

Acknowledgements

Support for this work was partially provided by PRIN 2004094015_002 "Thermo-hydraulic-mechanical and Electro-mechanical Modelling of ITER Superconducting Magnets". This support is gratefully acknowledged.

References

1. Boso D P, Litewka P, Schrefler B A, Wriggers P (2005), in press on Int J Numer Meth Engng, Published online in Wiley InterScience (www.interscience.wiley.com). DOI: 10.1002/nme.1427.
2. Young W C, Budynas R, (2001) Roark's Formulas for Stress and Strain, McGraw-Hill Professional
3. Wriggers P, Zavarise G. (1997) Comm Numer Meth Engng 13:429-438.
4. Zavarise G, Wriggers P (2000) Int J Numer Meth Engng 49:977-1006.
5. Litewka P, Wriggers P (2002) Int J Numer Meth Engng 53:2019-2041.

A study of symbolic description, numerical efficiency and accuracy of 2D and 3D contact formulations

J. Korelc[1], J. Lengiewicz[2], and S. Stupkiewicz[2]

[1] University of Ljubljana, Jamova 2, Ljubljana, Slovenia, jkorelc@fgg.uni-lj.si
[2] Institute of Fundamental Technological Research, Warsaw, Poland, jleng@ippt.gov.pl, sstupkie@ippt.gov.pl

Abstract. Complex 3D contact formulations, including various types of smoothing, advanced friction laws and sensitivity analysis have become computationally progressively expensive and, for moderate size problems, can be the main obstacle for practical application of these formulations. The aim of the paper is to present a general symbolic description of contact problems and to study the efficiency and the accuracy of several formulations of contact finite elements. For the study to be comprehensive, an approach is needed that enables derivation of the required formulas (e.g. element residual and tangent), their finite element coding, and running of some benchmark problems in an objective way. This has been achieved by using a symbolic approach to derivation of formulas and automatic code generation.

1 Symbolic Approach to Problem Solving in Mechanics of Solids

Abstract description of the governing equations and subsequent automatic generation of numerical code has opened a new way in coding of the finite element routines [9]. For complex problems, like frictional contact interactions, symbolic description on a high abstract level has enabled significant advantages, particularly considering the derivation of the residual vector and stiffness matrix. Accordingly, implementation of various formulations of frictional contact within a single FE environment is now possible with a moderate effort. It is thus possible to compare these formulations and to assess their efficiency and accuracy.

In this work, several contact smoothing techniques for finite-deformation multi-body contact problems are investigated. In particular, various approaches to symbolic description of contact, including the use of forward and backward modes of automatic differentiation procedures, are analyzed.

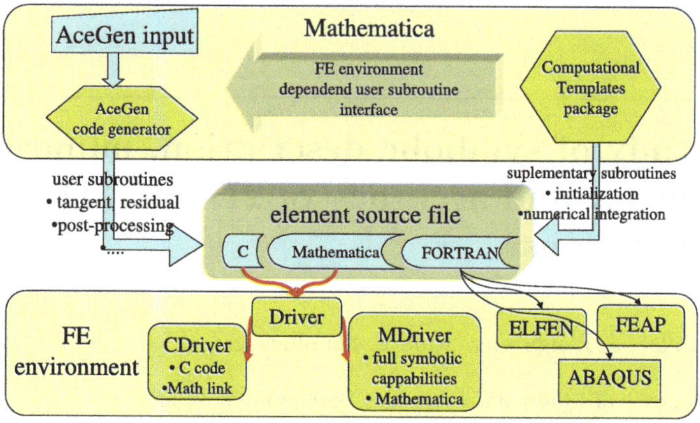

Fig. 1. Hybrid system for multi-language and multi-environment code generation.

Contact elements studied in the present paper have been derived using a hybrid system for automatic generation of numerical codes. The system consists of three major components. The *Mathematica* package *AceGen* is used for symbolic derivation of formulae needed in numerical procedures and for automatic generation of respective codes. The approach implemented in *AceGen* overcomes the problem of expression growth by combining several techniques: symbolic algebra system *Mathematica*, automatic differentiation, automatic code generation, and theorem proving. The second component, called *Computational Templates*, is a collection of prearranged modules for the automatic creation of the interface between the finite element code and the finite element environment. *AceGen* and *Computational Templates* together enable multi-language and multi-environment generation of nonlinear finite element codes from the same symbolic description. Currently supported are C, FORTRAN, and *Mathematica* languages and several research and commercial finite element environments (*FEAP*, *ELFEN*, *ABAQUS*, *Driver*).

The third component of the system is a model finite element environment called *Driver*. The advantage of this finite element environment is that it exists in two equivalent versions. The first version (*MDriver*) is written in *Mathematica*'s symbolic language, so when a particular problem is analyzed the advantages of *Mathematica* can be exploited. The second version (*CDriver*) is written in C language and is connected with *Mathematica* so that large-scale problems can be solved. The structure of the system is depicted in Fig. 1. General description of *AceGen* and *Computational Templates* together with numerous examples can be found in [9, 2].

2 Automatic differentiation procedures

Differentiation is an arithmetic operation that plays crucial role in the development of numerical procedures. The exact analytical derivatives are difficult to derive, which is why the numerical differentiation is often used instead. Automatic differentiation (AD) represents an alternative solution to the numerical differentiation as well as to the manually derived expressions. If one has a computer code which evaluates a function $f(a_1, a_2, a_3, \ldots, a_n)$ and needs to compute the gradient of f with respect to arbitrary variables, then automatic differentiation tools (see e.g. [3]) can be applied to generate the appropriate program code. Two approaches, often recalled as forward and backward mode of automatic differentiation, can be used for this purpose. The forward mode accumulates the derivatives of auxiliary variables with respect to the independent variables, while the backward mode propagates adjoins, i.e. the derivatives of the final values with respect to auxiliary variables. Consequently, the numerical work ratio w for the evaluation of gradient is proportional to the number of variables ($w(f) \leq \alpha\, n$) in the case of forward mode and constant in the case of backward mode ($w(f) \leq 5$), where

$$w(f) = \frac{computational\, work(f(a_1, a_2, a_3, ..., a_n), \nabla f = \frac{\partial f}{\partial a_i})}{computational\, work(f(a_1, a_2, a_3, ..., a_n))}$$

Traditional formulations in solid mechanics, including contact problems, involve partial derivatives ($\frac{\partial(\cdot)}{\partial(\cdot)}$), total derivatives ($\frac{D(\cdot)}{D(\cdot)}$), directional derivatives, consistent derivatives, etc. These mathematical formalisms can all be represented by AD procedures provided that the AD exceptions are handled in a proper way. Some typical situations are presented in Fig. 2 where function f depends on two sets of variables, $\mathbf{a}$ and $\mathbf{b}$.

In the first case (A), there exists an explicit algorithmic dependency of $\mathbf{b}$ with respect to $\mathbf{a}$, so that in principle the derivatives can be obtained automatically without users intervention simply by the chain rule. However, there also exists a profound mathematical relationship that enables evaluation of derivatives in a more efficient way. This is often the case when the evaluation of $\mathbf{b}$ involves iterative loops, inverse matrices, etc.

Case B represents the situation when variables $\mathbf{b}$ are independent variables and variables $\mathbf{a}$ implicitly depend on $\mathbf{b}$. This implicit dependency has to be considered for differentiation. In this case, the AD would not give the correct result without user intervention. A typical example for this situation is an arbitrary transformation of coordinates.

In the third case (C), there exists an explicit dependency between variables $\mathbf{b}$ and $\mathbf{a}$ that has to be neglected for differentiation. This is a special case of situation A that often appears in the formulation of mechanical problems where instead of total variation some arbitrary variation of the given quantity has to be evaluated.

type of "conditional" derivative	formalism	AceGen description
A	$\Delta f = \left.\frac{\partial f(\mathbf{a},\mathbf{b}(\mathbf{a}))}{\partial(\mathbf{a})}\right\vert_{\frac{\partial(\mathbf{b})}{\partial(\mathbf{a})}=\mathbf{M}}$	a ⊢ SMSReal[a$$] b ⊢ f_b[a] M ⊢ f_M[a, b] SMSDefineDerivative[b, a, M] f ⊢ f_f[a, b] Δf ⊢ SMSD[f, a]
B	$\Delta f = \left.\frac{\partial f(\mathbf{b})}{\partial(\mathbf{a}(\mathbf{b}))}\right\vert_{\frac{\partial(\mathbf{b})}{\partial(\mathbf{a})}=\mathbf{M}}$	b ⊢ SMSReal[b$$] a ⊢ SMSFreeze[f_a[b]] M ⊢ f_M[a, b] SMSDefineDerivative[b, a, M] f ⊢ f_f[b] Δf ⊢ SMSD[f, a]
C	$\Delta f = \left.\frac{\partial f(\mathbf{a},\mathbf{b}(\mathbf{a}))}{\partial(\mathbf{a})}\right\vert_{\frac{\partial(\mathbf{b})}{\partial(\mathbf{a})}=\mathbf{0}}$	a ⊢ SMSReal[a$$] b ⊢ f_b[a] f ⊢ f_f[a, b] Δf ⊢ SMSD[f, a, "Constant"→b]

Fig. 2. Automatic differentiation exceptions.

3 General symbolic description of contact problems

The general scheme appropriate for the direct symbolic description of 3D contact elements is given in Fig. 3. Out of the variety of available contact formulations, only the approach employing the node-to-segment formulation and augmented Lagrangian treatment of contact constraints [8] is presented. Similar schemes can be derived also for other formulations, e.g. penalty, segment-to-segment, mortar, etc., cf. [10, 3]. The present formulation is based on the general formulation of transient coupled non-linear systems [1].

In Figure 3, $\mathbf{a}$ denotes the vector of all element unknowns which is composed of the displacement $\mathbf{u}^s$ of the slave node, displacements $\mathbf{u}_i^m$ of the nodes defining the master segment, and the normal λ_N and tangential $\boldsymbol{\lambda}_T$ Lagrange multipliers. The position of the slave node is denoted by $\mathbf{x}^s$.

The master surface is parameterized by reference coordinates ξ^1 and ξ^2. For a general master surface interpolation, defined by the position vector $\mathbf{x}^m(\xi^1, \xi^2)$ and the surface normal $\mathbf{n}(\xi^1, \xi^2)$, an additional set of nonlinear equations $\boldsymbol{\Phi} = \mathbf{0}$ (the local problem) has to be solved in order to determine the closest point projection of the slave node onto the master surface. The projection defines also the normal gap g_N. Since $\mathbf{x}^m$ and $\mathbf{n}$ depend on the displacements of the nodes involved in the interpolation of the master surface, there exists an implicit dependence of $\mathbf{b} = \{g_N, \xi^1, \xi^2\}$ on $\mathbf{a}$ (through the local problem solution). An AD exception is thus introduced in order to avoid differentiation of the Newton loop of the local problem.

The approach proposed in [7] is adopted for calculation of the tangential slip increment $\Delta\xi^i$, namely

1. $\mathbf{a} = \{\mathbf{a}^u, \mathbf{a}^\lambda\}$, $\mathbf{a}^u = \{\mathbf{u}^s, \mathbf{u}_1^m, \mathbf{u}_2^m, \ldots\}$, $\mathbf{a}^\lambda = \{\lambda_N, \boldsymbol{\lambda}_T\}$
2. local problem (closest point projection):

$$\begin{cases} \mathbf{b} = \{\xi^1, \xi^2, g_N\} \\ \boldsymbol{\Phi}(\mathbf{b}) = \mathbf{x}^m(\xi^1, \xi^2) + g_N \mathbf{n}(\xi^1, \xi^2) - \mathbf{x}^s = \mathbf{0} \\ \text{local NR: } \begin{cases} {}_{\Phi}\mathbf{K}_j = \dfrac{\partial \boldsymbol{\Phi}_j}{\partial \mathbf{b}_j} \\ {}_{\Phi}\mathbf{K}_j \Delta \mathbf{b}_j + \boldsymbol{\Phi}_j = \mathbf{0} \\ \mathbf{b}_{j+1} := \mathbf{b}_j + \Delta \mathbf{b}_j \end{cases} \\ \text{define AD exception: } \dfrac{\partial \mathbf{b}}{\partial \mathbf{a}} = -\left[\dfrac{\partial \boldsymbol{\Phi}}{\partial \mathbf{b}}\right]^{-1} \dfrac{\partial \boldsymbol{\Phi}}{\partial \mathbf{a}} \end{cases}$$

3. $\Delta \boldsymbol{\xi} = \dfrac{\partial \boldsymbol{\xi}}{\partial \mathbf{a}^u}(\mathbf{a}^u - {}^n\mathbf{a}^u)$
4. $\hat{\lambda}_N = \lambda_N + \varrho g_N$, $\hat{\boldsymbol{\lambda}}_T = \boldsymbol{\lambda}_T + \varrho \Delta \boldsymbol{\xi}$
5. $\hat{k} = \begin{cases} -\mu \hat{\lambda}_N, & \hat{\lambda}_N \le 0 \\ 0, & \hat{\lambda}_N > 0 \end{cases}$
6. $\Pi_N = \begin{cases} (\lambda_N + \frac{\varrho}{2} g_N) g_N, & \hat{\lambda}_N \le 0 \text{ (contact)} \\ -\frac{1}{2\varrho}\lambda_N^2, & \hat{\lambda}_N > 0 \text{ (separation)} \end{cases}$
7. $\Pi_T = \begin{cases} \left\{ \begin{array}{ll} (\boldsymbol{\lambda}_T + \frac{\varrho}{2}\Delta\boldsymbol{\xi}) \cdot \Delta\boldsymbol{\xi}, & \|\hat{\boldsymbol{\lambda}}_T\| \le \hat{k} \text{ (stick)} \\ -\frac{1}{2\varrho}(\|\boldsymbol{\lambda}_T\|^2 - 2\hat{k}\|\hat{\boldsymbol{\lambda}}_T\| + \hat{k}^2), & \|\hat{\boldsymbol{\lambda}}_T\| > \hat{k} \text{ (slip)} \end{array} \right\}, & \hat{\lambda}_N \le 0 \\ -\frac{1}{2\varrho}\|\boldsymbol{\lambda}_T\|^2, & \hat{\lambda}_N > 0 \end{cases}$
8. $\boldsymbol{\Psi} = \left.\dfrac{\partial(\Pi_N + \Pi_T)}{\partial \mathbf{a}}\right|_{\frac{\partial \hat{k}}{\partial \mathbf{a}} = 0}$, define AD exception
9. $\mathbf{K} = \dfrac{\partial \boldsymbol{\Psi}}{\partial \mathbf{a}}$

Fig. 3. General symbolic description of augmented Lagrangian formulation of frictional contact.

$$\Delta\xi^i = \frac{\partial \xi^i}{\partial \mathbf{a}^u}(\mathbf{a}^u - {}^n\mathbf{a}^u),$$

where vectors $\mathbf{a}^u$ and ${}^n\mathbf{a}^u$ consist of the displacements of all nodes in the considered contact element at the current and at the previous time step, respectively.

The augmented Lagrange multipliers $\hat{\lambda}_N$ and $\hat{\boldsymbol{\lambda}}_T$, the normal ($\Pi_N$) and the tangential ($\Pi_T$) contributions to the contact Lagrangian are also provided in Fig. 3 for all possible contact states (contact or separation, stick or slip). The element residual vector $\boldsymbol{\Psi}$ and the tangent matrix $\mathbf{K}$ are then obtained by automatic differentiation of the contact Lagrangian $\Pi_N + \Pi_T$. Note that an AD exception

$$\boldsymbol{\Psi} = \left.\frac{\partial(\Pi_N + \Pi_T)}{\partial \mathbf{a}}\right|_{\frac{\partial \hat{k}}{\partial \mathbf{a}} = 0}$$

has to be defined in order to obtain proper frictional contact equations.

4 Study of efficiency and code size of 3d contact formulations

Following the procedure described in the previous section, various 3D contact elements have been derived. The details are beyond the scope of the paper and can be found elsewhere, e.g. [7, 9, 13]. Only a general comparison of code size and efficiency of the elements is provided below. In all the cases, the generated finite elements share some common characteristics: node-to-segment contact formulation; master surface formed by quadrilateral segments; Coulomb friction condition; augmented Lagrangian treatment of contact constraints; fully implicit solution of the closest point projection.

The following contact surface descriptions are considered (see Figure 4): "Bilinear" is the standard bilinear interpolation; "Bezier 9" is a smooth Bezier surface with 9 control nodes; "Bezier 16" is a smooth Bezier surface with 16 control nodes; "Gregory" is a smooth Gregory patch type surface [13] that can be applied also on unstructured meshes. For all the elements, two equivalent codes are generated. In the first case, all differentiations are performed using the forward AD mode and, in the second case, the backward AD mode is applied when appropriate.

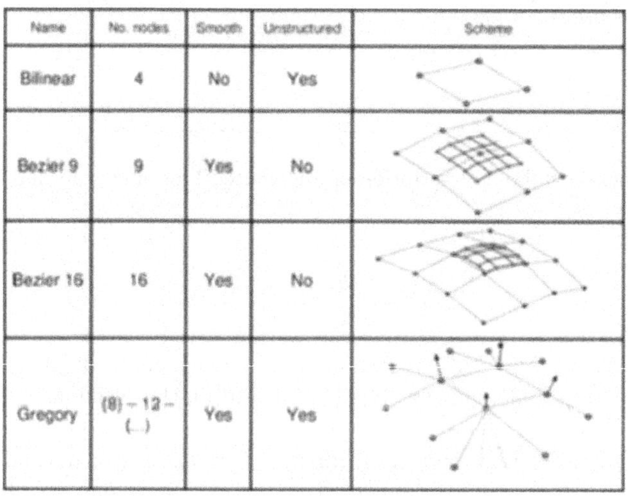

Name	No. nodes	Smooth	Unstructured	Scheme
Bilinear	4	No	Yes	
Bezier 9	9	Yes	No	
Bezier 16	16	Yes	No	
Gregory	(8) – 12 – (…)	Yes	Yes	

Fig. 4. Contact surface description.

In Table 1, the size of the generated "tangent and residual" user-subroutine codes are compared. The code size is in the range from 37 to 278 KB (kilobytes) with the significant increase for Gregory patch formulation. This is due to the complexity of the interpolating function and also due to the singularities that occur in the description of the Gregory patch. At the same time,

Table 1. Size of the contact element source code file.

Formulation	Forward AD (KB)	Backward AD (KB)
Bilinear	37	39
Bezier 9	91	94
Bezier 16	90	96
Gregory patch	278	292

Fig. 5. Example: sliding contact of two elastic blocks.

there is no significant difference in code size between forward and backward AD mode.

A simple numerical example, used to assess the numerical efficiency of the formulations, is depicted in Fig. 5. A solid block is pressed into the flat surface and dragged in the direction of the arrow. The evaluation time of the contact element residual vector and tangent matrix is compared with the total simulation time in Table 2. For forward AD mode, evaluation of contact elements takes from 11 per cent of the total time in the case of simple bilinear element to 77 per cent in the case of Gregory patch formulation. One can see that, for moderate size problems, contact evaluation can become a bottleneck of numerical simulation. It is also evident that the numerical efficiency is significantly improved when the backward mode is used. In the case of the most complicated Gregory patch formulation, speed up of 21 times is obtained as compared to the forward mode. Thus we can conclude that the use of backward mode of automatic differentiation is essential for derivation of numerically efficient 3D smooth contact elements.

Table 2. Relative evaluation time and convergence properties.

Formulation	Forward AD % total time	Backward AD % total time	Speed up	Total iterations
Bilinear	11.	0.9	13.	154
Bezier 9	29.	2.3	17.	75
Bezier 16	41.	3.5	16.	102
Gregory patch	77.	14.	21.	86

References

1. J. Korelc. Multi-language and multi-environment generation of nonlinear finite element codes. *Engineering with Computers*, 18:312 327, 2002.
2. J. Korelc. Automatic generation of numerical codes with introduction to AceGen 6.0 symbolic code generator. Available at http://www.fgg.uni-lj.si/Symech/, 2006.
3. A. Griewank. On Automatic Di erentiation. In M. Iri and K. Tanabe, editors, *Mathematical Programming: Recent Developments and Applications*, pages 83 108. Kluwer Academic Publishers, Amsterdam, 1989.
4. G. Pietrzak and A. Curnier. Large deformation frictional contact mechanics: continuum formulation and augmented Lagrangian treatment. *Comp. Meth. Appl. Mech. Engng.*, 177(3 4):351 381, 1999. 5.
5. T.A. Laursen. *Computational Contact and Impact Mechanics.* Springer-Verlag, Berlin, 2002. 6.
6. P. Wriggers. *Computational Contact Mechanics.* Wiley, Chichester, 2002. 7.
7. P. Michaleris, D.A. Tortorelli, and C.A. Vidal. Tangent operators and design sensitivity formulations for transient non-linear coupled problems with applications to elastoplasticity. *Int. J. Num. Meth. Engng.*, 37:2471 2499, 1994. 8.
8. G. Pietrzak. C*ontinuum mechanics modelling and augmented Lagrangian formulation of large deformation frictional contact problems.* PhD thesis, Lausanne, EPFL, 1997.
9. L. Krstulović-Opara. *A C^1-continuous formulation for finite deformation contact.* PhD thesis, Institute für Baumechanik und Numerische Mechanik, Universität Hannover, 2001.
10. M.A. Puso and T.A. Laursen. A 3D contact smoothing method using Gregory patches. *Int. J. Num. Meth. Engng.*, 54:1161 1194, 2002.

Part II

Mathematical analysis

Existence theorems for noncoercive incremental contact problems with Coulomb friction

L.-E. Andersson[1] and A. Rietz[2]

[1] Department of Mathematics, Linköping University, SE-581 83 Linköping, Sweden, leand@mai.liu.se
[2] Department of Mathematics, Linköping University, SE-581 83 Linköping, Sweden, anrie@mai.liu.se

1 Introduction

For static or incremental contact problems with Coulomb friction there are satisfactory and well known existence results for the coercive case, *i.e.*, when the elastic body is anchored so that rigid body motions are not possible, see [3, 1, 6, 7, 2]. The articles by Jaru**u**sek and Cocu, [7, 2] indeed contain results for the noncoercive case, *i.e.*, when rigid body motions are possible. However, the compatibility conditions which are used to ensure the existence of a solution, are the same that guarantee that the corresponding contact problem without friction has a solution. The condition is essentially that the applied force field should push the elastic body towards the obstacle. One of few previous articles containing friction-dependent compatibility conditions is [1].

By using a friction-less contact condition we cannot, for example, obtain any existence result for a two-dimensional problem with an elastic body being pushed towards a flat obstacle. It seems intuitively clear that a solution exists in this case, provided that the external forces are not applied at a too acute angle, so that the frictional reaction forces from the obstacle cannot balance them.

In this paper we shall give compatibility conditions for existence *which involve the coefficient of friction.* Cases which are treated are finite-dimensional systems, continuous elastic systems with Signorini contact conditions and a non-local friction law and continuous systems with a normal compliance conditions. It will also be indicated how to generalize to Signorini contact conditions with a local friction law.

The compatibility condition for existence of solution can be given a very simple mechanical interpretation, indicating that lack of solutions implies that a corresponding contact problem where the elastic body has been *made rigid*, is dynamically *unstable* for all possible mass distributions.

The results given in this paper are based on the thesis [9] by the second author.

2 A discrete system

Let us consider a truss in R^d, $d = 2$ or $d = 3$, consisting of finitely many nodes. The index set for the nodes is denoted by I. Some of these nodes can come into contact with rigid flat obstacles and we denote the set of these contact indices by I_c. Let u_i be the displacement of node i. We form a block vector u of the displacement vectors u_i, $i \in I$.

An external force f_i is acting at each node i. We denote the block vector of forces by f. When $i \in I_c$, the node may be acted upon by a contact force r_i from the obstacle. Defining $r_i = 0$ when $i \notin I_c$ we form the corresponding block vector r of reaction forces.

The linear elasticity of the truss is then expressed by

$$Ku = f + r \tag{1}$$

if all displacements are measured from equilibrium. The stiffness matrix, K, is symmetric and positive semidefinite. K is positive definite if a sufficiently large number of nodes have prescribed displacements, for example when the truss is rigidly anchored. The case which is of interest for us is when the null space $\mathcal{N}$ of K is a non-trivial subspace of the linear space of rigid body velocities.

When $i \in I_c$, we let n_i denote a unit normal of the obstacle directed into the obstacle. We decompose the displacement u_i into its normal component $u_{i,N} = u_i \cdot n_i$ and tangential component $u_{i,T} = u_i - u_{i,N} n_i$. Let us also decompose the contact forces r_i into components in the same way. This decomposition enables us to express the impenetrability of the body and obstacle according to Signorini,

$$u_{i,N} \le 0,\ r_{i,N} \le 0,\ r_{i,N} u_{i,N} = 0 \text{ for all } i \in I_c. \tag{2}$$

In order to simplify the presentation and notation we have assumed that the initial state is the equilibrium state with zero forces and displacements and that the initial gaps between contact nodes and obstacles are zero. The somewhat more general case where we have an incremental problem with a nonzero initial state and nonzero initial gaps can rather easily be reduced to the static case that we deal with here.

The Coulomb law of friction defines the tangential reaction force $r_{i,T}$ caused by the external force field f. Coulomb's law of friction requires

$$\begin{cases} |r_{i,T}| \le -\mu_i r_{i,N} & \\ u_{i,T} = 0 & \text{if } |r_{i,T}| < -\mu_i r_{i,N} \\ u_{i,T} = -\lambda_i r_{i,T} & \text{if } |r_{i,T}| = -\mu_i r_{i,N} > 0 \end{cases}$$

where $\lambda_i \geq 0$ and $i \in I_c$. The coefficient of friction μ_i is assumed to be positive, $\mu_i \geq \mu_{min} > 0$ for all $i \in I_c$, where μ_{min} is a constant. The norm $|\cdot|$ is the usual Euclidean norm of a vector.

We are now in position to formulate the discrete friction problem studied in this paper. Given a force field f, find an equilibrium u, r such that Coulomb's friction law and Signorini's contact law are fulfilled, *i.e.*, such that

$$\begin{cases} Ku = f + r & \\ r_i = 0 & \text{when } i \notin I_c \\ u_{i,N} \leq 0, r_{i,N} \leq 0, u_{i,N} r_{i,N} = 0 & \text{when } i \in I_c \\ |r_{i,T}| \leq -\mu_i r_{i,N} & \\ u_{i,T} = 0 & \text{if } |r_{i,T}| < -\mu_i r_{i,N} \\ u_{i,T} = -\lambda_i r_{i,T} & \text{if } |r_{i,T}| = -\mu_i r_{i,N} > 0 \end{cases} \tag{3}$$

where $\lambda_i \geq 0$.

Before proceeding we introduce the inner product

$$\langle f, v \rangle = \sum_{i \in I} f_i \cdot v_i \tag{4}$$

of block vectors representing force f and displacement v and let us define a norm

$$|v| = \sqrt{\sum_{i \in I} |v_i|^2}$$

for displacement block vectors v and a corresponding norm

$$|f| = \sup_{|v|=1} |\langle f, v \rangle| = \sqrt{\sum_{i \in I} |f_i|^2}$$

for block vectors of forces. Thus, the same notation, $|\cdot|$, will be used for the Euclidean norm of forces, displacements and for the norm of block vectors of forces and displacements.

2.1 Existence theorem for the discrete problem

Theorem 1. *Lack of solutions to the discrete static noncoercive friction problem implies that for any choice of the positive definite matrix D, there exist a rigid body velocity $w \in \mathcal{N}$ such that $w \neq 0$, $w_{i,N} \leq 0$ for $i \in I_c$, a contact force field r and a constant $\alpha \geq 0$ such that*

$$\alpha \langle Dw, v \rangle = \langle f, v \rangle + \langle r, v \rangle \tag{5}$$

for all $v \in \mathcal{N} = \{v : Kv = 0\}$, The contact forces r_i fulfill $|r_{i,T}| \leq -\mu_i r_{i,N}$ and are related to w by the conditions $w_{i,N} r_{i,N} = 0$ and

$$\begin{cases} w_{i,T} = 0 & \textit{if } |r_{i,T}| < -\mu_i r_{i,N} \\ w_{i,T} = -\lambda_i r_{i,T} & \textit{if } |r_{i,T}| = -\mu_i r_{i,N} > 0 \end{cases} \tag{6}$$

where $\lambda_i \geq 0$.

For the proof we consider a sequence of auxiliary coercive problems where the first equation in the noncoercive problem (3) is replaced by

$$(K + \varepsilon_n D)u = f + r \tag{7}$$

with $0 < \varepsilon_n \to 0$ and where the existence of solutions u^n, r^n is guaranteed by standard theorems for variational inequalities, see for example [4].

Now assume that the noncoercive problem (3) does not have a solution. Then for the sequence of solutions u^n to the coercive problems we must have $|u^n| \to \infty$ as $n \to \infty$. In fact, if there exists a bounded sequence of solutions u^n then it is easy to verify that some subsequence must converge to a solution of the noncoercive problem (3).

The proof of the main theorem may now be obtained using the following lemma for the case that $|u^n| \to \infty$.

Lemma 1. *Let us define $w^n = u^n/|u^n|$. Then*

I.

$$w_i^n \cdot r_i^n = -\mu_i |r_{i,N}^n||w_{i,T}^n|$$

and

$$\langle r^n, w^n \rangle = \sum_i w_i^n \cdot r_i^n \leq 0.$$

II. There is a convergent subsequence of w^n, such that $w^n \to w$, where $w \in \mathcal{N}$ and $|w| = 1$. Moreover

$$\lim_{n\to\infty} \varepsilon_n u^n = \alpha w$$

where $\alpha \geq 0$, and

$$|\varepsilon_n u^n| < C < \infty \text{ for all } n$$

where C is some constant.

III. The sequence of contact forces r^n obtained is bounded.

The proof of this lemma is rather straightforward. Only the third part requires some more careful analysis.

2.2 Mechanical interpretation

Assume that the applied force is such that the static friction problem (3) lacks solutions. Let us consider a rigid truss exposed to the force field, $f + r$. We shall show that the dynamical problem for this rigid truss has a solution where the truss is in motion, which indicates that the quasistatic modelling of the truss is inadequate in this case.

For the rigid body motion $u(t)$ we then have the following laws of motion

$$\langle M \frac{d^2u(t)}{dt^2}, v \rangle = \langle f + r, v \rangle \tag{8}$$

for all $v \in \mathcal{N}$, see [5]. Here M denotes a diagonal mass matrix for our rigid truss.

Since the incremental problem was assumed to lack solutions, Theorem 1 with D chosen as M yields the existence of a rigid body motion w and a contact force r satisfying (6) such that

$$\langle f, v\rangle + \langle r, v\rangle = \alpha\langle Mw, v\rangle.$$

The law of motion (8) then yields

$$\langle M\frac{d^2u(t)}{dt^2}, v\rangle = \langle f + r, v\rangle = \alpha\langle Mw, v\rangle \tag{9}$$

for all $v \in \mathcal{N}$. Since $\frac{d^2u(t)}{dt^2}$, w and v are vectors in $\mathcal{N}$ and M is positive definite we conclude from (9) that

$$\frac{d^2u(t)}{dt^2} = \alpha w$$

Solving this equation with the initial conditions

$$u(0) = u_0, \qquad \frac{du(0)}{dt} = v_0 w$$

where we assume that $u_0 \in V$, $(u_0)_{i,N} = 0$ for all $i \in I_c$ and that $v_0 > 0$ is a constant, we get,

$$u(t) = u_0 + (v_0 t + \alpha t^2/2)w. \tag{10}$$

We need to verify the solution (10) is compatible with the contact force field r under the assumption of contact with Coulomb friction, and also that the contact nodes do not penetrate the obstacles. In fact we have

$$\frac{du(t)_{i,T}}{dt} = (v_0 + \alpha t)w_{i,T} = -\lambda_i(v_0 + \alpha t)r_{i,T}$$

that is, $\frac{du(t)}{dt}$ and r have opposite direction, as required. Moreover,

$$\frac{du(t)_{i,N}}{dt} = (v_0 + \alpha t)w_{i,N} \leq 0$$

guaranteeing non-penetration.

3 Existence theorem for a continuous system with nonlocal friction law

Let us consider an elastic body, which in its undeformed state occupies a bounded domain Ω in R^d, $d = 2$ or $d = 3$, with a Lipschitz boundary Γ. The body Ω is subject to volume forces $f \in (L^2(\Omega))^d$ in the domain and surface tractions $t \in (H^{-1/2}(\Gamma))^d$ with support in $\Gamma_t \subset \Gamma$ on the boundary. A part

$\Gamma_c \subset \Gamma$ has possible contact with an obstacle. We assume that Γ_c and Γ_t are relatively open with disjoint closures, *i.e.*, with a positive distance. We further assume that some relative neighbourhood $\Gamma_s \subset \Gamma$ of the closure of Γ_c has $C^{1,\beta}$-regularity, with $0 < \beta$. In order to model the nonlocal friction law we assume that we are given a regularization operator

$$S : H^{-1/2}(\Gamma) \mapsto L^2(\Gamma)$$

which is linear, compact, such that the complement of supp$(S(p))$ has at most a distance ρ to supp(p) and such that $p \geq 0$ implies that $S(p) \geq 0$ for $p \in H^{-1/2}(\Gamma)$. We also require that $\rho < \mathrm{dist}(\Gamma_c, \Gamma_t)$ and introduce the convex closed set

$$V = \{v \in (H^1(\Omega))^d : v_N \leq 0 \text{ on } \Gamma_c\}.$$

For the coefficient of friction μ we require that $\mu \in C(\Gamma)$ and $\mu \geq \mu_{min} > 0$.

Using the nonpenetration condition $u_N \leq 0$ on Γ_c, the nonlocal friction law

$$|r_T| \leq -\mu S(r_N) \quad \text{on} \quad \Gamma$$

and contact and friction conditions analogous to those for the discrete case, (3), the variational formulation of the static noncoercive friction problem takes the following form.

Variational noncoercive problem
Find $u \in V$ such that

$$a(u, v-u) + \big[-\mu S(r_N), |u_T| - |v_T| \big]_\Gamma \geq (L, v-u) \tag{11}$$

for all $v \in V$.

Here

$$(L, v) = \langle f, v \rangle_\Omega + \big[t, v\big]_\Gamma,$$

$\langle \cdot, \cdot \rangle_\Omega$ denotes the inner product in $(L^2(\Omega))^d$, $\big[\cdot, \cdot\big]_\Gamma$ is the dual pairing between $(H^{-1/2}(\Gamma))^d$ and $(H^{1/2}(\Gamma))^d$ and $a(u, v)$ is the bilinear elastic energy functional.

Further r denotes the contact force field, *i.e.*, an element of $(H^{-1/2}(\Gamma))^d$ such that $r_i = \sigma_{ij}(u)n_j$ on $\bar{\Gamma}_c$ and $r = 0$ outside $\bar{\Gamma}_c$. We have used that $\sigma_{ij}(u)n_j$ is zero outside $\bar{\Gamma}_c \cup \bar{\Gamma}_t$.

Finally we introduce the following geometric condition on the shape of the contact surface Γ_c.

Geometric condition:

$$\left.\begin{array}{r} v \in \mathcal{N} \\ v_N(x) \leq 0 \text{ when } x \in \Gamma_c \end{array}\right\} \Rightarrow \begin{array}{c} v(x) = 0 \text{ at most at} \\ \text{one point } x \text{ of } \bar{\Gamma}_c. \end{array}$$

We now state the main theorem.

Theorem 2. *Under the previous assumptions, lack of solutions to the friction problem (11) implies that for every function* $\phi \in L^\infty(\Omega)$ *such that* $\phi \geq \phi_{min} > 0$ *there exists a rigid body displacement* $w \in V$*, a contact force field* r *of finite total variation on* Γ_s *such that* $-r_N$ *is a positive measure with support in* $\overline{\Gamma}_c$ *and a constant* $\alpha \geq 0$ *such that*

I.

$$\alpha\langle \phi w, v\rangle_\Omega = (L, v) + \big[r, v\big]_\Gamma \tag{12}$$

for all $v \in \mathcal{N}$.

II.

$$\big[r, v-w\big]_{\Gamma_S} - \big[\mu S(r_N), |v_T| - |w_T|\big]_\Gamma \geq 0 \tag{13}$$

for all $v \in V$*,* v *continuous on* Γ_s.

For the proof of the existence theorem we study the following sequence of auxiliary coercive problems where for the perturbation parameter ε_n we have $\varepsilon_n > 0$ and $\varepsilon_n \to 0$ as $n \to \infty$.

Auxiliary coercive problems:
Find $u \in V$ such that

$$\varepsilon_n\langle \phi u, v-u\rangle_\Omega + a(u, v-u) + \langle -\mu S(r_N), |u_T| - |v_T|\rangle \geq (L, v-u)$$

for all $v \in V$.

By standard theorems for variational inequalities the auxiliary problems have solutions u^n. If $\|u^n\| := \|u^n\|_{(H^1(\Omega))^d}$ are bounded as $n \to \infty$ then it is straightforward to show that some subsequence converges weakly to a solution u of the noncoercive problem. For the case when the noncoercive problem lacks solutions we must therefore have $\|u^n\| \to \infty$. Introducing $w^n = u^n/\|u^n\|$ one may show that $w^n \to w \in \mathcal{N}$ strongly in $(H^1(\Omega))^d$, that $r^n \to r$ weakly* as a sequence of functionals on $C(\Gamma)$ and that (12) and (13) are valid. The inequality (13) represents the Coulomb friction conditions and the equality (12) has the same mechanical interpretation as in the previous section, *i.e.*, that lack of solutions to the noncoercive problem implies that the corresponding setting with a rigid body and the same applied force field f is unstable under the nonlocal friction law.

One of the main difficulties in the proof is to show that the total masses of the positive measures $-r_N^n$ are uniformly bounded as $n \to \infty$. For this the geometrical condition on the shape of Γ_c is needed.

4 Existence theorem for a continuous system with a normal compliance friction law

Using the same assumptions and notation as in the previous section the normal compliance variational formulation of the coercive problem is the following.

Normal compliance variational problem
Find $u \in (H^1(\Omega))^d$ such that

$$a(u, v-u) + \varphi_\kappa(u, v-u) + j_\kappa(u, v) - j_\kappa(u, u) \geq (L, v-u) \tag{14}$$

for all $v \in (H^1(\Omega))^d$

Here $\varphi_\kappa(u, v) = \kappa \int_{\Gamma_c} [u_N]_+ v_N \, ds$ and $j_\kappa(u, v) = \kappa \int_{\Gamma_c} \mu [u_N]_+ |v_T| \, ds$

We next introduce a sequence of auxiliary coercive problems by adding a term $\varepsilon_n \langle \phi u, v-u \rangle_\Omega$ to the left hand side of (14).

Arguing similarly as in the previous section we conclude that lack of solutions for the coercive problems implies that $w^n = u^n / \|u^n\|$ converges to $w \in \mathcal{N}$ strongly in $(H^1(\Omega))^d$, that the normal contact forces $-r_N^n = \kappa [u_N]_+$ have uniformly bounded masses as $n \to \infty$. For some subsequence, $r^n \to r$ weakly* as a sequence of functionals on $C(\Gamma)$ and (12) and (13), without the regularization term S, are valid for the measure r and the displacement field w.

It is important to notice that in the normal compliance case, *the masses of $-r_N^n$ are also uniformly bounded in the penalty parameter κ.*

References

1. Bonfanti G, A noncoercive friction problem with tangential applied forces in three dimensions, Boll. Un. Mat. Ital. 7, 1993, pp. 149-165.
2. Cocu M., 1984, Existence of solutions of Signorini problems with friction, *Int. J. Engng. Sci.* **22**, 567-575.
3. Duvaut G., 1980 Équilibre d'un solide élastique avec contact unilatéral et frottement de Coulomb, *C.r. hebd. Séanc. Acad. Sci. Paris,* 290 A, 263-265.
4. Fan K, A minimax inequality and applications, in: Shisha (Ed.), Inequalities III, Academic Press, New York, 1972, pp. 103-113
5. H. Goldstein H, Classical Mechanics, 2:nd ed, Addison-Wesley, Massachusetts, 1980.
6. Jaru*u*sek J, Contact problems with bounded friction. Coercive case, Czechoslovak Math. J. 33 (1983) 237-261.
7. Jaru*u*sek J., 1984, Contact problems with bounded friction. Semicoercive case, Czechoslovak Math. J., 34, 619-629.
8. Ne*u*cas J., Jar*u*usek J., Haslinger J., 1980, On the solution of the variational inequality to the Signorini problem with small friction, Bolletino U.M.I., 5, 796-811.
9. Rietz A (2005) Existence theroems for noncoercive incremental contact problems with Coulomb friction. Thesis, Linköping university, Sweden. Linköping Studies in Science and Technology, Dissertation No. 7 925. Available at http://www.ep.liu.se/.

Local uniqueness results for the discrete friction problem

P. Hild[1] and Y. Renard[2]

[1] Laboratoire de Mathématiques de Besanccon, CNRS UMR 6623, Université de Franche-Comté, 16 route de Gray, 25030 Besanccon, France,
hild@math.univ-fcomte.fr

[2] MIP, CNRS UMR 5640, INSAT, Complexe scientifique de Rangueil, 31077 Toulouse, France,
renard@insa-toulouse.fr

Abstract. In this paper we consider the frictional contact problem involving the Signorini contact model and the Coulomb friction law in elastostatics. We focus on the behavior of the set of solutions in the finite-dimensional case where the friction coefficient is a parameter. We study local uniqueness and Lipschitz or C^1 continuation of solutions. We come to the conclusion that for any contact status there exists a generalized eigenvalue problem and that the solutions are locally unique if the friction coefficient is not an eigenvalue.

1 Introduction and problem set up

Our aim is to propose and to study a framework for the finite-dimensional approximation of the unilateral contact problem with Coulomb friction in elastostatics in order to obtain results ensuring local uniqueness and continuation of solutions. As far as we know the only existing results concerned with uniqueness in the finite-dimensional case are global and assume that the friction coefficient is small. This has been first obtained for a particular finite element approximation in [3] and many authors proved later the same kind of result for other finite-dimensional approximations. Nevertheless there does not exist any uniqueness result for large friction coefficients.

Roughly speaking our method can be summarized as follows. We use a formulation of the frictional contact conditions (without any regularization or smoothing) in which the discrete problem is written $\mathcal{H}(\mathcal{F}, Y) = 0$ where $\mathcal{F}$ is the friction coefficient and Y is a vector containing the displacement field as well as the normal and tangential loads on the contact zone. Having at our disposal a solution $(\mathcal{F}, Y)$ to the discrete problem we formulate an eigenvalue problem depending on the status of the nodes on the contact zone. The eigenvalue problem comes from the application of a generalized implicit function theorem involving Clarke's gradient which is well-adapted to our formulation of the frictional contact conditions. We write the eigenvalue problem both in the smooth case (when $\mathcal{H}$ is $\mathcal{C}^1$ near $(\mathcal{F}, Y)$) and in the nonsmooth case (when $\mathcal{H}$ is only Lipschitz-continuous near $(\mathcal{F}, Y)$, i.e., when some points on the contact zone satisfy grazing contact or vanishing slip conditions). The main result obtained in this paper is that the solution $(\mathcal{F}, Y)$ is locally unique if the friction coefficient is not an eigenvalue.

Let $\Omega \subset \mathbb{R}^d$ ($d = 2$ or 3) be a bounded domain representing the reference configuration of a linearly elastic body whose boundary $\partial\Omega$ consists of three nonoverlapping open parts Γ_N, Γ_D and Γ_C with $\overline{\Gamma_N} \cup \overline{\Gamma_D} \cup \overline{\Gamma_C} = \partial\Omega$. We assume that the measures of Γ_C and Γ_D are positive. The body is submitted to a prescribed density of loads on Γ_N, it is clamped on Γ_D and a unilateral contact condition with static Coulomb friction holds between the body and a flat rigid foundation on Γ_C (see [4]) where $\mathcal{F}$ is a Lipschitz-continuous nonnegative function on Γ_C representing the friction coefficient. We define the spaces:

$$V = \{v \in H^1(\Omega; \mathbb{R}^d), v = 0 \text{ on } \Gamma_D\}, X_N = \{v_N|_{\Gamma_C} : v \in V\}, X_T = \{v_T|_{\Gamma_C} : v \in V\}$$

and their dual spaces V', X'_N and X'_T. The set of admissible (i.e., nonpositive) normal stresses on Γ_C is given by:

$$\Lambda_N = \{f_N \in X'_N : \langle f_N, v_N \rangle_{X'_N, X_N} \geq 0, \;\; \forall v_N \in X_N, v_N \leq 0\}.$$

Given $\lambda_N \in \Lambda_N$ the set of admissible (i.e., belonging to Coulomb's cone) tangential stresses on Γ_C is defined as:

$$\Lambda_T(\mathcal{F}\lambda_N) = \{f_T \in X'_T : \langle f_T, w_T \rangle_{X'_T, X_T} + \langle \mathcal{F}\lambda_N, |w_T| \rangle_{X'_N, X_N} \leq 0, \forall w_T \in X_T\}.$$

Let $V^h \subset V$ be a family of finite-dimensional sub-vector spaces indexed by h coming from a regular finite element discretization of the domain Ω (the notation h represents the diameter of the largest element). We define:

$$X_N^h = \{v_N^h|_{\Gamma_C} : v^h \in V^h\}, \qquad X_T^h = \{v_T^h|_{\Gamma_C} : v^h \in V^h\}.$$

Let us denote also by $X_N'^h \subset X'_N \cap L^2(\Gamma_C)$ and $X_T'^h \subset X'_T \cap L^2(\Gamma_C; \mathbb{R}^{d-1})$ some finite element approximations of X'_N and X'_T respectively satisfying the Babuska-Brezzi "inf-sup" condition:

$$\inf_{\lambda^h = (\lambda_N^h, \lambda_T^h) \in X_N'^h \times X_T'^h} \; \sup_{v^h \in V^h} \frac{\displaystyle\int_{\Gamma_C} \lambda_N^h v_N^h \, d\Gamma + \int_{\Gamma_C} \lambda_T^h . v_T^h \, d\Gamma}{\|\lambda^h\|_{X_N'^h \times X_T'^h} \|v^h\|_V} \geq c > 0, \qquad (1)$$

where c may depend on h.

Then, we choose $\Lambda_N^h \subset X_N'^h$ and $\Lambda_T^h(\mathcal{F}\lambda_N^h) \subset X_T'^h$ as closed convex approximations of Λ_N and $\Lambda_T(\mathcal{F}\lambda_N)$ respectively (note that the conditions $\Lambda_N^h \subset \Lambda_N$ and $\Lambda_T^h(\mathcal{F}\lambda_N^h) \subset \Lambda_T(\mathcal{F}\lambda_N)$ are generally not satisfied). We set:

$$u^h(x) = \sum_{i=1}^{k_1} u^i \varphi_i(x), \quad \lambda_N^h(x) = \sum_{i=1}^{k_2} \lambda_N^i \psi_i(x), \quad \lambda_T^h(x) = \sum_{i=1}^{k_3} \lambda_T^i \xi_i(x),$$

where

$$U = (u_i)_{i=1..k_1}, \quad L_N = (\lambda_N^i)_{i=1..k_2}, \quad L_T = (\lambda_T^i)_{i=1..k_3},$$

and φ_i, ψ_i and ξ_i are the shape functions of the chosen finite element method. We introduce the following matrices: $(B_N)_{ij} = \int_{\Gamma_C} \psi_i \mathrm{n}.\varphi_j \, d\Gamma$, $1 \leq i \leq k_2, 1 \leq j \leq k_1$, $(B_T)_{ij} = \int_{\Gamma_C} \xi_i.\varphi_j \, d\Gamma$, $1 \leq i \leq k_3, 1 \leq j \leq k_1$, and $K = (K)_{ij}$, $1 \leq i, j \leq k_1$ denotes the stiffness matrix. Let $\overline{\Lambda}_N^h$ be the convex set of vectors whose components are the nodal values on Γ_C of functions in Λ_N^h:

$$\overline{\Lambda}_N^h = \left\{ L_N = (\lambda_N^i)_{1 \leq i \leq k_2} \in \mathbb{R}^{k_2} : \sum_{i=1}^{k_2} \lambda_N^i \psi_i \in \Lambda_N^h \right\}.$$

If $L_N \in \overline{\Lambda}_N^h$ is the vector containing the nodal values of $\lambda_N^h \in \Lambda_N^h$ we define:

$$\overline{\Lambda}_T^h(\mathcal{F} L_N) = \left\{ L_T = (\lambda_T^i)_{1 \leq i \leq k_3} \in \mathbb{R}^{k_3} : \sum_{i=1}^{k_3} \lambda_T^i \xi_i \in \Lambda_T^h(\mathcal{F} \lambda_N^h) \right\}.$$

Therefore the finite-dimensional frictional contact problem becomes: for a given $r > 0$, find $U \in \mathbb{R}^{k_1}, L_N \in \mathbb{R}^{k_2}$ and $L_T \in \mathbb{R}^{k_3}$ satisfying:

$$\begin{cases} KU = F + B_N^T L_N + B_T^T L_T, \\ L_N = P_{\overline{\Lambda}_N^h}(L_N - r B_N U), \\ L_T = P_{\overline{\Lambda}_T^h(\mathcal{F} L_N)}(L_T - r B_T U), \end{cases} \tag{2}$$

where F represents the external loads and the notation $P_{\overline{\Lambda}_N^h}$ (resp. $P_{\overline{\Lambda}_T^h(\mathcal{F} L_N)}$) stands for the projection operator onto $\overline{\Lambda}_N^h$ (resp. $\overline{\Lambda}_T^h(\mathcal{F} L_N)$) with respect to the Euclidean inner product in $\mathbb{R}^{k_2}$ (resp. $\mathbb{R}^{k_3}$).

Remark 1. In (2), the first equality describes the equilibrium, the second represents the unilateral contact conditions and the last one incorporates the frictional conditions.

Remark 2. Equations (2) do not involve any regularization of the frictional contact conditions. Therefore the solution U, L_N, L_T does not depend on r.

The advantage of this formulation is that contact and friction conditions are expressed without constraints and with Lipschitz-continuous expressions. The existence of solutions to Problem (2) is established in [6] for any friction coefficient. Although no specific regularity for the mapping $G \longmapsto \overline{\Lambda}_T^h(G)$ is necessary to prove existence and uniqueness except that $\overline{\Lambda}_T^h(G)$ has to be a closed convex set, we assume that the following additional property holds:

$$(X, G) \longmapsto P_{\overline{\Lambda}^h_T(G)}(X) \text{ is Lipschitz-continuous and piecewise } \mathcal{C}^1. \quad (3)$$

This is not a restricting hypothesis since all known discretizations of the Coulomb friction condition satisfy (3). In particular this ensures the existence of Clarke's generalized derivative.

2 The local uniqueness results

Given $r > 0$ and denoting $Y = (Y_U, Y_N, Y_T) \in \mathbb{R}^{k_1} \times \mathbb{R}^{k_2} \times \mathbb{R}^{k_3}$ and $k = k_1 + k_2 + k_3$, we consider the Lipschitz-continuous and piecewise $\mathcal{C}^1$ mapping $\mathcal{G} : \mathbb{R}^{k_2} \times \mathbb{R}^k \longmapsto \mathbb{R}^k$ defined by:

$$\mathcal{G}(G, Y) = \begin{pmatrix} KY_U - F - B_N^T P_{\overline{\Lambda}^h_N}(Y_N - rB_N Y_U) - B_T^T P_{\overline{\Lambda}^h_T(G)}(Y_T - rB_T Y_U) \\ \frac{1}{r}(P_{\overline{\Lambda}^h_N}(Y_N - rB_N Y_U) - Y_N) \\ \frac{1}{r}(P_{\overline{\Lambda}^h_T(G)}(Y_T - rB_T Y_U) - Y_T) \end{pmatrix}.$$

The problem $\mathcal{G}(G, Y) = 0$ corresponds to a friction problem with an *a priori* known friction threshold $-G$ (also called the Tresca friction problem). Now we introduce the map $\mathcal{H} : \mathbb{R} \times \mathbb{R}^k \longmapsto \mathbb{R}^k$ defined by $\mathcal{H}(\mathcal{F}, Y) = \mathcal{G}(\mathcal{F} Y_N, Y)$. It is straightforward that a solution to Problem (2) is a solution to:

$$\mathcal{H}(\mathcal{F}, Y) = 0 \quad (4)$$

and conversely. Since the maps $\mathcal{G}$ and $\mathcal{H}$ are Lipschitz-continuous, they admit a generalized gradient in Clarke's sense everywhere denoted $\nabla\mathcal{G} = (\nabla_1\mathcal{G}, \nabla_2\mathcal{G})$ and $\nabla\mathcal{H} = (\nabla_1\mathcal{H}, \nabla_2\mathcal{H})$. Let us recall a characterization of the generalized gradient (see [1]) for a Lipschitz-continuous function $\mathcal{J}(Z)$ which fails to be differentiable in a set Ω_f of zero Lebesgue measure: $\nabla\mathcal{J}(Z) = \text{co}\left\{\lim_{i\to\infty} \nabla\mathcal{J}(Z_i) : Z_i \to Z, Z_i \notin \Omega_f\right\}$, where the notation "co" stands for the convex hull. So we obtain:

$$\nabla\mathcal{G}(G, Y) = \Big(\nabla_1\mathcal{G}(G, Y), \nabla_2\mathcal{G}(G, Y)\Big) = \begin{pmatrix} A_{UG} & A_{UU} & A_{UN} & A_{UT} \\ 0 & A_{NU} & A_{NN} & 0 \\ A_{TG} & A_{TU} & 0 & A_{TT} \end{pmatrix}$$

$$= \begin{pmatrix} -B_T^T W_G & K + rB_N^T W_N B_N + rB_T^T W_T B_T & -B_N^T W_N & -B_T^T W_T \\ 0 & -W_N B_N & \frac{1}{r}(W_N - I_2) & 0 \\ \frac{1}{r} W_G & -W_T B_T & 0 & \frac{1}{r}(W_T - I_3) \end{pmatrix}$$

where W_N and (W_T, W_G) are selections of the generalized gradient of $X \mapsto P_{\overline{\Lambda}^h_N}(X)$ taken at $Y_N - rB_N Y_U$ and of the generalized gradient of $(X, G) \mapsto P_{\overline{\Lambda}^h_T(G)}(X)$ taken at $Y_T - rB_T Y_U$ respectively, and I_2 and I_3 denote the identity matrices of orders k_2 and k_3. We prove (see [4]):

Lemma 1. *Assume that the hypotheses (1) and (3) hold. Then the restriction* $\nabla_2\mathcal{G}(G,Y)$ *of the generalized gradient* $\nabla\mathcal{G}(G,Y)$ *is nonsingular in the sense that any selection of* $\nabla_2\mathcal{G}(G,Y)$ *is a nonsingular matrix.*

The generalized gradient of $\mathcal{H}$ is written:

$$\nabla\mathcal{H}(\mathcal{F},Y) = \begin{pmatrix} -B_T^T W_G Y_N & A_{UU} & A_{UN} & A_{UT} \\ 0 & A_{NU} & A_{NN} & 0 \\ \frac{1}{r}W_G Y_N & A_{TU} & 0 & A_{TT} \end{pmatrix} + \begin{pmatrix} A_{UG} \\ 0 \\ A_{TG} \end{pmatrix} \begin{pmatrix} 0 & 0_1 & \mathcal{F}I_2 & 0_3 \end{pmatrix}$$

where 0_1 (resp. 0_3) denotes the k_2-by-k_1 (resp. k_2-by-k_3) zero matrix. So

$$\nabla_2\mathcal{H}(\mathcal{F},Y) = \begin{pmatrix} A_{UU} & A_{UN} & A_{UT} \\ A_{NU} & A_{NN} & 0 \\ A_{TU} & 0 & A_{TT} \end{pmatrix} + \begin{pmatrix} A_{UG} \\ 0 \\ A_{TG} \end{pmatrix} \begin{pmatrix} 0_1 & \mathcal{F}I_2 & 0_3 \end{pmatrix}.$$

It is straightforward that $\nabla_2\mathcal{H}(\mathcal{F},Y)$ is singular if and only if there exist a selection of W_N and (W_T, W_G) and an eigenpair $(\frac{1}{\mathcal{F}}, V_N)$ such that:

$$-\begin{pmatrix} 0_1 & I_2 & 0_3 \end{pmatrix} \left(\nabla_2\mathcal{G}(\mathcal{F}Y_N, Y)\right)^{-1} \begin{pmatrix} A_{UG} \\ 0 \\ A_{TG} \end{pmatrix} V_N = \frac{1}{\mathcal{F}} V_N. \tag{5}$$

The next sufficient condition for local uniqueness and continuation of solutions is a direct application of the implicit function theorem for Lipschitz functions due to J.-B. Hiriart-Urruty (see [5] or [1]).

Theorem 1. *Assume that (1) and (3) hold. If* $(\mathcal{F}_0, Y_0)$ *solves (4) and if* $\mathcal{F}_0^{-1}$ *is not an eigenvalue to Problem (5) for any selection of* W_N, W_T *and* W_G, *then there exist a neighborhood* U *of* $\mathcal{F}_0$, *a neighborhood* V *of* Y_0 *and a Lipschitz-continuous function* $\zeta : U \longrightarrow V$ *such that*

$$\mathcal{H}(\mathcal{F},Y) = 0, \mathcal{F} \in U, Y \in V \Longleftrightarrow Y = \zeta(\mathcal{F}).$$

In the regular case we recover the result obtained from the standard implicit function theorem:

Corollary 1. *Assume that (1) holds. If* $(\mathcal{F}_0, Y_0)$ *solves (4), if* $\mathcal{H}$ *is* C^1 *near* $(\mathcal{F}_0, Y_0)$ *and if* $\mathcal{F}_0^{-1}$ *is not an eigenvalue to Problem (5) then there exist a neighborhood* U *of* $\mathcal{F}_0$, *a neighborhood* V *of* Y_0 *and a* C^1 *function* $\zeta : U \longrightarrow V$ *such that*

$$\mathcal{H}(\mathcal{F},Y) = 0, \mathcal{F} \in U, Y \in V \Longleftrightarrow Y = \zeta(\mathcal{F}).$$

Remark 3. The extension of these results to the continuous case is an open problem.

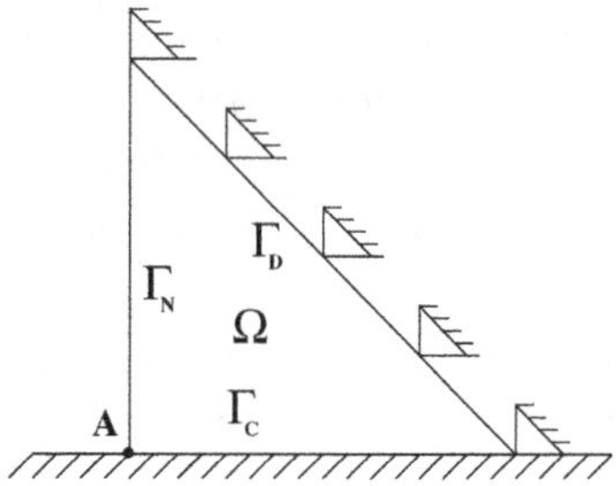

Fig. 1. Example of an elementary finite element mesh.

3 An elementary example

We consider the single linear triangular finite element depicted in Figure 1 (the lengths of Γ_C, Γ_N and Γ_D are ℓ, ℓ and $\sqrt{2}\ell$, respectively). We denote by $\lambda \geq 0$ and $\mu > 0$ the Lamé coefficients. The contact status is determined only by the behavior of point A. Since $(k_1, k_2, k_3) = (2, 1, 1)$ there are, for a given friction coefficient and a loading, four unknowns to the problem. We denote them by u_N and u_T (normal and tangential displacement at A), $\tilde{\lambda}_N$ and $\tilde{\lambda}_T$ (normal and tangential component of the generalized load at A). On Γ_C we set $\mathrm{n} = (0, -1)$ and the unit tangent vector is $\mathrm{t} = (1, 0)$. With $F^T = (F_N, F_T)$ the discrete frictional contact problem becomes:

$$\begin{cases} \dfrac{(\lambda+3\mu)}{2} u_N - \dfrac{(\lambda+\mu)}{2} u_T = \tilde{\lambda}_N + F_N, \\ -\dfrac{(\lambda+\mu)}{2} u_N + \dfrac{(\lambda+3\mu)}{2} u_T = \tilde{\lambda}_T + F_T, \\ \tilde{\lambda}_N = P_{]-\infty,0]}(\tilde{\lambda}_N - r u_N), \\ \tilde{\lambda}_T = P_{[\mathcal{F}\tilde{\lambda}_N, -\mathcal{F}\tilde{\lambda}_N]}(\tilde{\lambda}_T - r u_T), \end{cases} \tag{6}$$

where P_A denotes the standard projection operator from $\mathbb{R}$ onto $A \subset \mathbb{R}$. From (5) and Theorem 1, it can be easily checked that a solution $(\mathcal{F}, Y)$ is locally unique provided that for any selection $W_N \in \nabla P_{]-\infty,0]}(\tilde{\lambda}_N - r u_N)$ and $(W_T, -W_G) \in \nabla T(\tilde{\lambda}_T - r u_T, -\mathcal{F}\tilde{\lambda}_N)$ (where $T : \mathbb{R} \times \mathbb{R}_+ \to \mathbb{R}$ is such that $T(x, y) = P_{[-y,y]}(x)$), there is no nonzero solution $(V_{UN}, V_{UT}, V_N, V_T)$ to:

$$\begin{pmatrix} \frac{\lambda+3\mu}{2} + \frac{r\ell^2 W_N}{9} & -\frac{\lambda+\mu}{2} & -\frac{\ell W_N}{3} & 0 \\ -\frac{\lambda+\mu}{2} & \frac{\lambda+3\mu}{2} + \frac{r\ell^2 W_T}{9} & 0 & -\frac{\ell W_T}{3} \\ -\frac{\ell W_N}{3} & 0 & \frac{W_N - 1}{r} & 0 \\ 0 & -\frac{\ell W_T}{3} & 0 & \frac{W_T - 1}{r} \end{pmatrix} \begin{pmatrix} V_{UN} \\ V_{UT} \\ V_N \\ V_T \end{pmatrix} + \mathcal{F} \begin{pmatrix} 0 \\ -\frac{\ell W_G}{3} \\ 0 \\ \frac{W_G}{r} \end{pmatrix} V_N = 0. \tag{7}$$

The different situations are summarized hereafter where $a = (\lambda + 3\mu)/2$ and $b = (\lambda + \mu)/2$.

- Case a1: $u_N = 0, \tilde{\lambda}_T = \pm\mathcal{F}\tilde{\lambda}_N \neq 0, u_T \neq 0$. Then $W_N = 1, W_T = 0, W_G = \operatorname{sgn}(u_T)$. So $\mathcal{F} = -\operatorname{sgn}(u_T)\frac{a}{b}$ solves (7). Consequently we must have $\mathcal{F} = \frac{a}{b}$ and $u_T < 0$.
- Case a2: $u_N = 0, \tilde{\lambda}_T = \pm\mathcal{F}\tilde{\lambda}_N \neq 0, u_T = 0$. Then $W_N = 1, W_T = \alpha \in [0,1], W_G = -\operatorname{sgn}(\tilde{\lambda}_T)(1-\alpha)$. So $\mathcal{F} = \operatorname{sgn}(\tilde{\lambda}_T)\left(\frac{a}{b} + \frac{r\ell^2}{9b}\frac{\alpha}{1-\alpha}\right)$ solves (7). Consequently we must have $\mathcal{F} \in [\frac{a}{b}, +\infty[$ and $\tilde{\lambda}_T > 0$.
- Case a3: $|\tilde{\lambda}_T| < -\mathcal{F}\tilde{\lambda}_N$. So $W_N = 1, W_T = 1, W_G = 0$ and no $\mathcal{F}$ solves (7).
- Case a4: $u_N < 0, \tilde{\lambda}_T = \tilde{\lambda}_N = 0$. Then $W_N = 0$ and no $\mathcal{F}$ solves (7).
- Case a5: $u_N = 0, \tilde{\lambda}_T = \tilde{\lambda}_N = 0, u_T \neq 0$. Then $W_N = \gamma \in [0,1], W_T = 0, W_G = \operatorname{sgn}(u_T)$. So $\mathcal{F} = -\operatorname{sgn}(u_T)\left(\frac{a}{b} + \frac{1-\gamma}{\gamma}\frac{9(a^2-b^2)}{r\ell^2 b}\right)$ solves (7). Consequently we must have $\mathcal{F} \in [\frac{a}{b}, +\infty[$ and $u_T < 0$.
- Case a6: $u_N = 0, \tilde{\lambda}_T = \tilde{\lambda}_N = 0, u_T = 0$. Then $W_N = \gamma \in [0,1], W_T = \alpha \in [0,1], W_G = \beta \in [\alpha - 1, 1-\alpha]$. So $\mathcal{F} = \frac{9}{r\beta\ell^2}\left(\frac{b(1-\alpha)(1-\gamma)}{\gamma} - (\frac{a(1-\alpha)}{b} + \frac{r\alpha\ell^2}{9b})(\frac{a(1-\gamma)}{\gamma} + \frac{r\ell^2}{9})\right)$ solves (7). Consequently we must have $\mathcal{F} \in [\frac{a}{b}, +\infty[$.

Cases a1 and a2 deal with nonzero contact forces located on the boundary of Coulomb's cone. Case a1 corresponds to contact with slip and the solution is locally unique if $u_T > 0$. The solution is also locally unique if $u_T < 0$ and $\mathcal{F} \neq (\lambda+3\mu)/(\lambda+\mu)$. The remaining case (i.e., $u_T < 0$ and $\mathcal{F} = (\lambda+3\mu)/(\lambda+\mu)$) may correspond to a bifurcation case and it will be discussed hereafter. In case a2 there is sticking contact and the solution is locally unique when $\tilde{\lambda}_T < 0$. The solution is also unique when $\tilde{\lambda}_T > 0$ and $\mathcal{F} \in [0, (\lambda+3\mu)/(\lambda+\mu)[$. Bifurcation may occur for $\mathcal{F} \geq (\lambda+3\mu)/(\lambda+\mu)$ and $\tilde{\lambda}_T > 0$.

Case a3 corresponds to contact forces in the interior of Coulomb's cone and case a4 corresponds to a non contact state. In both cases the solution is locally unique for any positive friction coefficient.

Cases a5 and a6 correspond to vanishing contact forces (i.e., grazing contact). The solution is locally unique for $u_T > 0$ or for $u_T \leq 0$ with $\mathcal{F} \in [0, (\lambda+3\mu)/(\lambda+\mu)[$ (in case a6 we note that the friction coefficient value a/b is obtained when $\alpha = 0, \beta = -1, \gamma = 1$). Here also, bifurcations may be possible in the case $u_T \leq 0$ and $\mathcal{F} \geq (\lambda+3\mu)/(\lambda+\mu)$.

A more detailed study of (6) shows that there are only two loading situations, which we consider next, where local uniqueness is not observed.

First case: $(\lambda+3\mu)F_N + (\lambda+\mu)F_T < 0$ and $F_N > 0$.

An interesting phenomenon is that the bifurcation diagram is not connected (see Fig. 2). There is a solution corresponding to separation of point A which is valid for any friction coefficient. A stick solution solves the friction problem for any $\mathcal{F} \geq -F_T/F_N$ and a slip solution is valid for $\mathcal{F} > -F_T/F_N$. These two branches intersect each other at $\mathcal{F} = -F_T/F_N$ in a bifurcation point of type a2 relative to the previous classification.

Second case: $(\lambda+3\mu)F_N + (\lambda+\mu)F_T = 0$ and $F_N > 0$.

In this case a grazing contact solution with nonzero slip is valid for any friction coefficient. A stick solution is valid for $\mathcal{F} \geq (\lambda+3\mu)/(\lambda+\mu)$ and an

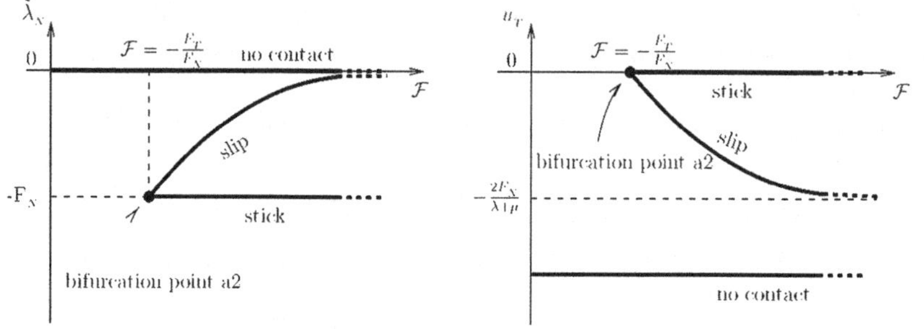

Bifurcation diagram $(\tilde{\lambda}_N, \mathcal{F})$ Bifurcation diagram $(u_T, \mathcal{F})$

Fig. 2. First case of bifurcation.

infinity of slip solutions occur for $\mathcal{F} = (\lambda + 3\mu)/(\lambda + \mu)$. In fact, it can be viewed as the limit case of the bifurcation diagram considered previously when $(\lambda + 3\mu)F_N + (\lambda + \mu)F_T$ goes to zero with negative values. Here the diagram is connected and it contains a vertical branch. The bifurcation points are of type a1, a2 and a5 following the previous classification (see Fig. 3).

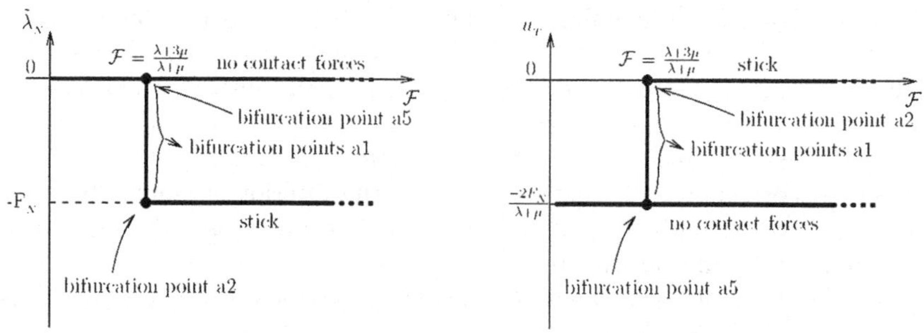

Bifurcation diagram $(\tilde{\lambda}_N, \mathcal{F})$ Bifurcation diagram $(u_T, \mathcal{F})$

Fig. 3. Second case of bifurcation.

References

1. Clarke FH (1983) Optimization and Nonsmooth Analysis. Wiley, New York
2. Duvaut G, Lions JL (1972) Les inéquations en mécanique et en physique. Dunod, Paris
3. Haslinger J (1983) Math Meth Appl Sci 5: 422–437
4. Hild P, Renard Y (2005) Quart Appl Math (to appear)
5. Hiriart-Urruty JB (1979) Math Oper Res 4: 79–97
6. Khenous H, Pommier J, Renard Y (2005) Appl Numer Math (to appear)

Analysis of a class of dynamic unilateral contact problems with friction for viscoelastic bodies

M. Cocou[1] and G. Scarella[2]

[1] Laboratoire de Mécanique et d'Acoustique - C.N.R.S., 31 chemin Joseph Aiguier, 13402 Marseille Cedex 20 and Université de Provence, U.F.R. - M.I.M., Marseille, France, cocou@lma.cnrs-mrs.fr

[2] INRIA Sophia-Antipolis, 2004 route des Lucioles - B.P. 93, 06902 Sophia Antipolis, France, Gilles.Scarella@sophia.inria.fr

Abstract. In this paper, we extend the existence results obtained in [1], for a dynamic unilateral contact problem with nonlocal friction between a Kelvin-Voigt viscoelastic body and a rigid obstacle, to the contact between two viscoelastic bodies and to a cracked viscoelastic body. We give classical and primal variational formulations for the problems. Penalized formulations are investigated by using an abstract existence and uniqueness result. Several estimates on the penalized solutions are established which allow to pass to the limit and to prove the existence of solutions for these dynamic unilateral contact problems with friction.

1 Introduction

In this work we study a class of dynamic contact problem for viscoelastic bodies, when Signorini's (or unilateral contact) conditions with nonlocal friction are considered. G. Duvaut and J.L. Lions [2] analyzed dynamic and quasistatic contact problems with Tresca's (or given) friction law involving linearly elastic and viscoelastic bodies. J.A.C. Martins and J.T. Oden [3] proved the existence and uniqueness, for a viscoelastic material, of a solution to dynamic contact problems with normal compliance and K.L. Kuttler and M. Shillor [4] have analyzed the dynamic bilateral contact with nonlocal friction for viscoelastic bodies. Existence and uniqueness results for the wave equation with unilateral boundary conditions for a half-space were proved by G. Lebeau and M. Schatzman [5], existence of solutions in the case of a smooth bounded domain subjected to unilateral conditions on his boundary was proved by J.U. Kim [6] for the wave equation and by J. Muñoz-Rivera and R. Racke [7] for a thermoelastic radially symmetric body. The dynamic Signorini contact problem with given friction for viscoelastic bodies has been analyzed by J. Jarušek [8].

In this paper, we extend the existence results obtained in [1], for a dynamic unilateral contact problem with nonlocal friction between a Kelvin-Voigt viscoelastic body and a rigid obstacle, to the contact between two viscoelastic bodies and to a cracked viscoelastic body, see also [9]. For the sake of simplicity, we consider in some detail only this second case, of a cracked body, which will clearly appear as including the case of the contact between two bodies. Firstly, we give classical and primal variational formulations of the problem. Secondly, penalized formulations are investigated by using an abstract existence and uniqueness result for the solution of some implicit evolution inequalities. Several estimates on the penalized solutions are established which allow to pass to the limit by using compactness arguments and to prove the existence of solutions for these dynamic unilateral contact problems with nonlocal friction. Particular unilateral contact problems and their numerical solutions based on fictitious domain formulations, with Lagrange multipliers representing the normal jump of the displacements, are presented in [10].

2 Classical and variational problems

We consider a cracked viscoelastic body characterized by a Kelvin-Voigt constitutive law which initially occupies a bounded domain Ω of $\mathbb{R}^d$, $d = 2$ or 3. In the sequel we will assume the small deformation hypothesis. We suppose that the boundary $\partial\Omega$ of Ω is composed of three parts such that $\partial\Omega = \overline{\Gamma_U} \cup \overline{\Gamma_F} \cup \overline{\Gamma_C}$, where Γ_U, Γ_F are sufficiently smooth and meas$(\Gamma_U) > 0$. The body is subjected to volume forces of density $\boldsymbol{f}$, zero displacements are imposed on Γ_U, prescribed tractions $\boldsymbol{F}$ are imposed on Γ_F and the crack is denoted by Γ_C.

2.1 Parametrization of the crack

We suppose that Γ_C is composed of two parts such that $\overline{\Gamma_C} = \overline{\Gamma^+} \cup \overline{\Gamma^-}$, which may be in unilateral contact with friction. To formulate boundary conditions on the crack, we suppose that we can decompose Ω into two open disjoint subsets Ω^+ and Ω^- with Lipschitz continuous boundaries $\partial\Omega^+$ and $\partial\Omega^-$, respectively, such that $\Omega = \Omega^+ \cup \Omega^- \cup \Gamma_V$, and $\Gamma_V \subset \partial\Omega^+ \cap \partial\Omega^-$, as it can be seen on Fig. 1. Γ_V may be considered as a virtual interface between Ω^+ and Ω^-. We choose a decomposition such that meas$(\Gamma_U^\alpha) > 0$, where $\Gamma_U^\alpha = \Gamma_U \cap \partial\Omega^\alpha$, and we denote $\Gamma^\alpha = \Gamma_C \cap \partial\Omega^\alpha$, $\alpha =+,-$.

Following a similar method to the one proposed by P. Boieri et al. [11], in order to express boundary conditions on the crack we introduce an open subset Ξ of $\mathbb{R}^{d-1}$ and we assume that the two faces of the crack can be parametrized by two C^1 functions, φ^+, φ^-, defined on Ξ. We suppose that $\varphi^+(\bar{x}) - \varphi^-(\bar{x}) \geq 0$, $\forall \bar{x} \in \Xi$ and we define each face Γ^α of the crack as the

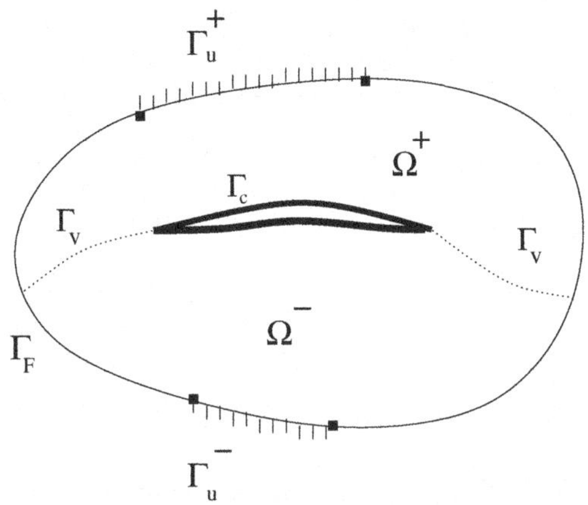

Fig. 1. A representation of Ω in 2D.

graph of φ^α on Ξ that is $\Gamma^\alpha = \{\ (\bar{x},\ \varphi^\alpha(\bar{x}))\ ;\ \bar{x} \in\ \Xi\}$, α =+,–. The gap between the two faces is given by $\tilde{g}(\xi) = \varphi^+(\xi) - \varphi^-(\xi)$ and the unit outward normal to Γ^α is denoted by $\boldsymbol{n^\alpha} : \Xi \to \mathbb{R}^d$. We adopt the following notations for the normal and tangential components of a displacement vector $\boldsymbol{v}$, of the corresponding relative displacement and of the stress vector corresponding to $\boldsymbol{\sigma}^\alpha$ with α =+,–,

$$\begin{cases} v_N^\alpha = v_N^\alpha(t,\xi) = \boldsymbol{v}^\alpha(t,\xi,\varphi^\alpha(\xi)).\boldsymbol{n}^\alpha(\xi), \boldsymbol{v}_T^\alpha = \boldsymbol{v}_T^\alpha(t,\xi) = \boldsymbol{v}^\alpha - v_N^\alpha \boldsymbol{n}^\alpha, \\ [v_N] = [v_N](t,\xi) = v_N^+ + v_N^-, \quad [\boldsymbol{v}_T] = [\boldsymbol{v}_T](t,\xi) = \boldsymbol{v}_T^+ - \boldsymbol{v}_T^-, \\ \sigma_N^\alpha = \sigma_N^\alpha(t,\xi) = (\boldsymbol{\sigma}^\alpha \boldsymbol{n}^\alpha).\boldsymbol{n}^\alpha, \boldsymbol{\sigma}_T^\alpha = \boldsymbol{\sigma}_T^\alpha(t,\xi) = \boldsymbol{\sigma}^\alpha \boldsymbol{n}^\alpha - \sigma_N^\alpha \boldsymbol{n}^\alpha. \end{cases}$$

Then one can show that the unilateral contact condition has the form

$$[u_N]\,(t,\xi) \leq g(\xi)\ \ \forall \xi\ \in\ \Xi,\ \text{where}\ g(\xi) = \frac{\varphi^+(\xi) - \varphi^-(\xi)}{\sqrt{1 + |\nabla\varphi^+|^2}}.$$

The term $g \geq 0$ is the normalized gap between the two faces of the crack.

2.2 Classical formulation

We denote by μ the friction coefficient and for the sake of simplicity we suppose that the density of the body is equal to 1. Let $\boldsymbol{\mathcal{A}} = (\mathcal{A}_{ijkl})$, $\boldsymbol{\mathcal{B}} = (\mathcal{B}_{ijkl})$ denote two fourth-order tensors, the elasticity tensor and the viscosity tensor. They satisfy the following classical symmetry and ellipticity conditions: $C_{ijkl} = C_{jikl} = C_{klij} \in W^{1,\infty}(\mathbb{R}^d)$, $\forall i,\ j,\ k,\ l = 1, \ldots, d$, $\exists \alpha_0 > 0$, $C_{ijkl}\tau_{ij}\tau_{kl} \geq \alpha_0\tau_{ij}\tau_{ij}\ \ \forall \boldsymbol{\tau} = (\tau_{ij})$ such that $\tau_{ij} = \tau_{ji}$, $\forall\ i,\ j = 1, \ldots, d$, where $\boldsymbol{C} = (C_{ijkl}) = \boldsymbol{\mathcal{A}}$, $\boldsymbol{\mathcal{B}}$ and the usual summation convention with respect to $i, j, k, l = 1, \ldots, d$, is used . A classical formulation of the unilateral contact problem with nonlocal friction is the following.**Problem $\mathbf{P_0}$**: Find $\boldsymbol{u} = \boldsymbol{u}(t, \boldsymbol{x})$ such that $\boldsymbol{u}(0) = \boldsymbol{u}_0, \;\; \dot{\boldsymbol{u}}(0) = \boldsymbol{u}_1$ in Ω and

$$\begin{cases} \ddot{u} - \operatorname{div} \boldsymbol{\sigma}(\boldsymbol{u}, \dot{\boldsymbol{u}}) = \boldsymbol{f}, \qquad \boldsymbol{\sigma} = \boldsymbol{\sigma}(\boldsymbol{u}, \dot{\boldsymbol{u}}) = \mathcal{A}\varepsilon(\boldsymbol{u}) + \mathcal{B}\varepsilon(\dot{\boldsymbol{u}}) \text{ in }]0,T[\times \Omega, \\ \boldsymbol{u} = \boldsymbol{0} \text{ in }]0,T[\times \Gamma_U, \qquad \boldsymbol{\sigma n} = \boldsymbol{F} \text{ in }]0,T[\times \Gamma_F, \\ [u_N] \leq g, \;\; \sigma_N^- = \sigma_N^+ \leq 0, \;\; \sigma_N^- \left([u_N] - g \right) = 0 \text{ in }]0,T[\times \Xi, \\ \boldsymbol{\sigma}_T^+ = -\boldsymbol{\sigma}_T^- \text{ in }]0,T[\times \Xi, \\ |\boldsymbol{\sigma}_T^+| \leq \mu \, |(\mathcal{R}\boldsymbol{\sigma})_N^+| \text{ and } \begin{array}{l} |\boldsymbol{\sigma}_T^+| < \mu \, |(\mathcal{R}\boldsymbol{\sigma})_N^+| \Rightarrow [\dot{\boldsymbol{u}}_T] = \boldsymbol{0}, \\ |\boldsymbol{\sigma}_T^+| = \mu \, |(\mathcal{R}\boldsymbol{\sigma})_N^+| \Rightarrow \exists \lambda \geq 0, [\dot{\boldsymbol{u}}_T] = -\lambda \boldsymbol{\sigma}_T^+, \end{array} \end{cases} \tag{1}$$

where $\varepsilon(\boldsymbol{v}) = \dfrac{1}{2}(\nabla \boldsymbol{v} + \nabla^t \boldsymbol{v})$ and $\mathcal{R}$ is a regularization which will be defined later.

Remark 1. If $\Gamma_V = \emptyset$, then the problem P_0 is a classical unilateral contact problem between two viscoelastic bodies.

2.3 Variational formulation

Let us first introduce two Hilbert spaces $(\boldsymbol{H}, |.|)$, $(\boldsymbol{V}, \|.\|)$ with the associated scalar product denoted, respectively, by $(.,.)$, $\langle .,. \rangle$, and the set $\boldsymbol{K}$ as follows: $\boldsymbol{H} = \left[L^2(\Omega)\right]^d$, $\boldsymbol{V} = \{\boldsymbol{v} \in \left[H^1(\Omega)\right]^d;\ \boldsymbol{v} = \boldsymbol{0} \text{ a.e. on } \Gamma_U\}$, $\boldsymbol{K} = \{\boldsymbol{v} \in \boldsymbol{V};\ [v_N] \leq g \text{ a.e. in } \Xi\}$. We assume that $\boldsymbol{u}_0 \in \boldsymbol{K}$, $\boldsymbol{u}_1 \in \boldsymbol{V}$, $\boldsymbol{F} \in W^{1,\infty}(0,T;[L^2(\Gamma_F)]^d)$, $\boldsymbol{f} \in W^{1,\infty}(0,T;\boldsymbol{H})$, $g \in H^{1/2}_{00}(\Xi)$, $g \geq 0$ almost everywhere on Ξ, $\mu \in L^\infty(\Xi)$ and $\mu \geq 0$ almost everywhere on Ξ. The particular case $g = 0$ corresponds to a cut. Let us define two bilinear forms a, b on $\boldsymbol{V} \times \boldsymbol{V}$, $J : \boldsymbol{V}^3 \to \mathbb{R}$ and $\boldsymbol{L} \in W^{1,\infty}(0,T;\boldsymbol{V})$ by

$$a(\boldsymbol{v}, \boldsymbol{w}) = \int_\Omega \mathcal{A}\varepsilon(\boldsymbol{v}) : \varepsilon(\boldsymbol{w}) \, dx, \quad b(\boldsymbol{v}, \boldsymbol{w}) = \int_\Omega \mathcal{B}\varepsilon(\boldsymbol{v}) : \varepsilon(\boldsymbol{w}) \, dx,$$

$$J(\boldsymbol{u}, \boldsymbol{v}, \boldsymbol{w}) = \int_\Xi \mu \, | \left((\mathcal{R}\boldsymbol{\sigma}) \, (\boldsymbol{u}, \boldsymbol{v}) \right)_N^+ | \; |[\boldsymbol{w}_T]| \, d\xi,$$

$$\langle \boldsymbol{L}, \boldsymbol{v} \rangle = \int_\Omega \boldsymbol{f} . \boldsymbol{v} \, dx + \int_{\Gamma_F} \boldsymbol{F} . \boldsymbol{v} \, ds \;\; \forall \, \boldsymbol{u}, \, \boldsymbol{v}, \, \boldsymbol{w} \in \boldsymbol{V}.$$

We suppose that $\mathcal{R} : [L^2_{\text{sym}}(\Omega)]^{d^2} \to [H^1(\Omega)]^{d^2}$ is a linear and continuous regularization of $\boldsymbol{\sigma} = \boldsymbol{\sigma}(\boldsymbol{u}, \boldsymbol{v}) \;\; \forall \, \boldsymbol{u}, \, \boldsymbol{v} \in \boldsymbol{V}$, such that $((\mathcal{R}\boldsymbol{\sigma}) \, (\boldsymbol{u}_0, \boldsymbol{u}_1))_N^+ = 0$ and $\exists C > 0$, $\|((\mathcal{R}\boldsymbol{\sigma}) \, (\boldsymbol{u}, \boldsymbol{v}))_N^+\| \leq C \left(|\boldsymbol{u}| + |\boldsymbol{v}| \right) \;\; \forall \, \boldsymbol{u}, \, \boldsymbol{v} \in \boldsymbol{V}$. The same type of regularization is considered in [4], where a particular case is also presented. We assume the following compatibility relation on the initial conditions: $\exists \, \boldsymbol{l} \in \boldsymbol{H}$ such that $(\boldsymbol{l}, \boldsymbol{v}) + a(\boldsymbol{u}_0, \boldsymbol{v}) + b(\boldsymbol{u}_1, \boldsymbol{v}) = \langle \boldsymbol{L}(0), \boldsymbol{v} \rangle \;\; \forall \, \boldsymbol{v} \in \boldsymbol{V}$ and we denote by $\langle .,. \rangle_{-1/2,\,1/2}$ the duality pairing between $[H^{-1/2}(\Omega)]^d$ and $[H^{1/2}(\Omega)]^d$. A variational formulation of problem P_0 is the following.

Problem $\mathbf{P_1}$: Find $\boldsymbol{u} \in W^{1,2}(0,T;\boldsymbol{V}) \cap C^1([0,T];[H^{-1/2}(\Omega)]^d)$, $\boldsymbol{u}(t) \in \boldsymbol{K}$ for all $t \in]0,T[$, such that

$$\begin{cases} \langle \dot{\boldsymbol{u}}(T), \boldsymbol{v}(T) - \boldsymbol{u}(T) \rangle_{-1/2,\,1/2} - (\boldsymbol{u}_1, \boldsymbol{v}(0) - \boldsymbol{u}_0) \\ \\ + \displaystyle\int_0^T \{ -(\dot{\boldsymbol{u}}, \dot{\boldsymbol{v}}) + a(\boldsymbol{u}, \boldsymbol{v}) + b(\dot{\boldsymbol{u}}, \boldsymbol{v}) + J(\boldsymbol{u}, \dot{\boldsymbol{u}}, \boldsymbol{v} + \dot{\boldsymbol{u}} - \boldsymbol{u}) \} \, dt \\ \\ \geq \displaystyle\int_0^T \langle \boldsymbol{L}, \boldsymbol{v} - \boldsymbol{u} \rangle \, dt + \int_0^T \{ -|\dot{\boldsymbol{u}}|^2 + a(\boldsymbol{u}, \boldsymbol{u}) + b(\dot{\boldsymbol{u}}, \boldsymbol{u}) + J(\boldsymbol{u}, \dot{\boldsymbol{u}}, \dot{\boldsymbol{u}}) \} \, dt \\ \forall\, \boldsymbol{v} \in L^\infty(0,T;\boldsymbol{V}) \cap W^{1,2}(0,T;\boldsymbol{H}),\ \boldsymbol{v}(t) \in \boldsymbol{K} \text{ a.e. } t \in]0,T[. \end{cases} \tag{2}$$

We adopt this primal variational formulation because it enables us to prove the existence of a solution by using only one implicit evolution inequality, but mixed formulations are also possible.

3 The penalized problem

3.1 Formulations of the penalized problem

We consider a penalized viscoelastic contact problem, with a solution denoted by $\boldsymbol{u}_\varepsilon$, verifying the same equations in Ω and the same boundary conditions on Γ_U, Γ_F, Γ as before, except for unilateral contact conditions. The penetration between the two faces of the crack is penalized. Let us define the mapping $\Phi_\varepsilon : \boldsymbol{V} \times \boldsymbol{V} \to \mathbb{R}$ by

$$\Phi_\varepsilon(\boldsymbol{v}, \boldsymbol{w}) = \frac{1}{\varepsilon} \int_\Xi ([v_N] - g)_+ [w_N] \, ds, \ \ \forall\, \boldsymbol{v}, \boldsymbol{w} \in \boldsymbol{V}, \text{ where } r_+ = \max(0, r).$$

We shall now use the decomposition of Ω into Ω^+ and Ω^- explained before which enables us to reduce the variational analysis to more classical domains. For $\boldsymbol{w} \in [L^2(\Omega)]^d$, we set $\hat{\boldsymbol{w}} = (\boldsymbol{w}^+, \boldsymbol{w}^-)$, where $\boldsymbol{w}^\alpha$ is the restriction of $\boldsymbol{w}$ on Ω^α. We shall introduce new problems, set on $\Omega^+ \times \Omega^-$, in the following functional framework: $\boldsymbol{H}^\alpha = \left[L^2(\Omega^\alpha)\right]^d$, $\hat{\boldsymbol{H}} = \boldsymbol{H}^+ \times \boldsymbol{H}^-$, $\boldsymbol{V}^\alpha = \{\boldsymbol{v}^\alpha \in \left[H^1(\Omega^\alpha)\right]^d ; \boldsymbol{v}^\alpha = \boldsymbol{0} \text{ a.e. on } \Gamma_U^\alpha\}$, $\alpha = +, -$, $\hat{\boldsymbol{V}} = \{\hat{\boldsymbol{v}} = (\boldsymbol{v}^+, \boldsymbol{v}^-) \in \boldsymbol{V}^+ \times \boldsymbol{V}^- ; \boldsymbol{v}^+ = \boldsymbol{v}^- \text{ a.e. on } \Gamma_V\}$. Let us define the mapping $\Psi : \boldsymbol{V} \to \hat{\boldsymbol{V}}$ by $\boldsymbol{v} \mapsto \Psi\boldsymbol{v} = \hat{\boldsymbol{v}} = (\boldsymbol{v}^+, \boldsymbol{v}^-)$. We set $\hat{\boldsymbol{K}} = \Psi(\boldsymbol{K}) = \{\Psi\boldsymbol{v};\ \boldsymbol{v} \in \boldsymbol{K}\}$, $[\hat{v}_N] = [v_N]$ and $[\hat{\boldsymbol{v}}_T] = [\boldsymbol{v}_T]$, where $\boldsymbol{v} = \Psi^{-1}\hat{\boldsymbol{v}}$ for all $\hat{\boldsymbol{v}} \in \hat{\boldsymbol{V}}$. We shall use the following notations: $\forall\, s \in \mathbb{R}$, $\boldsymbol{H}^s = [H^s(\Omega^+)]^d \times [H^s(\Omega^-)]^d$, $\langle \hat{\boldsymbol{u}}, \hat{\boldsymbol{v}} \rangle_{-s,s} = \langle \boldsymbol{u}^+, \boldsymbol{v}^+ \rangle_{\boldsymbol{H}^{-s}(\Omega^+), \boldsymbol{H}^s(\Omega^+)} + \langle \boldsymbol{u}^-, \boldsymbol{v}^- \rangle_{\boldsymbol{H}^{-s}(\Omega^-), \boldsymbol{H}^s(\Omega^-)}$, $\forall \hat{\boldsymbol{u}} \in \boldsymbol{H}^{-s}$, $\forall\, \hat{\boldsymbol{v}} \in \boldsymbol{H}^s$. Let us define the following mappings, with $\alpha =+,-$:

$$\langle \hat{\boldsymbol{L}}, \hat{\boldsymbol{v}} \rangle_{\hat{\boldsymbol{V}}} = \langle \boldsymbol{L}, \boldsymbol{v} \rangle,\ \hat{a}(\hat{\boldsymbol{u}}, \hat{\boldsymbol{v}}) = a(\boldsymbol{u}, \boldsymbol{v}) = a^+(\boldsymbol{u}^+, \boldsymbol{v}^+) + a^-(\boldsymbol{u}^-, \boldsymbol{v}^-),$$

$$\text{where } a^\alpha(\boldsymbol{u}^\alpha, \boldsymbol{v}^\alpha) = \int_{\Omega^\alpha} \boldsymbol{\mathcal{A}}\varepsilon(\boldsymbol{u}^\alpha) : \varepsilon(\boldsymbol{v}^\alpha)\, dx \ \ \forall\, (\boldsymbol{u}^\alpha, \boldsymbol{v}^\alpha) \in \boldsymbol{V}^\alpha \times \boldsymbol{V}^\alpha,$$

$$\hat{b}(\hat{\boldsymbol{u}}, \hat{\boldsymbol{v}}) = b(\boldsymbol{u}, \boldsymbol{v}) = b^+(\boldsymbol{u}^+, \boldsymbol{v}^+) + b^-(\boldsymbol{u}^-, \boldsymbol{v}^-),$$

$$\text{where } b^\alpha(\boldsymbol{u}^\alpha, \boldsymbol{v}^\alpha) = \int_{\Omega^\alpha} \boldsymbol{\mathcal{B}}\varepsilon(\boldsymbol{u}^\alpha) : \varepsilon(\boldsymbol{v}^\alpha)\, dx \ \ \forall\, (\boldsymbol{u}^\alpha, \boldsymbol{v}^\alpha) \in \boldsymbol{V}^\alpha \times \boldsymbol{V}^\alpha,$$

$$\hat{J}(\hat{\boldsymbol{u}}, \hat{\boldsymbol{v}}, \hat{\boldsymbol{w}}) = \int_{\Xi} \mu\, |\, (\mathcal{R}\boldsymbol{\sigma}\,)\,(\boldsymbol{u}, \boldsymbol{v}))_N^+|\ \, |\, [\boldsymbol{w}_T]\,|\, d\xi,$$

for all $\hat{\boldsymbol{u}}, \hat{\boldsymbol{v}}, \hat{\boldsymbol{w}} \in \hat{\boldsymbol{V}}$, where $\boldsymbol{u}$, $\boldsymbol{v}$, $\boldsymbol{w}$ satisfy $\Psi\boldsymbol{u} = \hat{\boldsymbol{u}}$, $\Psi\boldsymbol{v} = \hat{\boldsymbol{v}}$, $\Psi\boldsymbol{w} = \hat{\boldsymbol{w}}$. We consider an auxiliary penalized problem which corresponds to the previous decomposition of Ω. Let us define the mapping $\hat{\Phi}_\varepsilon : \hat{\boldsymbol{V}} \times \hat{\boldsymbol{V}} \to \mathbb{R}$ by $\hat{\Phi}_\varepsilon(\hat{\boldsymbol{u}}, \hat{\boldsymbol{v}}) = \Phi_\varepsilon(\Psi^{-1}\hat{\boldsymbol{u}}, \Psi^{-1}\hat{\boldsymbol{v}})$ for all $\hat{\boldsymbol{u}}$, $\hat{\boldsymbol{v}} \in \hat{\boldsymbol{V}}$.

Problem $\hat{P}_2^\varepsilon$: Find $\hat{\boldsymbol{u}}_\varepsilon = (\boldsymbol{u}_{\boldsymbol{\varepsilon}}^+, \boldsymbol{u}_{\boldsymbol{\varepsilon}}^-) \in W^{1,2}(0,T; \hat{\boldsymbol{V}}) \cap W^{2,2}(0,T; \hat{\boldsymbol{H}})$ such that $\hat{\boldsymbol{u}}_\varepsilon(0) = \hat{\boldsymbol{u}}_0$, $\dot{\hat{\boldsymbol{u}}}_\varepsilon(0) = \hat{\boldsymbol{u}}_1$, and

$$\begin{cases} (\ddot{\hat{\boldsymbol{u}}}_\varepsilon, \hat{\boldsymbol{w}} - \dot{\hat{\boldsymbol{u}}}_\varepsilon)_{\hat{\boldsymbol{H}}} + \hat{a}(\hat{\boldsymbol{u}}_\varepsilon, \hat{\boldsymbol{w}} - \dot{\hat{\boldsymbol{u}}}_\varepsilon) + \hat{b}(\dot{\hat{\boldsymbol{u}}}_\varepsilon, \hat{\boldsymbol{w}} - \dot{\hat{\boldsymbol{u}}}_\varepsilon) + \hat{\Phi}_\varepsilon(\hat{\boldsymbol{u}}_\varepsilon, \hat{\boldsymbol{w}} - \dot{\hat{\boldsymbol{u}}}_\varepsilon) \\ + \hat{J}(\hat{\boldsymbol{u}}_\varepsilon, \dot{\hat{\boldsymbol{u}}}_\varepsilon, \hat{\boldsymbol{w}}) - \hat{J}(\hat{\boldsymbol{u}}_\varepsilon, \dot{\hat{\boldsymbol{u}}}_\varepsilon, \dot{\hat{\boldsymbol{u}}}_\varepsilon) \geq \langle \hat{\boldsymbol{L}}, \hat{\boldsymbol{w}} - \dot{\hat{\boldsymbol{u}}}_\varepsilon \rangle_{\hat{\boldsymbol{V}}} \ \ \forall\, \hat{\boldsymbol{w}} \in \hat{\boldsymbol{V}}, \text{ a.e. } t \in\,]0,T[\,. \end{cases} \tag{3}$$

Theorem 1. *Under the above assumptions, there exists a unique solution* $\hat{\boldsymbol{u}}_\varepsilon$ *of problem* $\hat{P}_2^\varepsilon$.

To prove the existence and uniqueness of the penalized solution, we consider an abstract problem for which we shall establish an existence and uniqueness result.

3.2 Analysis of an abstract evolution inequality

Let $(H, |.|)$ and $(V, \|.\|)$ be two Hilbert spaces with corresponding scalar products denoted by $(.,.)$ and $\langle .,. \rangle$, respectively, such that V is dense in H and the imbedding from V into H is compact. We also introduce the set $\mathcal{K}$ as

$$\mathcal{K} = \{v \in W^{2,2}(0,T;H) \cap W^{1,2}(0,T;V);\, v(0) = u_0,\ \dot{v}(0) = u_1\},$$

where u_0, $u_1 \in V$. Let a and b be two bilinear, symmetric, continuous and V-elliptic forms defined on V^2, in the following sense:

$$\begin{aligned} &\exists\, m_a,\ m_b > 0,\ a(u,v) \leq m_a \|u\|\, \|v\|,\ \ b(u,v) \leq m_b \|u\|\, \|v\| \quad \forall\, u,\ v \in V, \\ &\exists\, A,\ B > 0\ a(v,v) \geq A\, \|v\|^2,\ \ b(v,v) \geq B\, \|v\|^2 \quad \forall\, v \in V. \end{aligned} \tag{4}$$

Let $\beta : V \to \mathbb{R}$ and $\phi : [0,T] \times V^3 \to \mathbb{R}$ two weakly continuous mappings such that

$$\phi(t,u,v,w_1 + w_2) \leq \phi(t,u,v,w_1) + \phi(t,u,v,w_2), \tag{5}$$

$$\phi(t,u,v,\theta w) = \theta\, \phi(t,u,v,w) \quad \forall t \in [0,T],\ \forall\, u,\ v,\ w,\ w_{1,2} \in V,\ \forall\, \theta \geq 0, \tag{6}$$

$$\phi(.,0,0,.) = 0, \quad \beta(0) = 0, \tag{7}$$

$$\begin{aligned} &\exists\, \eta_0 > 0 \text{ such that } \forall\, t_{1,2} \in [0,T],\ \forall\, u_{1,2}, v_{1,2}, w \in V, \\ &\quad |\phi(t_1,u_1,v_1,w) - \phi(t_2,u_2,v_2,w)| \\ &\quad \le \eta_0(|t_1-t_2| + |\beta(u_1-u_2)| + |v_1-v_2|)\, \|w\|, \end{aligned} \tag{8}$$

$$\begin{aligned} &\exists\, \eta > 0 \text{ such that } \forall\, t_{1,2} \in [0,T],\ \forall\, u_{1,2}, v_{1,2}, w_{1,2} \in V, \\ &\quad |\phi(t_1,u_1,v_1,w_1) - \phi(t_1,u_1,v_1,w_2) \\ &\quad +\ \phi(t_2,u_2,v_2,w_2) - \phi(t_2,u_2,v_2,w_1)| \\ &\quad \le \eta\,(\,|t_1-t_2| + \|u_1-u_2\| + |v_1-v_2|)\, \|w_1-w_2\|. \end{aligned} \tag{9}$$

We suppose $L \in W^{1,\infty}(0,T;V)$ and we assume the following compatibility condition on the initial data :

$$\exists\, l \in H \quad (l,w) + a(u_0,w) + b(u_1,w) + \phi(0,u_0,u_1,w) = \langle L(0), w\rangle \quad \forall w \in V \tag{10}$$

and we consider the following problem.
Problem $\mathcal{P}$: Find $u \in \mathcal{K}$ such that for almost all t $\in\,]0,T[$

$$\begin{aligned} (\ddot u, v-\dot u) + a(u, v-\dot u) + b(\dot u, v-\dot u) + \phi(t,u,\dot u,v) \\ -\phi(t,u,\dot u,\dot u) \ge \langle L, v-\dot u\rangle \quad \forall\, v \in V. \end{aligned} \tag{11}$$

Using an incremental approach the following existence result is proved in [12].

Theorem 2. *Under the assumptions on u_0, u_1, L, (4)-(9) and the compatibility condition (10), there exists a unique solution of the problem $\mathcal{P}$.*

4 Existence of a solution of unilateral contact problem with nonlocal friction

Using (3), it follows that there exists a positive constant M independent of ε, such that, for all $\varepsilon > 0$, the following estimates on $\hat{\boldsymbol{u}}_\varepsilon$ hold

$$\begin{cases} |\dot{\hat{\boldsymbol{u}}}_\varepsilon|_{\hat{\boldsymbol{H}}} \le M \quad \forall t \in\,]0,T[\,, \quad \displaystyle\int_0^T \|\dot{\hat{\boldsymbol{u}}}_\varepsilon\|^2_{\hat{\boldsymbol{V}}}\, ds \le M, \\ \|\hat{\boldsymbol{u}}_\varepsilon\|_{\hat{\boldsymbol{V}}} \le M, \quad \|([\hat u_{\varepsilon N}] - g)_+\|_{L^2(\Xi)} \le M\sqrt{\varepsilon} \quad \forall t \in\,]0,T[, \end{cases} \tag{12}$$

and that $\ddot{\hat{\boldsymbol{u}}}_\varepsilon$ is bounded in $L^2(0,T;\boldsymbol{H}^{-1})$ by a constant independent of ε. These estimates imply that there exists $\hat{\boldsymbol{u}} = (\boldsymbol{u}^+, \boldsymbol{u}^-)$ such that, up to a subsequence,

$$\begin{aligned} &\hat{\boldsymbol{u}}_\varepsilon \rightharpoonup \hat{\boldsymbol{u}} \quad \text{weak} * \text{ in } L^\infty(0,T;\hat{\boldsymbol{V}}),\ \ddot{\hat{\boldsymbol{u}}}_\varepsilon \rightharpoonup \ddot{\hat{\boldsymbol{u}}} \quad \text{in } L^2(0,T;\boldsymbol{H}^{-1}), \\ &\dot{\hat{\boldsymbol{u}}}_\varepsilon \rightharpoonup \dot{\hat{\boldsymbol{u}}} \quad \text{in } L^2(0,T;\hat{\boldsymbol{V}}) \quad \text{weak} * \text{ in } L^\infty(0,T;\hat{\boldsymbol{H}}). \end{aligned} \tag{13}$$

Using a compactness result of J. Simon [13], up to a subsequence we obtain

$$\dot{\hat{\boldsymbol{u}}}_\varepsilon \to \dot{\hat{\boldsymbol{u}}} \;\text{ in } L^2(0,T;\hat{\boldsymbol{H}}) \text{ and in } C([0,T];\boldsymbol{H}^{-1/2}),\;\; \hat{\boldsymbol{u}}_\varepsilon \to \hat{\boldsymbol{u}} \text{ in } C([0,T];\boldsymbol{H}^{1/2}). \tag{14}$$

For all $\hat{\boldsymbol{v}} \in L^\infty(0,T;\hat{\boldsymbol{V}}) \cap W^{1,2}(0,T;\hat{\boldsymbol{H}})$ such that $\hat{\boldsymbol{v}}(t) \in \hat{\boldsymbol{K}}$ for almost every $t \in]0,T[$, we choose in (3) $\hat{\boldsymbol{w}} = \dot{\hat{\boldsymbol{u}}}_\varepsilon + \hat{\boldsymbol{v}} - \hat{\boldsymbol{u}}_\varepsilon$. We integrate with respect to $t \in]0,T[$, integrate by parts the acceleration term and using an argument of monotonicity for $\hat{\Phi}_\varepsilon$ and relations (13) and (14) we obtain by passing to the lower limit

$$\begin{cases} \langle \dot{\hat{\boldsymbol{u}}}(T), \hat{\boldsymbol{v}}(T) - \hat{\boldsymbol{u}}(T)\rangle_{-1/2,1/2} - (\hat{\boldsymbol{u}}_1, \hat{\boldsymbol{v}}(0) - \hat{\boldsymbol{u}}_0)_{\hat{\boldsymbol{H}}} \\ + \displaystyle\int_0^T \left\{ -(\dot{\hat{\boldsymbol{u}}}, \dot{\hat{\boldsymbol{v}}})_{\hat{\boldsymbol{H}}} + \hat{a}(\hat{\boldsymbol{u}}, \hat{\boldsymbol{v}}) + \hat{b}(\dot{\hat{\boldsymbol{u}}}, \hat{\boldsymbol{v}}) + \hat{J}(\hat{\boldsymbol{u}}, \dot{\hat{\boldsymbol{u}}}, \hat{\boldsymbol{v}} + \dot{\hat{\boldsymbol{u}}} - \hat{\boldsymbol{u}}) \right\} dt \\ \geq \displaystyle\int_0^T \langle \hat{\boldsymbol{L}}, \hat{\boldsymbol{v}} - \hat{\boldsymbol{u}}\rangle_{\hat{\boldsymbol{V}}}\, dt + \int_0^T \left\{ -|\dot{\hat{\boldsymbol{u}}}|^2_{\hat{\boldsymbol{H}}} + \hat{a}(\hat{\boldsymbol{u}}, \hat{\boldsymbol{u}}) + \hat{b}(\dot{\hat{\boldsymbol{u}}}, \hat{\boldsymbol{u}}) + \hat{J}(\hat{\boldsymbol{u}}, \dot{\hat{\boldsymbol{u}}}, \dot{\hat{\boldsymbol{u}}}) \right\} dt \\ \forall\, \hat{\boldsymbol{v}} \in L^\infty(0,T;\hat{\boldsymbol{V}}) \cap W^{1,2}(0,T;\hat{\boldsymbol{H}}),\ \hat{\boldsymbol{v}}(t) \in \hat{\boldsymbol{K}} \text{ a.e. } t \in]0,T[. \end{cases}$$

Thus, $\boldsymbol{u}$ defined by $\boldsymbol{u} = \boldsymbol{u}^+$ on Ω^+ and $\boldsymbol{u} = \boldsymbol{u}^-$ on Ω^- is a solution of (2). Consequently, the following result holds.

Theorem 3. *Under the above assumptions, there exists a solution of problem* P_1.

The existence of a solution in the purely elastic case (i.e. with $\boldsymbol{\mathcal{B}} = \boldsymbol{0}$) in a general nonsymmetric domain remains an open problem to our knowledge even for frictionless conditions.

References

1. Cocou M (2002) Z Angew Math Phys (ZAMP) 53:1099–1109
2. Duvaut G, Lions JL (1972) Les inéquations en mécanique et en physique. Dunod, Paris
3. Martins JAC, Oden JT (1987) Nonlinear Anal Theory Methods 11:407–428
4. Kuttler KL, Shillor M (2001) Nonlinear Anal Theory Methods 45:309–327
5. Lebeau G, Schatzman M (1984) J Differential Equations 53:309–361
6. Kim JU (1989) Commun Partial Differential Equations 14:1011–1026
7. Muñoz-Rivera J, Racke R (1998) SIAM J Appl Math 58:1307–1337
8. Jarušek J (1996) Czechoslovak Math J 46 (121):475–487
9. Cocou M, Scarella G (2004) C R Acad Sci Paris, Série I 338:341–346
10. Scarella G (2004) Etude théorique et numérique de la propagation d'ondes en présence de contact unilatéral dans un milieu fissuré. PhD Thesis, Université Paris Dauphine
11. Boieri P, Gastaldi F, Kinderlehrer D (1987) Appl Math Optim 15:251–277
12. Cocou M, Scarella G (2005) Z Angew Math Phys (ZAMP), to appear
13. Simon J (1987) Ann Mat Pura Applic 146:65–96.

Sthenic incompatibilities in rigid bodies motion

M. Frémond[1] and P. Isabella-Valenzi[2]

[1] Laboratoire Lagrange, LCPC, 58 boulevard Lefebvre, 75732, Paris cedex 15, France, fremond@lcpc.fr

[2] Università di Roma Tor Vergata, via del Politecnico,1, 00133 Roma, Italy, Isabella.Valenzi@ing.uniroma2.it

Abstract. When a rigid body slides with friction on a surface, hopping motion is observed: this is an everyday phenomenon. In rigid bodies mechanics, this phenomenon appears when it is no longer possible to compute the reaction contact forces. The difficulty is overcome by a motion theory involving velocity discontinuities. Velocity discontinuities may result either from an obstacle which makes impossible to compute the acceleration: this is *a cinematic incompatibility* or from the impossibility to compute the reaction forces: this is a *sthenic incompatibility.* We describe two examples: the Klein and Painlevé sthenic incompatibilities.

1 Introduction

When a rigid body collides with a rigid plane, it is no longer possible to solve the smooth equations of motion because it is impossible to compute the acceleration. This is a *cinematic incompatibility.* Collision theory, assumes a time discontinuity of the velocity. This assumption associated with the basic laws of mechanics, i.e., equations of motions and constitutive laws satisfying the laws of thermodynamics, gives a solution to this problem by producing a predictive theory which takes into account the cinematic incompatibilities [2, 1].

The motion of rigid bodies may involve friction which introduces reaction forces. These forces depend on the velocities through algebraic or differential equations. It may happen that these equations have no solution whereas there is no cinematic incompatibility. Again it is no longer possible to solve the equations of motion. What occurs? May the predictive theory cope with this unexpected situation? Is it too schematic and has more sophistication to be added? In this case, one may think that the rigidity assumption has to be removed. We show that the above mentioned collision theory is rich enough to provide a solution and that there is no necessity to get rid of the rigidity assumption. We call this kind of incompatibility, a *sthenic incompatibility.* We describe two examples: the Klein and Painlevé sthenic incompatibilities, [4], [6], [2].

2 The Klein sthenic incompatibility

Let us consider a bar with length $2l$, mass m, the ends of which are moving in two slides which are fixed to a massive rigid support, (Fig. 1). One slide has a Coulomb friction, the other one is without friction. The state of the *system bar-slides* depends on a unique parameter: the abscissa $x(t)$ of the bar center of mass G. An horizontal force F is applied at distance $b = GA$ of the center of mass (b is positive downward, the b of Fig. 1 is positive). A torque $\hat{C}$ is also applied. The angle of the bar with the inferior slide is $\theta \in]0, \pi[$. Let the horizontal initial velocity $U = dx/dt$ be given and let us find the bar motion. For some values of the data, U, F and $\hat{C}$ it is not possible to solve the equations of motions, more precisely it is not possible to find the reaction forces of the slides, [4], [2]. What occurs? We show that it is possible to overcome the difficulty within the rigid body theory by describing carefully what occurs when velocities are discontinuous, [2], [1].

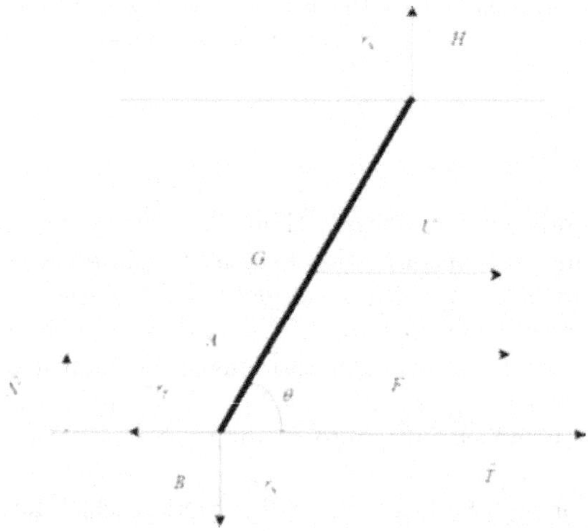

Fig. 1. Ends B and H of a bar are guided by two slides with normal vector $\mathbf{N}$ (direction N is normal to the slides, direction T is tangential). The superior slide is without friction. The inferior slide has a Coulomb friction. Force F is applied at distance $b = GA$ from center of mass G, θ is the angle between the bar and the horizontal axis. Inferior slide applies reaction force $(-r_T, -r_N)$ to the bar and superior slide force $(0, r_N)$.

2.1 The constitutive laws

The 2-D system is made of the rigid bar and of the two horizontal rigid slides which are fixed to a massive rigid support in such a way that we may assume their velocities are 0. Contact is bilateral in each slide. The superior slide is without friction. The inferior slide has a Coulomb friction. The interior forces of the system are the reaction force (r_T, r_N) of the slide at point B together with the reaction percussions (P_T, P_N) in case the velocities are discontinuous, i.e., there is a collision or a velocity jump. We assume Coulomb friction at point B. The Coulomb friction in case of discontinuity of velocity, involves the quantity $(U^+ + U^-)$ as shown in [2] and [1], (U^+ and U^- are the velocities after and before the velocity discontinuity). We choose the same friction coefficient μ for the smooth and non-smooth evolutions due to experimental evidence and theoretical results, [1].

2.2 The equations

The equations of motion and constitutive laws give the following equations [2]:

Smooth evolution

Almost everywhere in time

$$m\frac{dU}{dt} = -r_T + F, \tag{1}$$

$$0 = -r_T l \sin\theta + 2r_N l \cos\theta + Fb\sin\theta + C, \tag{2}$$

by denoting $C = \hat{C} - bF_N \cos\theta$. Second equation is due to the zero angular velocity. Constitutive law for the normal reaction force is

$$r_N \in \partial I_0(0) = \mathbf{R},$$

where I_0 is the indicator function of the origin of $\mathbf{R}$, [5]. Normal reaction force can be positive or negative because the contact is bilateral. The Coulomb constitutive law is

$$r_T \in \partial I_{r_N}(U), \tag{3}$$

with

$$I_{r_N}(x) = \mu\,|r_N|\,|x|\,.$$

Non smooth evolution

At any time

$$m\,[U] = -P_T, \tag{4}$$

$$0 = -P_T l \sin\theta + 2P_N l \cos\theta, \tag{5}$$

where $[U] = U^+ - U^-$. Second equation is due to the zero angular velocity. Constitutive law for the normal reaction percussion is

$$P_N \in \partial I_0(0) = \mathbf{R}.$$

Normal reaction percussion is positive or negative because the contact is bilateral. The dissipation function for the tangential velocity is

$$I_{P_N}(x) = \mu |P_N| |x|,$$

which gives the constitutive law, [2], [1]

$$P_T \in \partial I_{P_N}(U^+ + U^-). \tag{6}$$

2.3 An example of sthenic incompatibility

In a smooth evolution, the reaction forces or the interior forces satisfy two algebraic equations (2) and (3), when velocity U is known. If these equations have no solution, it is impossible to solve the differential equation (1), thus to solve the smooth equations of motion: for instance, in case

$$U > 0,\ bF + \frac{C}{\sin\theta} < 0,\ |\mathrm{tg}\theta| > \frac{2}{\mu},$$

as shown in Fig. 2. Impossibility does not result from the impossibility to compute the acceleration as when a solid collides a rigid plane. It results from the impossibility to compute the interior forces. We have a *sthenic incompatibility* whereas we have a cinematic incompatibility when a solid collides a rigid plane.

Let us note that with such initial velocity U, force F and torque C, the smooth evolution we are expecting because there is no obstacle, cannot exist. A difficulty seems to prevent to solve the equations of motion. It may be shown that this is not the case and that this situation is completely normal. The difficulty is overcome by the system by having a velocity discontinuity because in case

$$U^+ + U^- = 0 \text{ and } |\mathrm{tg}\theta| > \frac{2}{\mu},$$

there is an unique solution $U^+ < 0$ of the three algebraic equations (4), (5) and (6). Indeed, motion may go on with this new initial condition because in case

$$U < 0,\ bF + \frac{C}{\sin\theta} < 0,\ |\mathrm{tg}\theta| > \frac{2}{\mu},$$

there are two possible reaction forces r_T as it is shown in Fig.3.

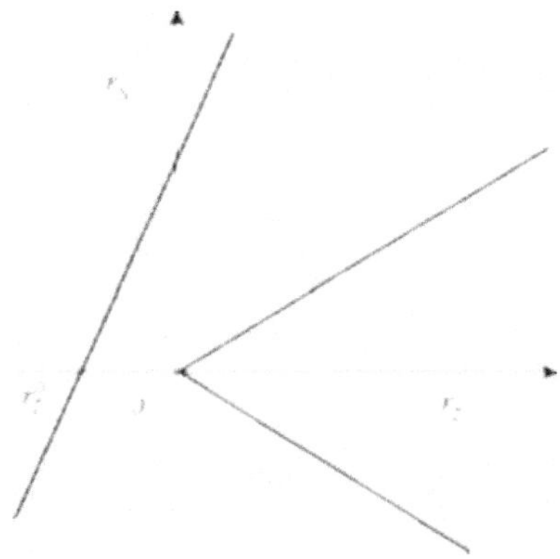

Fig. 2. For $U > 0$, $bF + \frac{C}{\sin\theta} < 0$, and $|\text{tg}\theta| > \frac{2}{\mu}$, there is no possible reaction force because half-lines $r_T = \mu |r_N|$ and line $0 = -r_T l \sin\theta + 2lr_N \cos\theta + bF \sin\theta + C$ do not intersect. This is a sthenic incompatibility: it is impossible to find reaction forces which satisfy the equations of motion.

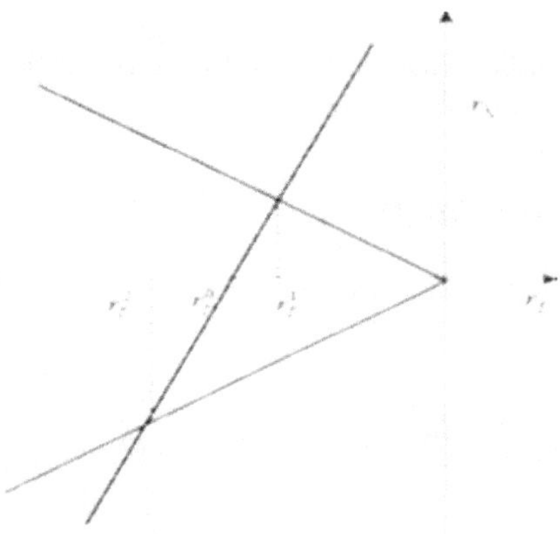

Fig. 3. For $U < 0$, $bF + \frac{C}{\sin\theta} < 0$, $|\text{tg}\theta| > \frac{2}{\mu}$, there are two reaction forces at intersection of half-lines $r_T = -\mu |r_N|$ and line $0 = -r_T l \sin\theta + 2lr_N \cos\theta + bF \sin\theta + C$.

3 The Painlevé sthenic incompatibility

Let us consider a rigid slender bar which is sliding with friction on an horizontal plane, as illustrated in Fig. 4. The mass of the bar is m, its length is $2l$, its mass moment of inertia is I. The coefficient of friction is μ. The bar, being pointed in the direction of motion, is sliding towards the left. The velocity of the center of mass is (U_T, U_N). Velocity U_T is the horizontal or tangential velocity and velocity U_N is the vertical or normal velocity. The velocity of the contact point A is $(V_T, V_N) = (U_T + \omega l sin\theta,\ U_N - \omega l cos\theta)$, where θ is the angle of the bar with respect to the horizontal and $\omega = d\theta/dt$ is the angular

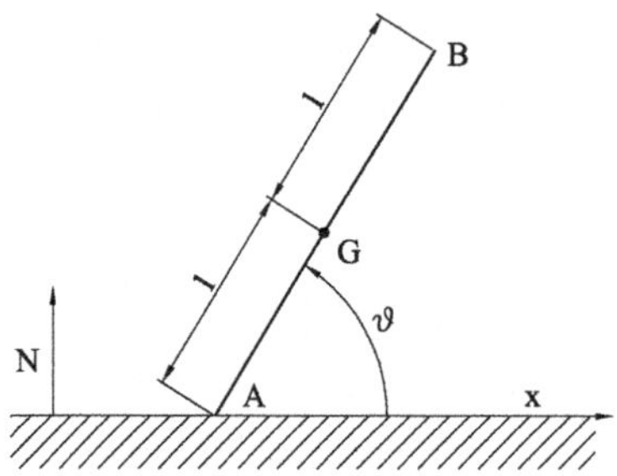

Fig. 4. Painlevé's example.

velocity. In some configurations, the sliding motion becomes impossible when the contact forces diverge to infinity.

3.1 The equations of motion

It is easy to get, *almost everywhere in time*:

$$m\frac{dU_N}{dt} = -r_N - g,\ m\frac{dU_T}{dt} = -r_T,$$
$$I\frac{d\omega}{dt} = l(-r_T sin\theta + r_N cos\theta), \tag{7}$$

where $-(r_T, r_N)$ is the reaction of the plane and $(0, -g)$ is the gravity force, and *at any time*:

$$m[U_N] = -P_N,\ m[U_T] = -P_T,$$
$$I[\omega] = l(-P_T sin\theta + P_N cos\theta), \tag{8}$$

where $-(P_T, P_N)$ is the percussion reaction of the plane on the bar. The constitutive laws for the reaction forces and percussions are still the Coulomb friction law.

We suppose that the bar is sliding towards the left: $y_a = 0$, $V_N = 0$, $V_T < 0$. We suppose that at the beginning of the motion $\omega^2 ml sin\theta - g < 0$ and $I - ml^2 cos\theta(\mu sin\theta - cos\theta) > 0$. The normal reaction is

$$r_N = \frac{I(\omega^2 ml sin\theta - g)}{I - ml^2 cos\theta(\mu sin\theta - cos\theta)} \geq 0.$$

We suppose that the evolution is such that either the denominator or both the numerator and the denominator of r_N go to zero with r_N going to infinity. In this situation the smooth evolution is no longer possible. What happens? It may be shown that, depending on the cinematic and geometric conditions, the bar may leave the plane either smoothly (i.e., without velocity discontinuity) or non smoothly (i.e. with a velocity discontinuity). This property is given by the algebraic and differential equations resulting from the non smooth and smooth Coulomb constitutive laws and equations of motion (8) and (7).

Consider now the data $m = l = 1, I = 1/12,\ \mu = 0.9,\ g = 1$, with the initial conditions $y_A(0) = 0, \theta(0) = 0.85, V_T(0) = -7, V_N(0) = 0, \omega(0) = 0$. The reaction r_N diverges to infinity. It can be proved that a discontinuity of velocity occurs, [2]. The future velocities (U_T^+, U_N^+, ω^+) depending on (U_T^-, U_N^-, ω^-) are given by the algebraic equations (8) and the Coulomb constitutive law. In this configuration there is *not uniqueness* of the solution. The angular velocity ω^+ is indeterminate. It depends on the parameter $[\omega]$ which verifies, [2]:

$$0 \leq [\omega] \leq \frac{-2V_T^-}{l(\mu cos\theta + sin\theta)} = [\omega]_{max}. \tag{9}$$

Figs. 5 and 6 show the effect of the sthenic incompatibility for $[\omega] = 2$.

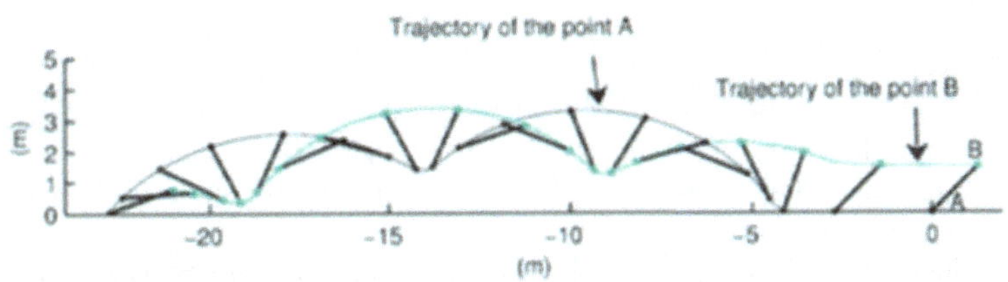

Fig. 5. Sthenic incompatibility for $[\omega] = 2 < [\omega]_{max} = 7.4398$. The sthenic incompatibility is responsible for the jump of the bar moving towards the left.

4 Conclusion

The predictive motion theory involving velocity discontinuities, takes into account both *cinematic and sthenic incompatibilities*, [2]. The velocity discontinuities result from two different reasons: the best-known *cinematic incompatibilities*, when it is impossible to compute the acceleration and the less-known *sthenic incompatibilities* when it is impossible to compute the reaction or interior forces. These two incompatibilities are equivalent: they are overcome by velocity discontinuities determined by the theory. The difficulties in modelling the frictional hopping motion disappear if one uses this collision theory that

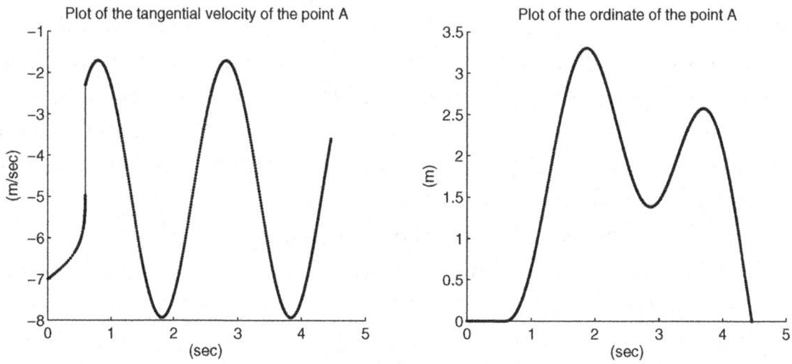

Fig. 6. Sthenic incompatibility for $[\omega] = 2$. After the jump of the horizontal velocity due to the sthenic incompatibility, the bar flies, makes two turns and falls again on the plane.

satisfies the basic requirements of mechanics. We prove that when a smooth evolution is not possible, a velocity discontinuity occurs. The converse is also true. Let us also note that as it is usual with Coulomb friction law, the solutions of the Painlevé and Klein problems are not always unique.

References

1. M. Frémond, (2001), *Non-smooth thermo-mechanics*. Springer.
2. M. Frémond, (2005), *Collisions*. Università di Roma Tor Vergata, Dipartimento di Ingegneria Civile.
3. J.H. Jellet, (1872), Treatise on the theory of friction, Hodges, Foster and Co.
4. F. Klein, (1910), Zu Painlevés Kritik des Coulombschen Reigbungsgsetze, *Zeitsch. Math. Phys.*, **58**, 186–191.
5. J.J. Moreau, (2003), *Fonctionnelles convexes, Séminaire sur les équations aux dérivées partielles*. Collège de France, Paris (1966) and, Università di Roma Tor Vergata, Dipartimento di Ingegneria Civile.
6. P. Painlevé, (1905), Sur les lois du frottement de glissement, *C. R. Acad. Sci.*, **121**, 112–115; **141**, 401–405; **141**, 546–552.

Study of two quasistatic viscoplastic contact problems with adhesion

M. Sofonea[1] and W. Han[2]

[1] Laboratoire de Mathématiques et Physique pour les Systèmes, Université de Perpignan, 52 Avenue Paul Alduy, 66860 Perpignan, France, sofonea@univ-perp.fr

[2] Department of Mathematics, University of Iowa, Iowa City, IA 52242, USA, whan@math.uiowa.edu

Abstract. We consider two quasistatic contact problems for viscoplastic materials. The contact is modeled with Signorini's condition in the first problem, and with normal compliance in the second one. In both problems the adhesion of the contact surfaces, caused by glue, is taken into account and the evolution of the bonding field is described by a first order differential equation. For each model, we provide the variational formulation, state a result on the existence of a unique weak solution, and indicate that the solution of the Signorini problem can be obtained as the limit of the solutions of the problem with normal compliance as the stiffness coefficient of the foundation tends to infinity. We also introduce and discuss a fully discrete scheme for solving the Signorini problem; under certain solution regularity assumptions, an optimal order error estimate holds.

1 Introduction

Processes of adhesion are important in many industrial settings where parts, usually nonmetallic, are glued together. For this reason, adhesive contact between bodies, when a glue is added to prevent the surfaces from relative motion, has recently received increased attention in the literature. General models with adhesion can be found in [1, 3, 4, 5, 6]. Results on the mathematical analysis of various adhesive contact problems can be found in [2, 6, 7, 8] and references therein. The main new idea in these papers is the introduction of a surface internal variable, the *bonding field* β, which describes the fractional density of active bonds on the contact surface. As a fraction its values are restricted to the range $0 \leq \beta \leq 1$. At a point on the bonding contact surface Γ_3, when $\beta = 1$, the adhesion is complete and all the bonds are active; when $\beta = 0$ all the bonds are inactive, severed, and there is no adhesion; when $0 < \beta < 1$ the adhesion is partial and only a fraction β of the bonds is active.

This paper describes two models of adhesive contact between a viscoplastic body and a foundation with emphasis in their variational and numerical analysis. It is organized as follows. In Section 2 we introduce the models, list the assumption on the data and present their variational formulations. In Section 3 we state existence and uniqueness results for the models as well as a convergence result. Finally, in Section 4 we consider a fully-discrete scheme for numerical solution and provide an optimal order error estimate.

2 The problems

The physical setting is as follows. A viscoplastic body occupies a bounded domain $\Omega \subset \mathbb{R}^d$ $(d = 2, 3)$ with a regular boundary Γ that is partitioned into three disjoint measurable parts Γ_1, Γ_2 and Γ_3, such that $\text{meas}(\Gamma_1) > 0$. Let $[0, T]$ denote the time interval of interest for some $T > 0$. The body is clamped on $\Gamma_1 \times (0, T)$ and thus the displacement field vanishes there. A volume force of density $\boldsymbol{f}_0$ acts in $\Omega \times (0, T)$ and a surface traction of density $\boldsymbol{f}_2$ acts on $\Gamma_2 \times (0, T)$. Moreover, the body is in adhesive contact with a foundation on $\Gamma_3 \times (0, T)$. We assume that the process is quasistatic, and therefore the inertial term is neglected in the equation of motion.

In the first problem the contact is modeled by Signorini's condition with adhesion. The classical formulation for the process is the following: find a displacement field $\boldsymbol{u} : \Omega \times [0, T] \to \mathbb{R}^d$, a stress field $\boldsymbol{\sigma} : \Omega \times [0, T] \to \mathcal{S}^d$, and a bonding field $\beta : \Omega \times [0, T] \to \mathbb{R}$ such that

$$\dot{\boldsymbol{\sigma}} = \mathcal{E}\boldsymbol{\varepsilon}(\dot{\boldsymbol{u}}) + \mathcal{G}(\boldsymbol{\sigma}, \boldsymbol{\varepsilon}(\boldsymbol{u})) \quad \text{in } \Omega \times (0, T), \tag{1}$$

$$\text{Div}\,\boldsymbol{\sigma} + \boldsymbol{f}_0 = \boldsymbol{0} \quad \text{in } \Omega \times (0, T), \tag{2}$$

$$\boldsymbol{u} = \boldsymbol{0} \quad \text{on } \Gamma_1 \times (0, T), \tag{3}$$

$$\boldsymbol{\sigma\nu} = \boldsymbol{f}_2 \quad \text{on } \Gamma_2 \times (0, T), \tag{4}$$

$$\begin{aligned} &u_\nu \le 0, \quad \sigma_\nu \le \gamma_\nu \beta^2 \widetilde{R}(u_\nu), \\ &\qquad (\sigma_\nu - \gamma_\nu \beta^2 \widetilde{R}(u_\nu)) u_\nu = 0 \quad \text{on } \Gamma_3 \times (0, T), \end{aligned} \tag{5}$$

$$-\boldsymbol{\sigma}_\tau = p_\tau(\beta) R^*(\boldsymbol{u}_\tau) \quad \text{on } \Gamma_3 \times (0, T), \tag{6}$$

$$\dot{\beta} = -(\gamma_\nu \beta \widetilde{R}(u_\nu)^2 + \gamma_\tau \beta \|R^*(\boldsymbol{u}_\tau)\|^2 - \epsilon_a)_+ \quad \text{on } \Gamma_3 \times (0, T), \tag{7}$$

$$\boldsymbol{u}(0) = \boldsymbol{u}_0, \quad \boldsymbol{\sigma}(0) = \boldsymbol{\sigma}_0 \quad \text{in } \Omega, \tag{8}$$

$$\beta(0) = \beta_0 \quad \text{on } \Gamma_3. \tag{9}$$

Here and below $\mathcal{S}^d$ represents the space of second order symmetric tensors on $\mathbb{R}^d$ whereas "$\cdot$" and "$\|\cdot\|$" denote the inner product and the Euclidean norms on $\mathcal{S}^d$ and $\mathbb{R}^d$; $\boldsymbol{\varepsilon}(\boldsymbol{u})$ denotes the linearized strain tensor, $\boldsymbol{\nu}$ is the unit outer normal on Γ, $u_\nu = \boldsymbol{u} \cdot \boldsymbol{\nu}$, $\boldsymbol{u}_\tau = \boldsymbol{u} - u_\nu \boldsymbol{\nu}$, $\sigma_\nu = (\boldsymbol{\sigma\nu}) \cdot \boldsymbol{\nu}$ and $\boldsymbol{\sigma}_\tau = \boldsymbol{\sigma\nu} - \sigma_\nu \boldsymbol{\nu}$; the dot above a quantity represents its derivative with respect to the time variable and $r_+ = \max\{r, 0\}$.

Equation (1) represents the constitutive law of the material in which $\mathcal{E}$ is a fourth-order elasticity tensor and $\mathcal{G}$ is a constitutive function, (2) represents the equilibrium equation, and (3), (4) are the displacement and traction boundary conditions, respectively.

Conditions (5) represent the Signorini condition with adhesion where $\widetilde{R}(u_\nu) = (-R(u_\nu))_+$ and R is the truncation operator

$$R(s) = R_L(s) = \max\{-L, \min\{s, L\}\},$$

$L > 0$ being the characteristic length of the bond, beyond which there is no any additional traction (see, e.g., [6]); by choosing L large enough, we can assume that $R(u_\nu) = u_\nu$; then from (5) we recover the contact condition

$$u_\nu \leq 0, \quad \sigma_\nu + \gamma_\nu \beta^2 u_\nu \leq 0, \quad (\sigma_\nu + \gamma_\nu \beta^2 u_\nu) u_\nu = 0 \quad \text{on } \Gamma_3 \times (0,T),$$

which was used in [1, 2, 6] in the study of adhesive contact problems with linearly elastic materials.

Condition (6) is the tangential boundary condition on the contact surface Γ_3 in which $p_\tau(\beta)$ is a given function and R^* is a truncation operator,

$$R^*(\boldsymbol{v}) = R^*_L(\boldsymbol{v}) = \begin{cases} \boldsymbol{v} & \text{if } \|\boldsymbol{v}\| \leq L, \\ L \dfrac{\boldsymbol{v}}{\|\boldsymbol{v}\|} & \text{if } \|\boldsymbol{v}\| > L. \end{cases}$$

This condition shows that the shear on the contact surface depends on the adhesion field and on the tangential displacement, but, again, only up to the bond length L.

Equation (7) describe the evolution of the bonding field with given material parameters γ_ν, γ_τ and ϵ_a. In (8) and (9), $\boldsymbol{u}_0, \boldsymbol{\sigma}_0$ and β_0 are given initial displacement, stress and bonding fields.

In the second problem the contact is modeled with normal compliance and the classical formulation for the process is the following: find a displacement field $\boldsymbol{u}^\lambda : \Omega \times [0,T] \to \mathbb{R}^d$, a stress field $\boldsymbol{\sigma}^\lambda : \Omega \times [0,T] \to \mathcal{S}^d$, and a bonding field $\beta^\lambda : \Omega \times [0,T] \to \mathbb{R}$ such that

$$\dot{\boldsymbol{\sigma}}^\lambda = \mathcal{E}\varepsilon(\dot{\boldsymbol{u}}^\lambda) + \mathcal{G}(\boldsymbol{\sigma}^\lambda, \varepsilon(\boldsymbol{u}^\lambda)) \quad \text{in } \Omega \times (0,T), \tag{10}$$

$$\operatorname{Div} \boldsymbol{\sigma}^\lambda + \boldsymbol{f}_0 = \boldsymbol{0} \quad \text{in } \Omega \times (0,T), \tag{11}$$

$$\boldsymbol{u}^\lambda = \boldsymbol{0} \quad \text{on } \Gamma_1 \times (0,T), \tag{12}$$

$$\boldsymbol{\sigma}^\lambda \boldsymbol{\nu} = \boldsymbol{f}_2 \quad \text{on } \Gamma_2 \times (0,T), \tag{13}$$

$$-\sigma^\lambda_\nu = \frac{1}{\lambda}\, p_\nu(u^\lambda_\nu) - \gamma_\nu (\beta^\lambda)^2 \widetilde{R}(u^\lambda_\nu) \quad \text{on } \Gamma_3 \times (0,T), \tag{14}$$

$$-\boldsymbol{\sigma}^\lambda_\tau = p_\tau(\beta^\lambda) R^*(\boldsymbol{u}^\lambda_\tau) \quad \text{on } \Gamma_3 \times (0,T), \tag{15}$$

$$\dot{\beta}^\lambda = -(\gamma_\nu \beta^\lambda \widetilde{R}(u^\lambda_\nu)^2 + \gamma_\tau \beta^\lambda \|R^*(\boldsymbol{u}^\lambda_\tau)\|^2 - \epsilon_a)_+ \quad \text{on } \Gamma_3 \times (0,T), \tag{16}$$

$$\boldsymbol{u}^\lambda(0) = \boldsymbol{u}_0, \quad \boldsymbol{\sigma}^\lambda(0) = \boldsymbol{\sigma}_0 \quad \text{in } \Omega, \tag{17}$$

$$\beta^\lambda(0) = \beta_0 \quad \text{on } \Gamma_3. \tag{18}$$

Here p_ν is a given function and $\lambda > 0$ is a penalization parameter which may be interpreted as a deformability coefficient of the foundation, and then $1/\lambda$ is the surface stiffness coefficient. Indeed, when λ is smaller the reaction force of the foundation to penetration is larger and so the same force will result in a smaller penetration, which implies that the foundation is less deformable. When λ is larger the reaction force of the foundation to penetration is smaller, and so the foundation is less stiff and more deformable.

Next, we introduce variational formulations of the above problems. We seek the displacement in the space $V = \{\boldsymbol{v} \in H^1(\Omega)^d : \boldsymbol{v} = \mathbf{0} \text{ on } \Gamma_1\}$, the stress field in the space $Q = \{\boldsymbol{\sigma} = (\sigma_{ij}) : \sigma_{ij} = \sigma_{ji} \in L^2(\Omega)\}$ and the bonding field in the space $B = L^2(\Gamma_3)$. These are Hilbert spaces with the inner products $(\boldsymbol{u}, \boldsymbol{v})_V = (\boldsymbol{\varepsilon}(\boldsymbol{u}), \boldsymbol{\varepsilon}(\boldsymbol{v}))_Q$, $(\boldsymbol{\sigma}, \boldsymbol{\tau})_Q = \int_\Omega \sigma_{ij}\tau_{ij}\,dx$, $(\zeta, \xi)_B = \int_{\Gamma_3} \zeta\xi\,da$ and the associated norm $\|\cdot\|_V$, $\|\cdot\|_Q$ and $\|\cdot\|_B = \|\cdot\|_{L^2(\Gamma_3)}$, respectively. Also, we use the set

$$\mathcal{Q} = \{\,\theta : [0,T] \to B : 0 \le \theta(t) \le 1 \ \forall\, t \in [0,T], \text{ a.e. on } \Gamma_3\,\}.$$

We assume the elasticity tensor $\mathcal{E}$ and the function $\mathcal{G}$ satisfy the following conditions.

$$\left.\begin{array}{l}
\text{(a) } \mathcal{E} = (\mathcal{E}_{ijkl}) : \Omega \times \mathcal{S}^d \to \mathcal{S}^d.\\
\text{(b) } \mathcal{E}_{ijkl} \in L^\infty(\Omega),\ 1 \le i,j,k,l \le d.\\
\text{(c) } \mathcal{E}\boldsymbol{\sigma}\cdot\boldsymbol{\tau} = \boldsymbol{\sigma}\cdot\mathcal{E}\boldsymbol{\tau},\ \forall\,\boldsymbol{\sigma},\boldsymbol{\tau} \in \mathcal{S}^d, \text{ a.e. in } \Omega.\\
\text{(d) There exists } \alpha_0 > 0 \text{ such that}\\
\qquad \mathcal{E}\boldsymbol{\tau}\cdot\boldsymbol{\tau} \ge \alpha_0\|\boldsymbol{\tau}\|^2\ \forall\,\boldsymbol{\tau} \in \mathcal{S}^d, \text{ a.e. in } \Omega.
\end{array}\right\} \tag{19}$$

$$\left.\begin{array}{l}
\text{(a) } \mathcal{G} : \Omega \times \mathcal{S}^d \times \mathcal{S}^d \to \mathcal{S}^d.\\
\text{(b) There exists } L_{\mathcal{G}} > 0 \text{ such that}\\
\qquad \|\mathcal{G}(\boldsymbol{x},\boldsymbol{\sigma}_1,\boldsymbol{\varepsilon}_1) - \mathcal{G}(\boldsymbol{x},\boldsymbol{\sigma}_2,\boldsymbol{\varepsilon}_2)\|\\
\qquad\quad \le L_{\mathcal{G}}\,(\|\boldsymbol{\sigma}_1 - \boldsymbol{\sigma}_2\| + \|\boldsymbol{\varepsilon}_1 - \boldsymbol{\varepsilon}_2\|)\\
\qquad\qquad \forall\,\boldsymbol{\sigma}_1,\boldsymbol{\sigma}_2,\boldsymbol{\varepsilon}_1,\boldsymbol{\varepsilon}_2 \in \mathcal{S}^d, \text{ a.e. } \boldsymbol{x} \in \Omega.\\
\text{(c) For any } \boldsymbol{\sigma},\boldsymbol{\varepsilon} \in \mathcal{S}^d,\ \boldsymbol{x} \mapsto \mathcal{G}(\boldsymbol{x},\boldsymbol{\sigma},\boldsymbol{\varepsilon})\\
\qquad \text{is measurable on } \Omega.\\
\text{(d) The mapping } \boldsymbol{x} \mapsto \mathcal{G}(\boldsymbol{x},\mathbf{0},\mathbf{0}) \text{ belongs to } Q.
\end{array}\right\} \tag{20}$$

Moreover, the normal compliance function p_ν and the tangential function p_τ satisfy the assumptions

$$\left.\begin{array}{l}
\text{(a) } p_\nu : \Gamma_3 \times \mathbb{R} \to \mathbb{R}_+.\\
\text{(b) There exists } L_\nu > 0 \text{ such that}\\
\qquad |p_\nu(\boldsymbol{x},r_1) - p_\nu(\boldsymbol{x},r_2)| \le L_\nu|r_1 - r_2|\\
\qquad \forall\, r_1,\ r_2 \in \mathbb{R}, \text{ a.e. } \boldsymbol{x} \in \Gamma_3.\\
\text{(c) } (p_\nu(\boldsymbol{x},r_1) - p_\nu(\boldsymbol{x},r_2))\,(r_1 - r_2) \ge 0\\
\qquad \forall\, r_1,\ r_2 \in \mathbb{R}, \text{ a.e. } \boldsymbol{x} \in \Gamma_3.\\
\text{(d) For any } r \in \mathbb{R},\ \boldsymbol{x} \mapsto p_\nu(\boldsymbol{x},r) \text{ is measurable on } \Gamma_3.\\
\text{(e) } p_\nu(\boldsymbol{x},r) = 0 \text{ for all } r \le 0.
\end{array}\right\} \tag{21}$$

$$\left.\begin{array}{l}\text{(a) } p_\tau : \Gamma_3 \times \mathbb{R} \longrightarrow \mathbb{R}_+.\\ \text{(b) There exists } L_\tau > 0 \text{ such that}\\ \qquad |p_\tau(\boldsymbol{x},\beta_1) - p_\tau(\boldsymbol{x},\beta_2)| \le L_\tau\, |\beta_1 - \beta_2|\\ \qquad \forall\, \beta_1,\, \beta_2 \in \mathbb{R},\ \text{a.e. } \boldsymbol{x} \in \Gamma_3.\\ \text{(c) There exists } M_\tau > 0 \text{ such that } |p_\tau(\boldsymbol{x},\beta)| \le M_\tau\\ \qquad \forall\, \beta \in \mathbb{R},\ \text{a.e. } \boldsymbol{x} \in \Gamma_3.\\ \text{(d) For any } \beta \in \mathbb{R},\ \boldsymbol{x} \mapsto p_\tau(\boldsymbol{x},\beta) \text{ is measurable on } \Gamma_3.\\ \text{(e) The mapping } \boldsymbol{x} \mapsto p_\tau(\boldsymbol{x},0) \text{ belongs to } L^2(\Gamma_3).\end{array}\right\} \tag{22}$$

The adhesion coefficients γ_ν, γ_τ and ϵ_a satisfy the conditions

$$\gamma_\nu,\ \gamma_\tau \in L^\infty(\Gamma_3), \quad \epsilon_a \in L^2(\Gamma_3), \quad \gamma_\nu,\ \gamma_\tau,\ \epsilon_a \ge 0 \quad \text{a.e. on } \Gamma_3. \tag{23}$$

We also suppose that the body forces and surface tractions satisfy

$$\boldsymbol{f}_0 \in W^{1,\infty}(0,T;L^2(\Omega)^d), \quad \boldsymbol{f}_2 \in W^{1,\infty}(0,T;L^2(\Gamma_2)^d) \tag{24}$$

and we denote by $\boldsymbol{f} : [0,T] \to V$ the function defined by

$$(\boldsymbol{f}(t),\boldsymbol{v})_V = \int_\Omega \boldsymbol{f}_0(t)\cdot\boldsymbol{v}\,dx + \int_{\Gamma_3} \boldsymbol{f}_2(t)\cdot\boldsymbol{v}\,da \quad \forall\,\boldsymbol{v} \in V,\ t \in [0,T].$$

We use the convex set of admissible displacements defined by

$$U = \{\boldsymbol{v} \in V : v_\nu \le 0 \text{ on } \Gamma_3\},$$

and the functionals $j_{ad} : L^\infty(\Gamma_3) \times V \times V \to \mathbb{R}$ and $j_{nc} : V \times V \to \mathbb{R}$,

$$j_{ad}(\beta,\boldsymbol{u},\boldsymbol{v}) = -\int_{\Gamma_3} \gamma_\nu \beta^2 \widetilde{R}(u_\nu) v_\nu\,da + \int_{\Gamma_3} p_\tau(\beta)\, R^*(\boldsymbol{u}_\tau)\cdot\boldsymbol{v}_\tau\,da,$$

$$j_{nc}(\boldsymbol{u},\boldsymbol{v}) = \int_{\Gamma_3} p_\nu(u_\nu) v_\nu\,da.$$

We assume that the initial data satisfy

$$\boldsymbol{u}_0 \in U, \quad \boldsymbol{\sigma}_0 \in Q, \quad \beta_0 \in B, \quad 0 \le \beta_0 \le 1 \ \text{ a.e. on } \Gamma_3. \tag{25}$$

$$(\boldsymbol{\sigma}_0, \boldsymbol{\varepsilon}(\boldsymbol{v}))_Q + j_{ad}(\beta_0,\boldsymbol{u}_0,\boldsymbol{v}) = (\boldsymbol{f}(0),\boldsymbol{v})_V \quad \forall\,\boldsymbol{v} \in V. \tag{26}$$

By a standard procedure we can derive the following variational formulation of the Signorini problem (1)–(9).

Problem 1. *Find a displacement field $\boldsymbol{u} : [0,T] \to V$, a stress field $\boldsymbol{\sigma} : [0,T] \to Q$, and a bonding field $\beta : [0,T] \to L^\infty(\Gamma_3)$ such that*

$$\dot{\boldsymbol{\sigma}}(t) = \mathcal{E}\boldsymbol{\varepsilon}(\dot{\boldsymbol{u}}(t)) + \mathcal{G}(\boldsymbol{\sigma}(t),\boldsymbol{\varepsilon}(\boldsymbol{u}(t))) \quad \text{a.e. } t \in (0,T), \tag{27}$$

$$\begin{aligned}\boldsymbol{u}(t) \in U, \quad & (\boldsymbol{\sigma}(t),\boldsymbol{\varepsilon}(\boldsymbol{v}-\boldsymbol{u}(t)))_Q + j_{ad}(\beta(t),\boldsymbol{u}(t),\boldsymbol{v}-\boldsymbol{u}(t))\\ & \ge (\boldsymbol{f}(t),\boldsymbol{v}-\boldsymbol{u}(t))_V \quad \forall\,\boldsymbol{v} \in U,\ t \in [0,T],\end{aligned} \tag{28}$$

$$\begin{aligned}\dot\beta(t) = -(\gamma_\nu \beta(t) \widetilde{R}(u_\nu(t))^2 + \gamma_\tau \beta(t) \|R^*(\boldsymbol{u}_\tau(t))\|^2 - \epsilon_a)_+\\ \text{a.e. } t \in (0,T),\end{aligned} \tag{29}$$

$$\boldsymbol{u}(0) = \boldsymbol{u}_0, \quad \boldsymbol{\sigma}(0) = \boldsymbol{\sigma}_0, \quad \beta(0) = \beta_0. \tag{30}$$

The variational formulation of the problem with normal compliance (10)–(18) is as follows.

Problem 2. *Find a displacement field* $\boldsymbol{u}^\lambda : [0,T] \to V$*, a stress field* $\boldsymbol{\sigma}^\lambda : [0,T] \to Q$*, and a bonding field* $\beta^\lambda : [0,T] \to L^\infty(\Gamma_3)$ *such that*

$$\dot{\boldsymbol{\sigma}}^\lambda(t) = \mathcal{E}\boldsymbol{\varepsilon}(\dot{\boldsymbol{u}}^\lambda(t)) + \mathcal{G}(\boldsymbol{\sigma}^\lambda(t), \boldsymbol{\varepsilon}(\boldsymbol{u}^\lambda(t))) \quad \text{a.e. } t \in (0,T), \tag{31}$$

$$(\boldsymbol{\sigma}^\lambda(t), \boldsymbol{\varepsilon}(\boldsymbol{v}))_Q + j_{ad}(\beta^\lambda(t), \boldsymbol{u}^\lambda(t), \boldsymbol{v}) + j_{nc}(\boldsymbol{u}^\lambda(t), \boldsymbol{v}) = (\boldsymbol{f}(t), \boldsymbol{v})_V \quad \forall \boldsymbol{v} \in V,\ t \in [0,T], \tag{32}$$

$$\dot{\beta}^\lambda(t) = -(\gamma_\nu \beta^\lambda(t) \widetilde{R}(u_\nu^\lambda(t))^2 + \gamma_\tau \beta^\lambda(t) \|R^*(\boldsymbol{u}_\tau^\lambda(t))\|^2 - \epsilon_a)_+ \quad \text{a.e. } t \in (0,T), \tag{33}$$

$$\boldsymbol{u}^\lambda(0) = \boldsymbol{u}_0, \quad \boldsymbol{\sigma}^\lambda(0) = \boldsymbol{\sigma}_0, \quad \beta^\lambda(0) = \beta_0. \tag{34}$$

In the next section we provide the unique solvability of Problems 1 and and 2. We also discuss the relationship between the two problems; it turns out that the solution of the problem with normal compliance converges to the solution of the problem with the Signorini condition, as the stiffness of the surface tends to infinity.

3 Existence, uniqueness and convergence results

For Problems 1 and 2, we have the following results concerning the solution existence, uniqueness, and relationship between the two problems.

Theorem 1. *Assume* (19)–(26).
1) There exists a unique solution $(\boldsymbol{u}, \boldsymbol{\sigma}, \beta)$ *to Problem* 1 *which satisfies*

$$\boldsymbol{u} \in W^{1,\infty}(0,T;V), \ \boldsymbol{\sigma} \in W^{1,\infty}(0,T;Q), \ \beta \in W^{1,\infty}(0,T,B) \cap \mathcal{Q}. \tag{35}$$

2) There exists a unique solution $(\boldsymbol{u}^\lambda, \boldsymbol{\sigma}^\lambda, \beta^\lambda)$ *to Problem* 2 *which satisfies the regularity expressed in* (35).
3) As $\lambda \to 0$*, the solution* $(\boldsymbol{u}^\lambda, \boldsymbol{\sigma}^\lambda, \beta^\lambda)$ *of Problem* 2 *converges to the solution* $(\boldsymbol{u}, \boldsymbol{\sigma}, \beta)$ *of Problem* 1*, that is,*

$$\|\boldsymbol{u}^\lambda(t) - \boldsymbol{u}(t)\|_V + \|\boldsymbol{\sigma}^\lambda(t) - \boldsymbol{\sigma}(t)\|_Q + \|\beta^\lambda(t) - \beta(t)\|_B \to 0$$

for all $t \in [0,T]$.

Theorem 1 can be proved based on results on variational inequalities and ordinary differential equations, and a fixed point argument, similar to those used in [8, Ch. 6]. We conclude that under the stated assumptions, the adhesive Signorini contact problem (1)–(9) has a unique weak solution with the regularity (35) and, for all $\lambda > 0$, the adhesive contact problem with normal compliance (10)–(18) has a unique weak solution with the same regularity. We also conclude that the weak solution of the adhesive contact problem with a rigid obstacle may be approached as closely as one wishes by the solution of the adhesive contact problem with a deformable foundation, with a sufficiently small deformability coefficient.

4 Numerical approximation

We turn now to the numerical approximation of the problems. We only discuss a fully discrete scheme for the adhesive Signorini problem; the discussion can be extended for the normal compliance problem. For definiteness and for the sake of simplicity, we assume that $\overline{\Omega} \subset \mathbb{R}^2$ is a planar polygonal domain and let $\{\mathcal{T}^h\}_h$ be a regular family of finite element triangulations of $\overline{\Omega}$. For $1 \le j \le 3$, $\overline{\Gamma_j} = \cup_{i=1}^{i_j} \Gamma_j^{(i)}$, and on each $\Gamma_j^{(i)}$, the unit outward normal vector is constant. The finite element partitions of $\overline{\Omega}$ are assumed to be compatible with the decomposition $\Gamma = \cup_{j=1}^{3} \cup_{i=1}^{i_j} \Gamma_j^{(i)}$, i.e., if S is an element side such that for some j and i, $S \cap \Gamma_j^{(i)}$ contains an interior point of S, then $S \subset \Gamma_j^{(i)}$. Corresponding to each partition $\mathcal{T}^h$, we use linear elements for $V^h \subset V$ and define a finite element set $U^h = V^h \cap U$. Note that for $\boldsymbol{v}^h \in U^h$, we have $v_\nu^h \le 0$ at each node on $\overline{\Gamma_3}$. We use piecewise constant functions for $Q^h \subset Q$. Let $\mathcal{T}_{\Gamma_3}^h$ be the partition of Γ_3 induced by $\mathcal{T}^h$. Then the space B is approximated by B^h, the space of piecewise constant functions corresponding to $\mathcal{T}_{\Gamma_3}^h$. Let $0 = t_0 < t_1 < \cdots < t_N = T$ be a partition of the time interval $[0, T]$, and denote $k_n = t_n - t_{n-1}$, $k = \max_{1 \le n \le N} k_n$. We use the notation $z_n = z(t_n)$ for a continuous function $z(t)$. For a sequence $\{f_n\}_{n=0}^N$, we introduce the backward divided difference $\delta f_n = (f_n - f_{n-1})/k_n$. We denote by $\mathcal{P}_{Q^h} : Q \to Q^h$ the orthogonal projection operator with respect to the inner product of Q, and by $\mathcal{P}_{B^h} : B \to B^h$ the orthogonal projection operator with respect to the inner product of B. Then a fully discrete approximation scheme for Problem 1 is the following.

Problem 3. *Find a discrete displacement field* $\boldsymbol{u}^{hk} = \{\boldsymbol{u}_n^{hk}\}_{n=0}^N \subset U^h$, *a discrete stress field* $\boldsymbol{\sigma}^{hk} = \{\boldsymbol{\sigma}_n^{hk}\}_{n=0}^N \subset Q^h$, *and a discrete bonding field* $\beta^{hk} = \{\beta_n^{hk}\}_{n=0}^N \subset B^h$ *such that*

$$\boldsymbol{u}_0^{hk} = \boldsymbol{u}_0^h, \quad \boldsymbol{\sigma}_0^{hk} = \boldsymbol{\sigma}_0^h, \quad \beta_0^{hk} = \beta_0^h, \tag{36}$$

and, for $n = 1, 2, \ldots, N$,

$$\delta\boldsymbol{\sigma}_n^{hk} = \mathcal{P}_{Q^h}\mathcal{E}\varepsilon(\delta\boldsymbol{u}_n^{hk}) + \mathcal{P}_{Q^h}\mathcal{G}\varepsilon(\boldsymbol{\sigma}_{n-1}^{hk}, \varepsilon(\boldsymbol{u}_{n-1}^{hk})), \tag{37}$$

$$\begin{aligned}(\boldsymbol{\sigma}_n^{hk}, \varepsilon(\boldsymbol{v}^h - \boldsymbol{u}_n^{hk}))_Q + j_{ad}(\beta_n^{hk}, \boldsymbol{u}_n^{hk}, \boldsymbol{v}^h - \boldsymbol{u}_n^{hk}) \\ \ge (\boldsymbol{f}_n, \boldsymbol{v}^h - \boldsymbol{u}_n^{hk})_V \quad \forall \boldsymbol{v}^h \in U^h,\end{aligned} \tag{38}$$

$$\delta\beta_n^{hk} = -\mathcal{P}_{B^h}(\gamma_\nu \beta_{n-1}^{hk} \widetilde{R}(u_{n-1,\nu}^{hk})^2 + \gamma_\tau \beta_{n-1}^{hk} \|R^*(\boldsymbol{u}_{n,\tau}^{hk})\|^2 - \epsilon_a)_+. \tag{39}$$

Here $u_{n,\nu}^{hk}$ and $\boldsymbol{u}_{n,\tau}^{hk}$ represent the normal component and the tangential part of the function $\boldsymbol{u}_n^{hk}$. By using the known solution $\boldsymbol{\sigma}_{n-1}^{hk}$, $\boldsymbol{u}_{n-1}^{hk}$, and β_{n-1}^{hk} at $t = t_{n-1}$, we first determine β_n^{hk} from (39). By a discrete analogue of the proof of Theorem 1, we can show that the system (37) and (38) determines a unique $\boldsymbol{u}_n^{hk}$. Then $\boldsymbol{\sigma}_n^{hk}$ is determined uniquely from (37). Thus, Problem 3 has a unique solution.

In the following, we focus on error estimation. For this purpose, we assume the following additional solution regularity:

$$\boldsymbol{u} \in W^{2,1}(0,T;V) \cap C([0,T];H^2(\Omega)^d). \tag{40}$$

$$\boldsymbol{\sigma} \in C([0,T];H^1(\Omega)^{d\times d}) \cap W^{1,1}(0,T;Q). \tag{41}$$

$$\beta \in W^{2,1}(0,T;B) \cap C^1([0,T];\tilde{H}^1(\Gamma_3)). \tag{42}$$

Here, $\tilde{H}^1(\Gamma_3)$ is the set of functions that belong to $H^1(\Gamma_3^{(i)})$, $1 \le i \le i_3$.

Theorem 2. *Assume* (19)–(26) *and* (40)–(42). *Then, the following error estimate holds:*

$$\max_{0\le n\le N} \left(\|\boldsymbol{\sigma}_n - \boldsymbol{\sigma}_n^{hk}\|_Q + \|\boldsymbol{u}_n - \boldsymbol{u}_n^{hk}\|_V + \|\beta_n - \beta_n^{hk}\|_B\right) \le c\,(h+k).$$

A similar result can be stated in the study of the adhesive contact problem with normal compliance. The proofs are based on arguments similar to those used in [8, Ch. 6].

References

1. Cangémi L (1997) Frottement et adhérence: modèle, traitement numérique et application à l'interface fibre/matrice. Ph.D. Thesis, Univ. Méditerranée, Aix-Marseille I
2. Cocou M, Rocca R (2000) Math. Model. Num. Anal. 34:981–1001
3. Frémond M (1982) C. R. Acad. Sci. Paris, Série II 295:913–916
4. Frémond M (1987) J. Mécanique Théorique et Appliquée 6:383-407
5. Frémond M (2002) Non-Smooth Thermomechanics. Springer, Berlin
6. Raous M, Cangémi L, Cocou M (1999) Comput. Methods Appl. Engrg. 177:383–399
7. Shillor M, Sofonea M and Telega J J (2004) Models and Analysis of Quasistatic Contact. Springer, Berlin
8. Sofonea M, Han W, Shillor M (to appear) Contact Problems with Adhesion or Damage. Chapman-Hall/CRC Press, New York

A uniqueness criterion for the Signorini problem with Coulomb friction

Y. Renard

MIP, CNRS UMR 5640, INSAT, Complexe scientifique de Rangueil, 31077 Toulouse, France,
Yves.Renard@insa-toulouse.fr

Summary. Some optimal *a priori* estimates are given for the solutions to the Signorini problem with Coulomb friction (the so-called Coulomb problem) and a uniqueness criterion is exhibited. Recently, nonuniqueness examples have been presented in the continuous framework. It is proven, here, that if a solutions satisfies an hypothesis on the tangential displacement and if the friction coefficient is small enough, it is the unique solution to the problem.

1 The Signorini problem with Coulomb friction

Recently, examples of nonunique solutions for the Signorini problem with Coulomb friction (or simply the Coulomb problem) have been given by P. Hild in [6] and [36] for a large friction coefficient (greater than one). As far as we know, for a fixed geometry, it is still an open question to know wheither or not there is uniqueness of the solution for a sufficiently small friction coefficient. The present paper is a partial answer to this preoccupation. The major result is given by Proposition 5. More details on the result presented here can be found in [11].

Let $\Omega \subset \mathbb{R}^d$ ($d = 2$ or 3) be a bounded Lipschitz domain representing the reference configuration of a linearly elastic body.

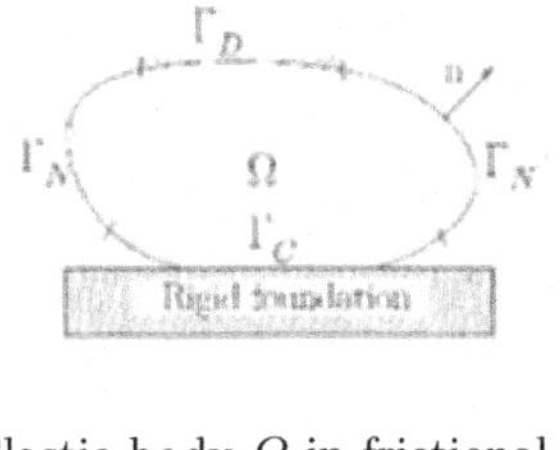

Fig. 1. Elastic body Ω in frictional contact.

This body is submitted to a Neumann condition on $\Gamma_N \subset \partial\Omega$, to a Dirichlet condition on $\Gamma_D \subset \partial\Omega$ and a unilateral contact with static Coulomb friction condition on the rest of the boundary Γ_C between the body and a flat rigid foundation. The problem consists in finding the displacement field $u(t,x)$ satisfying

$$-\operatorname{div} \sigma(u) = f \quad \text{in } \Omega, \tag{1}$$

$$\sigma(u) = \mathcal{A}\varepsilon(u) \quad \text{in } \Omega, \tag{2}$$

$$\sigma(u)\mathrm{n} = F \quad \text{on } \Gamma_N, \tag{3}$$

$$u = 0 \quad \text{on } \Gamma_D, \tag{4}$$

where $\sigma(u)$ is the stress tensor, $\varepsilon(u)$ is the linearized strain tensor, n is the outward unit normal to Ω on $\partial\Omega$, F and f are the given external loads, and $\mathcal{A}$ is the elastic coefficient tensor which satisfies classical conditions of symmetry and ellipticity.

On Γ_C, the displacement and the stress vector are decomposed into normal and tangential components as follows (we assume Γ_C to have the $\mathcal{C}^1$ regularity):

$$u_N = u.\mathrm{n}, \quad u_T = u - u_N\mathrm{n}, \quad \sigma_N(u) = (\sigma(u)\mathrm{n}).\mathrm{n}, \quad \sigma_T(u) = \sigma(u)\mathrm{n} - \sigma_N(u)\mathrm{n}.$$

The unilateral contact condition is expressed by

$$u_N \leq 0, \ \sigma_N(u) \leq 0, \ u_N\sigma_N(u) = 0. \tag{5}$$

Denoting $\mathcal{F} \geq 0$ the friction coefficient, the *static Coulomb friction* condition is

$$\text{if } u_T = 0 \text{ then } |\sigma_T(u)| \leq -\mathcal{F}\sigma_N(u), \tag{6}$$

$$\text{if } u_T \neq 0 \text{ then } \sigma_T(u) = \mathcal{F}\sigma_N(u)\frac{u_T}{|u_T|}. \tag{7}$$

1.1 Classical weak formulation

We present here the classical weak formulation proposed by G. Duvaut [3] [4]. Let us introduce the following Hilbert spaces

$$V = \{v \in H^1(\Omega;\mathbb{R}^d), v = 0 \text{ on } \Gamma_D\},$$

$$X = \{v_{|_{\Gamma_C}} : v \in V\} \subset H^{1/2}(\Gamma_C;\mathbb{R}^d),$$

$$X_N = \{v_{N|_{\Gamma_C}} : v \in V\}, \ \ X_T = \{v_{T|_{\Gamma_C}} : v \in V\},$$

and their topological dual spaces V', X', X'_N and X'_T. It is assumed that Γ_C is sufficiently smooth such that $X_N \subset H^{1/2}(\Gamma_C)$, $X_T \subset H^{1/2}(\Gamma_C;\mathbb{R}^{d-1})$, $X'_N \subset H^{-1/2}(\Gamma_C)$ and $X'_T \subset H^{-1/2}(\Gamma_C;\mathbb{R}^{d-1})$ (see [8] and [1]).

Now, the set of admissible displacements is defined as

$$K = \{v \in V, v_N \leq 0 \quad \text{a.e. on } \Gamma_C\}. \tag{8}$$

Let us define the following maps:

$$a(u,v) = \int_\Omega \mathcal{A}\varepsilon(u) : \varepsilon(v) dx, \qquad l(v) = \int_\Omega f.v dx + \int_{\Gamma_N} F.v d\Gamma,$$

$$j(\mathcal{F}\lambda_N, v_T) = -\langle \mathcal{F}\lambda_N, |v_T| \rangle_{X'_N, X_N}$$

Standard hypotheses are:

$a(\cdot,\cdot)$ is a bilinear symmetric V-elliptic and continuous form on $V \times V$:
$\exists\, \alpha > 0, \exists\, M > 0, a(u,u) \geq \alpha \|u\|_V^2, a(u,v) \leq M \|u\|_V \|v\|_V \; \forall u, v \in V,$ (9)
$l(\cdot)$ linear continuous form on V, i.e. $\exists\, L > 0, |l(v)| \leq L \|v\|_V, \; \forall v \in V$ (10)
$\mathcal{F} \in MX_N$ being a nonnegative multiplier in X_N. (11)

The latter condition ensure that $j(\mathcal{F}\lambda_N, v_T)$ is linear continuous on λ_N and convex lower semi-continuous on v_T when λ_N is a nonpositive element of X'_N (see for instance [2]). We refer to Maz'ya and Shaposhnikova [10] for the theory of multipliers. The set MX_N denote the space of functions $f : \Gamma_C \longrightarrow \mathbb{R}$ of finite norm $\|f\|_{MX_N} = \sup_{\substack{v_N \in X_N \\ v_N \neq 0}} \dfrac{\|f v_N\|_{X_N}}{\|v_N\|_{X_N}}$. Condition (9) implies that $a(\cdot,\cdot)$ is a scalar product on V and the associated norm $\|v\|_a = (a(v,v))^{1/2}$ is equivalent to the usual norm of V with $\sqrt{\alpha}\|v\|_V \leq \|v\|_a \leq \sqrt{M}\|v\|_V, \quad \forall v \in V$. The continuity constant of $l(\cdot)$ can also be given with respect to $\|\cdot\|_a$: $\exists\, L_a > 0, |l(v)| \leq L_a \|v\|_a, \; \forall v \in V$. Constants L and L_a can be chosen such that $\sqrt{\alpha} L_a \leq L \leq \sqrt{M} L_a$.

Problem (1) - (7) is formally equivalent to the following inequality formulation:

$$\begin{cases} \text{Find } u \in K \text{ satisfying} \\ a(u, v-u) + j(\mathcal{F}\sigma_N(u), v_T) - j(\mathcal{F}\sigma_N(u), u_T) \geq l(v-u), \quad \forall\, v \in K. \end{cases} \tag{12}$$

1.2 Neumann to Dirichlet operator

We intoduce here the Neumann to Dirichlet operator on Γ_C and its basic properties. This will allow to restrict the contact and friction problem to Γ_C. Let $\lambda = (\lambda_N, \lambda_T) \in X'$. Then, under hypotheses (9) and (10), the solution u to

$$\begin{cases} \text{Find } u \in V \text{ satisfying} \\ a(u,v) = l(v) + \langle \lambda, v \rangle_{X',X} \quad \forall\, v \in V, \end{cases} \tag{13}$$

is unique (see [4]). So it is possible to define the operator

$$\begin{aligned} \mathbb{E} : X' &\longrightarrow X \\ \lambda &\longmapsto u_{|\Gamma_C} \end{aligned}$$

This operator is affine and continuous. Moreover, it is invertible and its inverse is continuous. It is possible to express $\mathbb{E}^{-1}$ as follows: for $w \in X$, let u be the solution to the Dirichlet problem

$$\begin{cases} \text{Find } u \in V \text{ satisfying } u_{|\Gamma_C} = w \text{ and} \\ a(u,v) = l(v), \quad \forall\, v \in V, v_{|\Gamma_C} = 0, \end{cases} \tag{14}$$

then $\mathbb{E}^{-1}(w)$ is equal to $\lambda \in X'$ defined by

$$\langle \lambda, v \rangle_{X',X} = a(u,v) - l(v), \quad \forall\, v \in V.$$

It is also possible to define the following norms on Γ_C relatively to $a(\cdot\,,\cdot)$

$$\|v\|_{a,\Gamma_C} = \inf_{w \in V,\ w_{|\Gamma_C} = v} \|w\|_a \ , \quad \|\lambda\|_{-a,\Gamma_C} = \sup_{\substack{v \in X \\ v \neq 0}} \frac{\langle \lambda, v \rangle_{X',X}}{\|v\|_{a,\Gamma_C}} = \sup_{\substack{v \in V \\ v \neq 0}} \frac{\langle \lambda, v \rangle_{X',X}}{\|v\|_a}$$

which are equivalent respectively to the norms in X and X':

$$\frac{\sqrt{\alpha}}{C_1}\|v\|_X \le \|v\|_{a,\Gamma_C} \le \sqrt{M}\gamma\|v\|_X \ , \quad \frac{1}{\sqrt{M}\gamma}\|\lambda\|_{X'} \le \|\lambda\|_{-a,\Gamma_C} \le \frac{C_1}{\sqrt{\alpha}}\|\lambda\|_{X'}$$

where $\gamma = \sup_{\substack{v \in X \\ v \neq 0}} \frac{\|w\|_V}{\|v\|_X}$ and C_1 is the continuity constant of the trace operator on Γ_C. With these norms, the estimates are straightforward since

$$\|\mathbb{E}(\lambda^1) - \mathbb{E}(\lambda^2)\|_{a,\Gamma_C} = \|\lambda^1 - \lambda^2\|_{-a,\Gamma_C}, \tag{15}$$

$$\|\mathbb{E}^{-1}(u^1) - \mathbb{E}^{-1}(u^2)\|_{-a,\Gamma_C} = \|u^1 - u^2\|_a = \|u^1 - u^2\|_{a,\Gamma_C}. \tag{16}$$

1.3 Direct weak inclusion formulation

Let $K_N = \{v_N \in X_N : v_N \le 0 \text{ a.e. on } \Gamma_C\}$ be the set of admissible normal displacements on Γ_C. The normal cone in X'_N to K_N at $v_N \in X_N$ is $N_{K_N}(v_N) = \{\lambda_N \in X'_N : \langle \lambda_N, w_N - v_N \rangle_{X'_N, X_N} \le 0,\ \forall w_N \in K_N\}$. The sub-gradient of $j(\mathcal{F}\lambda_N, u_T)$ with respect to the second variable is given by

$$\begin{aligned} \partial_2 j(\mathcal{F}\lambda_N, u_T) = \{\lambda_T \in X'_T : j(\mathcal{F}\lambda_N, v_T) &\ge j(\mathcal{F}\lambda_N, u_T) \\ &+ \langle \lambda_T, v_T - u_T \rangle_{X'_T, X_T},\ \forall v_T \in X_T\}. \end{aligned}$$

Problem (12) is then equivalent to the following problem (see [9])

$$\begin{cases} \text{Find } u \in V, \lambda_N \in X'_N \text{ and } \lambda_T \in X'_T \text{ satisfying} \\ (u_N, u_T) = \mathbb{E}(\lambda_N, \lambda_T), \\ -\lambda_N \in N_{K_N}(u_N) \quad \text{in } X'_N, \\ -\lambda_T \in \partial_2 j(\mathcal{F}\lambda_N, u_T) \quad \text{in } X'_T. \end{cases} \tag{17}$$

2 Elementary estimates on the Coulomb problem

A solution (u, λ) to Problem (17) satisfies the complementarity relations

$$\langle \lambda_N, u_N \rangle_{X'_N, X_N} = 0 \quad , \qquad \langle \lambda_T, u_T \rangle_{X'_T, X_T} = \langle \mathcal{F}\lambda_N, |u_T| \rangle_{X'_N, X_N} .$$

This implies $\langle \lambda, u \rangle_{X', X} \leq 0$, which expresses the dissipativity of contact and friction conditions. A consequence is that any solution to Problem (12) can be bounded independently of the friction coefficient.

Proposition 1 *Assume hypotheses (9), (10), (11) are satisfied. Let (u, λ) be a solution to Problem (17). Then $\|u\|_a \leq L_a$ and $\|\lambda\|_{-a,\Gamma_C} \leq L_a$.*

Proof. One has $\|u\|_a^2 = a(u, u) = l(u) + \langle \lambda, u \rangle_{X', X} \leq L_a \|u\|_a$, which states the first estimates. The estimate on $\|\lambda\|_{-a,\Gamma_C}$ can be performed using the intermediary solution $u^{\mathcal{N}}$ to the following problem with a homogeneous Neumann condition on Γ_C

$$a(u^{\mathcal{N}}, v) = l(v), \quad \forall v \in V. \tag{18}$$

Since $\|u^{\mathcal{N}}\|_a \leq L_a$ for the same reason as for u, and using (16) one has

$$\begin{aligned} \|\lambda\|_{-a,\Gamma_C}^2 &= a(u - u^{\mathcal{N}}, u - u^{\mathcal{N}}) \\ &= \langle \lambda, u - u^{\mathcal{N}} \rangle_{X', X} \leq -\langle \lambda, u^{\mathcal{N}} \rangle_{X', X} \leq L_a \|\lambda\|_{-a,\Gamma_C} \end{aligned}$$

■

It is possible to compare $\|u\|_a$ to the corresponding norm of the solution u^c to the Signorini problem without friction defined as follows

$$\begin{cases} \text{Find } u^c \in K \text{ satisfying} \\ a(u^c, v - u^c) \geq l(v - u^c), \quad \forall\, v \in K. \end{cases} \tag{19}$$

Under hypotheses (9) and (10), this problem has a unique solution (see [8]).

Proposition 2 *Assuming hypotheses (9), (10), (11) are satisfied, let u be a solution to Problem (12), u^c be the unique solution to Problem (19) and $u^{\mathcal{N}}$ the solution to Problem (18), then $\|u\|_a \leq \|u^c\|_a \leq \|u^{\mathcal{N}}\|_a$.*

Proof. One has $a(u^{\mathcal{N}}, u^{\mathcal{N}}) = l(u^{\mathcal{N}})$, $a(u^c, u^c) = l(u^c)$ and $a(u, u) = l(u) + \langle \lambda_T, u_T \rangle_{X'_T, X_T}$. Since u^c is the solution to the Signorini problem without friction, it minimizes over K the energy functional $\dfrac{1}{2}a(v, v) - l(v)$. The solution $u^{\mathcal{N}}$ minimizes this energy functional over V. Thus, since also $u \in K$, one has

$$\frac{1}{2}a(u^{\mathcal{N}}, u^{\mathcal{N}}) - l(u^{\mathcal{N}}) \leq \frac{1}{2}a(u^c, u^c) - l(u^c) \leq \frac{1}{2}a(u, u) - l(u),$$

and the following relations allow to conclude

$$a(u^c, u^c) - a(u^{\mathcal{N}}, u^{\mathcal{N}}) = l(u^c - u^{\mathcal{N}}),$$

$$a(u, u) - a(u^c, u^c) = l(u - u^c) + \langle \lambda_T, u_T \rangle_{X'_T, X_T} \leq l(u - u^c).$$

■

It is also possible to estimate how far from u^c is a solution u to Problem (12). Let us introduce the following norms on Γ_C. For $v \in X$ and $\lambda \in X'$ let us define

$$\|v_T\|_{a,\Gamma_C} = \inf_{\substack{w \in V \\ w_T = v_T}} \|w\|_a = \inf_{\substack{z \in X \\ z_T = v_T}} \|z\|_{a,\Gamma_C},$$

$$\|v_N\|_{a,\Gamma_C} = \inf_{\substack{w \in V \\ w_N = v_N}} \|w\|_a = \inf_{\substack{z \in X \\ z_N = v_N}} \|v\|_{a,\Gamma_C}.$$

$$\|\lambda_T\|_{-a,\Gamma_C} = \sup_{\substack{v_T \in X_T \\ v_T \neq 0}} \frac{\langle \lambda_T, v_T \rangle_{X'_T, X_T}}{\|v_T\|_{a,\Gamma_C}},$$

$$\|\lambda_N\|_{-a,\Gamma_C} = \sup_{\substack{v_N \in X_N \\ v_N \neq 0}} \frac{\langle \lambda_N, v_N \rangle_{X'_N, X_N}}{\|v_N\|_{a,\Gamma_C}}.$$

The following equivalence of norms are immediate:

$$\frac{\sqrt{\alpha}}{C_1}\|v_N\|_{X_N} \leq \|v_N\|_{a,\Gamma_C} \leq \gamma\sqrt{M}\|v_N\|_{X_N},$$

$$\frac{\sqrt{\alpha}}{C_1}\|v_T\|_{X_T} \leq \|v_T\|_{a,\Gamma_C} \leq \gamma\sqrt{M}\|v_T\|_{X_T}.$$

$$\frac{1}{\gamma\sqrt{M}}\|\lambda_N\|_{X'_N} \leq \|\lambda_N\|_{-a,\Gamma_C} \leq \frac{C_1}{\sqrt{\alpha}}\|\lambda_N\|_{X'_N},$$

$$\frac{1}{\gamma\sqrt{M}}\|\lambda_T\|_{X'_T} \leq \|\lambda_T\|_{-a,\Gamma_C} \leq \frac{C_1}{\sqrt{\alpha}}\|\lambda_T\|_{X'_T}.$$

This allow also to define $\|\mathcal{F}\|_a = \sup_{\substack{v_N \in X_N \\ v_N \neq 0}} \frac{\|\mathcal{F}v_N\|_{a,\Gamma_C}}{\|v_N\|_{a,\Gamma_C}}$. an equivalent norm on MX_N.

Lemma 1. *There exists $C_3 > 0$ such that for all $\lambda \in X'$*
$\|\lambda_T\|_{-a,\Gamma_C} \leq C_3\|\lambda\|_{-a,\Gamma_C}, \quad \|\lambda_N\|_{-a,\Gamma_C} \leq C_3\|\lambda\|_{-a,\Gamma_C}.$

Lemma 2. *There exists $C_4 > 0$ such that*
$\|\mathcal{F}|v_T|\,\|_{a,\Gamma_C} \leq C_4\|\mathcal{F}\|_a\|v_T\|_{a,\Gamma_C}, \ \forall v_T \in X_T.$

Proof. One has $\|\mathcal{F}|v_T|\,\|_{a,\Gamma_C} \leq \|\mathcal{F}\|_a\|\,|v_T|\,\|_{a,\Gamma_C}$. Moreover, the norm $\|\cdot\|_{X_N}$ is equivalent to the norm

$\|v_N\|^2_{1/2,\Gamma_C} = \|v_N\|^2_{L^2(\Gamma_C)} + \int_{\Gamma_C}\int_{\Gamma_C} \frac{|v_N(x) - v_N(y)|^2}{|x-y|^d} dxdy$ (see [1]) and it is easy to verify that $\| \, |v_T| \, \|_{1/2,\Gamma_C} \leq \|v_T\|_{1/2,\Gamma_C}$ for any $v_T \in X_T$. Thus, the result can be deduced from the previously presented equivalence of norms. ■

The tangential stress corresponding to u can be estimated as follows. As $\lambda_T \in \Lambda_T(\mathcal{F}\lambda_N)$, one has

$$\begin{aligned} \|\lambda_T\|_{-a,\Gamma_C} &\leq \sup_{\substack{v_T \in X_T \\ v_T \neq 0}} \frac{-\langle \mathcal{F}\lambda_N, |v_T| \rangle_{X'_N, X_N}}{\|v_T\|_{a,\Gamma_C}} \\ &\leq C_4 \|\mathcal{F}\|_a \|\lambda_N\|_{-a,\Gamma_C} \leq L_a C_3 C_4 \|\mathcal{F}\|_a, \end{aligned}$$

and the following result holds.

Proposition 3 *Assuming hypotheses (9), (10), (11) are satisfied, let u be a solution to Problem (12) and u^c be the solution to Problem (19), then $\|u^c - u\|_a \leq L_a C_3 C_4 \|\mathcal{F}\|_a$.*

Proof. With $\lambda \in X'$ and $\lambda^c \in X'$ the corresponding stresses on Γ_C, because $-\lambda^c_N \in N_{K_N}(u^c_N - g)$ and $-\lambda_N \in N_{K_N}(u_N - g)$ and the fact that N_{K_N} is a monotone set-valued map, one has $\langle \lambda^c_N - \lambda_N, u^c_N - u_N \rangle_{X'_N, X_N} \leq 0$. Now, $\|u^c - u\|_a$ can be estimated as follows

$$\|u^c - u\|^2_a = \langle \lambda^c - \lambda, u^c - u \rangle_{X', X} \leq \|\lambda_T\|_{-a,\Gamma_C} \|u^c - u\|_a \leq L_a C_4 \|\mathcal{F}\|_a \|u^c - u\|_a.$$

■

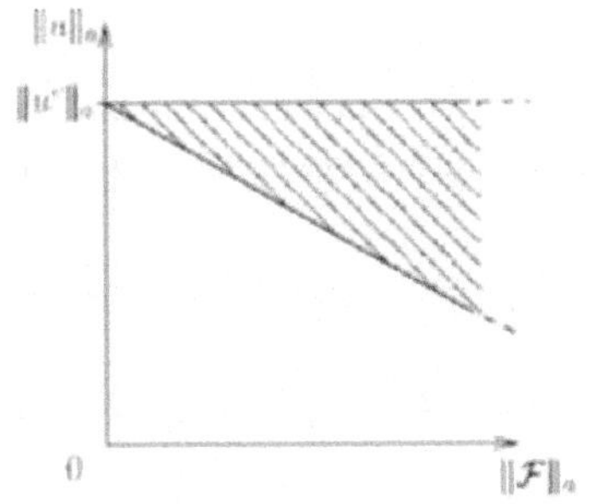

Fig. 2. *Admissibility zone for $\|u\|_a$.*

3 A uniqueness criterion

P. Hild in [6, 36] exhibit some multi-solutions for the Coulomb problem in triangular domains. These solutions have been obtained for a large friction coefficient ($\mathcal{F} > 1$) and for a tangential displacement having a constant sign. For the moment, it seems that no multi-solution has been exhibited for an arbitrary small friction coefficient in the continuous case. The result presented here is a partial uniqueness result which determines some cases where it is possible to say that a particular solution of the Coulomb problem is in fact the unique solution. The partial uniqueness results we present are deduced from the estimate given by the following lemma.

Lemma 3. *Assuming hypotheses (9), (10), (11) are satisfied, if u^1 and u^2 are two solutions to Problem (12) and λ^1 and λ^2 are the corresponding contact stresses on Γ_C, then one has the following estimate*

$$\|u^1-u^2\|_a^2 = \|\lambda^1-\lambda^2\|_{-a,\Gamma_C}^2 \leq \langle \zeta - \lambda_T^2, u_T^1 - u_T^2 \rangle_{X_T', X_T} \quad \forall \zeta \in -\partial_2 j(\mathcal{F}\lambda_N^1, u_T^2).$$

Proof. One has

$$\|u^1 - u^2\|_a^2 = \langle \lambda_N^1 - \lambda_N^2, u_N^1 - u_N^2 \rangle_{X_N', X_N} + \langle \lambda_T^1 - \lambda_T^2, u_T^1 - u_T^2 \rangle_{X_T', X_T}.$$

Due to the monotonicity of N_{K_N}, one has $\langle \lambda_N^1 - \lambda_N^2, u_N^1 - u_N^2 \rangle_{X_N', X_N} \leq 0$. Thus

$$\|u^1 - u^2\|_a^2 \leq \langle (\lambda_T^1 - \zeta) + (\zeta - \lambda_T^2), u_T^1 - u_T^2 \rangle_{X_T', X_T} \quad \forall \zeta \in -\partial_2 j(\mathcal{F}\lambda_N^1, u_T^2).$$

The result is implied by the monotonicity of the set-valued map $\partial_2 j(\mathcal{F}\lambda_N, u_T)$. ∎

In the case $d = 2$, it is possible to give the following result:

Proposition 4 *Assuming hypotheses (9), (10), (11) are satisfied and $d = 2$, if u is a solution to Problem (12) such that $u_T > 0$ and $C_3\|\mathcal{F}\|_a < 1$ then u is the unique solution to Problem (12) (When $\mathcal{F}$ is constant over Γ_C the condition reduces to $C_3\mathcal{F} < 1$).*

Proof. Let us assume that $\bar{u}$ is another solution to Problem (12), with $\bar{\lambda}_N$ and $\bar{\lambda}_T$ the corresponding contact stresses on Γ_C. Then from Lemma 3 one has

$$\|\bar{u} - u\|_a^2 \leq \langle \zeta - \lambda_T, \bar{u}_T - u_T \rangle_{X_T', X_T} \quad \forall \zeta \in -\partial_2 j(\mathcal{F}\bar{\lambda}_N, u_T).$$

Because $u_T > 0$, one has $\lambda_T = \mathcal{F}\lambda_N$ and $-\partial_2 j(\mathcal{F}\bar{\lambda}_N, u_T)$ contains $\mathcal{F}\bar{\lambda}_N$. Thus,

$$\begin{aligned} \|\bar{u} - u\|_a^2 &\leq \langle \mathcal{F}(\bar{\lambda}_N - \lambda_N), \bar{u}_T - u_T \rangle_{X_T', X_T} \\ &\leq C_3 \|\bar{\lambda} - \lambda\|_{-a,\Gamma_C} \|\mathcal{F}(\bar{u} - u)\|_a \leq C_3 \|\mathcal{F}\|_a \|\bar{u} - u\|_a^2, \end{aligned}$$

which implies $\bar{u} = u$ when $C_3\|\mathcal{F}\|_a < 1$. ∎

Of course, the same reasoning if valid for $u_T < 0$.

Let us now define the space of multipliers $M(X_T \to X_N)$ of the functions $\xi : \Gamma_C \to \mathbb{R}^d$ such that $\xi.\mathrm{n} = 0$ *a.e.* on Γ_C and such that the two following equivalent norms are finite:

$$\|\xi\|_{M(X_T \to X_N)} = \sup_{\substack{v_T \in X_T \\ v_T \neq 0}} \frac{\|\xi.v_T\|_{X_N}}{\|v_T\|_{X_T}}, \quad \text{and} \quad \|\xi\|_a = \sup_{\substack{v_T \in X_T \\ v_T \neq 0}} \frac{\|\xi.v_T\|_{a,\Gamma_C}}{\|v_T\|_{a,\Gamma_C}}.$$

It is possible to give the following more general result:

Proposition 5 *Assume hypotheses (9), (10), (11) are satisfied. Les u be a solution to Problem (12) such that $\lambda_T = \mathcal{F}\lambda_N\xi$, with $\xi \in M(X_T \to X_N)$, $\xi \in Dir_T(u_T)$ a.e. on Γ_C where $Dir_T(.)$ is the sub-derivative of the convex map $\mathbb{R}^d \ni x \longmapsto |x_T|$. and $C_3\|\mathcal{F}\|_a\|\xi\|_a < 1$. Then u is the unique solution to Problem (12).*

Proof. Let us assume that $\bar{u}$ is another solution to Problem (12), with $\bar{\lambda}_N$ and $\bar{\lambda}_T$ the corresponding contact stresses on Γ_C. Then from Lemma 3 one has

$$\|\bar{u} - u\|_a^2 \leq \langle \zeta - \lambda_T, \bar{u}_T - u_T \rangle_{X'_T, X_T} \qquad \forall \zeta \in -\partial_2 j(\mathcal{F}\bar{\lambda}_N, u_T).$$

Then a possible choice is $\zeta = \mathcal{F}\bar{\lambda}_N\xi$, and one has

$$\begin{aligned}\|\bar{u} - u\|_a^2 &\leq \langle \mathcal{F}\xi(\bar{\lambda}_N - \lambda_N), \bar{u}_T - u_T \rangle_{X'_T, X_T} \\ &\leq C_3\|\mathcal{F}\|_a\|\xi\|_a\|\bar{\lambda} - \lambda\|_{-a,\Gamma_C}\|\bar{u} - u\|_a \\ &\leq C_3\|\mathcal{F}\|_a\|\xi\|_a\|\bar{u} - u\|_a^2,\end{aligned}$$

which implies $\bar{u} = u$ when $C_3\|\mathcal{F}\|_a\|\xi\|_a < 1$. ■

References

1. R.A. Adams. *Sobolev spaces.* Academic Press, 1975.
2. L.-E. Andersson. Existence results for quasistatic contact problems with Coulomb friction. *Applied Mathematics and Optimisation*, 42 (2): pp 169–202, 2000.
3. G. Duvaut. Problèmes unilatéraux en mécanique des milieux continus. *in: Actes du congrès international des mathématiciens (Nice 1970), Tome 3, Gauthier-Villars, Paris*, pages pp 71–77, 1971.
4. G. Duvaut, J.L. Lions. *Les inéquations en mécanique et en physique.* Dunod Paris, 1972.
5. R. Hassani, P. Hild, I. Ionescu, N.-D. Sakki. A mixed finite element method and solution multiplicity for Coulomb frictional contact. *Comput. Methods Appl. Mech. Engrg.*, pages 4517–4531, 192 (2003).
6. P. Hild. An example of nonuniqueness for the continuous static unilateral contact model with Coulomb friction. *C. R. Acad. Sci. Paris*, pages 685–688, 337 (2003).
7. P. Hild. Non-unique slipping in the Coulomb friction model in two-dimensional linear elasticity. *Q. Jl. Mech. Appl. Math.*, pages 225–235, 57 (2004).
8. N. Kikuchi, J.T. Oden. *Contact problems in elasticity.* SIAM, 1988.
9. P. Laborde, Y. Renard. Fixed points strategies for elastostatic frictional contact problems. submitted.
10. V.G. Maz'ya, T.O. Shaposhnikova. *Theory of multipliers in spaces of differentiable functions.* Pitman, 1985.
11. Y. Renard. A uniqueness criterion for the Signorini problem with Coulomb friction. submitted.

Finite element/boundary element coupling for two-body elastoplastic contact problems with friction

A. Chernov, S. Geyn, M. Maischak, and E.P. Stephan

Institut für Angewandte Mathematik, Universität Hannover, Welfengarten 1, D-30167 Hannover, Germany,
chernov@ifam.uni-hannover.de

Abstract. We consider two-body contact problems in elastoplasticity (plasticity with isotropic hardening) with and without friction and present solution procedures based on the coupling of finite elements and boundary elements. One solution method consists in rewriting the problem with penalty terms taking care of the frictional contact conditions [4], see also [8]. Then, its discretized version is solved by applying the radial return algorithm for both friction and plastification. We perform a segment-to-segment contact discretization which allows also to treat friction. Another solution procedure uses mortar projections [2] together with a Dirichlet-to-Neumann (DtN) algorithm for the frictional contact part [6]; here we still use radial return for the plasticity part. Furthermore, extending the approach in [7] we can rewrite the contact problems with friction (given as variational inequalities without regularization) as saddle point problems and directly apply Uzawa's algorithm. Comments are given for adaptive procedures [5]. Numerical benchmarks are given for small deformations and demonstrate the wide applicability of the given methods.

1 Introduction

Multibody frictional contact processes, as well as processes with plastification are of great importance and have many applications in industry. Therefore, designing sufficiently accurate and efficient algorithms to solve them becomes one of the most successful and fast developing brunches of modern applied mathematics and engineering.

In the elastic case, the strong formulation of the contact problem can be reduced to a variational inequality over a convex set of admissible functions. Then a suitable discretization must be performed to achieve a convenient coupling of the solids.

[1] this research was supported by the DFG in GRK 615

[2] this research was supported by the DFG grant no. STE 573/5-1

The use of matching meshes on the contact interface is often a too strong and inconvenient restriction on the discretization. It will be natural to subordinate meshes to particular properties of the bodies, i.e. different geometries and material characteristics. The need to handle shifted meshes arises, for example, in case of finite deformations. If only conforming meshes are allowed, the bodies should be discretized again after each iteration step. Moreover, if the non-conforming meshes are forbidden, one has to give up such important tools as independent adaptive procedures in both bodies and independent automatic mesh generation. We discuss here the mortar approach to handle the contact interface, as well as segment-to-segment discretization for the penalty method. The mortar method is realized with DtN algorithm. The combination of Polyak and Uzawa methods is used to solve a contact subproblem with friction. The Newton method is used for the penalty approach; frictional contact conditions are realized with the return mapping algorithm.

As contact conditions represent the boundary interaction of the solids, plastic deformations as internal effects can be represented the same way for both mortar and penalty approaches. Plasticity with isotropic hardening is realized with the return mapping procedure and the Newton algorithm.

The use of FE and BE methods is possible for treating both contact and plastic behaviour. Therefore the FE/BE coupling combined with adaptive algorithms becomes a very flexible and powerful tool in modern mechanics.

2 The weak formulation for FE/BE coupling

Our model problem comes from the metal forming. Assume that we have an elastic stamp and an elastoplastic work piece. They are pushed together, so that plastic deformations occur in the work piece. Without loss of generality we call the stamp a slave body Ω^s, the work piece a master body Ω^m. Let $\Gamma^i = \Gamma^i_D \cup \Gamma^i_N \cup \Gamma^i_C$ be the boundary of $\Omega^i \subset \mathbb{R}^2$ for $i = s, m$. With symbols without superscripts we define the corresponding pairs consisting of slave and master, i.e. $\boldsymbol{u} := (\boldsymbol{u}^s; \boldsymbol{u}^m)$ in $\Omega^s \times \Omega^m$. Denote normal and tangential stress by

$$\sigma^i_n := \boldsymbol{n}^i \cdot \boldsymbol{\sigma}^i \cdot \boldsymbol{n}^i, \qquad \sigma^i_t \boldsymbol{t}^i := \boldsymbol{\sigma}^i \cdot \boldsymbol{n}^i - \sigma^i_n \boldsymbol{n}^i.$$

We identify the contact boundaries with a bijective mapping $R : \Gamma^s_C \longrightarrow \Gamma^m_C$, which acts in the direction of $\boldsymbol{n}^s$. Assume there is some initial gap g between the bodies given by R. Define normal and tangential jumps by

$$[u_n] := (\boldsymbol{u}^s(x) - \boldsymbol{u}^m(R(x))) \cdot \boldsymbol{n}^s, \qquad [u_t] := (\boldsymbol{u}^s(x) - \boldsymbol{u}^m(R(x))) \cdot \boldsymbol{t}^s.$$

Let $\hat{\boldsymbol{u}}_D$ be the prescribed displacements on $\Gamma^s_D \times \Gamma^m_D$ and $\hat{\boldsymbol{t}}_N$ be the prescribed tractions on $\Gamma^s_N \times \Gamma^m_N$. Denote the plastic behaviour of the work piece by the nonlinear operator $\mathcal{A}_{pl}$. We model it as plasticity with isotropic hardening. The explicit form of $\mathcal{A}_{pl}$ will be specified later. The strong formulation of the two body contact problem with Coulomb friction is given by

$$
\begin{aligned}
-\text{div } \sigma(\boldsymbol{u}^s) &= 0 && \text{in } \Omega^s,\\
-\text{div } \sigma(\mathcal{A}_{pl}\boldsymbol{u}^m) &= 0 && \text{in } \Omega^m,\\
\boldsymbol{u} &= \hat{\boldsymbol{u}}_D && \text{on } \Gamma_D^s \times \Gamma_D^m,\\
\sigma(\mathbf{u}) \cdot \mathbf{n} &= \hat{\boldsymbol{t}}_N && \text{on } \Gamma_N^s \times \Gamma_N^m,
\end{aligned}
\qquad (1)
$$

$$
\left.\begin{array}{c}
\sigma_n := \sigma_n^s = \sigma_n^m, \quad \sigma_t := \sigma_t^s = \sigma_t^m,\\
\sigma_n \leq 0, \quad [u_n] - g \leq 0, \quad \sigma_n([u_n] - g) = 0,\\
|\sigma_t| \leq \mu|\sigma_n|, \quad \sigma_t[u_t] + \mu|\sigma_n| \cdot |[u_t]| = 0
\end{array}\right\} \text{ on } \Gamma_C.
$$

The more simple Tresca friction model is given by

$$
|\sigma_t| \leq s, \quad \sigma_t[u_t] + s|[u_t]| = 0,
$$

where the function $s > 0$ is called "given friction" function. In order to obtain the BE weak formulation in the slave body we define the Steklov-Poincaré operator S, which is a Dirichlet-to-Neumann mapping. S is symmetric and positive definite on $\boldsymbol{H}^{1/2}(\Gamma^s)$ up to rigid body movements. It is defined as

$$
S := \frac{1}{2}(W + (K' + I)V^{-1}(K + I)) \, : \, \boldsymbol{H}^{1/2}(\Gamma^s) \to \boldsymbol{H}^{-1/2}(\Gamma^s).
$$

The boundary integral operators are given by

$$
\begin{aligned}
V\varphi(x) \quad &:= \int_{\Gamma^s} \varphi(y) G(x,y)\, ds_y, \quad Ku(x) \quad := \int_{\Gamma^s} u(y)(\mathcal{T}_y G(x,y))^T\, ds_y,\\
K'\varphi(x) &:= \mathcal{T}_x \int_{\Gamma^s} \varphi(y) G(x,y)\, ds_y, \; Wu(x) := -\mathcal{T}_x \int_{\Gamma^s} u(y) \mathcal{T}_y G(x,y)\, ds_y,
\end{aligned}
$$

where the fundamental solution for the Lamé equation in 2D is

$$
G(x,y) = \frac{\lambda + 3\mu}{4\pi\mu(\lambda + 2\mu)} \left\{ \log \frac{1}{|x-y|} I + \frac{\lambda+\mu}{\lambda+3\mu} \frac{(x-y)(x-y)^T}{|x-y|^2} \right\}.
$$

The variational formulation equivalent to (1) for FE/BE coupling is: Find $\boldsymbol{u} \in \boldsymbol{K}$:

$$
\beta(\boldsymbol{u}, \boldsymbol{w} - \boldsymbol{u}) + j([w_t]) - j([u_t]) \geq L(\boldsymbol{w} - \boldsymbol{u}) \quad \forall \boldsymbol{w} \in \boldsymbol{K}, \qquad (2)
$$

with the convex cone of admissible functions

$$
\boldsymbol{K} := \{\boldsymbol{u} := (\boldsymbol{u}^s; \boldsymbol{u}^m) \in \boldsymbol{H}_D^{1/2}(\Gamma^s) \times \boldsymbol{H}_D^1(\Omega^m) \, : \, [u_n] \leq g \text{ on } \Gamma_C^s \times \Gamma_C^m\}
$$

and

$$
\begin{aligned}
\beta(\boldsymbol{u}, \boldsymbol{w}) &:= \int_{\Gamma^s} S\boldsymbol{u}^s \cdot \boldsymbol{w}^s\, ds + \int_{\Omega^m} \sigma(\mathcal{A}_{pl}\boldsymbol{u}^m) : \nabla_s \boldsymbol{w}^m\, dx,\\
\boldsymbol{L}(\boldsymbol{w}) &:= \sum_{l=s,m} \int_{\Gamma_N^l} \hat{\boldsymbol{t}}_N \cdot \boldsymbol{w}^l\, ds, \qquad j([w_t]) := \int_{\Gamma_C} s|[w_t]|\, ds.
\end{aligned}
$$

A priori error estimates for pure FEM can be found in [1], numerical results for pure FEM in [6]. We studied a pure BEM discretization in [2].

As we mentioned in the introduction, it is often very convenient to have independent meshes in the slave and master body. In the two following sections we consider two methods allowing to preserve independent meshes.

3 Mortar approach

One possibility to perform data transfer through the contact interface is given by the mortar method. It was originally constructed for domain decomposition techniques, but suites to contact problems as well. The continuity of displacements and stresses is required in the weak sense and is done in terms of the so called normal and adjoint mortar projection, respectively.

In the following we need to define some functional spaces (see [1]). Assume that there are two trace meshes ξ_h^s, ξ_h^m induced by the independent discretizations of the stamp and the work piece. Let X_h^s, X_h^m be corresponding spaces of continuous piecewise linear functions on ξ_h^s, ξ_h^m. In the following we will not make any distinction between a function and the vector of its coefficients in the basis expansion of the discrete space. On ξ_h^s define a Lagrange multiplier space $M_h^s := \{\psi^s \in X_h^s : \forall T \in \xi_h^s,\ \partial\Gamma_C \in T \implies \psi_h|_T \in \mathcal{P}_0(T)\}$ and a mortar projection $\pi_h^s : X_h^m \to X_h^s$ such that $\forall \psi^s \in M_h^s$

$$\boldsymbol{u}^m - \pi_h^s(\boldsymbol{u}^m)|_{\partial\Gamma_C} = 0, \qquad \int_{\Gamma_C} (\boldsymbol{u}^m - \pi_h^s(\boldsymbol{u}^m))\, \psi^s\, ds = 0.$$

Its algebraic form is $\pi_h^s(\boldsymbol{u}^m) = D^{-1}B\boldsymbol{u}^m$, with a tridiagonal matrix D (as M_h^s and X_h^s are based on the same mesh ξ_h^s), and a sparse mass matrix B caused by testing with basis functions defined on ξ_h^s, ξ_h^m, which have local supports.

The boundary tractions are transferred by the adjoint operator π_h^{s*}, yielding for the adjoint mortar projection $\pi_h^{s*}(\boldsymbol{t}^s) = B^T D^{-T} \boldsymbol{t}^s$.

We use a **Dirichlet-to-Neumann algorithm** to solve problem (2).

1. Choose $\omega_D, \omega_N \in (0,1)$, set $\boldsymbol{X}_h^s \ni \boldsymbol{q}_0 := 0, \boldsymbol{X}_h^m \ni \boldsymbol{p}_1 := 0$
2. Solve elastoplastic problem with FEM:
   Find $\boldsymbol{u}^m \in \boldsymbol{H}_D^1$:
$$(\sigma(\mathcal{A}_{pl}\boldsymbol{u}^m), \nabla_s \boldsymbol{w}^m) = L^m(\boldsymbol{w}^m) - \langle \boldsymbol{p}_k, \boldsymbol{w}^m \rangle_{\Gamma_C} \qquad \forall \boldsymbol{w}^m \in \boldsymbol{H}_0^1 \tag{3}$$
3. Transfer obstacle, damping $\boldsymbol{q}_k = (1-\omega_D)\boldsymbol{q}_{k-1} + \omega_D \pi_h^s(\boldsymbol{u}_k^m)$
4. Solve elastic frictional contact problem with BEM:
   Find $\boldsymbol{u}^s \in \boldsymbol{K}_{\boldsymbol{q}_k} := \{\boldsymbol{u}^s : u_n^s - q_{kn} \le g\}$ such that $\forall \boldsymbol{w}^s \in \boldsymbol{K}_{\boldsymbol{q}_k}$
$$\langle S\boldsymbol{u}^s, \boldsymbol{w}^s - \boldsymbol{u}^s \rangle + j(w_t^s - q_{kt}) - j(u_t^s - q_{kt}) \ge L^s(\boldsymbol{w}^s - \boldsymbol{u}^s) \tag{4}$$
5. Compute residual $r_k^s \in X_h^s : \langle \boldsymbol{r}_k^s, \boldsymbol{w}^s \rangle := \langle S\boldsymbol{u}^s, \boldsymbol{w}^s \rangle - L^s(\boldsymbol{w}^s)$
6. Transfer scaled contact traction, damping
$$\langle \boldsymbol{p}_{k+1}, \boldsymbol{w}^m \rangle_{\Gamma_C} = (1-\omega_N)\langle \boldsymbol{p}_k, \boldsymbol{w}^m \rangle_{\Gamma_C} + \omega_N \langle \boldsymbol{\pi}_h^{s*}(r_k^s), \boldsymbol{w}^m \rangle_{\Gamma_C}$$
7. Repeat with 2, stop if $||p_{k+1} - p_k|| \le \mathrm{TOL}_{DtN} \cdot ||p_k||$

remIn the DtN algorithm we perform the separation of the nonlinearities. The two-body elastoplastic frictional contact problem is decomposed in two simpler one-body problems (elastic with friction/elastoplastic).

In order to solve the elastic problem with frictional contact we rewrite formulation (4) in an equivalent form with a Lagrange multiplier:
Find $\boldsymbol{u}^s \in \boldsymbol{K}_{\boldsymbol{q}_k}, \lambda_u \in \Lambda$:

$$\langle S\boldsymbol{u}^s, \boldsymbol{w}^s - \boldsymbol{u}^s\rangle + \int_{\Gamma_C} s\lambda_u (w_t^s - u_t^s)\, ds \geq L^s(\boldsymbol{w}^s - \boldsymbol{u}^s), \tag{5}$$

$$\lambda_u (u_t^s - q_{kt}) = |u_t^s - q_{kt}| \text{ a.e. on } \Gamma_C, \qquad \forall \boldsymbol{w}^s \in \boldsymbol{K}_{\boldsymbol{q}_k},$$

where $\Lambda = \{\lambda \in L^2(\Gamma_S) : |\lambda| \leq 1 \text{ a.e. on } \Gamma_C\}$. The product $s\lambda_u$ plays the role of the tangential contact traction. The Lagrange multiplier λ_u itself has the meaning of the sliding direction, if sliding occurs. Problem (5) is solved by the **Uzawa algorithm**.

1. Choose $\lambda^0 \in \Lambda$, $\rho > 0$
2. Solve frictionless contact with Polyak (modified CG) algorithm
   Find $\boldsymbol{u}_m^s \in \boldsymbol{K}_{\boldsymbol{q}_k}, \lambda_m \in \Lambda$:

$$\langle S\boldsymbol{u}_m^s, \boldsymbol{w}^s - \boldsymbol{u}_m^s\rangle \geq L^s(\boldsymbol{w}^s - \boldsymbol{u}_m^s) - \int_{\Gamma_C} s\lambda_k (w_t^s - u_{mt}^s)\, ds \qquad \forall \boldsymbol{w}^s \in \boldsymbol{K}_{\boldsymbol{q}_k}$$

3. Set $\lambda_{m+1} := P_\Lambda(\lambda_m + \rho\, s\, (u_{mt} - q_{kt}))$
4. Repeat with 2, stop if $||\lambda_{m+1} - \lambda_m|| \leq \mathrm{TOL}_U \cdot ||\lambda_m||$

As we have shown the Uzawa algorithm converges for sufficiently small ρ [3].

The elastoplastic problem in the work piece (3) is solved with the Newton-Raphson method using an incremental loading for the Neumann data on the contact boundary. The return mapping algorithm is used to keep the stress in the admissible range.

Our model problem represents the stamping procedure with isotropic hardening of the work piece (Section 2). In Fig. 1 we present results for FE/FE coupling. They should be compared with FE/BE results for the penalty approach (Fig. 2). We use bilinear basis functions on quadrilaterals and material parameters $E^s = E^m = 266926.0, \nu^s = \nu^m = 0.29$, $s = 0.22$; yield stress and hardening parameter for the work piece are $\sigma_Y^m = 4.0, h_Y^m = 450.0$. The bodies are coming into contact due to prescribed displacements on the bottom of the work piece $\hat{\boldsymbol{u}}_D = 0.96 \cdot 10^{-4}$. The damping parameters are $\omega_D = 0.5, \omega_N = 0.7$; the scaling parameter for Uzawa is $\rho = 8.264 \cdot 10^5$; the tolerances are $\mathrm{TOL}_{DtN} = \mathrm{TOL}_U = 10^{-6}, \mathrm{TOL}_N = 10^{-4}$.

The stress deviator norm is presented only for the work piece, as the stamp is assumed to be linear elastic. The brown region in it represents the maximum value of the stress deviator norm, i.e. the plastic region. Table 1 shows the number of DtN iterations depending on the damping parameters. The numbers of Uzawa iterations in the first DtN iteration are given in brackets. As the sliding direction is correctly recognized after the first DtN iteration, the Uzawa algorithm needs only 2 iterations in all later DtN iteration steps. The scaling parameter for Uzawa $\rho = 8.264 \cdot 10^5$ is chosen experimentally. It can be seen that the optimal values of the damping parameters ω_D, ω_N are between 0.5 and 0.7. For large damping parameters no convergence is observed.

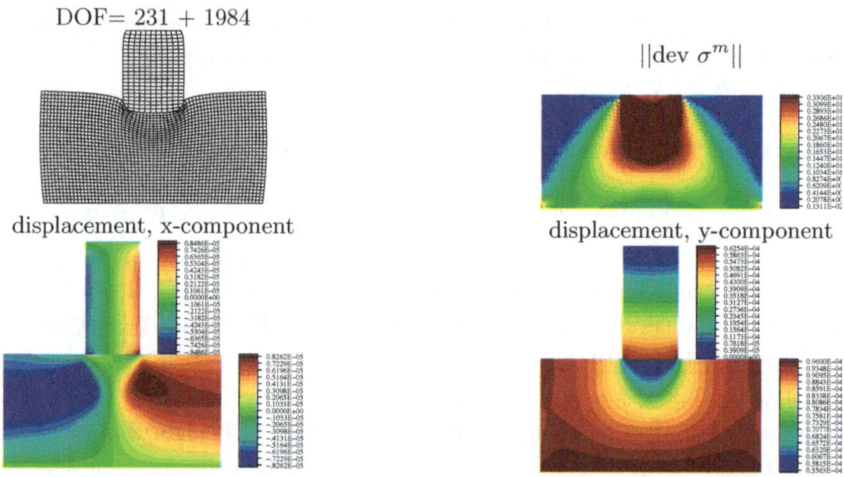

Fig. 1. Numerical experiments for mortar method with DtN algorithm.

Table 1. Mortar method: # DtN (Uzawa) iterations.

$\omega_D \setminus \omega_N$	0.3	0.5	0.7	0.9	1.0
0.3	27(34)	19(34)	13(34)	9(34)	9(34)
0.5	19(17)	13(17)	11(17)	-	-
0.7	12(13)	11(12)	-	-	-
0.9	13(9)	-	-	-	-

4 Penalty approach

The penalty approach allows us to rewrite variational inequality (2) as a variational equation. Setting the contact stress proportional to the penetration and penalizing with small parameters $\varepsilon_n, \varepsilon_t \ll 1$:

$$-\tilde{\sigma}_n(\boldsymbol{u}) := \frac{1}{\varepsilon_n}([u_n] - y)^+ \qquad \begin{aligned} -\tilde{\sigma}_t(\boldsymbol{u}) &:= sP_\Lambda(\frac{1}{\varepsilon_t}[u_t]) && \text{(Tresca)} \\ -\tilde{\sigma}_t(\boldsymbol{u}) &:= \mu|\sigma_n(\boldsymbol{u})|P_\Lambda(\frac{1}{\varepsilon_t}[u_t]) && \text{(Coulomb)} \end{aligned}$$

Now, the resulting weak formulation reads: Find $\boldsymbol{u} \in \boldsymbol{H}_D^{1/2}(\Gamma^s) \times \boldsymbol{H}_D^1(\Omega^m)$

$$\begin{aligned} \int_{\Gamma^s} S\boldsymbol{u}^s \cdot \boldsymbol{w}^s \, ds + \int_{\Omega^m} \sigma(\mathcal{A}_{pl}\boldsymbol{u}^m) : \nabla_s \boldsymbol{w}^m \, dx - \int_{\Gamma_C^i} \tilde{\sigma}(\boldsymbol{u}) \cdot [\boldsymbol{w}] \, ds \\ = \sum_{i=s,m} \int_{\Gamma_N^i} \hat{\boldsymbol{t}}_N \cdot \boldsymbol{w}^i \, ds \qquad \forall \boldsymbol{w} \in \boldsymbol{H}_D^{1/2}(\Gamma^s) \times \boldsymbol{H}_D^1(\Omega^m) \end{aligned}$$

This problem is solved by the Newton-Raphson method. The segment-to-segment discretization is used for computation of contact matrices. Note that

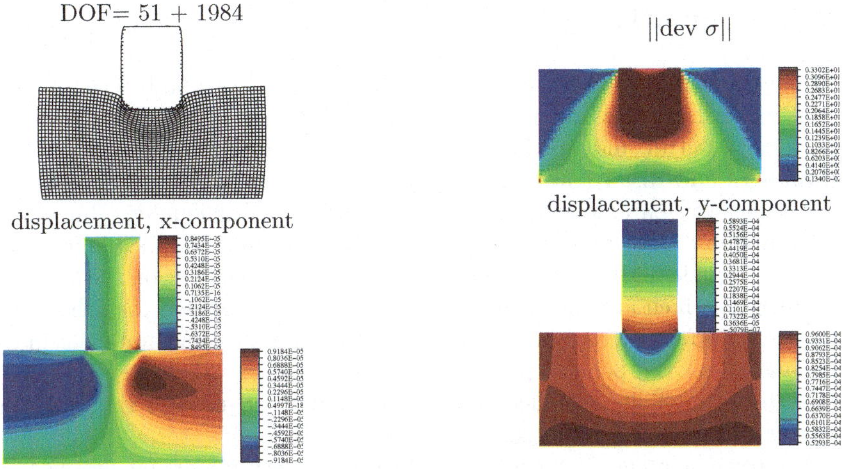

Fig. 2. Numerical experiments for penalty method with Newton-Raphson algorithm.

there are two nonlinearities in the problem due to frictional contact and due to plasticity. We use the return mapping algorithm for both plastification and frictional contact.

The numerical examples (Fig. 2) were done for the same parameters as for the mortar simulations, but the Coulomb friction law is used instead of given friction. The value $\mu = 0.2$ of the friction coefficient is chosen to have nearly the same maximal tangential displacements in the stamp. Piecewise linear continuous BEM discretizations are used in the stamp. Piecewise bilinear basis functions on quadrilaterals are used for FEM discretizations of the work piece. The tolerance is $\mathrm{TOL}_N = 10^{-4}$.

Table 2. Penalty method.

$1/\varepsilon_n$	$1/\varepsilon_t$	# Newton iterations	$\|\|([u_n] - g)^+\|\|_{L_2(\Gamma_C)}$
$20 \cdot E^m$	$10 \cdot E^m$	520	$0.8 \cdot 10^{-6}$
$10 \cdot E^m$	$5 \cdot E^m$	356	$0.15 \cdot 10^{-5}$
$5 \cdot E^m$	$2.5 \cdot E^m$	248	$0.29 \cdot 10^{-5}$
$2.5 \cdot E^m$	$1.25 \cdot E^m$	175	$0.55 \cdot 10^{-5}$

The mortar FE/FE and the penalty FE/BE approaches are in a good agreement: 5-8% difference in displacement, 2-3% difference in stress. The numbers of Newton iterations needed are given in Table 2. Note, smaller penalty parameters reduce the L_2 norm of the penetration function $([u_n]-g)^+$. But it increases the condition number of the Galerkin matrix as well as the number of Newton iterations.

5 Final remarks

We have considered two approaches for solving the two-body elastoplastic problem with frictional contact using the FE/BE coupling. Both methods are very powerful and are in a good agreement. Nethertheless, both have advantages and disadvantages. The mortar approach is independent of any kind of penalty parameters and solves the variational inequality. But a DtN iteration consists of nested (inner/outer) loops where the inner loops solve one-body problems. The convergence of DtN strongly depends on the damping parameters, which are not allowed to be sufficiently large. On the other hand the penalty approach consists of only one loop. Here the disadvantage is in the dependence on the penalty parameters. Their smaller values give physically more relevant results (smaller penetration) but they increase the condition number of the Galerkin matrix.

Adaptive strategies for contact problems are very important and use advantages of nonmatching meshes which are allowed here. In [5] we derive residual error indicators for FE and BE discretizations for linear elastic contact. That opens us a way to construct an adaptive approach for the two-body elastic problem with frictional contact using FE/BE coupling.

References

1. F. Ben Belgacem, P. Hild, P. Laborde, *Extension of the mortar finite element method to a variational inequality modelling unilateral contact*, Math. Mod. Meth. Appl. Sci., 9 (1999), pp. 287–303
2. A. Chernov, M. Maischak, E.P. Stephan, *A Mortar BEM for Two Body Contact Problems with Friction*, preprint no. 78, IfAM, 2005
3. A. Chernov, S. Geyn, M. Maischak, E.P. Stephan, *BEM for Contact Problems with Given Friction for Linear Elasticity*, preprint no. 79, IfAM, 2005
4. A. Chernov, S. Geyn, M. Maischak, E.P. Stephan, *FE/BE Procedures for Elastoplastic Contact Problems*, preprint no. 80, IfAM, 2005
5. A. Chernov, E.P. Stephan, *Residual A Posteriori Error Estimates for Contact Problems with Friction*, preprint no. 81, IfAM, 2005
6. R.H. Krause, B.I. Wohlmuth, *A Dirichlet-Neumann type algorithm for contact problems with friction*, Comput. Visual. Sci., 5 (2002), pp. 139–148
7. M. Maischak, E. P. Stephan, *A FEM-BEM coupling method for a nonlinear transmission problem modelling Coulomb Friction Contact*, Comp. Meth. Appl. Mech. Eng. 194 (2005), no. 2-5, 453-466
8. P. Wriggers, *Computational Contact Mechanics*, John Wiley & Sons, 2002

Part III

Contact models, results and applications

On the numerical simulation of non-smooth, resonant vibrations of delaminated structures

I. Müller, P. Vielsack, and K. Schweizerhof

Institut für Mechanik, Universität Karlsruhe (TH)
Kaiserstr. 12, D-76131 Karlsruhe,
mechanik@ifm.uni-karlsruhe.de

Abstract. Oscillation properties of delaminated structures are governed by dissipative impact-like contacts in the debonded region. The contribution focuses on the numerical simulation of this special type of contact. A robust and efficient contact strategy is presented mainly based on the theory of sudden impacts embedded in Finite Element methods.

1 Introduction

Composite laminates are being increasingly used as primary structural components in various engineering fields due to their inherently high specific mechanical properties. Correspondingly, a major interest lies in non-destructive testing of the structural integrity, particularly in relation to the occurrence of delaminations between adjacent plies as one of the most common failure modes in composite laminates.

The need for quantitative global damage detection methods has led to the development and continued research into vibration based methods [1]. Investigations of the nonlinear vibrational response are promising for localization and to quantify the size of the delamination on a global basis. The nonlinearity arises from a local contact phenomena, the clapping mechanism. The delaminated layer and the remaining part of a sandwich structure periodically strike against one another during the oscillation [2]. This dynamic contact-impact problem has as the simplest approximation two pendulums which are simultaneously excited by harmonic base excitation [2], [3]. Depending on the frequency and amplitude of excitation, a cascade of bifurcations with intermittent windows of irregular motions and different numbers of impacts during one excitation period can occur [2]. The essential point in this scenario is its dependency on the amount of energy dissipation at each impact [4].

In the following, a more realistic situation will be considered. The delaminated sandwich beam consists of two separated laminae with different cross sections. First, experiments are performed to provide a realistic reference for the oscillation behavior dominated by the continuously evolving lateral contact. Second, a mechanical model for this continuous system is proposed based on discretization into a system with multi degrees of freedom derived by Finite Element methods, whereas the fundamental challenge consists in the reliable capturing of the periodically appearing impact-like dynamic contacts within the stationary state of motion. It will be shown that standard procedures for contact modeling in FE methods as given by the penalty formulation are associated with a distinct sensitivity of the computed solution in regard to the regularization parameters to be chosen. In an effort to overcome these difficulties, a promising alternative approach of remarkable robustness is addressed for description of impact-like contacts which is mainly based on the theory of sudden impacts involving contact dissipation.

2 Experimental investigations

The mechanical system under consideration is depicted in figure 1 [a]. It consists of two laminae, whose shapes distinctly deflect from an ideal straight line in the statical stress-free state. Both laminae of length $685mm$ are clamped

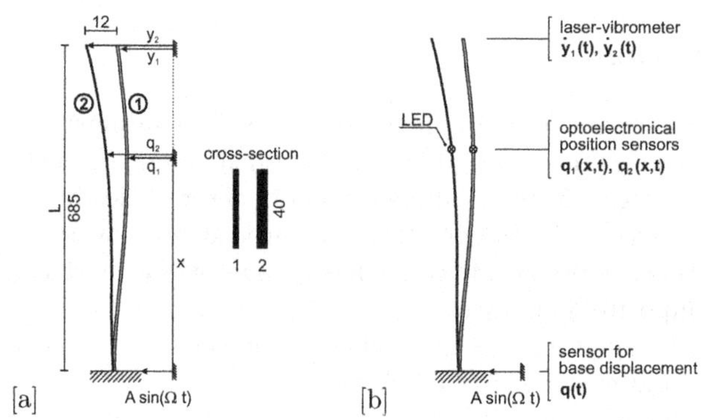

Fig. 1. Mechanical system: [a] geometry, [b] sensors.

at the lower ends. The upper ends are free. The maximum width of the gap at the upper ends at rest is $12mm$. Both laminae are made from aluminium. The cross-sections are $2 \times 40mm$ (thick lamina) and $1 \times 40mm$ (thin lamina), respectively. The displacements $q_1(x,t)$ and $q_2(x,t)$ at the distance x along the vertical at time t are absolute quantities. The base excitation $A\sin(\Omega t)$ is harmonic with amplitude A and driving angular frequency Ω. In the following, the amplitude $A = 2.5mm$ will be kept constant and only the frequency $\Omega = 2\pi f$ will be changed.

To characterize the properties of oscillations several sensors are used. Two optoelectronic position sensors give the distance of two corresponding LEDs to a vertical reference. In this way, the initial shape $q_{1,0}(x)$ and $q_{2,0}(x)$ of both laminae and the displacements $y_1(t) = q_1(L,t)$ and $y_2(t) = q_2(L,t)$ at the top will be monitored. For some selected excitation frequencies the deflection curves $q_1(x,t)$ and $q_2(x,t)$ during the oscillation are captured. Additionally, a laser vibrometer measures the velocity $\dot{y}_1(t)$ and $\dot{y}_2(t)$ at the upper end of both laminae. A position sensor at the lower end controls the amplitude of excitation A. Two stationary time-displacement responses with different

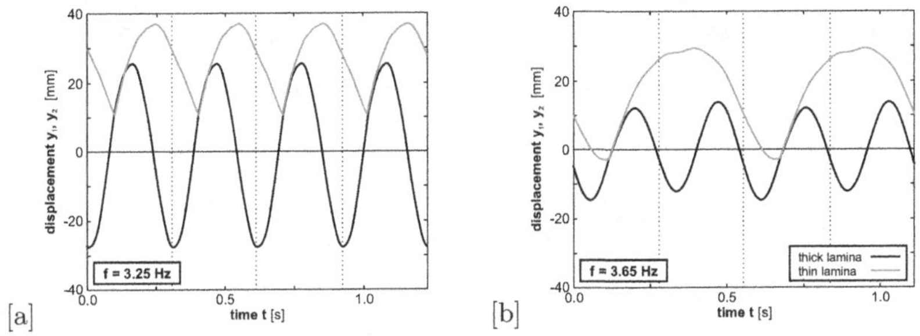

Fig. 2. Time-displacement curves for different excitation frequencies and constant amplitudes (experiment).

frequencies of excitation (the used values $f = \{3.25Hz, 3.65Hz\}$) are chosen from a frequency range of $0.5Hz \leq f \leq 4.0Hz$, where the lowest natural frequencies of both laminae can be found. The selected frequencies involve a strong contact interaction of both parts, due to the coupling of both subsystems.

The value $f = 3.25Hz$ lies in the vicinity of the resonance of the subsystem with coordinate y_1 (thick lamina). As can be seen in figure 2 [a], the thin lamina is beaten by the thick one. In contrast, an excitation frequency $f = 3.65Hz$ lies within a window of bifurcated oscillations. Period doubling occurs in such a way that one contact in two excitation periods can be found.

To explore the phenomenon of a continuously evolving contact line during the oscillation in more detail, Figure 3 depicts several states of the deflection shape during one period of excitation T. In fact, the contact duration is not constant for specific points along the beam axis. Despite this fact, in the following, the only considered point at the tip of the beams is supposed to be representative for the total motion. As can be seen (Fig. 2), the time interval of contact during one period of excitation is relatively large. Such a phase of common motion of both laminae is only possible if the initial impact, which initiates the contact phase, is highly dissipative. Otherwise, both laminae would separate immediately.

Figure 4 captures the velocities $\dot{y}_1(t)$ and $\dot{y}_2(t)$ of both laminae within two periods of excitation with $f = 3.25Hz$. The discontinuity (jump) in the velocities at the beginning of contact on the considered point is clearly visible. It is evident that this phenomenon can be described by classical impact theory.

Fig. 3. Operation deflection shape during one excitation period T, $f = 3.25Hz$ (experiment).

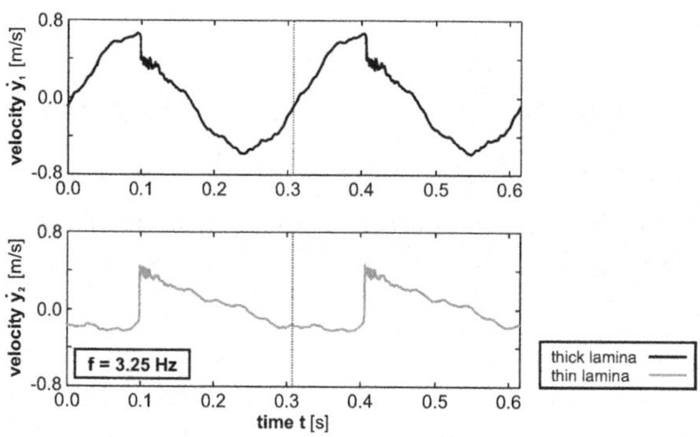

Fig. 4. Impact-dominated velocities of both laminae at the free ends in the delamination problem (experiment).

This theory links the relative velocity $\dot{Q}(t^{\langle c\rangle -}) = \dot{q}_u(t^{\langle c\rangle -}) - \dot{q}_v(t^{\langle c\rangle -})$ of the impacting masses m_u, m_v before contact and the relative velocity $\dot{Q}(t^{\langle c\rangle +}) = \dot{q}_u(t^{\langle c\rangle +}) - \dot{q}_v(t^{\langle c\rangle +})$ after contact by a coefficient of restitution e

$$\dot{Q}(t^{\langle c\rangle +}) = -e\,\dot{Q}(t^{\langle c\rangle -})\,. \tag{1}$$

The corresponding coefficient of restitution $0 \leq e \leq 1$ depends on the amount of energy dissipation at contact time $t^{\langle c\rangle}$. To get this coefficient, the velocities $\dot{y}_1(t^{\langle c\rangle -})$, $\dot{y}_2(t^{\langle c\rangle -})$ before impacting at the tip and $\dot{y}_1(t^{\langle c\rangle +})$, $\dot{y}_2(t^{\langle c\rangle +})$ after the impact are taken from the experiment. The ratio of masses is approximately given by the ratio of cross-sections as $m_1/m_2 = 1/2$. Within the scope of the experimentally gained data, the well-known formulas of NEWTON's impact law reveal a range of $0 \leq e \leq 0.1$ for the restitution coefficient. This result confirms the large contact damping arising at lateral impacts on beams.

3 Computation by finite element methods

For the numerical calculation, the one-dimensional continuum problem is transformed into two multi-DOF systems by FE methods. Both laminae are discretized by the same number of EULER-BERNOULLI beam elements. This leads to the semi-discrete equations of motion as two linear systems with $2n$ DOF

$$\begin{aligned} \mathbf{M_1}\ddot{\mathbf{q}}_1(t) + \mathbf{C_1}\dot{\mathbf{q}}_1(t) + \mathbf{K_1}\mathbf{q}_1(t) &= \mathbf{f}_1(t) \\ \mathbf{M_2}\ddot{\mathbf{q}}_2(t) + \mathbf{C_2}\dot{\mathbf{q}}_2(t) + \mathbf{K_2}\mathbf{q}_2(t) &= \mathbf{f}_2(t) \end{aligned} \quad . \tag{2}$$

The generalized coordinates given in eq. (2) can be subdivided into translatory and rotational DOFs leading to the displacements u and the rotations φ

$$\mathbf{q_i} = \left[u_{i,1}, \varphi_{i,1}, \ldots, u_{i,n/2}, \varphi_{i,n/2} \right]^T \quad \text{with } y_i = u_{i,n/2} \qquad i = 1(1)2 \, . \tag{3}$$

A prescribed displacement as a harmonic base excitation is applied to the lowest node of each subsystem. Additionally, a clamped support is assumed on these nodes. Coupling of both subsystems, represented by the two sets of equations (2), occurs because of several lateral contact events along the longitudinal axis of the beams. This fact causes the strong nonlinearity of the problem. The continuous evolution of the contact line must be separated into several node-to-node contacts of opposite nodes. During the oscillation the succession of contact points and the corresponding time when contacts occur are unknown *a priori*. These are typical features of a non-smooth dynamic system.

Dealing with non-smooth dynamic systems within FE methods, two fundamental challenges appear, namely to find an appropriate problem-oriented procedure of time integration and to capture the respective contact situation within numerical contact formulation. The first one is connected with the accuracy of detecting the switching times (contact times) of the non-smooth system to be integrated [5]. In general, in view of the numerical time integration with a certain discrete time step Δt only a limited degree of accuracy can be achieved in regard to determine the contact times. This leads to permanent numerical disturbances in the course of the motion. The undisturbed problem is orbitally stable as long as these disturbances remain below a critical limit. Thus, a sufficiently small time step Δt is needed for an orbital stable solution. The present contribution will not focus to the problem of an appropriate time integration. Here, the standard NEWMARK scheme ($2\alpha_N = \beta_N = 1/2$) is adopted. Only the problem of the contact formulation will be discussed in the following.

3.1 Contact Description via Penalty Method

Standard procedures in FE methods for contact modeling are the Augmented-LAGRANGE method and the penalty method. The main problem of both methods is the inclusion of the strongly dissipative character of the dynamic contact. Applications of the first procedure are not known in this field. The classical penalty method can be extended by introducing penalty damping in addition to the penalty stiffness. Thus, this procedure requires the choice of two penalty parameters, which have no physically specifiable quantity [6]. Moreover, these parameters are mutually dependent with the time step of integration [3]. A parameter estimation is only possible by comparing numerical results with an experimental reference for stationary oscillations [4].

Suppose that the system contains n_N possible pairs of contact nodes, then the actual state of contact is given by a set of indices $\mathcal{M}_C$ from all n_C contact points, which are closed at present time t. In the considered case of a time-discrete system with constant time steps Δt, a contact event is captured by the condition

$$g_{N,k_C} = u_{2,k_C} - u_{1,k_C} \leq 0\,, \quad \forall k_C \in \mathcal{M}_C\,. \tag{4}$$

Then the overall state of contact on actual time t reads

$$\begin{aligned} \mathcal{M}_N &= \{1(1)n_N\} \\ \mathcal{M}_C(t) &= \{k_C \in \mathcal{M}_N \mid g_{N,k_C} \leq 0\} \end{aligned}\,. \tag{5}$$

After each time step Δ the contact condition is checked. For all contact points contained in the index set $\mathcal{M}_C$ penalty stiffness c_k and penalty damper c_d must be applied. As a benchmark problem the time response of the experiment with excitation frequency $f = 3.25Hz$ (Fig. 2) is considered. Figure 5 contains a parametric plane that separates regions containing the correct type of solution from regions of distorted types of motion arising from an incorrect set of parameters. For this purpose, the numerical results computed for certain sets of parameters c_K, c_D have been categorized in comparison to the experimental results. Figure 5 reveals a distinct parameter sensitivity of the numerical

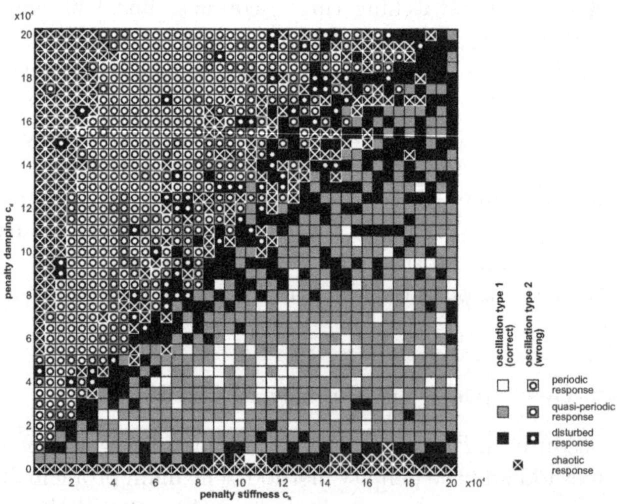

Fig. 5. Parametric plane for sets of penalty parameters c_K, c_D.

solution in regard to the regularization parameters. All sets of parameters taken from an upper left triangle in the parameter space (see Fig. 5) are associated with an incorrect numerical solution - a bifurcated oscillation. In the treated parameter range only in a few cases (Fig. 5: white bricks) a qualitatively correct result is obtained. Without information about the expected motion, a decision is not possible, whether the chosen set of penalty parameters is correct. Furthermore, treating another type of oscillation, a calibration of the penalty parameters c_K, c_D is required. Additionally, the parameter set is mutually dependent from the time step of integration. Due to these facts, the adopted penalty regularization could not be considered as an appropriate contact description for the treated type of contact, in particular with regard to the prognosis character of the simulation technique for purpose of damage identification .

3.2 Contact description via impact formulation

Sudden impacts accompanied by strong energy dissipation can be modeled by the classical theory of impact in which the velocities during a contact event are controlled by a law of impact, e.g. NEWTON's impact law. This procedure, which has been proven in the field of rigid body dynamics, can be adopted for description of dissipative contact within the Finite Element methods. Applying an impact law, sudden change of the velocities of the contact nodes within a vanishing time interval is assumed. At each pair of contact nodes the law of impact must be applied.

In contrast to applying a law of impact with vanishing contact duration, the penalty method supposes a phase of permanent contact with minimal duration of one time step Δt for each contact event. Thus, a partial state of permanent contact is included herein. As a main advantage for computation, this regularization allows a description of permanent contact while the number of DOFs is kept unaltered in comparison to the freely vibrating system.

In the following, the two methods are connected in a way that permits both a sudden impact and a motion in permanent contact. Therefore, the law of impact is employed to capture the impact-like contact phenomena showing a sudden change of the velocities as well as to observe energy dissipation during contact. Contact stiffness is only detached from the penalty method. In the case of permanent contact, this stiffness is adopted. Initially, two parameters are required (coefficient of restitution e, contact stiffness c_K), whereas the range of the main parameter e can be determined by experiments (see fig. 4). Since contact dissipation is solely captured by the law of impact, the two parameters are almost independent from one another, providing an additional advantage compared to the classical penalty method.

The system is governed by three partial states in motion, namely separated motion, sudden impact and permanent contact. Again, the actual state of contact at present time t can be observed by several sets of indicies. For this purpose, eq. 5 must be extended to capture all possible cases.

$$\begin{aligned}
\mathcal{M}_N &= \{1(1)n_N\} \\
\mathcal{M}_C &= \{k_C \in \mathcal{M}_N \mid g_{N,k_C} \leq 0\} \\
\mathcal{M}_M &= \{k_M \in \mathcal{M}_C \mid \dot{g}_{N,k_M} \geq \gamma \wedge (k_M \notin \mathcal{M}_{P0}(t_{j-1}) \vee k_M \notin \mathcal{M}_{pp}(t_{j-1}))\} \\
\mathcal{M}_{P0} &= \{k_{P0} \in \mathcal{M}_C \mid \dot{g}_{N,k_{P0}} < \gamma \wedge k_{P0} \notin \mathcal{M}_{PP}(t_{j-1})\} \\
\mathcal{M}_{PP} &= \{k_{PP} \in \mathcal{M}_C \mid k_{PP} \in \mathcal{M}_{P0}(t_{j-1}) \vee k_{PP} \in \mathcal{M}_{PP}(t_{j-1})\}
\end{aligned}$$

$$\text{with}\quad \begin{aligned}
\mathcal{M}_N &: \text{possible contact points } (n_N \text{ elements}) \\
\mathcal{M}_C &: \text{actually closed contact points } (n_C \text{ elements}) \\
\mathcal{M}_M &: \text{sudden impact with immediate separation } (n_M \text{ elements}) \\
\mathcal{M}_{P0} &: \text{beginning of permanent contact } (n_{P0} \text{ elements}) \\
\mathcal{M}_{PP} &: \text{persistent permanent contact } (n_{PP} \text{ elements})
\end{aligned} \tag{6}$$

The condition for the occurrence of a contact event $g_{N,k_C} \leq 0$ can be kept unmodified in relation to eq. (4). Therefore, the law of impact

$$\begin{aligned}
\dot{\mathbf{u}}_1^{\langle t_j \rangle}(t_j{}^+) &= \dot{\mathbf{u}}_1^{\langle t_j \rangle}(t_j{}^-) + \left(\mathbf{I} + \mathbf{e}^{\langle t_j \rangle}\right) \mathbf{M_2}^{\langle t_j \rangle} \left(\mathbf{M_1}^{\langle t_j \rangle} + \mathbf{M_2}^{\langle t_j \rangle}\right)^{-1} \dot{\mathbf{g}}_N^{\langle t_j \rangle}(t_j{}^-) \\
\dot{\mathbf{u}}_2^{\langle t_j \rangle}(t_j{}^+) &= \dot{\mathbf{u}}_2^{\langle t_j \rangle}(t_j{}^-) - \left(\mathbf{I} + \mathbf{e}^{\langle t_j \rangle}\right) \mathbf{M_1}^{\langle t_j \rangle} \left(\mathbf{M_1}^{\langle t_j \rangle} + \mathbf{M_2}^{\langle t_j \rangle}\right)^{-1} \dot{\mathbf{g}}_N^{\langle t_j \rangle}(t_j{}^-)
\end{aligned}$$
$$\text{with}\quad \begin{aligned}\mathbf{M_i}^{\langle t_j \rangle} &= \operatorname{diag}\left([\bar{m}_{i,kk}]\right) \\ \mathbf{e}^{\langle t_j \rangle} &= \operatorname{diag}\left([e_k]\right)\end{aligned} \qquad i = 1(1)2\,,\ \forall k \in \mathcal{M}_C(t_j) \vee \mathcal{M}_{P0}(t_j) \tag{7}$$

must be applied on all pairs of contact nodes contained in the sets of indicies $\mathcal{M}_M$, $\mathcal{M}_{P0}$. Herein $\bar{m}_{i,kk}$ denotes the masses affecting the impacting nodes at time $t^{\langle j \rangle}$, which can be estimated from the corresponding translatory DOFs in the diagonalized mass matrix. Furthermore, the diagonal matrix $\mathbf{e}^{\langle t_j \rangle}$ contains the coefficients of restitution e_k. In the following, the restitution coefficient is assumed to be constant for all pairs of contact nodes. Then, only one parameter e is needed, which is known from the preceding experimental investigation.

A sudden impact marks the beginning of each contact event. In the case of a vanishing translatory, relative velocity $\dot{g}_N = 0$, checked at each pair of contact nodes contained in $\mathcal{M}_c$, a partial state of permanent contact follows for these contact nodes. For all non-vanishing coefficients of restitution e, this switching condition can lead to a sequence of numerous impacts [7]. To avoid these phenomena, a small threshold γ for the relative velocity is introduced to assign the beginning of motion in permanent contact. Therefore, the weak inequality

$$\dot{g}_{N,k_{P0}} = \dot{u}_{2,k_{P0}} - \dot{u}_{1,k_{P0}} < \gamma\,, \quad \forall k_{P0} \in \mathcal{M}_{P0}\,,\ \gamma \ll 1 \tag{8}$$

allows the decision on whether a state of permanent contact follows on impact. In this case, the penalty stiffness c_K is added on the corresponding contact nodes indicated by the set of index $\mathcal{M}_{P0}$. Contact stiffness c_K for the corresponding pair of contact nodes remains in the system as long as the separation condition ($g_{N,k} = u_{2,k} - u_{1,k} > 0$, $k \in \mathcal{M}_{PP}$) is not satisfied. In the following, computations are presented for the two different types of oscillation whose

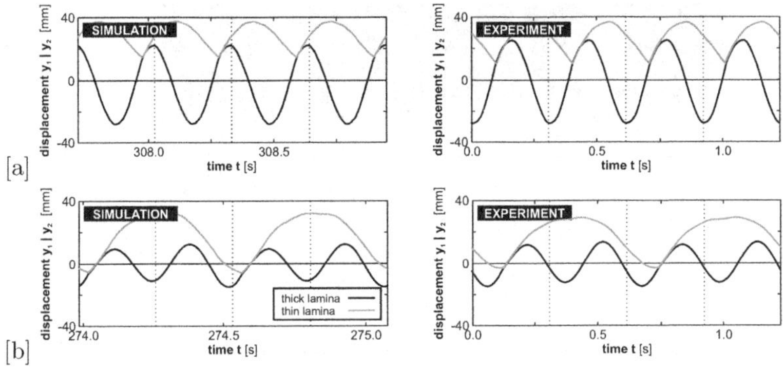

Fig. 6. Time-displacement diagram for the stationary motion of both lamina for the two excitation frequencies: [a] $f = 3.25Hz$, [b] $f = 3.65Hz$.

corresponding experimental results are given in Fig. 2. According to the experimental results, the coefficient of restitution as the main contact quantity is chosen as $e = 0.10$. Two additional numerical parameters have to be chosen: the penalty stiffness c_K and the threshold γ for permanent contact. Contact stiffness is set as $c_K = 100$, which is reasonable in this analysis. The threshold of the relative velocity is chosen in such a way that $\gamma = 0.0005$ is close to the thousandth part of the maximum relative velocity. These values are fixed for all cases treated above.

Considering the time histories (see Fig. 6), in all cases excellent agreement of experimental and numerical results can be noted in relation to both the amplitudes and the shape of oscillation. The main advantages over the classical penalty method are evident from this investigation. First, the main parameter e is experimentally determinable. The additionally required penalty stiffness c_K exhibits in this connection a very limited influence on the result. Moreover, the chosen set of parameters is applicable without modification to describe all treated types of motion that exhibit a wide variety. Such behavior cannot be expected from the spring-damper regularization by penalty method.

Furthermore, the described contact formulation shows a remarkable robustness relating to the choice of parameters e, c_K which can be found from the parametric plane of Figure 7. However, apart from a few exceptions, the basic type of response denoted by one contact event in one excitation period is preserved for all cases considered. When estimating the influence of the two parameters, the restitution coefficient mainly affects the solution in the treated range. Compared to the high sensitivity observed on classical penalty-regularization (see Fig. 5), the presented approach provides a promising alternative due to its robustness.

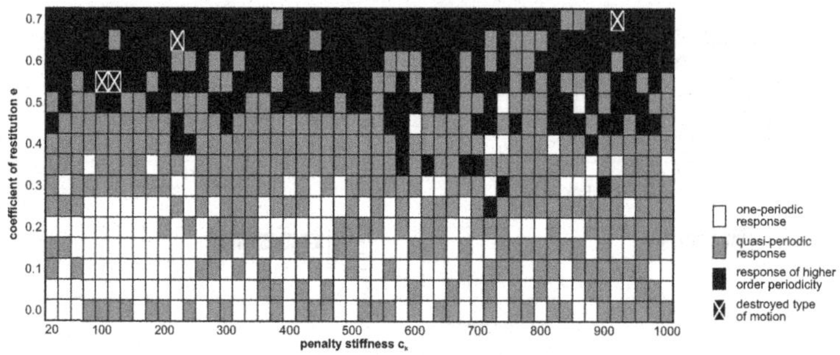

Fig. 7. Parametric plane for sets of parameters e, c_K.

4 Conclusions

Forced oscillations of delaminated structures are dominated by continuously evolving impact-like contacts. First, the arising contact phenomena are studied experimentally on a realistic model situation for the delamination problem. Depending on the frequency of excitation, a bifurcation scenario occurs. The experiments reveal a distinct amount of energy dissipation during the contact.

Armed with this knowledge, an appropriate contact description for numerical simulation using Finite Element methods is addressed mainly based on the theory of sudden impacts involving contact dissipation. In contrast to the classical penalty method, the main parameter, namely the coefficient of restitution, can be determined easily by experimental means. Computational results document the robustness of the procedure in relation to the choice of parameters. In most all if the cases considered above, the type of the experimental reference solution was preserved numerically within a widely open range of parameters. Furthermore, the set of parameters can be kept constant for all of the very different types of motion previously considered.

References

1. Zou L, Tong L, Steven GP (2000) Journal of Sound & Vibration 230(2):357–378
2. Vielsack P (2002) Journal of Sound & Vibration 253(2):347–358
3. Müller I, Vielsack P (2003) Journal of Computational & Applied Mechanics 4(2):173–186
4. Müller I, Konyukhov A, Vielsack P, Schweizerhof K (2005) Engineering Structures, 27(2):191–201
5. Vielsack P, Hartung A (1999) ZAMM 79(6):389–397
6. Engleder T, Vielsack P, Schweizerhof K (2002) Computational Mechanics 28:162–168
7. Valente AX, McClamroch NH, Mezi I (2003) International Journal of Non-Linear Mechanics 38(5):677–689

Mesoscopic particles – a new approach for contact and friction dynamics

G.-P. Ostermeyer

Institute of Dynamics and Vibrations, Technical University of Brunswick, P.O. Box 3329, 38 023 Braunschweig, Germany

Abstract. Within the classical dynamics of mechanical systems physical effects in boundary layers of contacting bodies are fundamental. Examples are impact motion or frictional contact. Usually the dynamics in boundary layers is given on other time and length scales than the dynamics of the system itself.

This paper deals with a mesoscopic particle discretisation of such contact zones or surfaces. These particles have hidden degrees of freedom from macroscopic point of view, which are separated only on a microscale. To be correct within the thermodynamic properties we have to use different time and length scale dynamics in these particle systems.

In frictional contact this particle discretisation can be used for detecting the heat generation and even wear properties. The research on frictional effects in boundary layers of contacting bodies was the root of developing the mesoscopic particle method.

1 Introduction

The atomic structure of material generates in natural way a multi scale problem. Usually methods as Multibody Systems or Finite Element discretisation are used for macroscopic models. The dynamic of microscopic structures like crystallites or even atoms in the material bodies take care of different atomic forces and describes for instance Brownian motion. This micro- or nanoscopic motion in detail is not of interest for macroscopic applications. But in macroscopic models the microscopic cosmos is taken into account approximately by introduction of elastic or dissipative forces or by introduction of thermodynamic variables, to describe microscopic dynamics in terms of heat and heat transfer. As the history shows, this is a very successful way to describe and to analyse macroscopic models in our world of experience.

But these approximations are not very successful for some questions. This is true for dissipative (plastic material) behavior and especially for frictional effects. On the atomic scale there is no energy dissipation but several mechanism of energy distribution mechanisms. There are very powerful tools to analyse the molecular dynamics (NEMD) even far away from the thermodynamic equilibrium in the model. But coupling these simulation methods with macroscopic simulation tools is not possible in principle. There is a gap of about 10^8 with respect to length and time scales. Long-time simulations of MD programs results in dynamics of some cubic nanometer for some nanoseconds. This is the reason for disjunctive developments and investigations in macrotribology and microtribology up to now.

A possible way out is the introduction of superatoms, to describe in parts or totally macroscopic models by particle methods. This has been firstly done by Boltzmann [1] 1890. He did investigations on elastic lattices. In 1917 Tomlinson [2] published a well known paper on the interaction of elastic point lattices explaining friction phenomena by for instance elastic hysteretic effects. Throughout the seventies and eighties especially Greenspan [3],[4] did many numerical investigations on macroscopic particle systems. The forces between these particles where generated by generalized Lennard – Jones potentials, allowing even the simulation of plastic material behavior, crack growing effects and fluidal motion beside the classical vibrational dynamics.

Main problem of these macroscopic particle discretisations are thermodynamic effects. The interpretation of local kinetic energy on a macroscopic scale as heat quantity leads to "heat waves ", which are reflected at the limit surface of the system. Of course the distribution of temperature in the particle field has to be parabolic rather than hyperbolic. In MD - simulations usually statistical methods are used to connect irreversible heat and particle motion and to avoid violations of thermo dynamical laws. But this is not applicable for macroscopic particle systems because the necessary time constants of simulation and numerically measurements would enforce only static systems.

The method of mesoscopic particle systems [5,6,7,8] uses a very new way of description of thermodynamical variables and their connection with mechanical degrees of freedom in the system. It uses hidden mechanical degrees of freedom, so called sub mechanical degrees of freedom and inner variables in each particle.

2 Heating of a rod by beating

This is a very simple example to uncover the methods of mesoscopic particles. Let us assume, there is a metallic cantilever. A well known experiment is the following. When the cantilever is beaten at the free end in longitudinal direction, there will occur small plastic deformations but mainly a heating up of the free end. The task is to describe this process.

A minimal model has to take care only for the heat quantity, which is brought in by some beatings with the hammer. No mechanical degrees of freedom in the rod are needed, because wave phenomena or plastic deformations in the rod are not of interest in this example (Fig. 1).

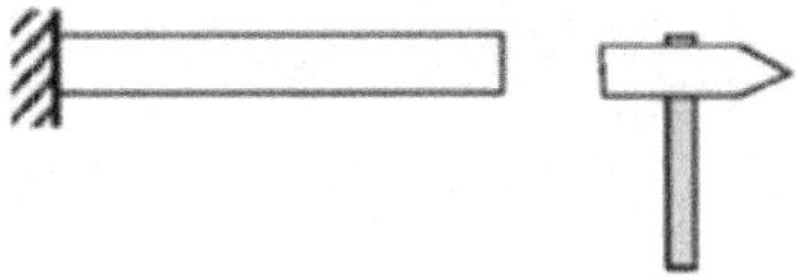

Fig. 1. Heating a rod.

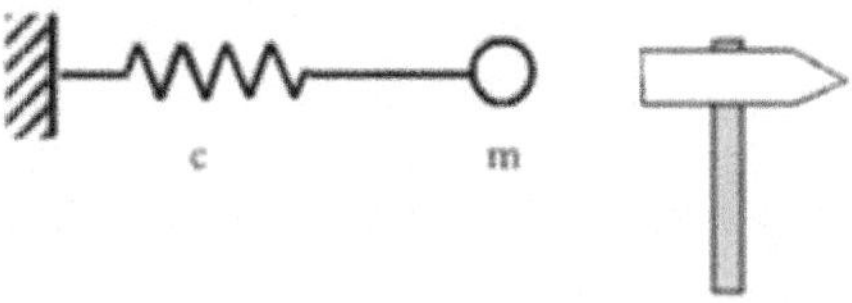

Fig. 2. A particle discretisation.

So this system is modeled by a one dimensional particle, elastically coupled to the wall. The vibrational degree of freedom of this particle should be a submechanical degree of freedom (Fig. 2). Only the energy is a macroscopic observable in the system.

The system starts with the particle at rest. The energy E is zero.

$$m\ddot{x} + cx = 0 \tag{1}$$

A beat results in vibrational motion of the particle. The energy quantity ΔE added to the system should be interpretated as heat W_1 after one beat:

$$W_1 = \Delta E. \tag{2}$$

When the same beat is repeated some time later, the total energy W_2 depends of course essentially on the phase and amplitude of the particle, when the hammer touch it. So after the second beat the energy could be

$$W_2 \in [0, 2\Delta E] \tag{3}$$

The second beat could make the rod cold again. This is not true and violates the thermodynamical laws. To avoid this situation, techniques of MD simulations can be used. So the hammer would be transformed to a statistical hammer, which can't see the phase of the particle vibration any more and interact only with the energy of the particle.

Another possibility is to transform the vibrational energy into heat by a viscous damper. With the mass m, the specific heat capacity c_v and the temperature T the heat energy W is

$$W = mc_vT. \tag{4}$$

The damper dissipates the vibrational energy. This energy could be summarized to the heat energy.

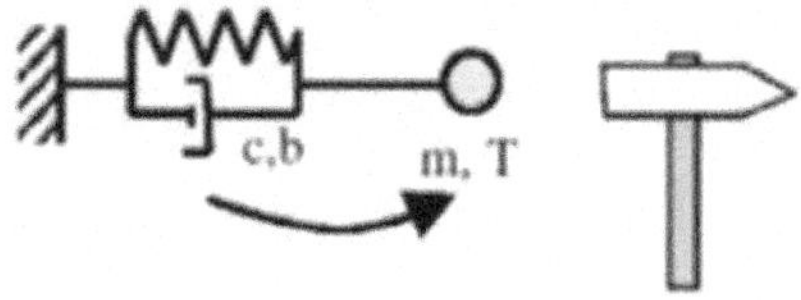

Fig. 3. A mesoscopic particle discretisation.

$$m\ddot{x} + b\dot{x} + cx = 0 \qquad \dot{T} = -\tfrac{1}{mc_v} b\dot{x}^2 \tag{5}$$

The temperature T is an additional variable of the particle. By this technique of damping out the vibrational energy the time to detect the heat quantity added by the next beat is of order of the oscillation period of this system. This is the minimal time for occurrence of beats to separate them as different or single beats. Because the vibration motion should be a submechanical degree of freedom, this frequency should be smaller than the highest frequency of interest in the macroscopic system.

This system of third order owns a first integral, which is the sum of kinetic, potential and heat energy:

$$\frac{1}{2}m\dot{x}^2 + \frac{1}{2}c\dot{x}^2 + mc_vT = const. \tag{6}$$

Usually the system is not adiabatic, heat flow and radiation has to be taken into account. Another problem occurs, when elastic waves should be detected too in this rod.

3 Energy separation in a linear chain

The rod in Fig. 1 is now discretizised in 100 points with equal mass, where each direct neighbors are coupled by a linear spring.

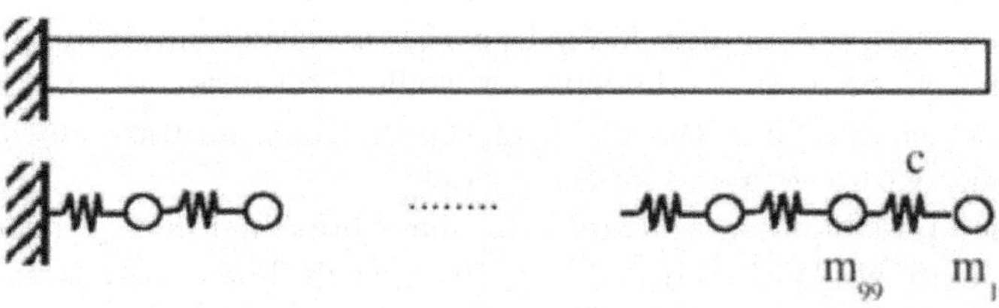

Fig. 4. rod, discretized with 100 discrete mass points.

The task now is to detect not only the heat but also the mechanical vibrations. Of course the problem hereby is to separate the heat from the elastic waves in the mass point-lattice. A "natural" isolation of the energies is possible using the dispersion-properties of the particle models. In the measurement technology this effect is regarded by Shannon's theorem for the selection of meaningful signals on channels with limited bandwidth.

Similar to that, elastic waves can be detected on the discrete mass pointlattice. This basic idea is illustrated by the linear chain in Fig. 4. In the following the displacements in the longitudinal direction are diagrammed perpendicular to the longitudinal axis.

The initial displacement of only the 50th particle applies the energy especially to the highest eigenforms. The dispersion results in stochastic vibrations with the diffuse energy propagation, corresponding with Schrödinger's computations [10] (Fig 5).

An initial sinusoidal displacement configuration produces, as expected, d'Alembert waves (see Fig. 6). The mechanical energy thereby spreads out hyperbolically as well (see Fig. 7). This is not the case starting with a Dirac displacement (Fig. 8). That diffuse energy propagation can approximatively depicted with the heat conduction. A temperature variable can analytically be given to each particle in the chain, thereby an energy integral as shown above is attached to the particle.

Beside the particles' interaction via the vibration they also interact by the inner temperature variables. According to its macroscopic character the classical parabolic thermal conduction is assumed here.

Half of the energy that is dissipated in the dampers will be added to each of the both neighbored particles (see Fig.9).

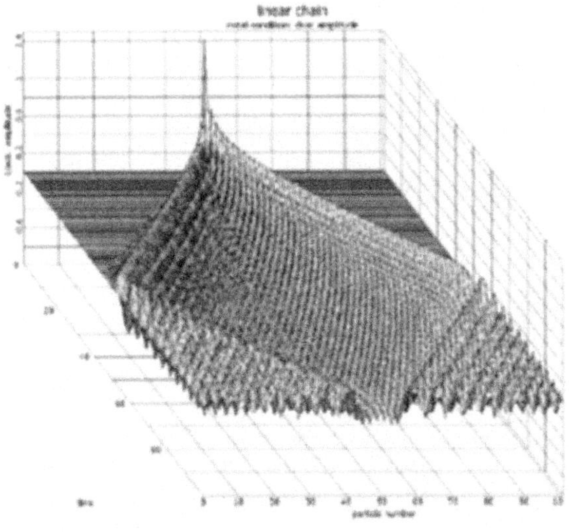

Fig. 5. Amplitudes after an initial Dirac amplitude.

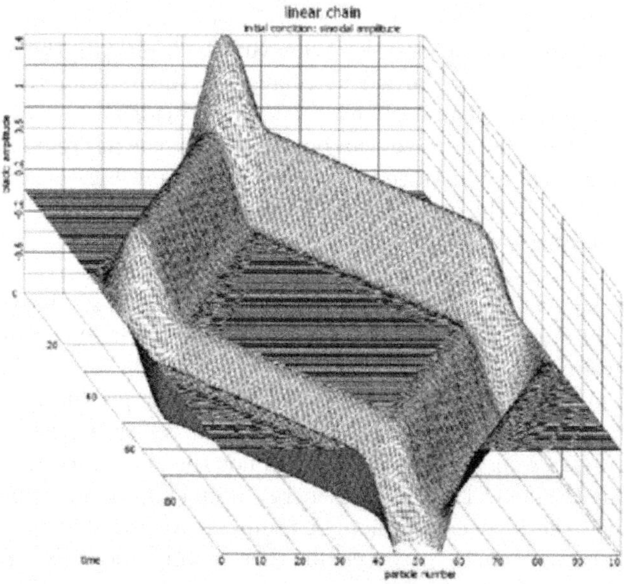

Fig. 6. d'Alembert waves after an initial sinusoidal distribution.

With these dampers mainly the high-frequency vibrations with the diffuse energy propagation will be damped. The low-frequency vibrations with their hyperbolic energy distribution will remain in the system for a long time.

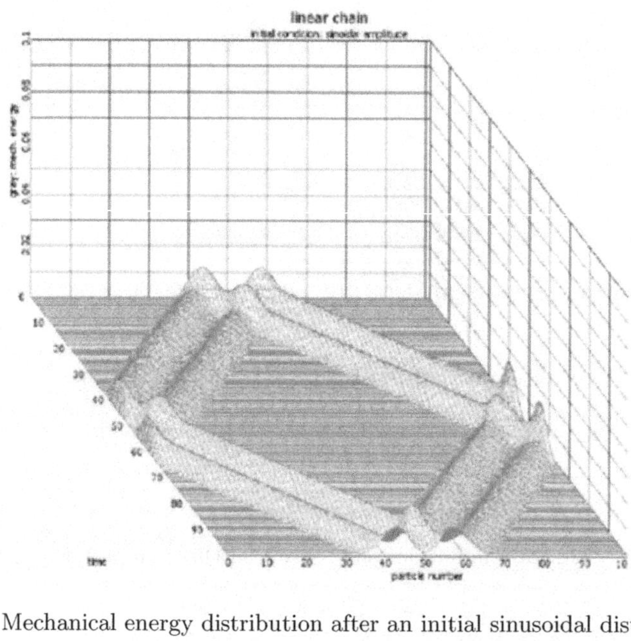

Fig. 7. Mechanical energy distribution after an initial sinusoidal distribution.

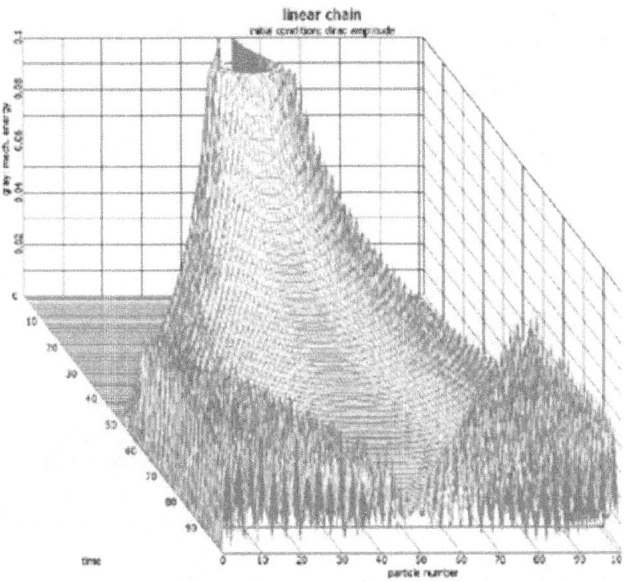

Fig. 8. Mechanical energy distribution after an initial Dirac distribution.

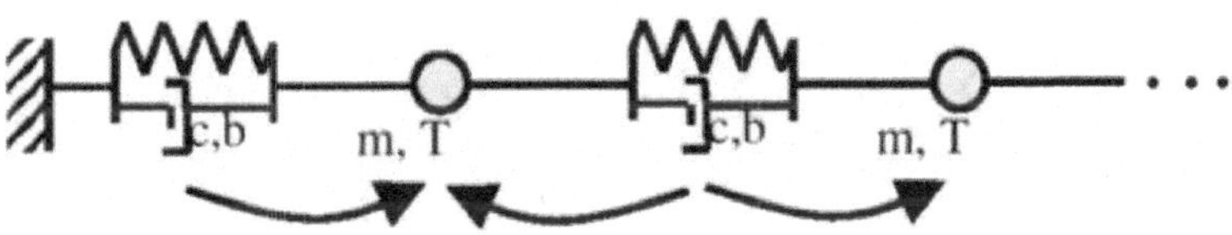

Fig. 9. Half of the dissipated energy is added to each of the neighbored particles.

The damping is an accepted technique for the smoothing of results, no matter if a real wave machine or a numerical experiment is considered. The innovative idea here is that this energy will remain in the system in the form of heat energy.

Fig. 10 shows the energy distribution in the rod caused by a Dirac - like stimulation. The diffuse energy propagation is herewith approximated by a parabolic distributed heat energy and a hyperbolic distributed energy remain.

A natural generalization of the particle model of the upper sections is a system of many particles. The number of particles depends on the number of macroscopic vibration modes which have to be taken into account.

Let $\mathbf{r}_i$ be the position vector of particle i, i = 1,...,n. Then the system can be described by the Lagrangian L and the Dissipation function D as follows

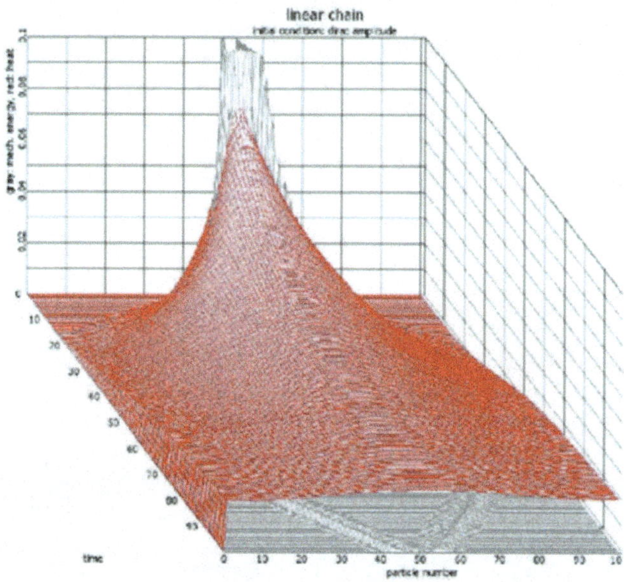

Fig. 10. Mechanic and thermal energy after an initial Dirac amplitude.

$$L = \sum_i \frac{1}{2} m \dot{r}_i^2 - \sum_{i,j} V_{ij} + V^E, \; V_{ij} = V(|r_i - r_j|), \tag{7}$$

$$D = \sum D_{ij} + D^E, \; D_{ij} = D(|\dot{r}_i - \dot{r}_j|), \tag{8}$$

where V^E and D^E denote interaction potential and dissipation function of the particle system with surroundings. Each particle has its own internal variable T_i, and each dissipation function D_{ij} assigns half of the dissipated energy to the heat of each particle i and j.

The heat carried by a particle describes a macroscopic property of the system which is generated by the temporary vibration energy in the hidden degrees of freedom. The macroscopic heat flow in the system should follow Fourier's law. This can be described by the non-stationary heat equation

$$c_v m \dot{T}_i := \frac{\partial}{\partial t} \int_V \rho c T dV = - \int_{\partial V} \underline{q} d\underline{A} + \int_V s dV, \tag{9}$$

where V denotes a small imaginary control volume around the particle i. The second term on the right side of the equation summarizes the heat production by the separation mechanism with all neighbor particles as described above. The first term on the right side represents the heat flow within the discretisized body.

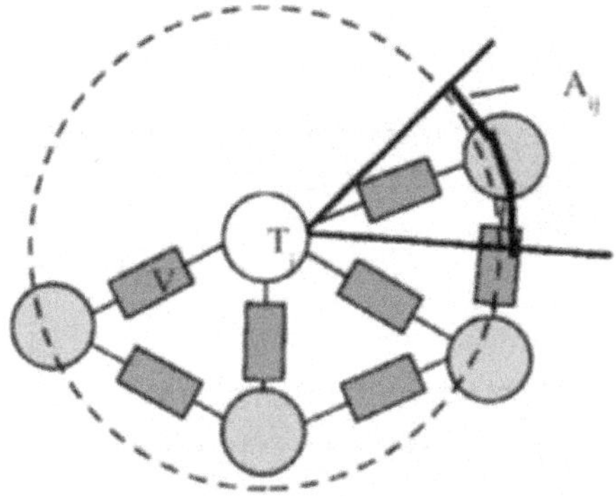

Fig. 11. The heat flow in the particle system.

Taking into account only the next neighbors, the equations of motion for each particle - simplified by finite volume methods and the assumption of homogeneity of material - are

$$\mathbf{m}\ddot{r}_i + \sum_j \left(\frac{\partial \mathbf{V_{ij}}}{\partial r_i} + \frac{\partial \mathbf{D}_{ij}}{\partial \dot{r}_{ij}} \right) + \frac{\partial \mathbf{V^E}}{\partial r_i} + \frac{\partial \mathbf{D^E}}{\partial \dot{r}_i} = 0, \tag{10}$$

$$\dot{T}_i = \frac{1}{c_v m} \left\{ \sum_{j=1}^{m} \left(\lambda A_{ij} \frac{T_i - T_j}{|r_j - r_i|} + \frac{1}{2} \frac{\partial D_{ij}}{\partial \dot{\mathbf{r}}_i} (\dot{\mathbf{r}}_i - \dot{\mathbf{r}}_j) \right) - (n - m)\gamma (T_i^4 - T^4) \right\} \tag{11}$$

where n is the maximum number and m is the effective number of neighbors. A_{ij} represents the part of surface of V that contributes to the heat flow between the particles i and j.

A many-particle model of a beam is shown in Fig. 12.

From a macroscopic point of view the beam model describes elastic waves and heat generation and distribution at the same time. Choosing the energy separation for simplification as a viscous damper, the resulting inner damping of the material damps out the high frequency vibrations – the hidden degrees of freedom - in the mesoscopic model in short time. The energy of the low frequency eigenmodes of the model which belong to the macroscopic degrees of freedom will stay almost unchanged for long time intervals compared to the time constant of the mesoscopic model.

In general, these mesoscopic particle models will generate two time scales. The macro time scale is related to the macroscopic behaviour of the model hiding the high frequency range of vibration in the model. This range is only visible on a micro time scale. The mathematical background for partitioning

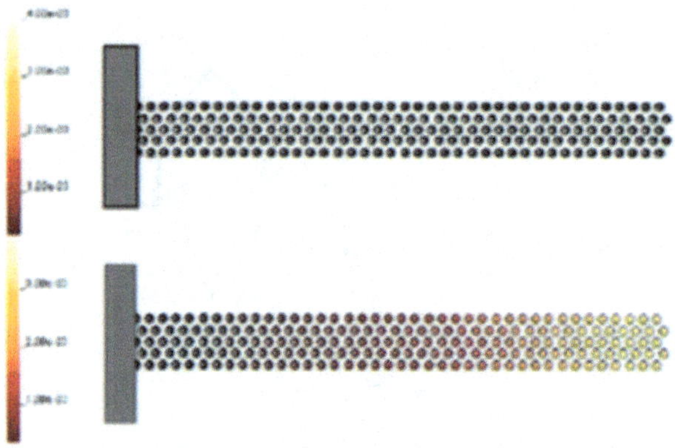

Fig. 12. The discretized beam after some beats.

the low and high frequency motion is dispersion which can be found in every discretized medium. A boundary between these frequencies can be explained by the well-known Shannon theorem. This will be shown elsewhere.

4 Elastoplastic behaviour of mesoscopic particle systems

The modified model of the beam now generates a body, which can be heated to an arbitrary high temperature. This is not very realistic, so that further modifications have to be done. We would like to take into account that the maximal heat energy which can be stored in the system is given by the thermal energy needed to bring the material to melting. We also would like to treat not only elastic, but also the elastoplastic behaviour of material. Both can be done by replacing the linear spring force by non-linear forces known from molecular dynamics, however a non-linear coupling between the potential and heat energy has to be introduced. An example of such coupling can be described rather simple by Lennard-Jones type forces. Let us define an interaction potential that depends explicitly on the stored heat energy $W - c_v mT$:

$$V_{LJW} = 4\varepsilon \left(\left(\frac{\sigma}{r}\right)^{12} - \sqrt{1 - \frac{W^2}{\varepsilon^2}} \left(\frac{\sigma}{r}\right)^{6} \right) . \tag{12}$$

Here the attractive forces turn imaginary, when W reaches the melting energy. When the melting point of the beam is reached, phase transition from the solid state to the liquid or gas state takes place as there are no attractive forces any more.

When there is more than one particle in the system, each particle i holds its own inner variable heat W_i. The influence of the heat on the potential V_{ij} depends on the number and heat of the neighbors of each particle.

$$V_{ij} = 4\varepsilon \left(\left(\frac{\sigma_{ij}}{r_{ij}} \right)^{12} - \sqrt{1 - \frac{\left(\frac{W_i}{n_n} + \frac{W_j}{n_j} \right)^2}{4\varepsilon^2}} \left(\frac{\sigma_{ij}}{r_{ij}} \right)^6 \right), r_{ij} = |r_i - r_j| \quad (13)$$

A simple model for testing such mesoscopic particle systems is a melting block, a specimen of a certain material of macroscopic size heated on the bottom side by a heating plate. This results in a heat transfer from the bottom to the top until the point is reached, where the body starts melting.

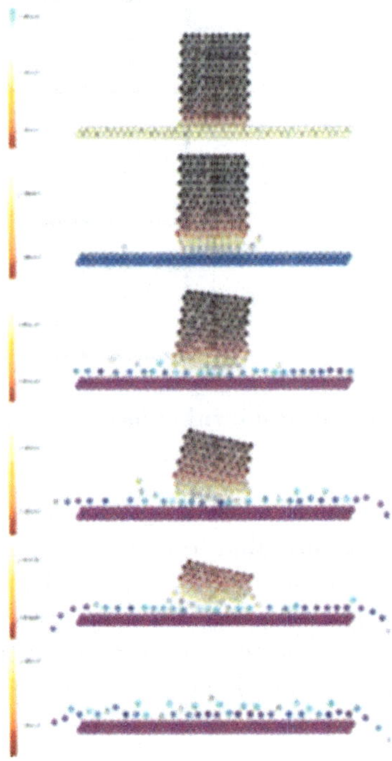

Fig. 13. A melting block.

The block is discretized into 160 two-dimensional particles. The total order of the differential system describing this model is 800. Each particle has two second order equations of motion and in addition one first order equation for its heat. The distance the modified Lennard-Jones forces and the dissipation forces are taken into account is $1.57\sigma_{ij}$. The gravitational force acts down in the normal direction to the heating plate. Heat radiation is taken into account for the boundary particles depending on the distance from the heating plate. The equations of motion are highly coupled and change, whenever particles

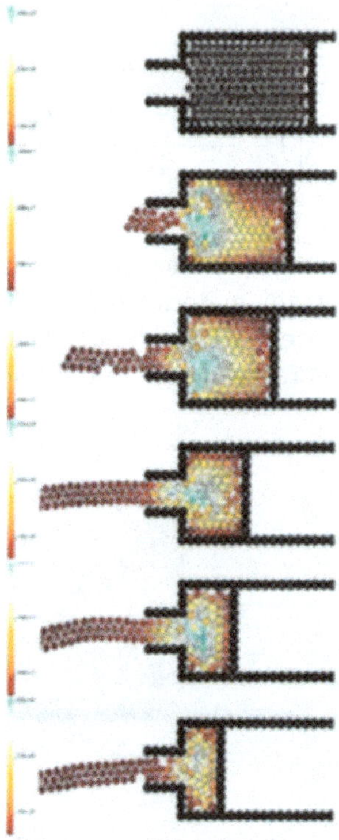

Fig. 14. Extruder build up with a mesoscopic particle system.

change their position and number of neighbour particles. Using this mesoscopic particle approach we are able to detect temperature and plastic flow. Such effects can be found in a simple example of an extruder, see Fig. 14. Pressing the plastic material through the tube produces heat, which leads to plastic flow and finally generates the extruder output. The mathematical model is equivalent to the model of the melting block.

5 Mesoscopic model of a friction process

The mesoscopic particle approach described above leads to a completely new method for discretizised layers in friction models.

During the friction process submechanical vibration energy is generated in the layers. Within the layers this energy is transformed partially to heat energy and elastic energy which is given in macroscopic quantities to the mechanical bodies outside the discretizised layers. Therefore there is a dissipative behaviour in the model from a macroscopic point of view and energy distribution flows from a microscopic point of view.

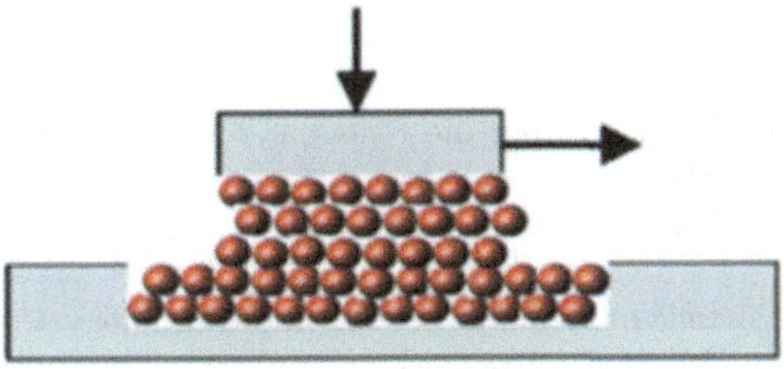

Fig. 15. Mesoscopic discretization of friction layer.

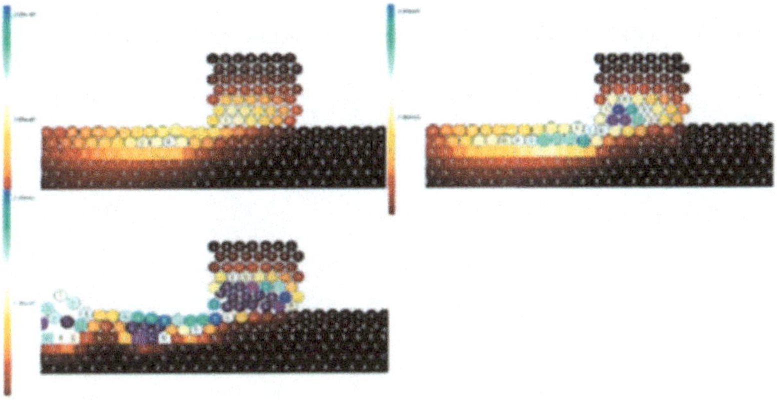

Fig. 16. Mesoscopic model of friction with different tangential velocities.

The author would like to emphasize that there are no a-priori assumptions on local dissipation of mechanical energy in these models.

The simulation in Fig. 16 shows the heat production in the friction process, some hot spot generations and destroying mechanisms (wear) on the surface. The resulting friction forces of the model have a falling characteristic related to sliding velocity.

6 Conclusion

We have shown, that the coupled thermomechanical dynamics of macroscopic systems can be described by introducing two-particle Lennard Jones interaction potentials depending on the stored heat energy. The two-particle potentials used in this paper gives only a rough approximation of the phase transition from solid body to gas and fluid state. These states of aggregation can not be completely described by two-particle interactions. In [9] many-particle forces are introduced, making possible a correct description of solids with different transversal rigidity, as well as transformation solid-liquid.

References

1. Bolzmann, L., Vorlesungen über die Principe der Mechanik, Verlag von Johann Anbrosius Barth, 1897
2. Tomlinson, G.A.: A molecular theory of friction. Phil. Mag. Ser. 7, Vol. 7, 905 ff (1929)
3. Greenspan, D.: Particle modeling in science and technology. Coll. Math. Societatis János Bolyai, Vol. 50, 51 ff (1988)
4. Greenspan, D.: A discrete numerical approach to fluid dynamics. Numerical Mathematics, Differential Equations, Information Processing 71, North Holland (1972)
5. Ostermeyer, G.P.: Friction Models with discrete Layers, ZAMM 1999

On wedged configurations with Coulomb friction

J.R. Barber[1] and P. Hild[2]

[1] Department of Mechanical Engineering, University of Michigan, Ann Arbor, MI48109-2125, USA,
jbarber@umich.edu

[2] Laboratoire de Mathématiques, Université de Franche-Comté/CNRS UMR 6623, 16 route de Gray, 25030 Besanc con, France,
hild@math.univ-fcomte.fr

Abstract. If the coefficient of friction is sufficiently large, elastic contact problems can exhibit 'wedged' solutions in which the body remains in a state of stress in the absence of applied loads. In this paper, we demonstrate that the critical coefficient of friction for wedging to occur is also the lowest real eigenvalue of a certain non-linear eigenvalue problem. Possible strategies for solving this eigenvalue problem are discussed.

1 Introduction

The elementary Coulomb friction law is a fruitful source of challenging mathematical questions in contact mechanics. Existence theorems for continuous static and quasi-static elastic contact problems involving Coulomb friction can be proved only subject to the condition that the coefficient of friction be sufficiently small (see [1, 2, 8, 4] for the static case and [26, 6] for the quasi-static case). Existence and uniqueness results for small friction coefficients were reported for finite element discretizations of the problem (with or without regularization techniques of the frictional contact conditions) by Oden and Pires [7], Haslinger [25], and Kikuchi and Oden [24]. The critical coefficient of friction below which the solution is unique depends upon the geometry and elastic properties of the system and no technique is presently available for determining it.

In this paper we shall be concerned with a particular category of non-unique solution associated with the phenomenon of *wedging*, in which the contacting bodies may remain in a self-sustaining stressed state even when the externally applied forces are removed. This is of concern in automated assembly operations [22, 23] where design requirements often necessitate close tolerances for the mating components, but slight misalignment in the assembly operation can then lead to an incorrectly assembled wedged state. In particular, we shall discuss possible methods of determining the critical coefficient of friction above which wedged solutions can exist.

2 Klarbring's model

Early insight into the physical basis behind the non-uniqueness of Coulomb friction solutions was provided by Klarbring [29, 28], who investigated the behavior of an elastically supported block in frictional contact with a rigid plane. He considered the case where an external force $\mathbf{F}$ increases monotonically with time, whilst maintaining the same direction, so the incremental quasi-static solution should be identical with the static solution. By assuming each possible state (stick, forward slip, backward slip, separation) in turn and solving the resulting linear equations of motion, he was able to identify a critical coefficient of friction f_c such that only one state was possible for a given load direction for $f < f_c$. For $f \geq f_c$ three solutions are obtained, corresponding to stick, separation and one direction of slip respectively [30].

Elementary calculations show that wedged states are also possible for Klarbring's model if and only if $f \geq f_c$ [37]. Thus, for Klarbring's model, wedging is always possible if the monotonic quasi-static solution is non-unique. Suppose now that the coefficient of friction is exactly equal to f_c. The multiple wedged states will still be solutions for the unloaded system, but they will now correspond to states of 'impending slip'. This suggests an alternative strategy for determining the critical coefficient of friction for wedging. We assume slip throughout the contact area and determine the conditions where non-trivial solutions exist with no external loading.

3 The eigenvalue problem

This method of determining the critical friction coefficient is actually a degenerate case of an eigenvalue problem defined by Hassani *et al.* [7] and Hild [17]. These authors demonstrated that discrete [7] and continuous [17] two-dimensional systems can posess an infinity of quasi-static solutions all involving slip for certain particular values of the coefficient of friction and that these critical coefficients can be determined by solving a linear eigenvalue problem.

Fig.1 shows a fairly general two-dimensional system consisting of an elastic body Ω, with boundary $\Gamma = \Gamma_D \cup \Gamma_N \cup \Gamma_C$, which is supported (displacement $\mathbf{U}= \mathbf{0}$) in Γ_D, loaded by prescribed tractions $\mathbf{F}$ in Γ_N, and which makes frictional contact with a rigid plane surface in Γ_C. We define the vector of normal contact forces as $\mathbf{P}$ and the corresponding vector of tangential nodal forces (positive when acting to the right) as $\mathbf{Q}$. The corresponding normal and tangential displacements in Γ_C are written $\mathbf{u}$, $\mathbf{v}$ respectively and the vector $\mathbf{w}$ represents all the remaining nodal displacements in the discretization.

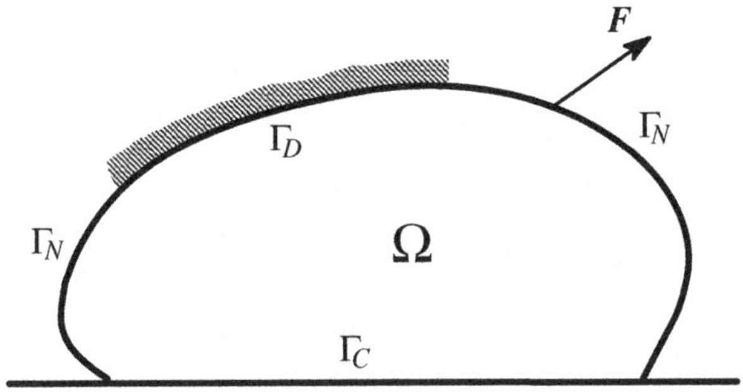

Fig. 1. Two-dimensional elastic contact problem.

If there is sliding friction in the same direction throughout Γ_C, we must have

$$\mathbf{u} = \mathbf{0}\,; \qquad \mathbf{Q} = f\mathbf{P}\,, \tag{1}$$

where f may be of either sign depending on the direction of slip. Suppose there exist two quasi-static solutions to this problem for given applied loads $\mathbf{F}$ and identify the deformation and stress fields by $\mathbf{S}_1$, $\mathbf{S}_2$, respectively. The boundary conditions (1) are linear and hence we can use linear superposition to show that the field $\mathbf{S} = \mathbf{S}_1 - \mathbf{S}_2$ also satisfies (1) and corresponds to zero applied tractions $\mathbf{F} = \mathbf{0}$. Thus, the existence of distinct multiple solutions involving slip also implies the existence of non-trivial solutions to the corresponding homogeneous (unloaded) problem.

With a suitable finite element discretization, there will be a finite number of nodes N in Γ_C and if the external loads $\mathbf{F} = \mathbf{0}$, there will be a linear relation between the contact tractions and displacements which can be written

$$\mathbf{K}\begin{Bmatrix}\mathbf{u}\\ \mathbf{v}\\ \mathbf{w}\end{Bmatrix} = \begin{bmatrix}\mathbf{K}_{nn} & \mathbf{K}_{nt} & \mathbf{K}_{ni}\\ \mathbf{K}_{nt} & \mathbf{K}_{tt} & \mathbf{K}_{ti}\\ \mathbf{K}_{ni} & \mathbf{K}_{ti} & \mathbf{K}_{ii}\end{bmatrix}\begin{Bmatrix}\mathbf{u}\\ \mathbf{v}\\ \mathbf{w}\end{Bmatrix} = \begin{Bmatrix}\mathbf{P}\\ \mathbf{Q}\\ \mathbf{0}\end{Bmatrix}\,, \tag{2}$$

or

$$\mathbf{C}\begin{Bmatrix}\mathbf{P}\\ \mathbf{Q}\\ \mathbf{0}\end{Bmatrix} = \begin{bmatrix}\mathbf{C}_{nn} & \mathbf{C}_{nt} & \mathbf{C}_{ni}\\ \mathbf{C}_{nt} & \mathbf{C}_{tt} & \mathbf{C}_{ti}\\ \mathbf{C}_{ni} & \mathbf{C}_{ti} & \mathbf{C}_{ii}\end{bmatrix}\begin{Bmatrix}\mathbf{P}\\ \mathbf{Q}\\ \mathbf{0}\end{Bmatrix} = \begin{Bmatrix}\mathbf{u}\\ \mathbf{v}\\ \mathbf{w}\end{Bmatrix}\,, \tag{3}$$

where $\mathbf{K}$ is the stiffness matrix and $\mathbf{C} = \mathbf{K}^{-1}$ is the compliance matrix.

Equations (1:(i), 3) then imply that $\mathbf{C}_{nt}\mathbf{Q} = -\mathbf{C}_{nn}\mathbf{P}$ and since $\mathbf{C}_{nn}$ is symmetric and positive definite (Γ_D is not null), we obtain $-(\mathbf{C}_{nn})^{-1}\mathbf{C}_{nt}\mathbf{Q} = \mathbf{P}$. Note that this equation depends on the entire finite element discretization, not only on the discretization near the contact zone, since it involves the coefficients of $\mathbf{K}^{-1}$, the inverse of the stiffness matrix. Substituting in (1:(ii)), we then obtain

$$-(\mathbf{C}_{nn})^{-1}\mathbf{C}_{nt}\mathbf{P} = \frac{1}{f}\mathbf{P}\,. \tag{4}$$

In other words, slip solutions of the assumed form exist if and only if $1/f$ is an eigenvalue of equation (4).

3.1 Relation to the wedging problem

If the configuration of Fig.1 can exist in a wedged state, the corresponding values of $\mathbf{u}$, $\mathbf{P}$, $\mathbf{Q}$ must satisfy the conditions

$$|Q_i| \leq fP_i\,; \quad u_i \geq 0\,; \quad P_i \geq 0\,; \quad u_iP_i = 0\,, \tag{5}$$

where $\mathbf{u}$, $\mathbf{P}$, $\mathbf{Q}$ are related through equations (1:(i), 2, 3).

Now suppose that the eigenvalue problem (4) has at least one real eigenvalue $1/f_c$ and that the corresponding eigenfunction satisfies the condition that the P_i $(i = 1, N)$ all have the same sign, so that we can choose the sign of the arbitrary multiplier on the eigenfunction so as to satisfy the condition $P_i \geq 0$. It follows that the same set of displacements and contact tractions will satisfy the conditions (5) for wedging if $f \geq f_c$.

Equation (4) is an $N \times N$ linear eigenvalue equation and it will always have N eigenvalues and associated eigenfunctions. However, the matrix $(\mathbf{C}_{nn})^{-1}\mathbf{C}_{nt}$ is not generally symmetric and hence there is no guarantee that all or indeed any of the eigenvalues will be real. Since the eigenvalues represent the coefficient of friction, complex values have no physical significance. Furthmore, even if real eigenvalues exist, there is no guarantee that the associated eigenfunctions satisfy the condition that the P_i $(i = 1, N)$ all have the same sign. It is fairly easy to generate examples in which these restrictions prevent the eigenvalue problem from defining a physically meaningful value of f_c. A simple example is the square block of Fig.2, discretized such that there exist only two nodes in Γ_C located at the two lower corners A, B [37].

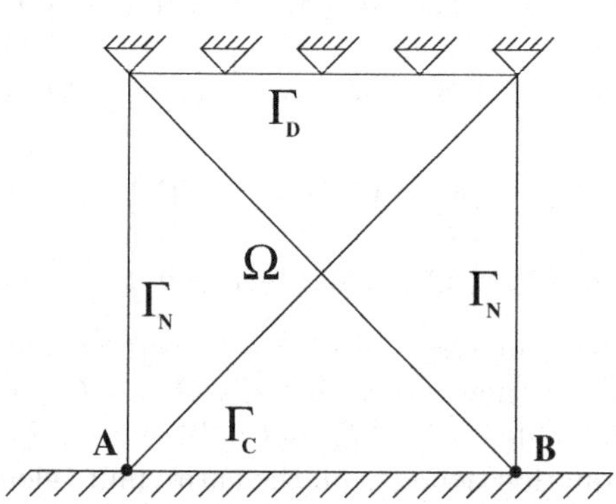

Fig. 2. Rectangular geometry.

3.2 A generalized non-linear eigenvalue problem

These restrictions do not necessarily imply that wedged states are impossible when the corresponding eigenvalue problem fails to yield a physically acceptable eigenfunction. For example, a simple investigation of the example of Fig. 2 shows that physically acceptable wedged states can occur (a) if both contact nodes are displaced in the same direction (say to the right) and the right hand node is allowed to separate, or (b) if both nodes are displaced towards the centre of the contact region and remain in contact [37].

These considerations suggest the definition of a generalized non-linear eigenvalue problem through the conditions

$$|Q_i| = f_c P_i \; ; \quad u_i \geq 0 \; ; \quad P_i \geq 0 \; ; \quad u_i P_i = 0 \; . \tag{6}$$

Comparison with (5) shows that the existence of a real eigenvalue f_c to this problem is a sufficient condition for wedged states to exist for $f \geq f_c$ and hence the minimum real eigenvalue $f_c^{\min}$ represents at least an upper bound for the critical coefficient of friction required for wedging to occur.

It is less clear whether this condition is also necessary — i.e. whether the existence of a wedged state with a given value of f implies the existence of a real eigenvalue for the problem (6) with $f_c \leq f$. However, the following thought experiment suggests that this is in fact the case. Suppose we start with the given wedged state and imagine some mechanism such as the gradual infusion of lubricant whereby the coefficient of friction can be gradually reduced. Eventually a value of f will be reached at which one node will start to slip and further reduction in f will transfer load to other nodes causing them to slip.

To place this argument on a more rigorous footing, consider an intermediate state in which a subset Γ_S of Γ_C is in a state of slip with coefficient of friction f, whilst the remainder $\Gamma_C - \Gamma_S$ remains stuck. In other words, we suppose the existence of a non-trivial solution to the problem

$$Q_i = \pm f P_i \quad \in \Gamma_S \tag{7}$$

$$|Q_i| < f P_i \quad \in \Gamma_C - \Gamma_S \tag{8}$$

$$u_i \geq 0 \; ; \quad P_i \geq 0 \; ; \quad u_i P_i = 0 \quad \in \Gamma_C \; . \tag{9}$$

where the sign taken in (4) depends on the direction of slip and may be different at different nodes.

Suppose we now reduce the coefficient of friction by an infinitesimal increment Δf. If the conditions after this increment are to satisfy (4, 9) with f replaced by $f - \Delta f$ and no change in state at any node, the corresponding incremental changes in $\mathbf{P}$,$\mathbf{Q}$,$\mathbf{u}$,$\mathbf{v}$ must satisfy the conditions

$$Q_i - \Delta Q_i = \pm(f - \Delta f)(P_i - \Delta P_i) \quad \in \Gamma_S \tag{10}$$

$$\Delta \mathbf{v} = \mathbf{0} \quad \in \Gamma_C - \Gamma_S \tag{11}$$

$$\Delta \mathbf{u} = \mathbf{0} \quad \in \Gamma_C \tag{12}$$

and since Δf is infinitesimal, we can drop second order small quantities in (10), giving

$$\Delta Q_i \mp f \Delta P_i = \pm \Delta f P_i \quad \in \Gamma_S \, . \tag{13}$$

Equations (11–13) define a linear problem for the incremental fields, since P_i is assumed known. We can therefore solve this problem and add the resulting increment in **P**,**Q**,**u**,**v** to the original state to obtain the new state, which will not violate the inequalities in (9), provided Δf is sufficiently small.

By making a succession of small but finite reductions Δf, we can monitor each inequality and determine which is the first to be violated as f is reduced. When a violation is detected, we can also solve for the exact value of Δf at which this node(s) just reaches the limiting condition ($P_i = 0$ for a 'slip' node or $|Q_i| = fP_i$ for a 'stick' node). We can then change the assumption at the node(s) in question and proceed.

Using this algorithm, as long as any nodes remain stuck, equations (4–9) define a wedged state. The limiting condition arises when the last node(s) in $\Gamma_C - \Gamma_S$ just reaches the slip condition. At this point, all nodes are either separated or slipping and hence the final wedged state is a solution of the generalized eigenvalue problem (6). This procedure could be applied for any initial wedged configuration and hence we conclude that the smallest eigenvalue f_c of (6) is also the critical coefficient of friction above which wedged states can occur.

3.3 Solution of the generalized eigenvalue problem

We now turn our attention to possible strategies for solving the non-linear eigenvalue problem (6). We first note that each contact node must be in one of three states (forward slip, backward slip or separation) and hence the number of possible combinations of states with N contact nodes is 3^N. Each of these combinations defines a classical linear eigenvalue problem, so we might solve each in turn and search the resulting solutions for the minimum real eigenvalue whose eigenfunction satisfies the inequality constraints with an appropriate choice for the sign of the multiplying constant.

This method proceeds by exhaustion and will generate the critical coefficient of friction for wedging from a finite number of numerical calculations. However, it is likely to be prohibitively computer intensive for large systems. An alternative approach is to develop an iterative method using strategies drawn from the corresponding unilateral contact problem. For example, suppose we solve the linear eigenvalue problem assuming unidirectional slip at all nodes and obtain a real eigenvalue associated with an eigenfunction that fails the requirement that the normal tractions be all of the same sign. We might then reformulate the linear eigenvalue problem making the assumption of separation at the nodes where the tractions are tensile. There is of course no guarantee that such an iterative procedure would be convergent, but any solutions so obtained could at least be used to define an upper bound for the critical coefficient of friction.

3.4 Numerical experiments

Yet another option, which we explore here numerically, is to program the 'relaxation' algorithm of Section 3.2. We first determine an initial wedged condition using prescribed values of $\mathbf{v}$ and a suitably high coefficient of friction and then try to reduce the friction coefficient using the following procedure:-

(i) Identify the node j at which the maximum ratio between tangential and normal traction is attained.
(ii) Relax the tangential displacement v_j at this node and solve the elasticity problem with no loads, a prescribed $\mathbf{v}$ and unilateral contact conditions (6) until another node reaches the same limiting condition.
(iii) Continue relaxation until all the nodes are at the same limiting condition, corresponding to slip and identify the limiting value of f.

We first consider the triangle Ω of vertices $A = (0,0)$, $B = (1,0)$ and $C = (0.6, 0.3)$ and we set $\Gamma_D =]B,C[$, $\Gamma_N =]A,C[$, $\Gamma_C =]A,B[$. The body Ω lies on a rigid foundation, the half-space delimited by the straight line (A,B). Moreover we assume that $\nu = 0.2$. We consider a mesh comprising 10 contact elements and 10 contact nodes (the node B is supposed to belong to Γ_D in the computations). We choose an initial 'constant' tangential displacement $\mathbf{v} = (0.05, ..., 0.05)$ (the 10 nodes move to the right with a displacement of 0.05) and we solve the elasticity problem with no loads and unilateral contact conditions (6). The range of the ratio between tangential and normal traction for this configuration is [0.715,4.021], so it corresponds to a wedged state for $f \geq 4.021$. We apply the above procedure with a relaxation coefficient of 0.999. After 1000 iterations the range of the ratio is [0.715,3.146] and after 10000 iterations it is [1.991,2.006], showing that wedging occurs if $f \geq 2.006$. We observe that the limiting value is $f = 2$ which precisely solves the associated linear eigenvalue problem corresponding to slip in the continuous case [17]. Moreover the limiting field $\mathbf{v}$ is no longer constant on the 10 nodes (and vanishing at B) but linear and the stress field is constant in Ω.

Next we consider the geometry of a tapered joint (with $\nu = 0.1$) which is a geometry conducive to wedging. Using symmetry conditions, the tapered joint can be divided into two parts and the geometry to be considered is a trapezium with two right angles. As before we choose a constant initial displacement $\mathbf{v} = (0.1, ..., 0.1)$, for which the range of the tangential/normal traction ratio is found to be [-0.0418,0.2128]. We used a relaxation coefficient of 0.99999. After 10000 iterations we obtain a wedged configuration for $f \geq 0.1002$ and the range of the ratio is [0.0995,0.1002]. It seems that the limiting value is $f = 0.1$ which is precisely equal to $\tan(\theta)$ (2θ is the angle of the entire joint).

The Von Mises stress fields after 1 and 10000 iterations are depicted in Fig. 3. The limiting stress field increases from the left to the right and admits physically relevant isolines which are parallel to the sides of the trapezium.

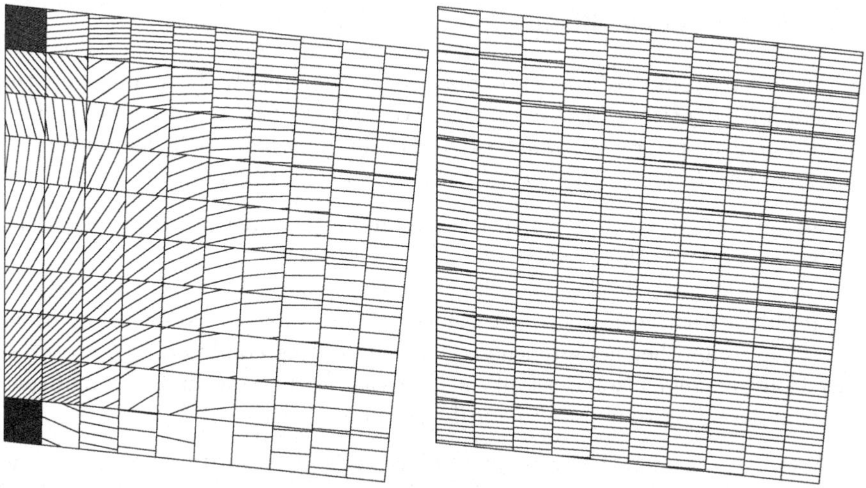

Fig. 3. Initial and limiting Von-Mises stress fields.

4 Conclusions

This paper represents a first attempt to quantify the coefficient of friction above which wedging is possible for a two-dimensional elastic contact problem. We have demonstrated that this is also the solution of a certain non-linear eigenvalue problem. Future work will be directed at testing various suggestions for the solution of this problem.

References

1. Nev cas J, Jaruv sek J, Haslinger J (1980) Boll Unione Mat Ital 17:796–811.
2. Jaruv sek J (1983) Czechoslovak Math J 33:237–261.
3. Kato Y (1987) Japan J Appl Math 4:237–268.
4. Eck C, Jaruv sek J (1998) Math Models Methods Appl Sci 8:445–468.
5. Andersson L-E (2000) Appl Math Optim 42:169–202.
6. Rocca R, Cocu M (2001) Int J Engrg Sci 39:1233–1255.
7. Oden J, Pires E (1983) ASME J Appl Mech 50:67–76.
8. Haslinger J (1983) Math Meth Appl Sci 5:422–437.
9. Kikuchi N, Oden JT (1988) Contact problems in elasticity: A study of variational inequalities and finite element methods, SIAM, Philadelphia.
10. Mosemann H, Wahl FM (2001) IEEE Trans Robotics Automation 17:709–718.
11. Sturges RH, Laowattana S (1996) J Mech Des 118:99–105.
12. Klarbring A (1984) Contact problems with friction, Doctoral Dissertation, Linköping University Linköping Sweden
13. Klarbring A (1990) Ing Arch 60:529–541
14. Cho H, Barber JR (1998) Math Comp Modeling 8:37–53
15. Barber JR, Hild P (2004) Proc. ASME/STLE Int. Joint Tribology Conference, Long Beach, CA.

16. Hassani R, Hild P, Ionescu I, Sakki N-D (2003) Comput Methods Appl Mech Engrg 192:4517–4531.
17. Hild P (2004) Q J Mech Appl Math 57:225–235
18. Huang SG, Schimmels JM, (2003) Int J Robotics Res 22:733–756
19. Kao I, Cutkosky MR (1992) Int J Robotics Res 11:20–40
20. Pang JS, Trinkle JC (1996) Math Programming 73:199–226
21. Bilkay O, Anlagan O (2004) Tribology Int 37:347–351
22. Mosemann H, Wahl FM (2001) IEEE Trans Robotics Automation 17:709–718
23. Sturges RH, Laowattana S (1996) J Mech Des 118:99–105
24. Kikuchi N, Oden JT (1988) Contact problems in elasticity: A study of variational inequalities and finite element methods, SIAM, Philadelphia
25. Haslinger J (1983) Math Meth Appl Sci 5:422–437
26. Andersson L-E (2000) Appl Math Optim 42:169–202
27. Dundurs J, Comninou M (1983) ASME J Appl Mech 50:77-84
28. Klarbring A (1990) Ing Arch 60:529–541
29. Klarbring A (1984) Contact problems with friction, Doctoral Dissertation, Linköping University Linköping Sweden
30. Cho H, Barber JR (1998) Math Comp Modeling 8:37–53
31. Hild P (2002) Int J Appl Math Comput Sci 12:41–50
32. Hassani R, Hild P, Ionescu IR, Sakki N-D (2003) Comput Methods Appl Mech Engrg 192:4517–4531
33. Johnson KL (1985) Contact Mechanics, Cambridge University Press, Cambridge 407
34. Zavarise G, Wriggers P (1999) Engng Comp 16:88–119
35. Anitescu M, Potra FA (1997) Nonlinear dynamics 14:231-247
36. Hild P (2004) Q J Mech Appl Math 57:235–245
37. Barber JR, Hild P (2004) Proc. ASME/STLE Int. Joint Tribology Conference, Long Beach, CA.

A genetic algorithm approach for wedged configurations with Coulomb friction

R. Hassani[1], I.R. Ionescu[2] and E. Oudet[2]

[1] Laboratoire de Géophysique Interne et Tectonophysique, CNRS & Université de Savoie, Campus Scientifique, 73376 Le Bourget du Lac, France, hassani@univ-savoie.fr

[2] Laboratoire de Mathématiques, CNRS & Université de Savoie, Campus Scientifique, 73376 Le Bourget du Lac, France, ionescu@univ-savoie.fr, oudet@univ-savoie.fr

Abstract. The wedged configuration with Coulomb friction (i.e. a nontrivial equilibrium state of a linear elastic body in a frictional unilateral contact with a rigid body under vanishing external loads), which is considered here in a 3-D continuous framework, was firstly studied in [1]. A supremal functional defined on the set of admissible normal displacement and tangential stresses is introduced. The infimum of this functional μ^w defines the critical friction coefficient for the wedged problem (WP). For friction coefficients μ with $\mu > \mu^w$ (WP) has at least a solution and for $\mu < \mu^w$ (WP) has no solution.

The (WP) problem is stated in a discrete framework using a mixed finite element approach and the (discrete) critical friction coefficient is introduced as the minimum of a specific functional. A genetic algorithm is used for the global minimization problem involving this non differentiable and non-convex functional. Finally, the analysis is illustrated with some numerical experiments.

1 Introduction

Wedged configurations appears to be of industrial interest in problems associated with automated assembly and manufacturing processes. The theoretical interest of wedged configurations is related to the non uniqueness of the equilibrium problem with Coulomb friction in linear elasticity (see for instance [8, 5, 7]). As far as we know, the first study on the subject was done by Barber and Hild [1], who have related it to the eigenvalue analysis of Hassani et al. [5, 7].

By a "wedged configuration with Coulomb friction" we mean a nontrivial equilibrium state of a linear elastic body under vanishing external loads which is in frictional contact with a rigid body. Our goal is to find the relation between the geometry of the elastic body (including the boundaries distribution) and the friction coefficient for which wedged configurations exist. It is beyond of the scope of the present work to discuss the quasi-static or dynamic trajectory of the body from the reference configuration to the wedged equilibrium. The (dynamic) stability conditions of the wedged configurations, which are not considered here, are the same as for the stability of any equilibrium state under Coulomb friction (see [9] for a recent study).

We consider the deformation of an elastic body occupying, in the initial unconstrained configuration a domain Ω in $\mathbf{R}^d$, with $d = 3$ in general and $d = 2$ in the in-plane configuration. The Lipschitz boundary $\partial\Omega$ of Ω consists of Γ_D, Γ_N and Γ_C. We assume that the displacement field $\boldsymbol{u}$ is vanishing on Γ_D and that the boundary part Γ_N is traction free (i.e. the density of surface forces is vanishing). In the initial configuration, the part Γ_C is considered as the candidate contact surface on a rigid foundation (see Figure 1) which means that the contact zone cannot enlarge during the deformation process. The contact is assumed to be frictional and the stick, slip and separation zones on Γ_C are not known in advance. In order to simplify the problem, and without any loss of generality we will suppose that the body Ω is not acted upon by a volume forces (i.e. the given density of volume forces are vanishing).

Denoting by $\mathbf{n}$ the unit outward normal vector of $\partial\Omega$ and by $\mu > 0$ the friction coefficient on Γ_C the wedged problem (WP) can be formulated as

Wedged problem (WP). Find $\mathbf{\Phi} : \Omega \to \mathbf{R}^d$ and μ with $\mathbf{\Phi} \neq 0$ and $\mu > 0$ such that

$$\boldsymbol{\sigma}(\mathbf{\Phi}) = \mathcal{C}\,\varepsilon(\mathbf{\Phi}), \qquad \mathbf{div}\,\boldsymbol{\sigma}(\mathbf{\Phi}) = \mathbf{0} \quad \text{in } \Omega, \tag{1}$$

$$\mathbf{\Phi} = \mathbf{0} \quad \text{on } \Gamma_D, \qquad \boldsymbol{\sigma}(\mathbf{\Phi})\mathbf{n} = \mathbf{0}, \quad \text{on } \Gamma_N, \tag{2}$$

$$\Phi_n \leq 0, \quad \sigma_n(\mathbf{\Phi}) \leq 0, \quad \Phi_n \sigma_n(\mathbf{\Phi}) = 0, \quad |\sigma_t(\mathbf{\Phi})| \leq -\mu\sigma_n(\mathbf{\Phi}) \quad \text{on } \Gamma_C, \tag{3}$$

where $\varepsilon(\mathbf{\Phi}) = (\nabla\mathbf{\Phi} + \nabla^T\mathbf{\Phi})/2$ denotes the linearized strain tensor field, $\mathcal{C}$ is a fourth order symmetric and elliptic tensor of linear elasticity and we adopted the following notation for the normal and tangential components: $\mathbf{\Phi} = \Phi_n\mathbf{n} + \mathbf{\Phi}_t$ and $\boldsymbol{\sigma}(\mathbf{\Phi})\mathbf{n} = \sigma_n(\mathbf{\Phi})\mathbf{n} + \boldsymbol{\sigma}_t(\mathbf{\Phi})$.

The function $\mathbf{\Phi}$ is determined up to a positive multiplicative constant, i.e. if $\mathbf{\Phi}$ is a solution then $t\mathbf{\Phi}$ is also a solution for all $t > 0$. Let also remark that if $\mathbf{\Phi}$ is a solution of (WP) for a friction coefficient μ then it is is also a solution for all friction coefficients $\bar{\mu} \geq \mu$.

Definition. Let $s \geq 1/2$, $p, q \in [1, +\infty]$ with $p \geq q$ be given. By a solution of the wedged problem (with the regularity (s, p, q)) we mean a nontrivial function $\mathbf{\Phi} \in H^1(\Omega)^d$ which satisfies (1)-(3), such that $\Phi_n \in H^s(\Gamma_C)$, $\boldsymbol{\sigma}_t(\mathbf{\Phi}) \in L^p(\Gamma_C)^d$ and $\sigma_n(\mathbf{\Phi}) \in L^q(\Gamma_C)$.

2 Critical friction as an infimum of a supremal functional

Let $\boldsymbol{\Sigma}_t$ and Σ_n be the spaces of the tangential and normal stresses and let denote by and S_n the space of normal displacements on Γ_C

$$\boldsymbol{\Sigma}_t = \{\boldsymbol{\tau} \in L^p(\Gamma_C)^d \ ; \ \boldsymbol{\tau} \cdot \mathbf{n} = 0\}, \quad \Sigma_n = L^q(\Gamma_C), \quad S_n = H^s(\Gamma_C),$$

with $s \geq 1/2$ and $p, q \in [1, +\infty]$.

For all $\boldsymbol{\tau} \in \boldsymbol{\Sigma}_t$ and $v \in S_n$ we consider the solution $\mathcal{U}(\boldsymbol{\tau}, v) = \boldsymbol{u} \in H^1(\Omega)^d$ of the following elasto-static problem :

$$\boldsymbol{\sigma}(\boldsymbol{u}) = \mathcal{C}\, \boldsymbol{\varepsilon}(\boldsymbol{u}), \qquad \mathbf{div}\, \boldsymbol{\sigma}(\boldsymbol{u}) = \mathbf{0} \quad \text{in } \Omega, \tag{4}$$

$$\boldsymbol{u} = \mathbf{0} \quad \text{on } \Gamma_D, \qquad \boldsymbol{\sigma}(\boldsymbol{u})\mathbf{n} = \mathbf{0} \quad \text{on } \Gamma_N, \tag{5}$$

$$u_n = v \quad \text{on } \Gamma_C, \qquad \sigma_t(\boldsymbol{u}) = \boldsymbol{\tau} \quad \text{on } \Gamma_C. \tag{6}$$

Since the stress $\boldsymbol{\sigma}(\boldsymbol{u}) \in H(\mathbf{div}\ ; \Omega)^d$ and $\sigma_n(\boldsymbol{u}) \in H^{-1/2}(\Gamma_C)$ we can define the operator $\mathcal{L} : \boldsymbol{\Sigma}_t \times S_n \to H^{-1/2}(\Gamma_C)$ by $\mathcal{L}(\boldsymbol{\tau}, v) =: \sigma_n(\boldsymbol{u})$.

Let $\mathcal{S}$ be a cone in the space of tangential stresses and normal displacements $\boldsymbol{\Sigma}_t \times S_n$ defined by

$$\mathcal{S} =: \{(\boldsymbol{\tau}, v) \in \boldsymbol{\Sigma}_t \times S_n \ ; \ (\boldsymbol{\tau}, v) \neq 0, \quad v \leq 0, \ v|\boldsymbol{\tau}| = 0, \quad \text{on } \Gamma_C\},$$

and we define the cone of admissible states $\mathcal{S}^{adm}$ (tangential stresses and normal displacements) by

$$\mathcal{S}^{adm} =: \{(\boldsymbol{\tau}, v) \in \mathcal{S}\ ; \ \mathcal{L}(\boldsymbol{\tau}, v) \in \Sigma_n, \ \mathcal{L}(\boldsymbol{\tau}, v) \leq 0, \ v\mathcal{L}(\boldsymbol{\tau}, v) = 0 \text{ on } \Gamma_C\}.$$

We consider now the supremal functional $J : \mathcal{S} \to \mathbf{R} \cup \{+\infty\}$ defined by

$$J(\boldsymbol{\tau}, v) = \operatorname*{ess\ sup}_{x \in \Gamma_C} Q(|\boldsymbol{\tau}(x)|, \mathcal{L}(\boldsymbol{\tau}, v)(x)),$$

where $Q : \mathbf{R}_+ \times \mathbf{R}_- \to \mathbf{R}_+ \cup \{+\infty\}$ is a quotient given by

$$Q(t, r) = \begin{cases} -\dfrac{t}{r}, & \text{if } r < 0 \\ 0, & \text{if } t = 0 \\ +\infty, & \text{if } r = 0, t > 0, \end{cases} \tag{7}$$

The following lemma (for the proof see [6]) gives the connection between the supremal functional J and the wedged problem.

Lemma 1. *For all $(\boldsymbol{\tau}, v) \in \mathcal{S}^{adm}$ with $J(\boldsymbol{\tau}, v) < +\infty$ the field $\boldsymbol{\Phi} = \mathcal{U}(\boldsymbol{\tau}, v)$ is a solution of (WP) with $\mu = J(\boldsymbol{\tau}, v)$.*

Let μ^w be the infimum of J on $\mathcal{S}^{adm}$,

$$\mu^w =: \inf_{(\boldsymbol{\tau}, v) \in \mathcal{S}^{adm}} J(\boldsymbol{\tau}, v),$$

which can be considered as a *critical friction coefficient (for the wedged problem)*. To see this we recall the following theorem from [6].

Theorem 1. *Suppose that $\mathcal{S}^{adm}$ is not empty and μ^w is finite. Then we have*
i) For all $\mu > \mu^w$ the problem (WP) has at least a solution.
ii) If $\mu < \mu^w$ then (WP) has no solution.

As it follows from the above theorem for a given geometry and for some given elastic coefficients (Poisson ratio in the case of isotropic elastic materials), wedged configurations exist only if the friction coefficient is larger than the critical value μ^w.

3 Mixed finite element approach of the critical friction

The body Ω is discretized by using a family of triangulations $(\mathcal{T}_h)_h$ made of finite elements of degree $k \geq 1$ where $h > 0$ is the discretization parameter representing the greatest diameter of a triangle in $\mathcal{T}_h$. The space of finite elements approximation is:

$$\mathbf{V}_h = \Big\{\boldsymbol{v}_h;\ \boldsymbol{v}_h \in (C(\overline{\Omega}))^d,\ \boldsymbol{v}_h|_T \in (P_k(T))^d\ \ \forall T \in \mathcal{T}_h,\ \boldsymbol{v}_h = \mathbf{0} \text{ on } \Gamma_D\Big\},$$

where $C(\overline{\Omega})$ stands for the space of continuous functions on $\overline{\Omega}$ and $P_k(T)$ represents the space of polynomial functions of degree k on T. On the boundary of Ω, we still keep the notation $\boldsymbol{v}_h = v_{hn}\mathbf{n} + \boldsymbol{v}_{ht}$ for every $\boldsymbol{v}_h \in \mathbf{V}_h$ and we denote by $(T_h)_h$ the family of $(d-1)$-dimensional mesh on Γ_C inherited by $(\mathcal{T}_h)_h$. Set

$$S_{hn} = \Big\{\nu; \nu = \boldsymbol{v}_h|_{\Gamma_C} \cdot \mathbf{n},\ \boldsymbol{v}_h \in \mathbf{V}_h\Big\},$$

the space of normal displacements which is included in the space of continuous functions on Γ_C which are piecewise of degree k on $(T_h)_h$. For the tangential and normal stresses we put

$$\boldsymbol{\Sigma}_{ht} = \Big\{\boldsymbol{\tau}_h; \boldsymbol{\tau}_h \in (C(\overline{\Gamma}_C))^{d-1}, \boldsymbol{\tau}_h|_T \in (P_k(T))^{d-1}\ \ \forall T \in T_h\Big\},$$

$$\Sigma_{hn} = \Big\{\sigma_h; \sigma_h \in C(\overline{\Gamma}_C), \sigma_h|_T \in P_k(T)\ \ \forall T \in T_h\Big\},$$

The discrete problem issued from the continuous wedged problem (WP) becomes:

Discrete wedged problem (WP)$_h$. Find $(\boldsymbol{\Phi}_h, \lambda_{hn}, \boldsymbol{\lambda}_{ht}) \in \mathbf{V}_h \times \Sigma_{hn} \times \boldsymbol{\Sigma}_{ht}$ such that

$$\int_\Omega \mathcal{C}\varepsilon(\boldsymbol{\Phi}_h) : \varepsilon(\boldsymbol{v}_h)\, d\Omega = \int_{\Gamma_C} \lambda_{hn} v_{hn}\, d\Gamma + \int_{\Gamma_C} \boldsymbol{\lambda}_{ht} \cdot \boldsymbol{v}_{ht}\, d\Gamma, \tag{8}$$

$$(\Phi_n)_i \leq 0,\ (\lambda_{hn})_i \leq 0,\ (\Phi_n)_i(\lambda_{hn})_i = 0, \quad |(\boldsymbol{\lambda}_{ht})_i| \leq -\mu(\lambda_{hn})_i, \tag{9}$$

for all $\boldsymbol{v}_h \in \mathbf{V}_h$ and $1 \leq i \leq p$, where $(\Phi_n)_i$, $(\lambda_{hn})_i$ and $(\boldsymbol{\lambda}_{ht})_i$ with $1 \leq i \leq p$, denote the nodal values on Γ_C of Φ_{hn}, λ_{hn} and $\boldsymbol{\lambda}_{ht}$ respectively.

Let us define now the discrete version of the operator $\mathcal{L}$ by $\mathcal{L}_h : \boldsymbol{\Sigma}_{ht} \times S_{hn} \to \Sigma_{hn}$ as follows. For all $\boldsymbol{\tau}_h \in \boldsymbol{\Sigma}_{ht}$ and $w_h \in S_{hn}$ we consider the solution $\boldsymbol{u}_h = \mathcal{U}_h(\boldsymbol{\tau}_h, w_h) \in \mathbf{V}_h$ of the following elasto-static problem

$$u_{hn} = w_h \text{ on } \Gamma_C, \quad \int_\Omega \mathcal{C}\varepsilon(\boldsymbol{u}_h) : \varepsilon(\boldsymbol{v}_h)\, d\Omega = \int_{\Gamma_C} \boldsymbol{\tau}_h \cdot \boldsymbol{v}_{ht}\, d\Gamma, \quad \forall \boldsymbol{v}_h \in \mathbf{W}_h,$$

where

$$\mathbf{W}_h =: \Big\{ \boldsymbol{v}_h \in \mathbf{V}_h \;;\; \boldsymbol{v}_h \cdot \mathbf{n} = 0, \quad \text{on } \Gamma_C \Big\}.$$

Let $\mathcal{L}_h(\boldsymbol{\tau}_h, w_h) \in \Sigma_{hn}$ be the normal stress associated to $\boldsymbol{u}_h = \mathcal{U}_h(\boldsymbol{\tau}_h, w_h)$, i.e.

$$\int_\Omega \mathcal{C}\varepsilon(\boldsymbol{u}_h) : \varepsilon(\boldsymbol{v}_h)\, d\Omega = \int_{\Gamma_C} \mathcal{L}_h(\boldsymbol{\tau}_h, w_h) v_{hn}\, d\Gamma + \int_{\Gamma_C} \boldsymbol{\tau}_h \cdot \boldsymbol{v}_{ht}\, d\Gamma, \quad \forall \boldsymbol{v}_h \in \mathbf{V}_h.$$

If p is the dimension of S_{hn} then the (discrete) linear operator $\mathcal{L}_h$ is a $p \times 3p$ matrix for the 3-D problem and a $p \times 2p$ matrix for the in-plane problem.

Let $\mathcal{S}_h$ be the cone (in the space of tangential stresses and normal displacements $\boldsymbol{\Sigma}_{ht} \times S_{hn}$) given by

$$\begin{aligned} \mathcal{S}_h =: \{ & (\boldsymbol{\tau}_h, v_h) \in \boldsymbol{\Sigma}_t \times S_n \;; (\boldsymbol{\sigma}, v_h) \neq 0, \quad (v_h)_i \leq 0, \\ & (v_h)_i |(\boldsymbol{\tau}_h)_i| = 0, \quad 1 \leq i \leq p \}, \end{aligned}$$

and $\mathcal{S}_h^{adm}$ the cone of admissible states

$$\begin{aligned} & \mathcal{S}_h^{adm} =: \{ (\boldsymbol{\tau}_h, v_h) \in \mathcal{S}_h \;; \\ & (\mathcal{L}_h(\boldsymbol{\tau}_h, v_h))_i \leq 0, \; (v_h)_i (\mathcal{L}_h(\boldsymbol{\tau}_h, v_h))_i = 0 \;\; 1 \leq i \leq p \}. \end{aligned}$$

We define the (discrete) supremal functional $J_h : \mathcal{S}_h^{adm} \to \mathbf{R} \cup \{+\infty\}$ as follows

$$J_h(\boldsymbol{\tau}_h, v_h) = \max_{1 \leq i \leq p} Q(|(\boldsymbol{\tau}_h)_i|, (\mathcal{L}_h(\boldsymbol{\tau}_h, v_h))_i),$$

with Q given by (7) and we put μ_h^w as

$$\mu_h^w =: \inf_{(\boldsymbol{\tau}_h, v_h) \in \mathcal{S}_h^{adm}} J_h(\boldsymbol{\tau}_h, v_h).$$

which turns out to be (see the following theorem of [6]) the *(discrete) critical frictional coefficient.*

Theorem 2. *Suppose that $\mathcal{S}_h^{adm}$ is not empty and μ_h^w is finite. Then we have*

i) There exists $(\boldsymbol{\tau}_h^*, v_h^*) \in \mathcal{S}_h^{adm}$ such that $J_h(\boldsymbol{\tau}_h^*, v_h^*) = \mu_h^w$. Moreover, $(\boldsymbol{\Phi}_h^*, \lambda_{hn}^*, \boldsymbol{\lambda}_{ht}^*)$ given by $\boldsymbol{\Phi}_h^* = \mathcal{U}_h(\boldsymbol{\tau}_h^*, v_h^*), \lambda_{hn}^* = \mathcal{L}_h(\boldsymbol{\tau}_h^*, v_h^*), \boldsymbol{\lambda}_{ht}^* = \boldsymbol{\tau}_h^*$, is a solution of $(WP)_h$ for $\mu \geq \mu_h^w$.

ii) If $\mu < \mu_h^w$ then the problem $(WP)_h$ has no solution.

4 Genetic algorithm approach

For the sake of simplicity, only the plane problem will be considered here but the extension to the 3-D problem can be done without any difficulty.

We give in the next some details of application of the Genetic algorithm to the plane problem for $k = 1$. For all $(\tau_h, v_h) \in \Sigma_{ht} \times S_{hn}$ with $\tau_h(x) = \sum_{i=1}^{p} T_i \psi_i(x), v_h(x) = \sum_{i=1}^{p} V_i \psi_i(x)$, we have $(\tau_h)_i = T_i, (v_h)_i = V_i$. First we have to compute the matrix L_{ij} of $\mathcal{L}_h$, i.e.

$$\mathcal{L}_h(\tau_h, v_h)(x) = \sum_{i=1}^{p} \left(\sum_{j=1}^{p} L_{ij} T_j + \sum_{k=1}^{p} L_{i,p+k} V_k \right) \psi_i(x).$$

Since the functional J_h is positively homogenous of degree 0 (i.e. $J_h(t(\boldsymbol{\tau}, v)) = J_h(\boldsymbol{\tau}, v)$ for all $t > 0$) we can normalize $\mathcal{S}_h$ through the "maximum" norm to get

$$\mathcal{S}_h^1 =: \{(\tau_h, v_h) \in \Sigma_t \times S_n \; ; \; V_i \in [-1, 0], \; T_i \in [-1, 1], \; V_i |T_i| = 0, \quad 1 \leq i \leq p\}.$$

The genetic algorithm is a technique of global optimization which can be useful if the computation time for J_h is small and if the dimension of $\mathcal{S}_h^1$ is not too large. In order to increase the efficiency of the algorithm, we reduce the dimension of $\mathcal{S}_h^1$ from $2p$ to p as follows. Firstly we remark that if $(\tau_h, v_h) \in \mathcal{S}_h^1$ then $(T_i, V_i) \in D =: \{0\} \times [-1, 0] \cup [-1, 1] \times \{0\}$. After that we construct $\boldsymbol{\theta} = (T, V) : [-1, 1] \to D$ as a continuous and surjective function (for details of this choice see [6]). We notice that the application $\Psi : (s_1, .., s_p) \to (\sum_{i=1}^{p} T(s_i)\psi_i, \sum_{i=1}^{p} V(s_i)\psi_i)$ is surjective from $[-1, 1]^p$ to $\mathcal{S}_h^1$. We can define now the set

$$K =: \left\{ (s_1, .., s_p) \in [-1, 1]^p \; ; \; \Psi(s_1, .., s_p) \in \mathcal{S}_h^{adm} \right\}$$

and $\mathcal{J} : [-1, 1]^p \to \mathbf{R}_+ \cup \{+\infty\}$ such that $\mathcal{J}(s_1, .., s_p) = J_h(\Psi(s_1, .., s_p))$

$$\mathcal{J}(s_1, .., s_p) = \begin{cases} \max\limits_{i=1,..,p} Q(T(s_i), \sum\limits_{j=1}^{p} L_{ij} T(s_j) + \sum\limits_{k=1}^{p} L_{i,p+k} V(s_k)), \\ \qquad \text{if } (s_1, .., s_p) \in K \\ +\infty, \quad \text{otherwise,} \end{cases}$$

to get the following minimization problem for $\mathcal{J}$ on $[-1, 1]^p$

$$\mu_h^w = \min_{(s_1, .., s_p) \in [-1,1]^p} \mathcal{J}(s_1, .., s_p).$$

From the definition of our optimization problem it is intuitively clear that the supremal functional has a great number of local minima. On the other hand, the functional is very smooth almost everywhere with respect to the parameters s_i inside of the admissible set. With such a local regularity it is straightforward to implement an efficient procedure of local optimization (with Newton's like methods for instance).

Considering those two aspects of our problem we used a stochastic algorithm based on the so called "genetic hybrid technique" (see for instance [4, 3] for the theoretical details of such aglorithms). The main idea of those methods is to manage in the same time a global random exploration of the search space and some local optimization steps. More precisely, we used an implementation of this stochastic method very close from the one proposed in the EO library (see [2]).

5 Numerical results

For the first test we wanted to give an example when the wedged problem and the (linear) spectral problem, introduced in [5, 7], has the same solution. For that we have chosen the wedged geometry of Figure 1, where we do not expect a non contact zone. Here the contact surface Γ_C is represented by the solid line and the other part of the boundary is stress free. We have found a very good agreement ($\mu_h^w = 0.300001$ and $\mu_h^s = 0.300005$) between the two solutions (i.e. between $(\mathbf{\Phi}_h^*, \mu_h^w)$ and $(\mathbf{\Phi}_h^s, \mu_h^s)$).

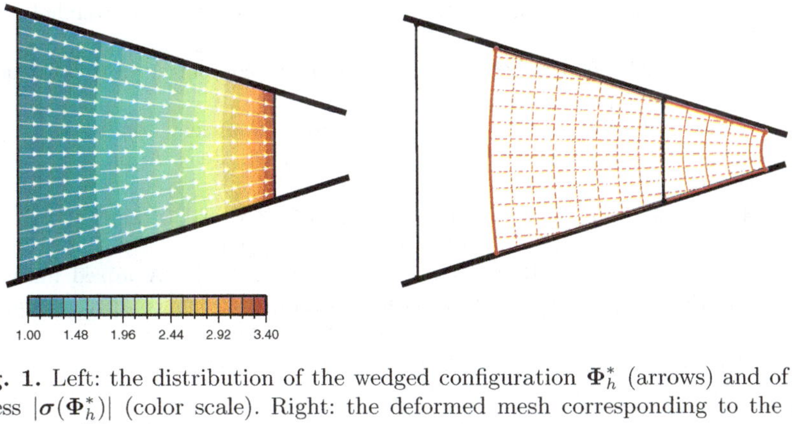

Fig. 1. Left: the distribution of the wedged configuration $\mathbf{\Phi}_h^*$ (arrows) and of the stress $|\sigma(\mathbf{\Phi}_h^*)|$ (color scale). Right: the deformed mesh corresponding to the displacement $\mathbf{\Phi}_h^*$.

In the second test we have considered the geometry drawn in Figure 3. As before the contact surface Γ_C is represented by the solid line and the other part of the boundary is stress free. Since the founded wedged solution $\mathbf{\Phi}_h^*$ exhibits two no contact zones, we cannot even formulate the spectral problem to do the comparison.

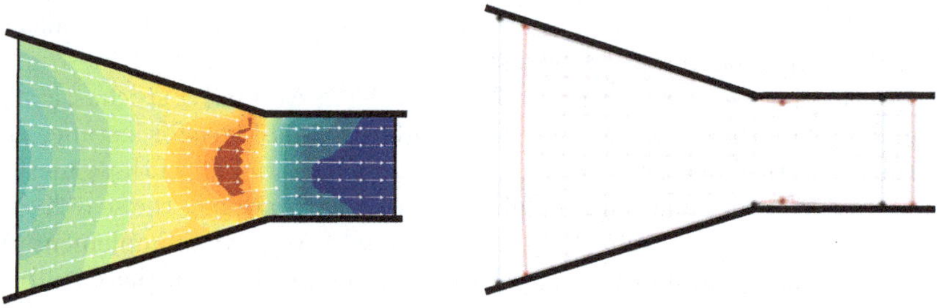

Fig. 2. Left: the distribution of the wedged configuration $\mathbf{\Phi}_h^*$ (arrows) and of the stress $|\boldsymbol{\sigma}(\mathbf{\Phi}_h^*)|$ (color scale). Right: the deformed mesh corresponding to the displacement $\mathbf{\Phi}_h^*$.

References

1. J. Barber and P. Hild (2004), Non-uniqueness, eigenvalues solutions and wedged configurations involving Coulomb frivtion, Proceedings of IJTC 2004, ASME/STLE International Joint Tribology Conference Long Beach California, USA, October 24–27
2. S. Cahon, N. Melab, E.G. Talbi, M. Schoenauer (2003), ParaDisEO-Based Design of Parallel and Distributed Evolutionary Algorithms, Artificial Evolution, 216–228.
3. L. Davis (1991), Handbook of Genetic Algorithms, Van Nostrand Reinhold, New York
4. A. E. Eiben, M. Schoenauer (2002), Evolutionary computing, Special issue on evolutionary computing , Inform. Process. Lett., 82, 1–6.
5. R. Hassani, P. Hild and I. R. Ionescu (2004), "Sufficient conditions of non-uniqueness for the Coulomb friction problem", Math. Meth. Appl. Sciences, Vol. 27, No. 3: 47–67
6. R. Hassani, I. R. Ionescu and E. Oudet (2005), Critical friction and wedged configurations.A genetic algorithm approach, submitted
7. R. Hassani, P. Hild, I. R. Ionescu and N. Sakki (2003), "A mixed finite element method and solution multiplicity for Coulomb frictional contact", *Computer Methods in Appl. Mech. and Engrg.*, vol. 192:4517–4531
8. A. Klarbring (1990), Examples of non-uniqueness and non-existence of solutions to quasistatic contact problems with friction, Ing. Archiv 60 529–541.
9. Martins J, Pinto da Costa A, (2004), Bifurcations and instabilities in frictional contact problems: theoretical relations, computational methods and numerical results, European Congress on Computational Methods in Applied Sciences and Engineering, P. Neittaanmaki, T. Rossi, S. Korotov, E. Onate, J. Périaux, and D. Knorzer (eds.)

Friction and contact between rough surfaces based on elastic-plastic sphere and rigid flat interaction

Y. Kligerman, I. Etsion, V. Brizmer, and Y. Kadin

Faculty of Mechanical Engineering, Technion – Israel Institute of Technology, Technion City, Haifa 32000 Israel,
mermdyk@technion.ac.il

Abstract. The review of a very broad research program carried out at Technion on different aspects of the fundamental contact and friction problem is presented.

Sliding inception is treated as a material failure mechanism. This is different from the conventional local Coulomb friction law approach that is based on a certain arbitrary friction coefficient or some modified forms that assume power law dependency between the normal and tangential stresses at the contact interface. The reviewed models show strong effect of the external normal load and nominal contact area on the static friction coefficient contrary to the classical laws of friction.

Contact of rough surfaces plays an important roll in friction modeling. Surface roughness can be modeled by multitude asperities having spherical summits of uniform curvature but a statistical height distribution. A single asperity interaction can be modeled by the contact between an elastic-plastic sphere and a rigid flat. Hence, the results obtained for the contact of a sphere and flat can be extended to the case of rough surfaces contact and friction.

Since the static friction is considered as a yield mechanism the problem of the elasticity terminus of a spherical contact is also discussed concerning perfectly slip and full stick contact conditions, and failure inception of ductile and brittle materials. The multiple load-unload contact cycles are also considered.

1 Introduction

The elastic-plastic contact and friction between a sphere and a flat is a fundamental problem in contact mechanics and tribology. It is applicable, for example, in areas such as particle handling, sealing, friction, wear, thermal and electrical conductivity, MEMS micro switches, and head-disk interaction in magnetic storage systems to name just a few.

Static friction was considered by the pioneers of friction research: Leonardo da Vinci, Guillame Amontons, Leonard Euler, Charles Augustin de Coulomb, George Rennie, Arthur-Jules Morin, Robert Hooke and others (see [1] and

[2]). In early experimental work it was observed that the proportionality of the force opposing relative motion to the force holding the bodies together seemed to be constant over a range of conditions.

Static friction coefficients are conveniently tabulated and incorporated into engineering handbooks. However, these tabulated values represent average coefficients of friction determined over a broad spectrum of test conditions and rough surface properties. While these numbers provide a general guideline for the sensitivity of the coefficient of friction to the materials in contact, they may not necessarily be representative of the coefficient of friction that will result between actual contact pairs. The friction coefficient is presently recognized as both material- and system-dependent [3] and is definitely not an intrinsic property of two contacting materials.

Chang et al. [4] presented a model (CEB friction model) for predicting the static friction coefficient of rough metallic surfaces. The CEB friction model uses a statistical representation of surface roughness [5] and calculates the static friction force that is required to fail all of the contacting asperities, taking into account their normal preloading. This approach is completely different from the classical Coulomb friction law and shows that the latter is a limiting case of a more general behavior where static friction coefficient actually decreases with an increasing applied load or decreasing nominal contact area. The CEB friction model actually treats the static friction as a plastic yield mechanism corresponding to the first occurrence of plastic deformation in the contacting asperities. This can severely underestimate friction coefficient values for contacting rough surfaces since it neglects the ability of an elastic-plastic deformed asperity to resist additional loading as was demonstrated by Kogut and Etsion [6, 2].

An accurate characterization of the contact between rough surfaces is important for analyzing many tribological problems. These include sealing, friction, performance and life of machine elements, etc.

Several approaches exist in modeling contacting rough surfaces. However, it seems that the most convenient one is the probabilistic approach, which is based on incorporating the behavior of a single asperity in a statistical model of multiple asperity contact.

The first probabilistic model for the contact of rough surfaces was presented by Greenwood and Williamson [5] (the GW model). This classic model is based on the Herz solution for a single elastic spherical asperity. Some pure plastic contact models emerged from the work of Abbott and Firestone [7] that relates the bearing area of a rough surface to its geometrical intersection with a flat.

The models for either pure elastic or pure plastic contact of rough surfaces overlook a wide intermediate range of interest where elastic-plastic contact prevails. An attempt to bridge this gap was made by Chang, et al. [8] (the CEB contact model). In this model each contacting asperity remains in elastic Herzian contact until a critical interference is reached, above which volume conservation of the asperity tip is imposed and a uniform average contact pressure is assumed. The original CEB model, because of its simplicity, was extended to different contacting surface geometries, asperity shapes and asperity height probability density functions (see, e.g. [9-11]).

The model presented in [12] by Kogut and Etsion offers an accurate finite element solution for the elastic perfectly plastic contact of a deformable sphere and a rigid flat by using constitutive laws appropriate to any mode of deformation, be it elastic or plastic. The contact was assumed frictionless. This more accurate FEM solution enables a revision of the CEB elastic-plastic model of contacting rough surfaces. The paper [13] offers such a revision.

New models for contact between an elastic-plastic sphere and a rigid flat are presented in this review. These include issues concerning perfectly slip and full stick contact conditions, failure inception of ductile and brittle materials, and multiple load cycles. A ductile material of the sphere is modeled as an elastic linear hardening one and follows the von Mises yield criterion. The failure of a brittle material follows the maximum tensile stress criterion.

2 The spherical contact model

A schematic representation of the contact problem is shown in Fig 1. A compliant hemispherical body of an un-deformed radius R comes into contact with a rigid flat surface.

The dashed and solid horizontal lines in Fig. 1 show the rigid flat positions before and after loading, respectively. The displacement, ω, of the rigid flat (which will be termed "interference" in the following), and the contact area with a radius, a, (see Fig. 1) correspond to a normal load, P, applied to the contact.

It is assumed that the contact interference, ω, and the contact radius, a, are much smaller than the un-deformed sphere radius, R. Hence, the analysis is carried out assuming small strain theory.

The problem is axisymmetric (2D) about the z axis. Therefore, it is sufficient to consider one half of the axial hemisphere section. The boundary conditions consist of constrain in the vertical direction at the hemisphere base and in the radial direction at its z axis.

The spherical surface is free elsewhere except from the axial constriction enforced by the contacting rigid flat.

3 Elasticity terminus of a spherical contact

In the present course of study the static friction is considered as a yield mechanism corresponding to the plastic deformation in the contacting asperities. Therefore the problem of the elasticity terminus of a spherical contact is important in contact mechanics and friction theory. This problem is analyzed in the work of Brizmer et al. [14] under both perfect slip and full stick conditions

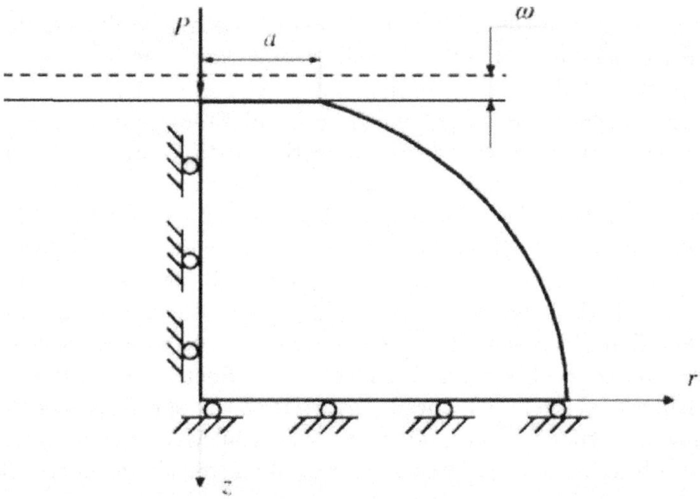

Fig. 1. Schematic model of the contact problem.

for a wide range of the sphere mechanical properties. The effect of these properties on failure inception is investigated by finding the critical interference and normal loading of the first plastic yield or brittle failure. The analysis is based on the analytical Hertz solution under frictionless slip condition and on a numerical solution under stick condition. The failure inception is determined by using either the von Mises criterion of plastic yield or the maximum tensile stress criterion of brittle failure.

An extensive review of the literature on spherical and cylindrical contacts under normal load with and without friction was presented by Adams and Nosonovsky [15] including the subject of yielding inception of normally loaded spherical contact. Most of the existing literature on elasticity terminus of spherical contact concern ductile materials and perfect slip contact condition. Very little work was done so far on the case of spherical contact under stick condition, and the few published solutions are either approximate or very complicated. An accurate analysis that predicts failure inception under full stick condition is still missing.

Applying the von Mises yield criterion for the Hertz solution (see Johnson, [16]) for perfect slip contact (with a rigid flat), one can obtain the critical interference, ω_c, and the critical load, P_c, corresponding to the yield inception of a ductile material:

$$\omega_c = \left[C_\nu \frac{\pi\,(1-\nu^2)}{2}\left(\frac{Y}{E}\right)\right]^2 R \tag{1}$$

$$P_c = \frac{\pi^3}{6} C_\nu^3 Y \left(R\,(1-\nu^2)\frac{Y}{E}\right)^2 \tag{2}$$

$$C_\nu = 1.234 + 1.256\ \nu \text{ for } 0.2 \leqslant \nu \leqslant 0.5 \tag{3}$$

where E and Y are the Young modulus and yield strength of the sphere material, respectively, and ν is its Poisson's ratio.

Similarly using the maximum tensile stress criterion for a perfect slip contact, the critical interference and the critical load corresponding to the failure inception of a brittle material is expressed respectively as follows:

$$\omega_{cf} = \left(\frac{F}{E}\right)^2 \left[\frac{3\pi\ (1-\nu^2)}{2\ (1-2\nu)}\right]^2 R \tag{4}$$

$$P_{cf} = \frac{9\pi^3 F}{2\ (1-2\nu)} \left(R \left(\frac{1-\nu^2}{1-2\nu}\right) \frac{F}{E}\right)^2 \tag{5}$$

where F is the failure tensile strength of the sphere material.

The failure inception for full stick condition was studied numerically, using a commercial ANSYS 8.0 package.

For ductile materials the ratio E/Y covered a range from 200 to 1000, and the Poisson's ratio varied from 0.2 to 0.5. For brittle materials the ratio E/F varied from 4000 to 16000 and the Poisson's ratio from 0.1 to 0.25.

The ratios of the critical interferences, δ_c/ω_c, and critical loads, L_c/P_c, in full stick over perfect slip for ductile materials were found to be functions of the Poisson's ratio only. For small values of the Poisson's ratio the behavior in stick, when high tangential stresses prevail in the contact interface, is much different than in slip. For high values of the Poisson's ratio the tangential stresses under stick condition are low and the behavior of the failure inception in stick and slip is similar.

For brittle materials the critical interference in stick, δ_{cf}, was found to be much larger than that in slip, ω_{cf}. This ratio varies from 34 for $\nu = 0.1$to 78 for $\nu = 0.25$.

4 Repeated loading of elastic-plastic spherical contact

Repeated loading and unloading of rough surfaces is an important problem, especially in the technology of micro/nano-systems, such as MEMS RF micro switches, for example, or head-disk interaction in magnetic storage system.

The process of first unloading of an elastic-plastic loaded sphere in contact with a rigid flat was studied by Etsion et al., [17], for a wide range of material properties and sphere radii. It was found that, depending on the maximum external load, plastic zones may remain locked inside the sphere upon completion of the unloading process. The behavior of an elastic-plastic spherical contact upon multiple loading-unloading cycles was studied by Kadin et al. [18].

The contact was assumed frictionless (perfect slip condition). Cyclic loading was performed for a range of maximum external loads. At each case the same maximum external load was applied periodically following a complete unloading. The sphere material is modeled as an elastic linear hardening and follows the von Mises yield criterion.

After the first loading a plastic core is located beneath the contact close to the sphere axis of symmetry. Another small plastic volume can be observed at the periphery of the contact zone.

The second stage consists of the unloading process. When the unloading process is completed, the contact load, and contact radius fall to zero however, the original un-deformed spherical geometry is not fully recovered. Residual stresses and strains remain locked in, and result in a deformed shape of the unloaded sphere. The equivalent von Mises stresses in certain peripheral zones can increase rather than decrease during the unloading stage.

When the maximum external load of the first loading is relatively moderate, the VM stresses at these peripheral zones after the first loading and then during the first unloading do not exceed the yield strength. In this case upon the second loading the VM stresses follow exactly the first unloading path. In this case the first unloading is reversible and the second loading is elastic.

When the maximum external load is high enough the stresses outside of the contact area after the second loading differ from those after the first loading. In these peripheral zones the von Mises (VM) stresses increase upon first unloading beyond the yield strength. During the second loading the VM stresses decrease, however, contrary to the previous case the first unloading is irreversible. Subsequent loading-unloading cycles are however, elastic.

5 Conclusions

A review of the research studies on contact and friction between rough surfaces based on elastic-plastic sphere and rigid flat interaction that are developed in the Tribology Laboratory of the Technion – Israel Institute of Technology was presented.

Recent results on elasticity terminus of a spherical contact showed that for small values of the Poisson's ratio the behavior in stick is much different than in slip. For high values of Poisson's ratio the behavior of the failure inceptions in stick and slip is similar.

Multiple loading-unloading of elastic-plastic spherical contact was analyzed showing that, depending on the maximum external load, the first unloading may be irreversible. However, subsequent loading-unloading cycles are elastic in a large range of practical maximum external loads.

References

1. Dowson, D. (1979) History of Tribology. Longman Inc., New York.
2. Kogut, L., and Etsion, I. (2004) A Static Friction Model for Elastic-Plastic Contacting Rough Surfaces. *J. Tribol.-T. ASME*, **126**, 34–40.
3. Blau, P. J. (2001) The Significance and Use of the Friction Coefficient. *Tribol. Int.*, **34**, 585–591.
4. Chang, W. R., Etsion, I., and Bogy, D. B. (1988) Static Friction Coefficient Model for Metallic Rough Surfaces. *J. Tribol.-T. ASME*, **110**, 57–63.
5. Greenwood, J. A., and Williamson, J. B. P. (1966) Contact of Nominally Flat Surfaces. *Proc. R. Soc. London, Ser. A*, **295**, 300–319.
6. Kogut, L., and Etsion, I. (2003) A Semi-Analytical Solution for the Sliding Inception of a Spherical Contact. *J. Tribol.-T. ASME*, **125**, 499–506.
7. Abbot, E. J. and Firestone, F. A. (1933) Specifying Surface Quality – a Method Based on Accurate Measurement and Comparison. *Mech. Eng.*, **55**, 569-572.
8. Chang, W. R., Etsion, I., and Bogy, D. B. (1987) An Elastic-plastic Model for the Contact of Rough Surfaces. *J. Tribol.-T. ASME*, **109**, 257–263.
9. Kogut, L., and Etsion, I. (2000) The Contact of a Compliant Curved and a Nominally Flat Rough Surfaces. *Tribol.T.*, **43**, 507–513.
10. Horng, J. H. (1998) An Elastic-Plastic Asperity Microcontact Model for Rough Surfaces. *J. Tribol.-T. ASME*, **120**, 82–88.
11. Yu, N. and Polycarpou, A. A. (2002) Contact of Rough Surfaces with Asymmetric Distribution of Asperity Heights. *J. Tribol.-T. ASME*, **124**, 367–376.
12. Kogut, L., and Etsion, I. (2002) Elastic-Plastic Contact Analysis of a Sphere and a Rigid Flat. *J. Appl. Mech.-T. ASME*, **69**, 657–662.
13. Kogut, L., and Etsion, I. (2003) A Finite Element Based Elastic-Plastic Model for the Contact of Rough Surfaces. *Tribol.T.*, **46**, 383–390.
14. Brizmer, V., Kligerman. Y., and Etsion, I. (2005) The Effect of Contact Conditions and Material Properties on the Elasticity Terminus of a Spherical Contact. *Int. J. Solids Struct.*, (submitted for publication)
15. Adams, G. G., Nosonovsky, M. (2000) Contact Modeling – Forces. *Tribol. Int.*, **33**, 431-442.
16. Johnson, K. L. (1985) Contact Mechanics. Cambridge University Press, Cambridge, MA.
17. Etsion, I., Kligerman, Y., and Kadin, Y. (2005) Unloading of an elastic-plastic loaded spherical contact. *Int. J. Solids Struct.*, **42**, 3716-3729.
18. Kadin, Y., Kligerman, Y., and Etsion, I. (2005) Multiple Loading-Unloading of an Elastic-Plastic Spherical Contact. *J. Mech. Phys. Solids* (submitted for publication).

TEDI (ThermoElasto-Dynamic Instability): a new mechanism for squeal & TEI

L. Afferrant and M. Ciavarella

CEMEC - PoliBa, Via Re David, 200, 70124 Bari, Italy,
luciano@poliba.i, mciava@poliba.it

Abstract. A new mechanism of dynamic instability is found, generated by the thermo-elastic deformations. In particular, it is found that even if coupling between dynamics and heat transfer seems apparently very weak (due to the very different time scales involved), the dynamic modes become unstable for arbitrarily small speeds, in a simple model involving an elastic layer compressed between two rigid plates and sliding out-of-plane. The present analysis neglects the effect of out-of-plane deformations and possible stick-slip in that direction.

1 Introduction

Instabilities in the sliding of elastic bodies are of interest in a wide range of scientific and industrial applications, including for example, the sliding of tectonic plates during earthquakes (Adams, 1995, Ben-Zion 2001, Rice *et al.* 2001), the mechanism of sliding for rubber-like materials (Barquins *et al.* 1996), the generation of noise and vibration in automotive brakes (Kincaid *et al.* 2003, Spurr, 1961) and stick-slip vibrations in machine tool slides and other linear positioning devices (Popp and Rudolph, 2004). These complex phenomena result from the interaction between relatively simple physical processes, notably the elastic deformation of the contacting bodies, the development of frictional forces at the interface opposing the motion. These processes become unstable generally at a critical value of the friction coefficient.

Another category of instability arises because of an additional process, the consequent generation of frictional heat. In fact, frictional heating induces a potentially unstable feedback process, for which thermoelastic deformations induce higher pressures in certain regions which in turn will receive more heating and will expand more, leading to a local increase in the contact pressure. Elementary reasoning shows that the product fV acts as the gain in the feedback process. Thus, we anticipate that for a given friction coefficient, there will be some sliding speed, V_{cr}, above which the system will be unstable. The corresponding eigenmode will then grow exponentially with time, leading eventally to localization of the contact load in a small region of the nominal contact area and to high local temperatures, known as hot spots. This phenomenon is known as *frictionally-excited thermoelastic instability* or *TEI* and is of critical importance in the design of brakes and clutches (Barber, 1969, Anderson and Knapp, 1989, Dow and Burton, 1972, Kennedy and Ling, 1974, Thoms, 1988).

The interaction of these two categories of instability has never been studied before, perhaps under the tacit assumption that the time scales of thermal effects is so much larger than that of dynamics effects, that quasi-static solutions are accurate enough. Dynamics in TEI models (i.e. inertia terms) is always neglected, hence neglecting the possibility that the localization of hot spots may induce additional braking force and stabilize stick-slip (or viceversa, enhance it). Viceversa, thermal effects are generally neglected in the friction dynamic studies, except perhaps indirectly at the asperity scales, where flash temperatures are well known to produce sliding weakening.

However, there is evidence that friction dynamics induces unexpected frictional heat flow behaviour (like the heat flow paradox in the St.Andreas's fault, in the discrepancy between the friction coefficient assumed in the calculations typical of adjacent rocks with the measured value of heat flow, Lachenbruch and Sass, 1973, 1980). On the other hand, TEI is suspected to give rise to low-frequency vibrations (thermal or hot judder, to distinguish it from cold judder). Finally, there are various other unsolved paradoxes in friction dynamics which perhaps would be better understood from a coupled theory. For example, the normal vibrations induced by sliding, and the fact that externally induced high frequency normal vibrations tend to lower friction or suppress stick-slip (Tolstoi, 1967), or the slow fronts in the beautiful recent experimental results by Fineberg's group (Rubinstein Cohen and Fineberg, 2004). The mathematics in principle is not that different in the two classes of instabilities. Both instabilities by definition involve a growth rate term with positive real part, which can be called the exploding eigenvalue. Both are based on some feedback mechanism. Since the problem are mathematically very similar, it would be nice to have a unified treatment.

In order to make a start, it is convenient to think about 1D models. Models as such have been found in the context of TEI for their simplicity to explain key essential features of existance, uniqueness and stability of thermoelastic contact (Barber *et al.* 1980), for the static case, and Ciavarella *et al.* (2003), for the full case with sliding. Hence, in this paper, the first attempt is made to consider the two effects coupled together, by considering the simplest possible system comprising a one-dimensional elastic layer bonded to a rigid half space and sliding against a second rigid half space, in order to see under which conditions, if any, the two classes of instabilities interact.

2 Formulation

Fig. 1 shows an elastic layer $0 < x < h$ which is bonded to a stationary rigid plane A at $x = 0$. A second rigid body B moves out-of-plane at velocity V and its plane surface makes sliding contact with the layer at $x = h$.

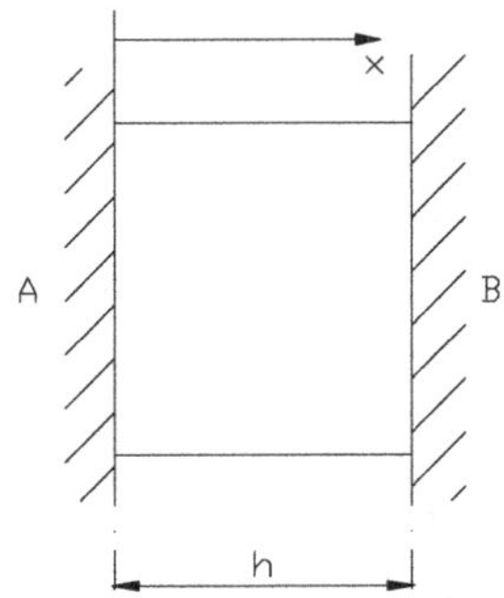

Fig. 1. Layer model.

The undeformed thickness of the layer at temperature $T(x) = 0$ is assumed to exceed h by a small amount Δ so as to ensure an initial contact pressure at the interface, where Coulomb friction conditions are assumed with a coefficient of friction f. We also assume that body A is maintained at temperature $T = 0$ and that the sliding body B is a non-conductor, so that all the heat generated by friction flows through the layer. We restrict attention to the one-dimensional plane strain problem, so the only non-zero displacement is u_x.

These conditions can be summarized in the boundary conditions

$$u_x = 0 \; ; \quad T = 0 \; ; \quad x = 0 \tag{1}$$

$$u_x = -\Delta \; ; \quad K\frac{\partial T}{\partial x} = -fV\sigma_{xx} \; ; \quad x = h \; , \tag{2}$$

where $\sigma_{xx}(x)$ is the tensile stress and K is the thermal conductivity.

The governing equations are the heat conduction equation

$$\frac{\partial^2 T}{\partial x^2} - \frac{1}{k}\frac{\partial T}{\partial t} = 0 \; , \tag{3}$$

the equation of motion

$$\frac{\partial \sigma_{xx}}{\partial x} - \rho\frac{\partial^2 u_x}{\partial t^2} = 0 \tag{4}$$

and the one-dimensional plane strain Hooke's law

$$\sigma = \frac{2\mu(1-\nu)}{(1-2\nu)}\frac{\partial u_x}{\partial x} - \frac{2\mu(1+\nu)\alpha T}{(1-2\nu)}, \tag{5}$$

where $\mu, \nu, \rho, k, \alpha$ are respectively the modulus of rigidity, Poisson's ratio, density, thermal diffusivity and coefficient of expansion for the material of the layer.

2.1 Dimensionless formulation

A convenient dimensionless formulation can be developed by defining the quantities

$$\xi = \frac{x}{h}\,;\quad \hat{u} = \frac{u_x}{h}\,;\quad \hat{t} = \frac{kt}{h^2}\,;\quad \hat{\sigma} = \frac{(1-2\nu)\sigma_{xx}}{2\mu(1-\nu)}\,;\quad \hat{T} = \frac{\alpha(1+\nu)T}{(1-\nu)}, \tag{6}$$

in which case equations (1–5) reduce to

$$\hat{u} = 0\,;\quad \hat{T} = 0\,;\quad \xi = 0 \tag{7}$$

$$\hat{u} = -\hat{\Delta}\,;\quad \frac{\partial \hat{T}}{\partial \xi} = -\hat{V}\hat{\sigma}\,;\quad \xi = 1\,, \tag{8}$$

$$\frac{\partial^2 \hat{T}}{\partial \xi^2} - \frac{\partial \hat{T}}{\partial \hat{t}} = 0\,, \tag{9}$$

$$\frac{\partial \hat{\sigma}}{\partial \xi} - \gamma^2 \frac{\partial^2 \hat{u}}{\partial \hat{t}^2} = 0 \tag{10}$$

$$\hat{\sigma} - \frac{\partial \hat{u}}{\partial \xi} + \hat{T} = 0\,, \tag{11}$$

where the dimensionless parameters

$$\hat{\Delta} = \frac{\Delta}{h}\,;\quad \hat{V} = \frac{2\mu\alpha(1+\nu)fVh}{K(1-2\nu)}\,;\quad \gamma = \frac{k}{ch} \tag{12}$$

and

$$c = \sqrt{\frac{2\mu(1-\nu)}{\rho(1-2\nu)}} \tag{13}$$

is the dilatational wave speed. The parameter γ defines the ratio between the time scales for elastic wave propagation and for heat conduction. The ratio k/c has dimensions of length and for most materials it is extremely small. Some typical values are 1.8 nm (steel), 23 nm (copper), 11 nm (aluminium), 0.8 nm (aluminium oxide). Thus, $\gamma \ll 1$ unless the layer is extremely thin.

2.2 Perturbation analysis

Following Burton *et al.* (1973) and Barber *et al.* (1980), we investigate the stability of the steady state by considering the possibility that a small perturbation in the temperature and displacement fields can grow exponentially with time. Thus we write

$$\hat{T}(\xi,\tau) = \hat{T}_0(\xi,\hat{t}) + \Theta(\xi)e^{b\hat{t}}; \tag{14}$$

$$\hat{u}(\xi,\tau) = \hat{u}_0(\xi,\hat{t}) + U(\xi)e^{b\hat{t}}; \tag{15}$$

$$\sigma(\xi,\tau) = \hat{\sigma}_0(\xi,\hat{t}) + S(\xi)e^{b\hat{t}}, \tag{16}$$

where $\hat{T}_0, \hat{u}_0, \sigma_0$ represent the unperturbed solution, which satisfies (7–11). Substituting (14–16) into (7–11), we find that the perturbation must satisfy the equations

$$\frac{d^2\Theta}{d\xi^2} - b\Theta = 0 \tag{17}$$

$$\frac{dS}{d\xi} - (\gamma b)^2 U = 0 \tag{18}$$

$$S - \frac{dU}{d\xi} + \Theta = 0\,, \tag{19}$$

with homogeneous boundary conditions

$$U = 0\,; \quad \Theta = 0\,; \quad \xi = 0 \tag{20}$$

$$U = 0\,; \quad \frac{d\Theta}{d\xi} + \hat{V}S = 0\,; \quad \xi = 1\,. \tag{21}$$

Solving (17, 18 and 19) with the above boundary conditions the following characteristic equation can be obtained

$$\hat{V} = \frac{\left(1-\gamma^2 z^2\right)\cosh\left(z\right)\sinh\left(\gamma z^2\right)}{\gamma[\cosh\left(\gamma z^2\right)\cosh\left(z\right) - 1 - \gamma z\sinh(z)\sinh\left(\gamma z^2\right)]}\,. \tag{22}$$

where $z = \sqrt{b}$. The zeros of (22) define the eigenvalues, b, at which non-trivial perturbations can exist.

2.3 Zeros of the characteristic equation

Equation (22) can be written in the symbolic form

$$\hat{V} = \mathcal{F}(\gamma,\xi,\eta)\,, \tag{23}$$

where $b = z^2 = \xi + \imath\eta$. Thus, ξ represents the exponential growth rate of the disturbance and η the frequency of the associated oscillation. We are particularly interested in the effect of $\hat{V}$ on ξ, since the system is unstable if there exist any zeros with $\xi > 0$. The dimensionless sliding speed $\hat{V}$ must be real, so (23) requires that

$$\Re(\mathcal{F}(\gamma, \xi, \eta)) = \hat{V} \tag{24}$$

$$\Im(\mathcal{F}(\gamma, \xi, \eta)) = 0 . \tag{25}$$

For a given value of γ, the zeros of equation (22) can be determined parametericaly by (i) selecting a value of ξ, (ii) solving (25) for η and then (iii) substituting ξ, η into (24) to determine $\hat{V}$.

Fig. 2 shows the relationship between $\hat{V}$ and $\xi \equiv \Re(b)$ for $\gamma = 0.1$.

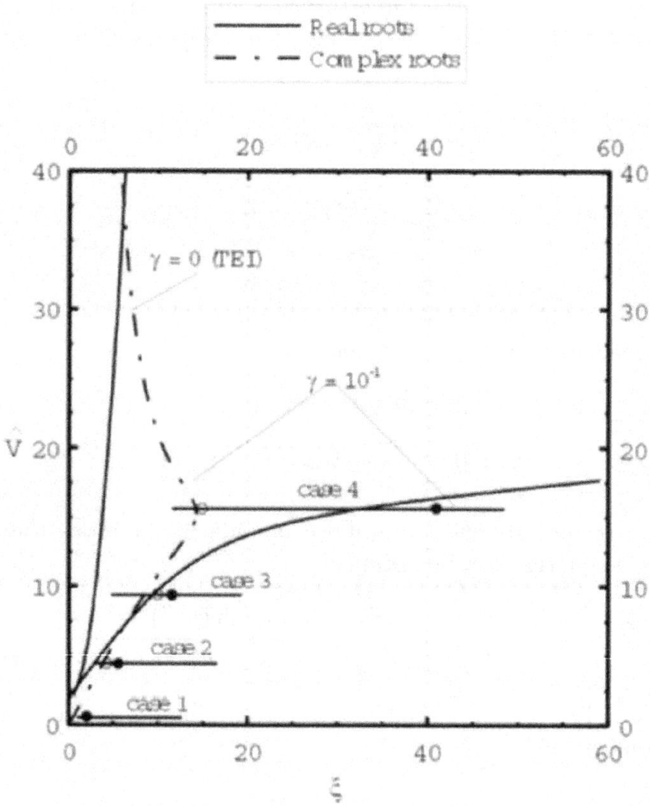

Fig. 2. Variation of the dimensionless sliding speed $\hat{V}$ with the dimensionless real part ξ of the growth rate for $\gamma = 0.1$.

Notice that the real zero passes into the complex plane at the value $\hat{V} = 2$, exactly as in the limiting case $\gamma = 0$. This is to be expected, since a real zero with $\xi = 0$ involves no variation of the stress or temperature perturbations (14–16) in time and hence the acceleration terms in the governing equations make no contribution. Thus, the elastodynamic properties of the system have no effect on the stability boundary, if this is determined by a real zero. However, Figure 2 uncovers the surprising conclusion that complex zeros exist with positive real parts for all values of $\hat{V} > 0$, showing that the system is unstable for all sliding speeds.

3 Numerical study of the transient problem

To explore the transient behaviour of the system, a numerical analysis was performed. The problem is solved (after discretization) as a system of equations at each timestep. We consider the following dimensionless equations to be solve

$$\frac{\partial^2 \hat{T}}{\partial \xi^2} - \frac{\partial \hat{T}}{\partial \hat{t}} = 0 \tag{26}$$

$$\frac{\partial^2 \hat{u}}{\partial \xi^2} - \gamma^2 \frac{\partial^2 \hat{u}}{\partial \hat{t}^2} - \frac{\partial \hat{T}}{\partial \xi} = 0 \tag{27}$$

$$\left.\frac{\partial \hat{T}}{\partial \xi}\right|_{\xi=1} + \frac{1}{2}\left(1 - sign\left(\hat{p}\right)\right)\hat{V}\left(\left.\frac{\partial \hat{u}}{\partial \xi}\right|_{\xi=1} - \left.\hat{T}\right|_{\xi=1}\right) = 0 \tag{28}$$

where $\hat{p}$ is the contact pressure and $sign(\hat{p}) = 1$ if $\hat{p} > 0$ and $sign(\hat{p}) = -1$ if $\hat{p} \leq 0$.

Additionally, the temperature and displacement at $\xi = 0$, the initial temperature, displacement and velocity distributions are prescribed. Further, the displacement at the sliding end ($\xi = 1$, *i.e.* $x = h$) is fixed to $-\Delta$ if the layer is in contact ($\hat{p} > 0$), the displacement is choosen in such way which the pressure venishes if the layer is separated. In this last case (*i.e.* when a negative pressure is detected) the displacement is choosen so that the following condition is satisfied

$$\left.\frac{\partial \hat{u}}{\partial \xi}\right|_{\xi=1} - \left.\hat{T}\right|_{\xi=1} = 0 \tag{29}$$

For the discretization, the layer is divided into n elements so that there are $n - 1$ internal nodes. A sequence of times $[\hat{t}^{start}, \ldots, \hat{t}^k, \hat{t}^{k+1}, \ldots, \hat{t}^{final}]$ is introduced. Equations (1), (2) and (3) will be discretized to form a system of equations for the $2n-2$ temperatures and displacements at the internal nodes and the temperature at $\xi = 1$, all at time $\hat{t}^{k+1}$ and assuming that everything is known at time $\hat{t}^k$.

In particular, equation (1) is discretized by using the Crank-Nicolson implicit scheme (in this way the system of finite difference equations obtained is unconditionally stable) and equation (2) is discretized using the Lax-Wendroff scheme for $\partial^2\hat{u}/\partial\xi^2 - \gamma^2\partial^2\hat{u}/\partial\hat{t}^2$ (which is second-order accurate with truncation error of $O[(\Delta\hat{t})^2, (\Delta\xi)^2])$ and a two-point finite difference approximation for $\partial\hat{T}/\partial\xi$. In all the examples the layer was divided into 100 finite elements of equal length. In the first example, we considered for the parameters γ and $\hat{V}$ the values 0.1 and 1, respectively, corresponding to case 1 of Fig. 2, for which there are only complex eigenvalues.

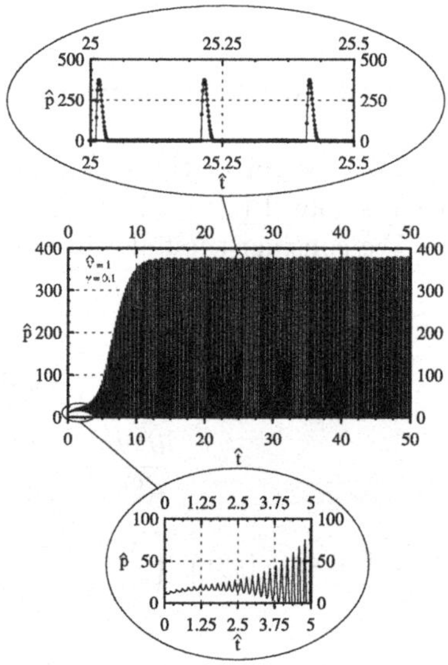

a: case 1

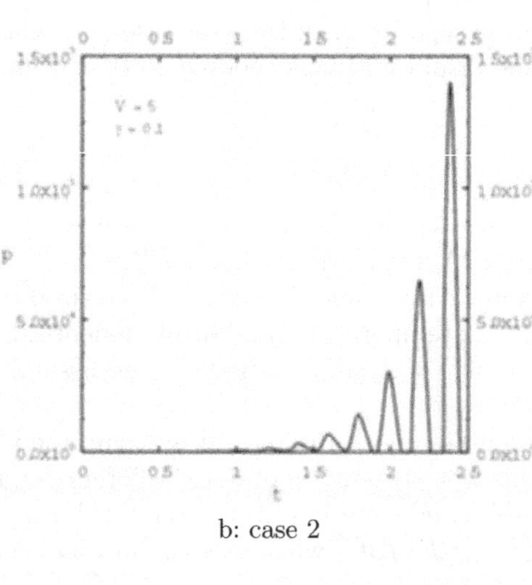

b: case 2

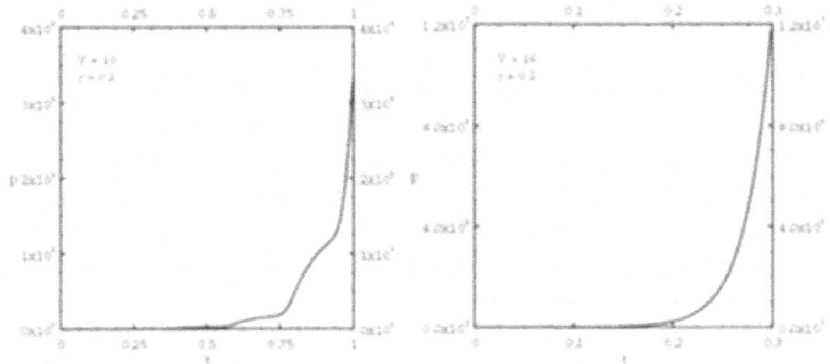

(c: case 3) (d: case 4)

Fig. 3: Evolution of the dimensionless contact pressure $\hat{p}$ at the layer end B ($\xi = 1$) with dimensionless time $\hat{t}$ for $\gamma = 10^{-1}$ and: $\hat{V} = 1$ (case 1); $\hat{V} = 5$ (case 2); $\hat{V} = 10$ (case 3); $\hat{V} = 16$ (case 4).

Fig. 3a shows the evolution of the dimensionless contact pressure $\hat{p}$ with dimensionless time $\hat{t}$. The results confirm that the existence of complex growth rates involves oscillating behavior of the system. In fact, after an initial small oscillation the layer separates and the following transient evolution is characterized by a continuous alternating of contact and separation conditions. Notice the mean contact pressure initially grows and then stabilize to a constant value.

The second example ($\gamma = 0.1$ and $\hat{V} = 5$) corresponds to case 2 of Fig. 2, for which both complex and real eigenvalues exist, but the complex ones dominate.

Fig. 3b shows that the layer experiences alternating contact and separation states. Also in this case the amplitude of the contact pressure oscillations continuously increases (periods longer than those shown in Fig. 3b were also considered).

Fig. 3c shows the transient evolution of the contact pressure when $\gamma = 0.1$ and $\hat{V} = 10$, corresponding to case 3 of Fig. 2 (real growth rates dominate on the complex ones). The pressure grows without limit, but the existence of complex eigenvalues involves a very small oscillation (the frequencies are very low).

Similar behavior can be noted in Fig. 3d where the evolution with time of the contact pressure is shown for $\gamma = 0.1$ and $\hat{V} = 16$.

4 Conclusions

We have shown a new mechanism for dynamic instability generated by TEI in a simple (1D) model of a layer sliding out-of-plane against a rigid plate. In this model, since we've neglected, as a preliminary investigation, the possibility of shear perturbations (i.e. bending out-of-plane), the dynamic modes are pure imaginary growth corresponding to simple waves of tension-compression travelling along the layer in the normal direction, whereas TEI instability gives a real growth rate above a critical sliding speed. The coupling of the two instabilities gives a complex growth rate at arbitrarily low speeds, which in practise will lead eventually to a non-linear regime of separation and impacts (possibly corresponding to noise). At larger speeds, the TEI modes continue to exist leading to seizure but are enhanced.

References

1. Y. Ben-Zion, Dynamic ruptures in recent models of earthquake faults, *J.Mech.Phys.Solids*, Vol.49 (2001), pp.2209–2244.
2. J.R. Rice, N.Lapusta and K.Ranjith, Rate and state dependent friction and the stability of sliding between elastically deformable solids, *J.Mech.Phys.Solids*, Vol.49 (2001), pp.1865–1898.
3. G.G. Adams, G.G., Self-excited oscillations of two elastic half-spaces sliding with a constant coefficient of friction, *ASME J. Appl. Mech.*, Vol. 62 (1995) pp. 867–872.
4. M. Barquins, A.A.Koudine and D.Vallet, On the existence of Schallamach waves during surface friction in rubber-like material as a function of thickness, *Comptes Rendu Mecanique,* Vol.323 (1996), pp.433–438.
5. N.M. Kinkaid, O.M.O'Reilly and P.Papadopoulos, Automotive disc brake squeal, *J. Sound Vib.*, Vol.267 (2003), pp.105-166.
6. R.T. Spurr, A Theory of Brake Squeal. Proceedings of the Automotive Division. Institute of Mechanical Engineers (AD) 1, (1961) pp. 33-40.
7. K. Popp and M. Rudolph, Vibration control to avoid stick-slip motion, *J.Vib.Control*, Vol.10 (2004), pp.1585–1600.
8. J.R. Barber, 1969 Thermoelastic instabilities in the sliding of conforming solids, *Proc. Roy.Soc.*, Vol. A312, pp. 381–394.
9. A.E. Anderson, & R.A. Knapp, 1989 Hot spotting in automotive friction systems, *Intl. Conf. on Wear of Materials*, Vol. 2, pp. 673–680.
10. T.A. Dow, & R.A. Burton, 1972 Thermoelastic instability of sliding contact in the absence of wear, *Wear*, Vol. 19, pp. 315–328.
11. F.E. Kennedy, & F.F. Ling, 1974 A thermal, thermoelastic and wear simulation of a high energy sliding contact problem, *ASME J.Lub.Tech.*, Vol. 96, pp. 497–507.
12. E. Thoms, 1988 Disc brakes for heavy vehicles, *Inst. Mech. Eng., Intl. Conf. on Disc Brakes for Commercial Vehicles*, C464/88, pp. 133–137.
13. A. Lachenbruch, J., Sass, 1973. Thermo-mechanical aspects of the San Andreas fault system. In: Proceedings of the Conference on Tectonic Problems of the San Andreas Fault System. pp. 192–205.

14. A. Lachenbruch, J., Sass, 1980. Heat flow and energetics of the San Andreas fault zone. Journal of Geophysical Research 85, 6185–6223.
15. D.M. Tolstoi, Signifiance of the normal degree of freedom and natural normal vibrations in contact friction, Wear 10 (1967) 199–213.
16. SM Rubinstein, G Cohen, J. Fineberg (2004) Detachment fronts and the onset of dynamic friction, Nature, 430 (7003): 1005-1009
17. J.R. Barber, J. Dundurs and M. Comninou, Stability considerations in thermoelastic contact, *ASME J. Appl. Mech.*, Vol.47 (1980), pp.871–874.
18. M. Ciavarella, L. Johansson, L.Afferrante, A.Klarbring and J.R.Barber, Interaction of thermal contact resistance and frictional heating in thermoelastic instability, *Int. J. Solids Structures*, Vol 40 (2003), pp. 5583–5597.

Contact behaviour of a sliding rubber element

W. Sextro[1], P. Moldenhauer[2], M. Wangenheim[2], M. Lindner[2], and M. Kröger[2]

[1] Institute of Mechanics, Graz University of Technology, Kopernikusgasse 24/III, A-8010 Graz, sextro@tugraz.at
[2] Institute of Mechanics, University of Hannover, Appelstr. 11, D-30167 Hannover, moldenhauer@ifm.uni-hannover.de

Abstract. Friction occurs in the contact between tyre and road. The friction of rubber material on dry surfaces is dominated by hysteresis and adhesion effects. Hysterisis friction is characterised by the energy dissipation within the visco-elastic material, which is caused by its deformation while passing the surface roughness. Hysteresis effects are modelled by an extended linear visco-elastic material with several MAXWELL elements. The development of a model in the time domain allows to consider nonlinear effects. Additionally temperature effects are taken into account based on the WLF-transformation. Adhesion forces originate from molecular bindings between the contact partners. This effect is simulated by applying a modified model of ACHENBACH on real surfaces. The temperature distribution within the friction contact region is investigated experimentally as well. Furthermore global stick-slip vibrations of a rubber block element are investigated using a global contact model. Numerical results are compared with experiments performed on a tribometer test rig.

1 Introduction

In many technical applications, e. g. seals, belts and tyres, the contact of rubber elements influences the dynamical behaviour. It can cause stick-slip-vibrations or squeal and hence acoustic problems. The friction contact behaviour depends on many parameters like relative velocity, surface roughness, normal pressure, lubrication, material properties and temperature. The friction of rubber can be classified according to four different physical phenomena in hysteresis, adhesion, cohesion and viscous friction. Hysteresis originates from internal material damping i.e. the deformation energy put into the elastomer can only partially be regained, see [1]. Adhesion friction is caused by intermolecular links between the contact partners that have to be separated. All the experiments mentioned in the following have been performed on the tribometer test rig, which is shown in Fig. 1. The rubber sample is pressed

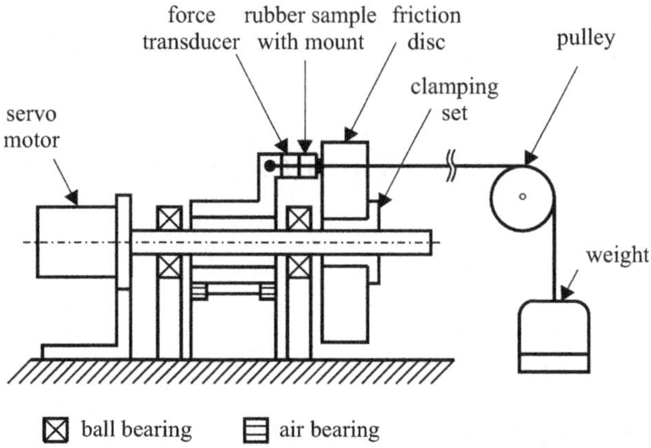

Fig. 1. Drawing of the tribometer test rig.

onto the rotating friction disk either by weights or by a magnetic actuator. Normal forces up to $F_N = 150$ N can be realised. The disc can be equipped with a variety of friction surfaces, e.g. steel, glass, aluminium, wood, grinding paper or asphalt and is driven by a servo motor. The relative velocity in the friction contact is adjustable between $v = 0.1$ mm/s and 3000 mm/s. The contact forces are measured by two three-component piezoelectric transducers. By means of a special chamber which encloses the friction disc and the rubber sample experiments at environmental temperatures between $T = 20°$C and 80°C can be realised. The rubber compound used for the experiments in this work is throughout a styrene butadiene rubber (SBR) filled with 60 phr carbon black.

To model large dynamic systems such as tyres it is necessary to develop effective numerical methods. A further reduction of computation time can be realised by a modal reduction of the investigated system.The contact has to be discretised due to the in general spatial motion and the non-linearity of the contact behaviour with friction. A point contact model can be applied which considers the friction contact, see [2]. Here, the friction contact with rough surfaces is described by nonlinear point contacts which consider the friction characteristic with respect to the sliding velocity or the normal force. This characteristic has to be identified either by means of physical models or experiments, see chap. 2. In chap. 3 a reduced model is applied to a rubber element to calculate stick-slip vibrations. To achieve a reduction of the computation time the rough surface is modelled as smooth and possesses the identified friction characteristic.

2 Modelling of rubber friction

2.1 Hysteresis friction

The presented hysteresis model is a generalised MAXWELL model, see Fig. 2. It is assumed that the deformations of the rubber are relatively small so that a linear model can be applied. However, material nonlinearities can be included by nonlinear spring and damper elements or MASING elements which allow to consider nonlinear hysteresis loops.

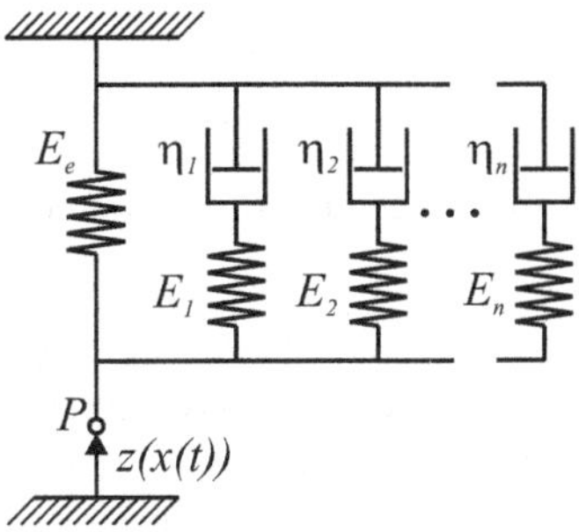

Fig. 2. Generalised MAXWELL model for hysteresis simulation.

The system parameters are identified by a fitting process of the frequency dependent storage E' and loss modulus E'', cp. [3]. Fig. 3 shows the approximation for a model with $n = 20$ MAXWELL elements. This number seems to be adequate to describe the dynamic behaviour of real elastomers. Each MAXWELL element implies a characteristic time constant $\tau_i = \eta_i / E_i$, $(i = 1, ..., n)$.

In the simulation the foot point P of the system is excited by a two-dimensional surface profile $z(x(t))$ which is time-dependent due to different sliding velocities v. The hysteretic friction coefficient μ_{hys} is given by

$$\mu_{hys} = \frac{W_{diss}}{\sigma_m A_0 x_{fric}} \tag{1}$$

with the dissipated energy W_{diss}, the average absolute stress at the foot point σ_m, the nominal contact area A_0 and the sliding distance x_{fric}. The dissipated energy W_{diss} is determined numerically by

$$W_{diss} = V \int_0^{t_{fric}} \sigma(t) \dot{\varepsilon}(t) \, dt, \tag{2}$$

while the relevant terms of the current stress $\sigma(t)$ and the time derivative of the strain $\dot{\varepsilon}(t)$ are output values from the simulation. Here, t_{fric} denotes the sliding time and V the volume of the excited rubber.

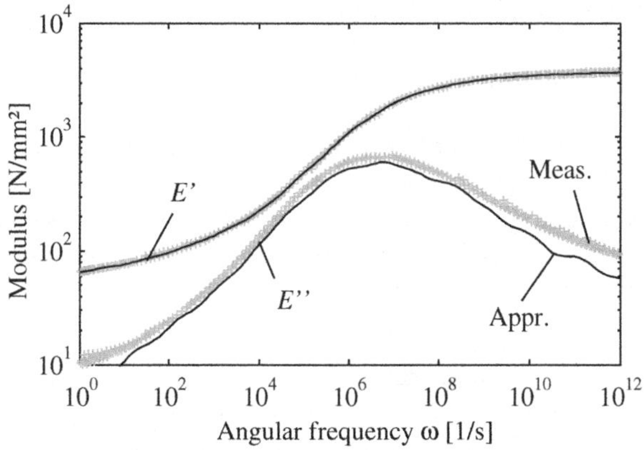

Fig. 3. Measured and approximated moduli from DMTA.

For a harmonic excitation of the foot point P the numerical results for the dissipated energy W_{diss} can be compared to the analytical solution, which show a very good agreement, see [1] and [4]. To estimate the friction coefficient on real rough surfaces measured height profiles are used as input parameter for the hysteresis simulation. Investigations with various roughness profiles show a typical maximum with respect to the sliding velocity and a dependency on the contained amplitudes and wavelengths, see [1].

The former remarks are limited to isothermal rubber conditions. However, in practice the rubber heats up with respect to the sliding velocity and the normal pressure due to the frictional power. This temperature increase changes the dynamic properties of the material significantly as it is formulated in the well-known WLF-equation [5]. In the simulation this effect can be considered by a modification of the time constants τ_i of the MAXWELL elements, which are only valid for the reference temperature T_{ref}. The time constants $\tau_{T,i}$ for a particular temperature T are determined by the shift factor a_T which can be found by applying the WLF-equation and the original time constants τ_i for the reference temperature $\tau_{T,i} = a_T \tau_i$. For the estimation of the temperature in the friction contact experiments with a thermography camera have been carried out for sliding velocities between 10...3000 mm/s. The measured temperature characteristic is used as input parameter for the calculation of the temperature dependent friction coefficient.

2.2 Adhesion friction

In the following a model for adhesion friction is presented which is based on the theory of ACHENBACH. In [6] an exemplary roughness is characterised as a superposition of two sinusoids which represent a macro and a micro roughness. The micro roughness has a characteristic amplitude H and wavelength L both in the nanometer scale. ACHENBACH defines a condition for the wetting of the micro cavities. They are spontaneously filled if the adhesion energy is greater or equal than the elastic energy of the rubber. The excitation frequency ω_m due to the micro roughness is given by $\omega_m = 2\pi v/L$ with v as the sliding velocity. To separate the surfaces the resulting friction force of adhesion F_{ad} is necessary. ACHENBACH introduces a wetting degree η which is a measure for the filling of the micro cavities. It depends on the relaxed modulus E_e and the dynamic storage modulus $E'(\omega_m)$,

$$\eta = \sqrt{\frac{E_e}{E'(\omega_m)}}. \tag{3}$$

For $v \to 0$ the wetting degree yields 1 and lessens with increasing sliding velocity due to the dynamic stiffening of the rubber. The macroscopic area of contact determines the potential regions for adhesive wetting which leads to a description of the adhesion force F_{ad} as the product of the macroscopic contact area A_{macr}, the wetting degree η, the surface energy $\Delta\gamma$ and an adhesion constant k_{adh},

$$F_{ad} = k_{adh} \Delta\gamma \eta(\omega_m) A_{macr}. \tag{4}$$

Fig. 4 shows the wetting degree η for a variation of the wavelength L of the micro roughness. Note that the wetting degree has diminished decisively at a velocity of 1 mm/s, where the macroscopic area of contact can be assumed as nearly constant.

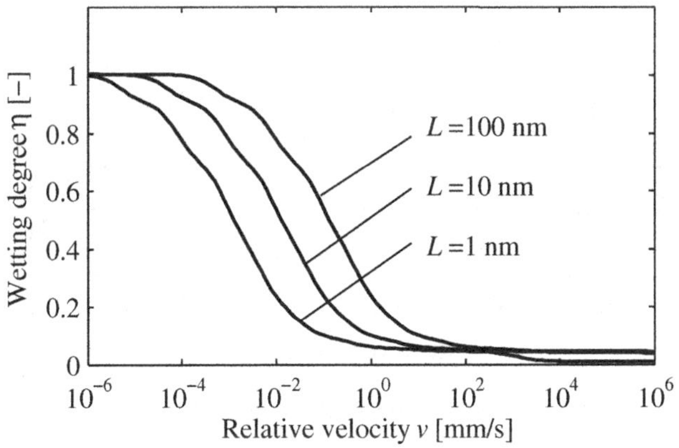

Fig. 4. Influence of the micro roughness on the wetting degree.

2.3 Comparison with experiments

The results of the hysteresis and adhesion modelling are now being compared to measurements conducted on the tribometer test rig. The compounds of the friction test sample and the dynamic modulus test sample, cp. Fig. 3, are identical. The friction surface is an aluminium oxide grinding paper (grit 400). The test sample has a diameter of 10 mm and a thickness of 2 mm and is loaded with an average normal pressure of $\sigma_m = 0.25$ N/mm^2. In the adhesion model the wavelength of the micro roughness has been set to $L = 50$ nm. The simulations for hysteresis and adhesion were performed with a model of 20 MAXWELL elements. The temperature has been considered in the hysteresis simulation by an approximation of the measurements from thermography. Fig. 5 depicts the measured friction coefficient, the hysteresis and adhesion simulation and the superposition. The measurements show an explicit maximum at a relative velocity of about $v = 200$ mm/s. In a higher velocity range the friction coefficient falls due to the viscoelastic character of the rubber. Up to a velocity of $v = 20$ mm/s the friction coefficient decreases which can result from adhesion friction. The comparison of the simulations with the experiments is very good.

These results from local rubber friction investigations can be used as an input for larger systems like tread blocks as a part of tyres.

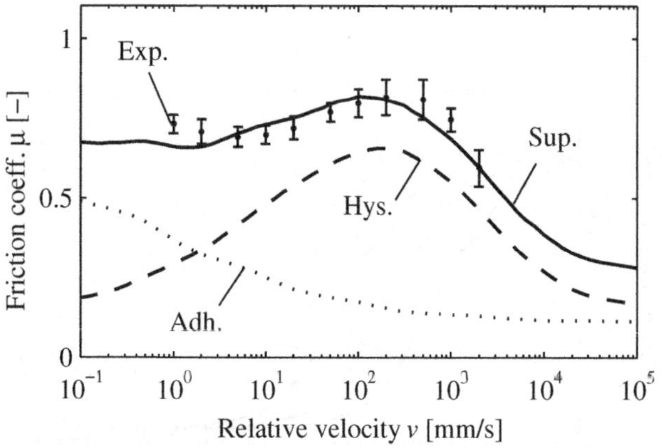

Fig. 5. Comparison Experiment - Simulation.

3 Rubber block investigations

Further experiments have been performed with rubber blocks. The dimension of these blocks are 15 x 15 mm^2 in contact area and 10 mm in height. The rubber material composition is SBR filled with 60 phr carbon black. The average contact pressure is maintained at $\sigma_m = 0.25$ N/mm^2.

3.1 Friction temperature

The temperature in the friction contact has great impact on the mechanical characteristics and friction behaviour of the participating materials. Therefore investigations on the temperature distribution in the contact region are of interest. Friction experiments have been performed with a rubber block on different surfaces. The temperature distribution within the contact area has been observed with a thermography camera. Fig. 6 shows the temperature distribution recorded with the thermography camera through an elongated hole in the friction surface. The width of the hole is 5 mm. The experiment parameters have been chosen to a sliding velocity of $v = 200$ mm/s on an aluminium oxide grinding paper (grit 400). As expected the largest temperatures occur at the trailing edge of the block.

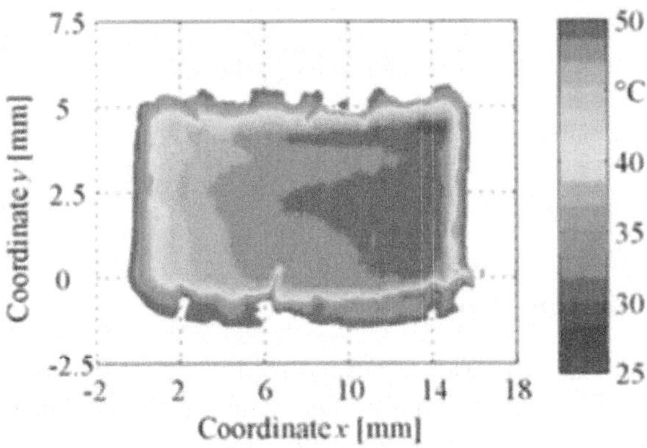

Fig. 6. Temperature distribution in friction contact.

Johnson [7] published solutions describing the temperature distribution in sliding contacts founded on the equation of heat conduction by Jaeger. For the temperature distribution within the contact area of the rubber element results the following expression:

$$\Delta\theta(X,Y) = \frac{q}{v\pi\rho c_p} \int\limits_{X-L}^{X+L} e^{-\xi} \int\limits_{Y-B}^{Y+B} \frac{exp\left[-\sqrt{(\xi^2+\zeta^2)}\right]}{\sqrt{(\xi^2+\zeta^2)}} \cdot d\xi \cdot d\zeta. \qquad (5)$$

with the width B, the lenght L, the temperature θ, density ρ, sliding velocity v, heat per time and area unit q, and the specific heat c_p. Hence, the temperature distribution in the investigated friction contact can be calculated. Fig. 7 shows the calculated and measured temperature at the centerline ($y = 2.5$ mm) of the contact area for a sliding velocity of $v = 100$ mm/s. Thermography experiment and calculation results of the temperature distribution show very good agreement as well.

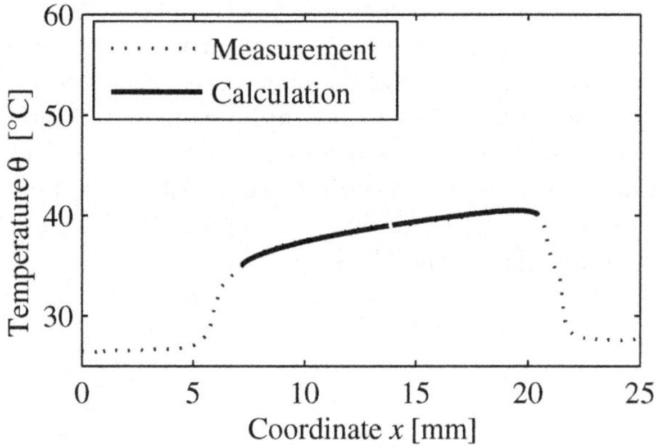

Fig. 7. Comparison of results with regard to temperature distribution.

3.2 Global vibration investigation

In this paragraph the attention is turned to dynamical friction phenomena of a rubber block. The global dynamic behaviour of the rubber element in the sliding direction is simulated by a 1-DOF oscillator using a linear standard model with one MAXWELL element, see Fig. 8. In this approach the actual flexural mode of the block is modelled as a longitudinal mode of the linear standard model with an identified mass m. Initially the normal force F_N and the sliding velocity v are required to calculate the tangential friction force F_T. The system parameters, i.e. spring and damping coefficients c_e, c_m, and d_e, the mass m and the sliding velocity dependent friction characteristics $\mu(v)$, have been identified with the experimental set-up shown in Fig. 1. The displacement x in sliding direction and its derivatives with respect to time are simulated by the model as the resulting output parameters. Here for small tangential displacements the rubber parameters' dependency on the exciting

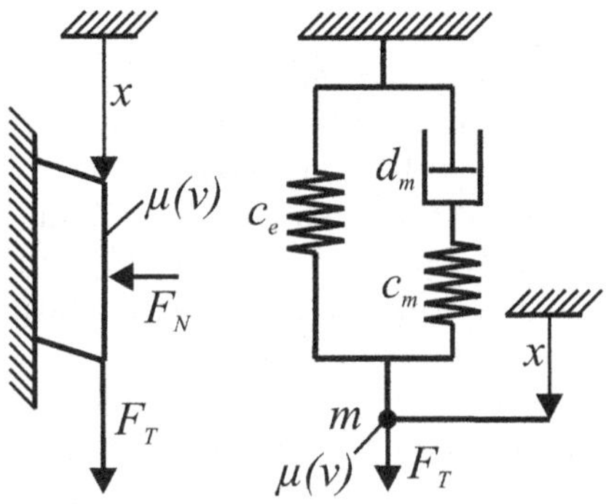

Fig. 8. Linear standard model for tangential displacement of rubber element.

amplitude can be neglected . Exemplary stick-slip vibrations are investigated in simulation and experiment.

Fig. 9 shows the phase diagram of the stick-slip vibrations of the rubber element on a glass surface. In the friction contact, the tangential velocity of the disc is $v = 14$ mm/s. The frequency of the stick-slip vibrations is $f = 31.5$ Hz. Especially during the slip phase simulation and measurement agree well. The velocity and displacement signals have been measured with a LASER vibrometer. As a matter of principle, the movement of the rubber element can only be measured with a distance of about 2 mm to the contact area. Therefore, the overshoot of the velocity at the beginning of the stick phase is not covered by the model.

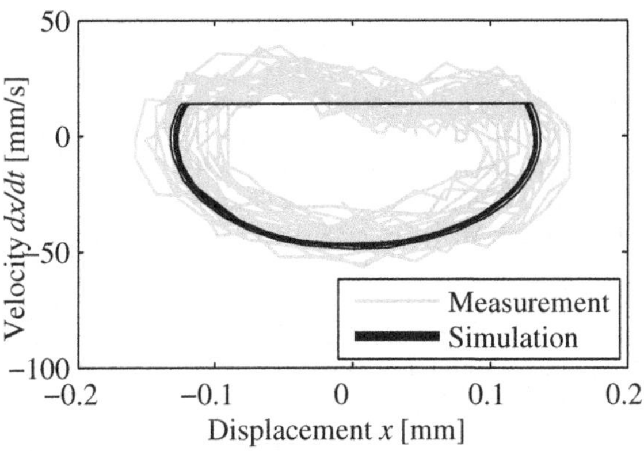

Fig. 9. Comparison of results.

4 Conclusion

Within this work a generalised MAXWELL model has been presented to simulate the hysteretic component of rubber friction which is strongly dependent on the visco-elastic character of the material. Temperature effects have been taken into account by the WLF transformation. A modified adhesion model of ACHENBACH has been adopted. A comparison with experiments shows a satisfactory agreement. Experiments on the temperature distribution in the friction contact of a rubber block showed good agreement to the corresponding numerical results. Global stick-slip vibrations have been simulated. Here simulation and measurements match relatively good as well.

In future investigations it is intended to consider temperature effects in the adhesion mode and to calculate the contact temperature of the rubber block by the frictional power. Furthermore the developed theory for a sliding rubber element will be applied to the rolling contact of tyres.

Acknowledgements

The authors want to dedicate this contribution to Prof. Karl Popp who initiated this work and advised its progress. Sadly, he passed away before it could be finished.

The investigations published in this report received financial support from the Deutsche Forschungsgemeinschaft (DFG) as a part of the collaborate research unit DFG-Forschergruppe "Dynamische Kontaktprobleme mit Reibung bei Elastomeren", to whom we extend our thanks. We also thank Dr. M. Klüppel from Deutsches Institut für Kautschuktechnologie (DIK) for providing rubber samples and experimental data.

References

1. Lindner M, Sextro W, Popp K (2004) Proc Appl Math Mech 4:101-102
2. Sextro W (2002) Dynamical contact problems with friction. Lecture notes in applied mechanics. Springer, Berlin Heidelberg New York
3. Achenbach M, Herdy M (2003) Kautschuk Gummi Kunststoffe 1-2/2003:24-31
4. Moldenhauer P, Lindner M, Kröger M, Popp K (2005) Modelling of hysteresis and adhesion friction of rubber in time domain. In: Austrell P E, Kari L (eds) Constitutive models for rubber IV:515-520. Balkema, Rotterdam
5. Williams M L, Landel R F, Ferry J D (1955) J Amer Chem Soc 77:3701-3707
6. Achenbach M, Frank E (2001) Tribologie und Schmierungstechnik 4:43-47
7. Johnson K L (1985) Contact Mechanics, Cambridge University Press, Cambridge

Sliding friction and contact mechanics of elastomers on rough surfaces

A. Le Gal, X. Yang, and M. Klüppel

Deutsches Institut für Kautschuktechnologie e.V., Eupener Str. 33, D-30519 Hannover

1 Introduction

The modeling of rubber frictional properties on rough surfaces is of considerable importance for the prediction of tire traction properties. The presented work consists in a physically based model describing the dynamic contact and the sliding friction of elastomers on rough surfaces. The self-affine character of rough substrates is analytically treated with correlation functions which depend on three surface descriptors. This allows a generalization of the concepts introduced by Greenwood and Williamson and a quantitative estimation of the velocity dependent real area of contact. The hysteresis friction can be modelled in the frame of this theory, originating from the energy losses due to the local deformation of the rubber from the surface asperities during dynamic contact. The experimental results show that the frictional behaviour under wet conditions is fairly described by the hysteresis friction. The transition from wet to dry friction is explained via an adhesion component assumed to be related with the real area of contact. This approach is also confirmed by the experiment and gives a deeper insight in the relationship between the material viscoelastic properties, the surface roughness and the frictional behavior.

2 Modeling of dynamic contact on rough surfaces

2.1 Self-affinity of rough surfaces

Recently, the self-affine behavior of road surfaces was observed, revealing the fractal nature of such objects [1]. This refers to a statistical invariance of the morphology under anisotropic dilations. In this case, the roughness is described by two closely related correlation functions, the height-difference correlation function $C_z(\lambda) =< (z(x+\lambda) - z(x))^2 >$ and the auto-correlation function $\Gamma_z(\lambda) =< z(x+\lambda)z(x) > - < z(x) >^2$, respectively, where the average < ... > is taken over all realizations of the profile $z(x)$ of the surface. For self-affine surfaces, characteristic scaling assumptions can be applied for the correlation functions. In particular, C_z (λ) follows a power law on small length scales:

$$C_z(\lambda) = \xi_\perp^2 \left(\frac{\lambda}{\xi_\parallel}\right)^{6-2D} \qquad for\ \lambda < \xi_\parallel \tag{1}$$

and approaches a constant value $\xi_\perp^2$ for $\lambda > \xi_\parallel$. The cross-over length scales $\xi_\parallel$ and $\xi_\perp$ describe the maximum roughness parallel and perpendicular to the profile and the exponent is governed by the surface fractal dimension $D(2 < D < 3)$.

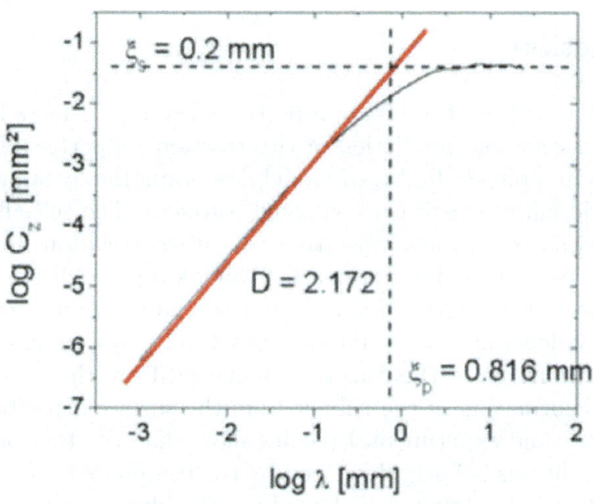

Fig. 1. Stylus measurement of the profile (left) and corresponding height difference correlation function (right) of the granite surface. The fractal descriptors $D \equiv 3 - H = 2.17$, $\xi_\perp = 0.20mm$, $\xi_\parallel = 0.81mm$ are indicated.

A similar behavior can be derived for the Fourier transform of the auto-correlation function, i. e. the spectral power density $S(f)$.. An example of the height-difference correlation of a rough granite surface, obtained from fifteen different profile measurements, is shown in Fig. 1.

2.2 Extension of the Greenwood-Williamson concepts

The modeling the dynamic contact of elastomers with self-affine interfaces is based on the Greenwood-Williamson (GW) approach of elastic contact with spheres of constant radius R and height distribution $\varphi(z)$. The load F_N is decomposed as a sum of N distinct contact forces $F_{n,i}$ of Hertzian type in dependence of the distance d between the rubber surface and the mean height $< z >$ of the surface formed by the spherical summits:

$$F_N = \sum_{i=1}^{N} F_{n,i} = \frac{16}{9} N \; |E^*(\omega)| \; R^{1/2} \int_d^\infty (z-d)^{3/2} \; \varphi(z)\, dz \qquad (2)$$

Here, $|E^*(\omega)|$ is the norm of the frequency dependent complex modulus that provides an extension of the model to dynamic contact conditions. The true contact area A is found as the sum of individual contact patches πa_i^2:

$$A = \pi \sum_{i=1}^{N} a_i^2 = 2\pi \; N \; R \int_d^\infty (z-d)\, \varphi(z)\, dz \qquad (3)$$

For an extension of the GW-concept to self-affine surfaces we consider an approximation of the rough surface by an arrangement of spherical summits, representing the asperities of the rough track on the largest length scale. The spherical summits are assumed to have a fixed curvature radius R given by the second derivative of the modulation of the surface profile on the largest roughness scale, which yields $R = \xi_{\|}^2/(4\pi^2\xi_\perp)$ [2].

The height distribution $\varphi_s(z)$ of the spherical summits can be obtained by comparing an analytical affine transformation of the original profile with a numerical distribution of the local maxima. This permits to identify an affine parameter s and the region of the profile in contact with the elastomer [3].

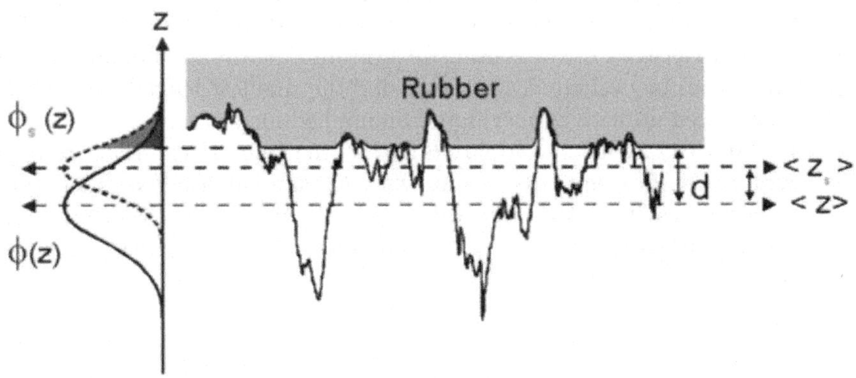

Fig. 2. Schematic view of the two height distributions $\varphi(z)$ of the surface profile and $\varphi_s(z)$ of the spherical summits on the largest roughness scale. The mean profile height $< z >$ and the summit height $< z_s >$ as well as the distance d between $< z >$ and the rubber surface are indicated.

The load F_N or the apparent stress $\sigma_o = F_N/A_o$ can now be evaluated similar to Equ. (2), if the summit distribution $\varphi_s(z)$ and the corresponding distance d are inserted [2,3]

$$\sigma_o \approx \frac{0.53\ \xi_\perp\ \left|E^*\left(\omega_{\min}\right)\right|}{\pi\ s^{3/2}\ \xi_{\|}}\ F_{3/2}\left(\frac{d_s}{\tilde{\sigma}_s}\right) \tag{4}$$

Here, we have used the relation $\tilde{\sigma}^2 = 1/2\ \xi_\perp^2$ and the abbreviation:

$$F_{3/2}(t) = \int_t^\infty (z-t)^{3/2}\ \varphi_s(z)\ dz \tag{5}$$

3 Real area of contact

The real area of contact is of considerable importance, as it determines the amount of cavities filled by the rubber as well as the number and size of local contact spots involved in the frictional process. In the case of self-affine surfaces, the interval of contact is found to be distributed over many length scales. It can be derived from an energy condition, stating that the elastic deformation work necessary to fill up a cavity should be larger that the elastically stored energy in the stress field of the rubber [2]:

$$\sigma(\lambda)\lambda^2 h(\lambda) \geq |E^* \cdot (\lambda)| h^3(\lambda) \tag{6}$$

Here, $|E^*(\lambda)|$ is the norm of the complex dynamic modulus at frequency $\omega = 2\pi\nu/\lambda$, ν is the sliding velocity and $h(\lambda)$ is the deformation of the rubber while filling up a cavity of horizontal size λ. From Equ. (6) a minimum contact length λ_{min} can be evaluated, characterising the smallest length scale where cavities are filled up with rubber under an applied load $\sigma(\lambda)$.

The real area of contact is then derived as the sum of "external" contact spots with the surface summits, scaling with F_0, and "internal" contact spots with the surface cavities, depending on $F_{3/2}$. It can be evaluated via the scaling behaviour exhibited by the microscopic summits and cavities of the profile as following [2,3]:

$$A_c(\lambda_{min}) = A_0 \left(\frac{1}{808\pi} \frac{(2D-4)^2\, \xi_{\|} F_0^2\left(\frac{d}{\tilde{\sigma}}\right) F_{3/2}\left(\frac{d}{\tilde{\sigma}_s}\right) \left|E^*\left(\frac{2\pi\,\nu}{\xi_{\|}}\right)\right|}{s^{3/2}\,(2D-2)^2\,\xi_\perp \left|E^*\left(\frac{2\pi\,\nu}{\lambda_{min}}\right)\right|} \right)^{\frac{1}{3}} \tag{7}$$

The parameter s is the affine scaling factor of order one, which relates the height distribution $\varphi(z)$ of the roughness profile to the summit height distribution $\varphi_s(z)$ at the largest roughness scale.

Fig. 4 shows simulations of the real area of contact in the range of low sliding velocities on the rough granite surface, calculated on the basis of the frequency dependent complex modulus at 3.5 % strain for the various S-SBR samples. The load is fixed at σ_o= 12.3 kPa. The curves are simulated within the velocity range of stationary sliding friction (10^{-6} to 1 m/s) for the silica and carbon black filled S-SBR in comparison with the same polymer without any filler system. The results indicate that the real area of contact is much higher in the case of unfilled material and lies around 0.3% of the nominal area at very low velocities. This effect is attributed to the soft character of unfilled rubber that promotes an improved ability for the rubber to follow the asperities of the rough substrate. A monotone decrease of A_c with rising velocity is observed, which can be related to an increase of the material stiffness while approaching the glassy domain, limiting the penetration of the surface corrugations in the rubber.

4 Modeling of rubber sliding friction

4.1 Hysteresis friction

The modeling of rubber friction on self-affine surfaces has been recently treated by several authors [2,4]. It is based on the hysteretic energy losses arising from the rubber deformation by surface asperities. Indeed, a rough surface characterized by a cavity on length scale λ in contact with an elastomer solid, gives rise to oscillating forces on the rubber with a frequency $\omega = 2\pi\nu/\lambda$.

Then, the frictional force resulting from hysteresis effects of the rubber can be estimated via the dissipated energy:

$$\Delta\tilde{E}_{diss} = \int\limits_0^V \int\limits_0^T d^3x \; dt \; \sigma \cdot \dot{\varepsilon} \tag{8}$$

where V is excitation volume, T is time of sliding and $\dot{\varepsilon}$ is time derivative of strain.

Due to the fractal nature of rough surfaces, the dynamic contact can be expressed as the sum of elementary processes distributed over many length scales. This leads to the following expression for the hysteresis friction coefficient [2]:

$$\mu_H \equiv \frac{F_H}{F_N} = \frac{1}{2\left(2\pi\right)^2} \frac{<\delta>}{\sigma_o \, \nu} \int\limits_{\omega_{min}}^{\omega_{max}} d\omega \; \omega \; \mathrm{E}'\left(\omega\right) \cdot S\left(\omega\right) \tag{9}$$

where E$''$ is the frequency dependent loss modulus of the rubber. Also, $<\delta>\equiv V/A_o$ is the layer thickness of the excitation volume. This layer thickness is assumed to increase linear with the mean penetration depth $<z_p>$ of the rubber ($<\delta> = b <z_p>$), which leaves a free parameter b. A contact analysis carried out via photogrammetry measurements showed that this assumption is verified on model indentors, leading to the conclusion that high strain amplitudes should be considered for the characterization of the dynamic modulus E*.

The two boundary frequencies in the integral of Equ. (9) indicate the frequency range of excitation of the rubber. They correspond to characteristic cut-off lengths of the interval of contact in which the rubber fills out the cavities of the rough substrate, with $\omega_{min} = 2\pi\nu/\xi_{||}$ and $\omega_{max} = 2\pi\nu/\lambda_{min}$.

Experimental friction tests were performed for two S-SBR systems filled with the same amount of carbon black and silica on a rough granite surface. In order to prevent the effects due to abrasion and heat build-up at the interface, the test were carried out up to a maximal sliding velocity of v = 1 cm/s and at a fixed load $\sigma_o = 12.3$ kPa. Moreover, under wet conditions, the surface was lubricated with a stabilized water detergent film in order to prevent interfacial effects due to the adhesion and identify the hysteresis contribution. Fig. 3 shows that the hysteresis friction fairly describes the frictional behavior under wet conditions for both compounds. It also appears that the level of dry friction is strongly influenced by the type of filler, whereby the silica filled compound exhibits higher dry friction values over the whole range of velocity.

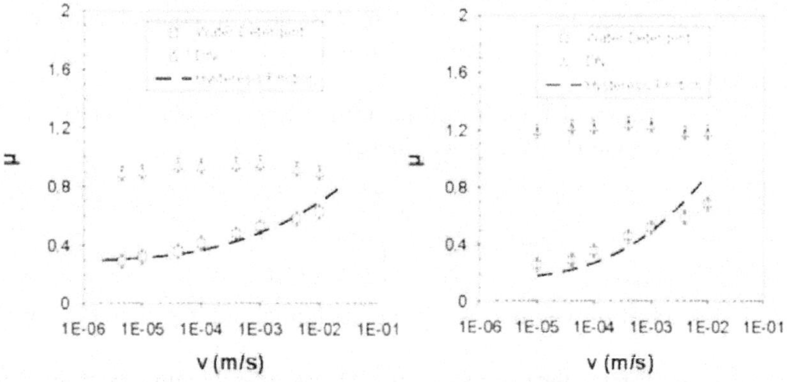

Fig. 3. Stationary friction curves (symbols) for the carbon black (left) and silica (right) filled S-SBR sample on a granite surface at load $\sigma_o = 12.3$ kPa under dry and wet conditions. For the simulated hysteresis, it yields $<\delta> / <z_p> = 42$ and $<\delta> / <z_p> = 50$, respectively.

5 Formulation of the adhesion friction

The additional contribution under dry conditions is interpreted in terms of adhesion effects occurring at the rubber-substrate interface, which scale with the real area of contact A_c. Accordingly, the friction coefficient can be written [5]:

$$\mu = \mu_A + \mu_H \qquad with \quad \mu_A = \frac{F_A}{F_n} = \frac{\sigma_s}{\sigma_o} \cdot \frac{A_c}{A_o} \tag{10}$$

The adhesion force F_A is determined by the interfacial true shearing stress σ_s, required to break the contact junctions. This quantity is treated as an open parameter that can be estimated from a fit to experimental data.

Fig. 4 shows the difference of the friction coefficients $\Delta\mu$ obtained under dry and lubricated (water detergent) conditions on the granite surface for the two filler systems in the range of low sliding velocities. Obviously, the data can be well adapted with the adhesion friction coefficient shown as dashed lines, demonstrating that a proper description of rubber friction on dry surfaces is possible with Equ. (10) and lead to satisfying correlation in the range of low sliding velocities. As expected, the adhesive component decreases successively with increasing sliding velocity.

It is interesting to note that the different levels of $\Delta\mu$ for the carbon black and silica systems depicted in Fig. 4 correlates with that of the real area of contact. Indeed, the same difference of about 40% is found for the silica and carbon black filled systems. The examination of the fitting parameters

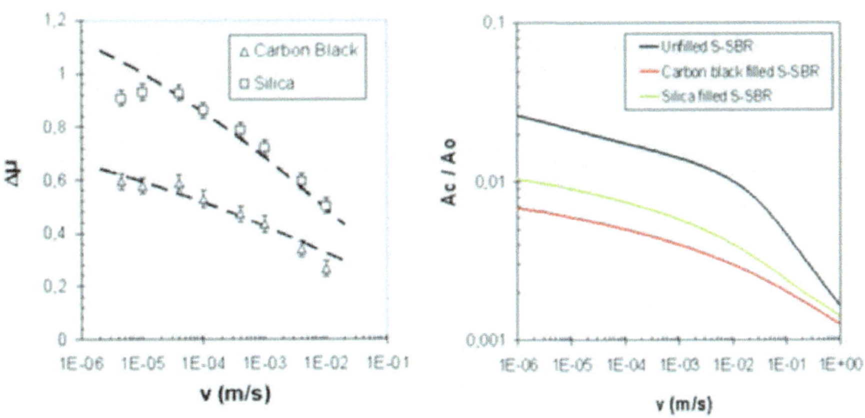

Fig. 4. Difference $\Delta\mu$ (symbols) of the experimental friction data under dry and lubricated (water detergent) conditions on granite at $\sigma_o = 12.3$ kPa and adaptations (lines) with Equ. (10) for the carbon black and silica filled S-SBR (left). Parameters are $\sigma_s = 13.2$ MPa and $\sigma_s = 13.8$ MPa, respectively. Simulated real area of contact (right).

σ_s shows that it is not the cause of the observed experimental difference. It confirms that the level of adhesion friction is governed by the real area of contact and indirectly by the viscoelastic properties of the filler materials [5].

6 Conclusion

The presented model proposes a formulation for the frictional behavior of rubber on rough surfaces under different contact conditions. The hysteresis friction arising from the energy losses associated with the local indentation of the rubber appears to be the dominating effect under wet conditions. The transition from wet to dry friction is assumed to be related with adhesion effects proportional to the real area of contact. Experimental data show a fair agreement with this formulation within the range of low sliding velocity and point out the role played by the viscoelastic properties during the frictional process.

References

1. Z. Radó, PhD-Thesis, Pennsylvania State University, USA (1994)
2. M. Klüppel and G. Heinrich, Rubber Chem. Technol. **73**, 578 (2000); ibid. Paper No. 43, ACS Rubber Division Meeting, Chicago, 13.-16. May (1999)
3. M. Klüppel, A. Müller, A. Le Gal and G. Heinrich, "Dynamic contact of tires with road tracks", Paper No. 49, ACS Meeting, San Francisco, April 28-30 (2003)
4. B. N. J. Persson, J. Chem. Phys. **115**, 3840 (2001); ibid. Paper No. 24, ACS Rubber Division Meeting, Cleveland, Ohio, 16. - 19. Oct. (2001)
5. A. Le Gal, X. Yang and M. Klüppel, Journal of Chemical Physics, in press (2005)

Unsteady rolling contact of rubber wheels

F. Gutzeit[1], W. Sextro[2], and M. Kröger[1]

[1] Institute of Dynamics and Vibrations, University of Hannover, Appelstr. 11, D-30167 Hannover,
gutzeit@ifm.uni-hannover.de

[2] Institute of Mechanics, Graz University of Technology, Kopernikusgasse 24/III, A-8010 Graz,
sextro@tugraz.at

Abstract. This paper presents a model for unsteady longitudinal slippage of a loaded rubber wheel. The aim of modelling was to calculate the dynamical contact behavior in a numerical efficient way. Further on, simulation results are compared to experimental data.

1 Introduction

Steady models for rolling contact loose accuracy, if the slippage excitation contains dynamical components. Exactly this kind of excitation is caused by the interventions of driving stability systems like ABS or ESP. A numerical efficient model for unsteady rolling contact would improve controlling results. The presented model was realized for a small solid rubber wheel (Ω 80 mm, width 18 mm). The developed algorithms are to be applied to tires in future. The measurements were done with an autonomous vehicle, see [1]. In the experiments, the velocity v_{abs} of the vehicle is hold constant, while the time depending slippage

$$\nu(t) = \frac{v_{abs} - \Omega(t)\, r_w}{v_{abs}} \tag{1}$$

of the wheel is forced. Figure 1 shows the orientation of the reference velocity v_{abs} and the angular velocity Ω, as well as the radius r_w of the undeformed wheel.

To calculate the contact behavior of a loaded wheel, the motion is divided in the rigid body motion and the relative deformations, see [2]. The relative deformations are composed of great static deformations, described by a FE-discretisation, and small dynamic deformations, which are approximated by a modal condensation of the FE-grid model. The dynamic contact behavior is calculated by point contact elements, attached to the shell surface of the wheel structure. The numerical efficiency of the model is gained by the combination

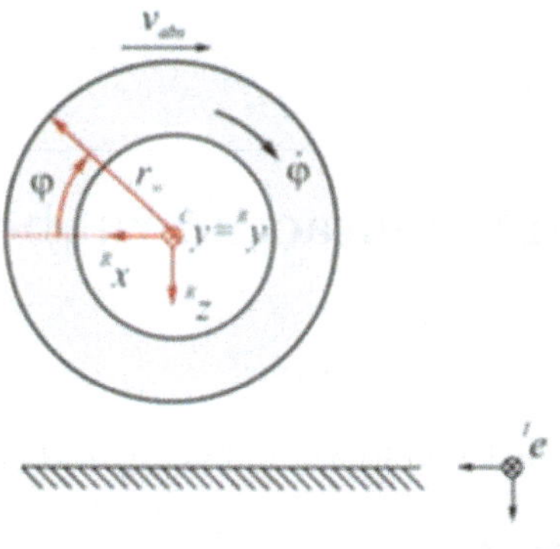

Fig. 1. Used coordinate systems and kinematic parameters.

of the modal approximated structure dynamics and the discretized description of the contact patch, see [3].

2 Model overview

For the description of the model, three coordinate systems are introduced, see Figure 1. The cylindrical ${}^{C}\mathbf{e}$ and cartesian coordinate system ${}^{R}\mathbf{e}$ accompany the wheel, the inertial system ${}^{I}\mathbf{e}$ is fixed to the ground.

Before the time step integration starts, the deformation of a still-standing wheel under a normal load F_N is calculated by a FE-approximation, where the contact nodes are free within the contact area. The normal force F_N corresponds to a static displacement r_{stat} in the normal direction. The nodes of the wheel center are displaced against the ground. The contact computation is carried out using the penalty method. Due to the nonlinear behavior of the system, a Newton-Raphson procedure is applied for the calculation of the large static displacements. The cylindrical coordinates ${}^{C}\mathbf{r}_{N,\,i}$ of each contacting node N_i are retrieved. The two dimensional cartesian static displacements ${}^{R}\mathbf{u}_{stat}({}^{C}\mathbf{r}_{N,\,i})$ of the contact nodes are arranged to the interpolation matrix ${}^{R}\mathbf{H}_{stat}({}^{C}\mathbf{r}_{N})$. The static contact pressure distribution is available in an appropriate interpolation matrix ${}^{R}\mathbf{H}_{p}({}^{C}\mathbf{r}_{N})$.

In the simulation, the structure dynamics are approximated by a linear modal approach and the nonlinear contact behavior is described by using a point contact model, see Figure 3. The contact patch is discretized and the local tangential forces $\mathbf{F}_{T,\,k}$ for passing material points M_k are obtained. Thus, the interpolation matrix of the contact shear stress distribution ${}^{R}\mathbf{H}_{\tau}({}^{C}_{R}\mathbf{r}_{M},\,t)$ is assembled to act on the wheel, in addition to the actuation moment M_{act}, which is generated by an actuator realizing the given reference angular velocity $\Omega_{ref}(t)$.

The wheel dynamics consist of the rigid body model and the modal condensed structure. With the modal condensed structure, the comparatively

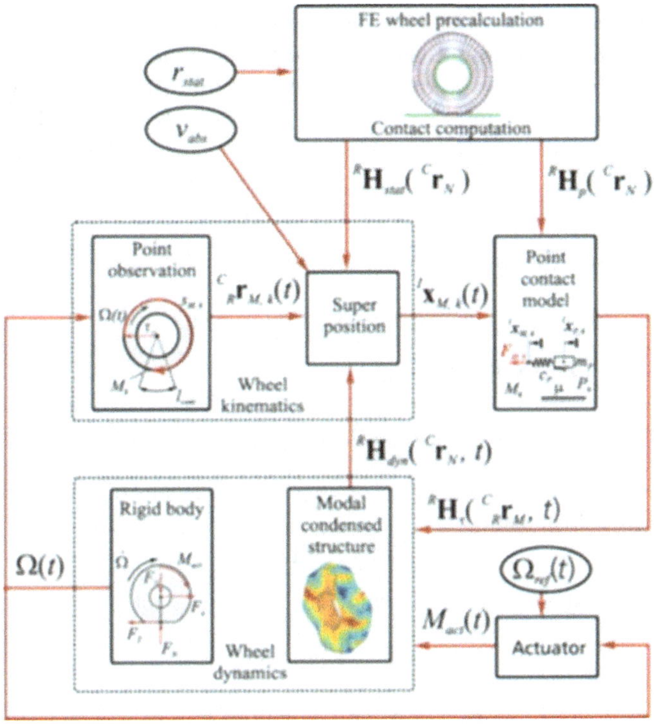

Fig. 2. Model for unsteady rolling contacts (input parameters are framed ovally).

small dynamic displacements ${}^{R}\mathbf{u}_{dyn}({}^{C}\mathbf{r}_{N,i}, t)$ are computed, which are composed to the interpolation matrix ${}^{R}\mathbf{H}_{dyn}({}^{C}\mathbf{r}_{N}, t)$, where gyroscopic effects are neglected. The rigid body system is used to gain the actual angular velocity $\Omega(\mathrm{t})$ of the wheel.

In the wheel kinematics, firstly the cylindrical coordinates ${}^{C}_{R}\mathbf{r}_{M,k}(t)$ of the material points M_k are observed within an intermediate, rigid body configuration R. For these coordinates, the static and dynamic displacements are gained by using the interpolation matrices ${}^{R}\mathbf{H}_{stat}({}^{C}\mathbf{r}_{N})$ and ${}^{R}\mathbf{H}_{dyn}({}^{C}\mathbf{r}_{N}, t)$. The calculated coordinates ${}^{R}\mathbf{x}_{M,k}(t)$ of M_k within the deformed configuration are transformed into the inertial system with the help of the velocity v_{abs}.

3 Wheel dynamics

As mentioned in Chapter 2, the dynamic displacements of a rolling, preloaded wheel structure are approximated by a modal condensation of the still standing FE-model. Before the Eigenvalue Problem is solved, the system's boundary conditions have to be changed. During the contact computation, the displacements of the wheel center nodes were impressed. For the time step integration, the displacements of the wheel center nodes are only fixed in radial direction to transduce the actuation moment to the wheel structure. Furthermore, the vertical displacements of the contact nodes are inhibited. The boundary conditions are realized by means of penalty matrices $\mathbf{K}_{cent}$ and $\mathbf{K}_{cont}$, which are added on the stiffness matrix $\mathbf{K}$ of the basic system. Thus, the resulting stiffness matrix

$$\mathbf{K}^* = \mathbf{K} + \mathbf{K}_{cent} + \mathbf{K}_{cont} \tag{2}$$

is introduced and the undamped eigenvalue problem

$$\left(-\omega_0^2\,\mathbf{M} + \mathbf{K}^*\right)\hat{\mathbf{u}}_{dyn} = \mathbf{0} \tag{3}$$

can be solved numerically. The modal matrix

$$\mathbf{T} = \left[\,\hat{\mathbf{u}}_{dyn,\,1}\;\hat{\mathbf{u}}_{dyn,\,2}\;\cdots\;\hat{\mathbf{u}}_{dyn,\,m}\,\right] \tag{4}$$

is composed of the first m eigenvectors $\hat{\mathbf{u}}_{dyn,\,r}$. By means of the modal matrix, the modal transformation

$$\mathbf{u}_{dyn} = \mathbf{T}\mathbf{q} \tag{5}$$

can be accomplished. Then, the equations of motion get decoupled with

$$\mathbf{M}_M = \mathbf{T}^T\mathbf{M}\mathbf{T} = \mathrm{diag}[m_r], \tag{6}$$

$$\mathbf{K}_M = \mathbf{T}^T\,\mathbf{K}^*\mathbf{T} = \mathrm{diag}[k_r]. \tag{7}$$

To derive the modal damping matrix, the Rayleigh hypothesis

$$\mathbf{D}_M = \mathbf{T}^T(\alpha\mathbf{M} + \beta\,\mathbf{K}^*)\mathbf{T} \tag{8}$$

is applied. Finally, the equation of motion is given by

$$\mathbf{M}_M\ddot{\mathbf{q}} + \mathbf{D}_M\dot{\mathbf{q}} + \mathbf{K}_M\mathbf{q} = \mathbf{T}^T(\mathbf{L}_{act}M_{act}(t) + \mathbf{A}_T\mathbf{F}_T(t)), \tag{9}$$

with the lever matrix $\mathbf{L}_{act}$, distributing the actuation moment $M_{act}(t)$ on the wheel center nodes, and the allocation matrix $\mathbf{A}_T$, assigning the contact DoFs to the DoFs of the entire system. The system is excited by the actuation moment $M_{act}(t)$ and the nodal contact forces $\mathbf{F}_{T,\,i}$, which are obtained by the interpolation matrix of the shear stress distribution ${}^R\mathbf{H}_\tau({}^C_R\mathbf{r}_M,\,t)$ and the corresponding contact area.

During the simulation, the modal coordinates $\mathbf{q}(t)$ of the modal differential equation are calculated for every time step. The needed dynamic displacements of the contact nodes are obtained by

$${}^R\mathbf{u}_{dyn} = \mathbf{A}_T^T\mathbf{T}\mathbf{q}, \tag{10}$$

which are finally arranged to the interpolation matrix ${}^R\mathbf{H}_{dyn}({}^C\mathbf{r}_N,\,t)$.

The actual angular velocity $\Omega(t)$ of the rigid body system is derived with the principle of conservation of angular momentum

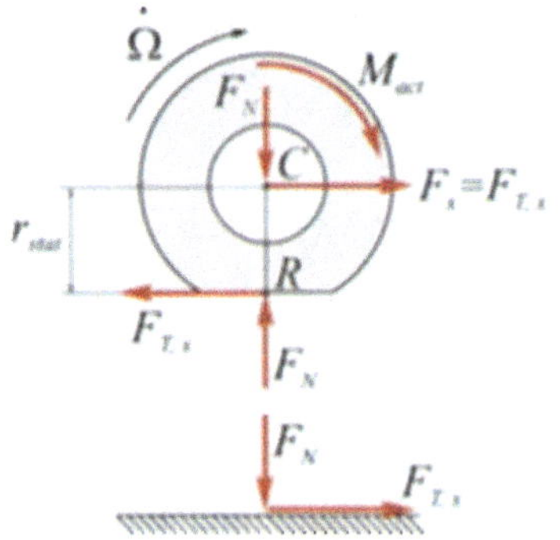

Fig. 3. Equilibrium of forces to obtain $\Omega(t)$.

$$\Omega(t) = \frac{1}{J_{yy}^{(C)}} \int_0^t (M_{act}(\bar{t}) + F_{T,x}(\bar{t})\, r_{stat})\, \mathrm{d}\bar{t} + \Omega_0, \tag{11}$$

with the angular momentum $J_{yy}^{(C)}$ of the wheel referring to the center point C of the wheel and the radius of the static deformed wheel r_{stat}, see Figure 4. Because of symmetry, the normal pressure distribution has no influence in this equation. The angular momentum $J_{yy}^{(C)}$ of the wheel can be calculated analytically due to the simple geometry.

4 Wheel kinematics

To obtain the motion of the material points M_k lying on the shell surface of the moving, deforming wheel, the points are firstly observed within an intermediate rigid body configuration R, expressed by the coordinates of an accompanying cylindrical coordinate system ${}^{C}\mathbf{e}$. These coordinates ${}^{C}_{R}\mathbf{r}_{M,k}(t)$ are cylindrical to make use of the axially symmetry of the wheel. In the point contact model, only those material points currently passing the contact patch are observed. To describe the positions of these points, it is beneficial to use cartesian coordinates. The cartesian coordinates ${}^{R}\mathbf{x}_{M,k}(t)$ of the points M_k within the deformed configuration are gained by superposing the cartesian coordinates ${}^{R}_{R}\mathbf{x}_{M,k}(t)$ of the points M_k within the rigid body configuration and the interpolated cartesian displacements

$${}^{R}\mathbf{x}_{M,k}(t) = {}^{R}_{R}\mathbf{x}_{M,k}(t) + {}^{R}\mathbf{u}_{stat}({}^{C}_{R}\mathbf{r}_{M,k}(t)) + {}^{R}\mathbf{u}_{dyn}({}^{C}_{R}\mathbf{r}_{M,k}(t), t). \tag{12}$$

The displacements ${}^{R}\mathbf{u}_{stat}({}^{C}_{R}\mathbf{r}_{M,k}(t))$ and ${}^{R}\mathbf{u}_{dyn}({}^{C}_{R}\mathbf{r}_{M,k}(t), t)$ of the points M_k are derived by applying the interpolation matrices ${}^{R}\mathbf{H}_{stat}({}^{C}\mathbf{r}_N)$ and ${}^{R}\mathbf{H}_{dyn}({}^{C}\mathbf{r}_N, t)$, respectively. Finally, the calculated coordinates are transformed into the inertial system

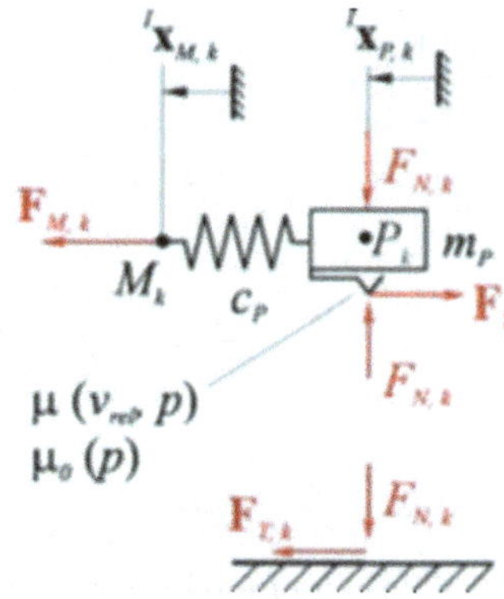

Fig. 4. Point contact element P_k.

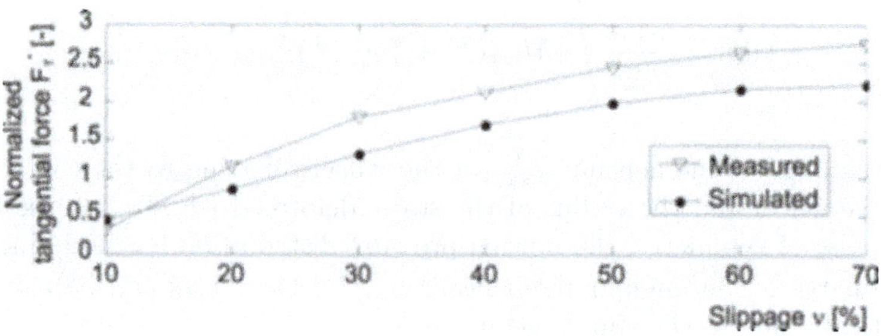

Fig. 5. Comparison of the tangential force characteristic $F_T^*(\nu)$ for steady rolling with $v_{abs} = 40\,\frac{mm}{s}$ and $F_N = 40\,N$, Natural Rubber (NR) on glass.

$$ {}^{I}\mathbf{x}_{M,\,k}(t) = {}^{I}\mathbf{x}_R(t) + {}^{R}\mathbf{x}_{M,\,k}(t), \tag{13} $$

where ${}^{I}\mathbf{x}_R(t)$ denotes the position vector of the reference point R, lying in the middle of the contact patch, which moves with the velocity v_{abs}, see Figure 4.

5 Point contact model

The position ${}^{I}\mathbf{x}_{M,\,k}(t)$ of the point M_k is the input value for the point contact model. Fig. 6 shows an exemplary point contact element P_k, with the spring constant c_P the mass m_P and the friction contact. The spring force is given by

$$ \mathbf{F}_{M,\,k} = c_P({}^{I}\mathbf{x}_{M,\,k} - {}^{I}\mathbf{x}_{P,\,k}). \tag{14} $$

The normal force $F_{N,\,k}$ acting on the element is derived by

$$ F_{N,\,k} = p({}^{C}_{R}\mathbf{r}_{M,\,k})\,A_P. \tag{15} $$

The normal pressure $p({}^{C}_{R}\mathbf{r}_{M,\,k})$ of the point M_k is gained by using the interpolation matrix ${}^{R}\mathbf{H}_{stat}({}^{C}\mathbf{r}_N)$. The representative area A_p of the contact

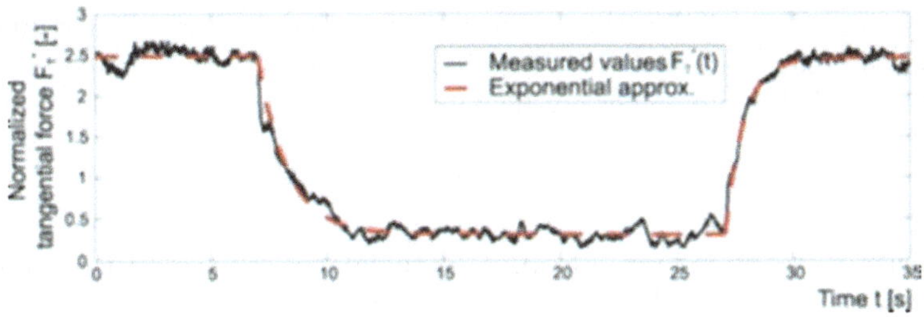

Fig. 6. Tangential force $F_T^*(t)$ during a slippage jump from $\nu_{start} = 50\,\%$ to $\nu_{target} = 10\,\%$ and back.

element is a fraction of the entire contact area A computed by the static precalculations.

If the contact element is within the sticking mode, the state variables of the contact element are derived as follows:

$$^I\dot{\mathbf{x}}_{P,\,k} = 0; \qquad \mathbf{F}_{T,\,k} = \mathbf{F}_{M,\,k}\,. \tag{16}$$

The sticking mode is left, if the condition

$$|\,\mathbf{F}_{M,\,k}\,| > \mu_0\, F_{N,\,k} \qquad \text{with} \qquad F_{N,\,k} \geq 0 \tag{17}$$

is fulfilled. Within the sliding mode, the contact force $\mathbf{F}_{T,\,k}$ is depending on $^I\dot{\mathbf{x}}_{P,\,k}$ and $F_{N,\,k}$, the position of P_k is derived using the impulse conservation principle

$$\begin{aligned} \mathbf{F}_{T,\,k} &= \mu\, F_{N,\,k}\, {}^I\mathbf{e}_{\dot{\mathbf{x}}}\,; \\ {}^I\dot{\mathbf{x}}_{P,\,k} &= \frac{1}{m_P}\int\limits_0^t (\mathbf{F}_{M,\,k}(\bar{t}) - \mathbf{F}_{T,\,k}(\bar{t}))\,\mathrm{d}\bar{t} + {}^I\dot{\mathbf{x}}_{P,\,k}(t=0), \end{aligned} \tag{18}$$

where $^I\mathbf{e}_{\dot{\mathbf{x}}}$ denotes the coordinates of the unit vector with the same direction as $^I\dot{\mathbf{x}}_{P,\,k}$. The sliding mode is interrupted, if the condition

$$|{}^I\,\dot{\mathbf{x}}_{P,\,k}\,| = 0 \tag{19}$$

is fulfilled. Finally, the shear stress at the position $^C_R\mathbf{r}_{M,\,k}$

$$\boldsymbol{\tau}_{M,\,k} = \frac{\mathbf{F}_{T,\,k}}{A_P} \tag{20}$$

can be calculated and for all points M_k the interpolation matrix $^R\mathbf{H}_\tau(^C_R\mathbf{r}_M,\,t)$ is assembled. At the start of the simulation $t = 0$ the variables are set to the initial conditions $^I\mathbf{x}_{P,\,k} = {}^I\,\mathbf{x}_{M,\,k}$ and $^I\dot{\mathbf{x}}_{P,\,k} = 0$, the friction mode is set to sticking.

The stiffness c_P of one of the k point contacts is a fraction of the global stiffness $c_0 = k\, c_P$. This stiffness c_0 is interpreted as a residual stiffness of the modal condensation. It is connected in series to the modal condensated structure and thus can generally be gained by comparing it to the basic FE-model

$$\frac{1}{c_n} = \frac{1}{c_m} + \frac{1}{c_0} \tag{21}$$

for a certain steady reference load. c_n is the global stiffness of the basic FE-System and c_m is a global stiffness of the modal system, which is approximated by the first m modes.

The partial mass m_P of the contact element P_k is introduced to increase the numerical stability of the system. Thus, the mass of the entire contact layer $m_0 = k\, m_P$ is chosen to be much smaller than the mass of the structure.

The characteristics of sliding friction $\mu(v_{rel}, p)$ and static friction $\mu_0(p)$ are gained experimentally, see [3].

6 Comparison with experiments

The presented model has been realized within the MATLAB/SIMULINK environment. The local static friction characteristic $\mu_0(p)$ was gained experimentally and was also used to approximate the sliding friction characteristic $\mu(p) = \mu_0(p)$. The experimentally identified sliding friction characteristic $\mu(v_{rel}, p)$ is much to low, because the standing contact of experimental set up heats up during the accomplishment and the temperature decreases the local friction coefficient. Therefore, the static friction characteristic was chosen as a first approach. Due to the low time constants only the first mode was used for the approximation of the structure dynamics. The modal damping was set to $D_{M,1} = 0.5$. For the calculation of the shear stress distribution 20 contact elements were used, arranged in a row in the center of the contact zone. All experiments were accomplished at constant normal force $F_N = 40\, N$ on the wheel and constant reference velocity $v_{abs} = 40\, mm/s$ of the wheel center. Figure 7 shows the comparison between simulation and experiments for steady slippage $\nu = const.$ For $\nu \geq 30$, the two graphs differ with $\Delta F_T^* \approx 0.5$. The reason for this difference is probably the lack of an appropriate local sliding friction characteristic. The qualitative progression of the simulated characteristic already fits relatively well to the experimental results. Hence, the distribution of sticking and sliding area within the contact zone is reproduced by the model for steady rolling.

In the following jump excitations for the slippage were chosen to investigate the dynamical behavior of the system. For such a slippage jump $\nu(t)$ the normalized tangential force

$$F_T^*(t) = \frac{F_T(t)}{F_N}, \tag{22}$$

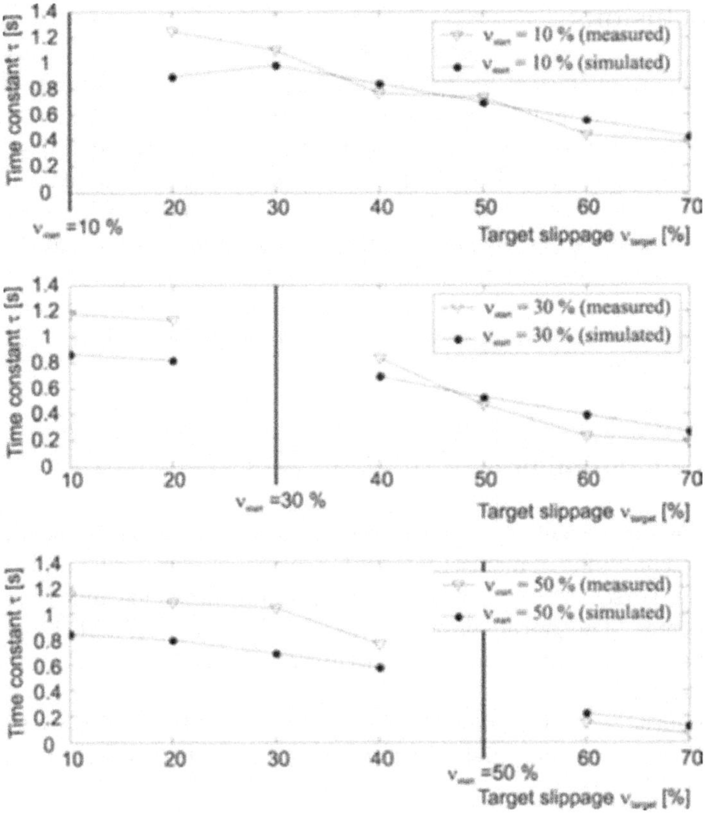

Fig. 7. Comparison of time constant τ for jumps from $\nu_{start} = 10\,\%$, $30\,\%$, $50\,\%$ to ν_{target} at $v_{abs} = 40\,\frac{mm}{s}$ and $F_N = 40N$, NR on glass.

where F_N denotes the normal contact force, is shown in Figure 3. To obtain a reference magnitude for the model verification, an exponential characteristic is fit to the experimental data. The time constant τ of this exponential approximation represents the delay of the dynamic contact behavior. The comparison of simulation and experiments for an unsteady excitation is shown in the diagrams of Figure 8. In the diagrams, three different values for the start value of the slippage jump ν_{start} are used as a parameter. The time constant τ is plotted versus the values of the target slippage ν_{target}. Regarding the experimental data, the characteristics of the time constants show two tendencies. There is a reciprocal dependency on the target value ν_{target} of the jump. Additionally, less distinctive, there is also a reciprocal dependency on the start value ν_{start} of the jump. For up-jumps $\nu_{target} > \nu_{start}$ the results of the model fit well to the experimental data. For down-jumps $\nu_{target} < \nu_{start}$, the simulation results differ stronger from the experimental data.

7 Conclusions

In this paper, an efficient model for unsteady rolling contact is presented. The model is experimentally validated by measurements of an autonomous vehicle. For the excitation with slippage jumps $\nu(t)$, an exponential saturation behavior is approximated to both, experimental data and simulation results. The characteristic time constant τ is used to verify the model. For increasing target values of the slippage ν_{target}, the time constant τ decreases. There also is a slight decreasing characteristic of the time constant τ for increasing start values ν_{start} of the slippage jump. For up-jumps, the time constant characteristics are reproduced quiet well by the model. In general, the comparison fits relatively well.

Acknowledgements

The investigations published in this report received financial support from the Deutsche Forschungsgemeinschaft (DFG) as a part of the collaborate research unit DFG-Forschergruppe "Dynamische Kontaktprobleme mit Reibung bei Elastomeren". The precalculations were done with an MATLAB FE-program, which has been built up at the Institute of Mechanics and Computational Mechanics (IBNM), University of Hannover.

The authors want to dedicate this contribution to Prof. Karl Popp who initiated this work and advised its progress. Sadly, he passed away before it could be finished.

References

1. Gutzeit, F., Kröger, M., Lindner, M., Popp, K. (2004): Experimental investigations on the dynamical friction behaviour of rubber, Proc. 6th Rubber Fall Colloquium Hannover, 523 pp.
2. Nackenhorst, U. (2000): Rollkontaktdynamik. Dynamische Analyse der Dynamik rollender Körper mit der Finite Elemente Methode. Institut für Mechanik, Universität der Bundeswehr Hamburg. ISSN 0939-2343.
3. Sextro, W. (2002): Dynamical Contact Problems with Friction. Models, Methods and Applications. Lecture Notes in Applied Mechanics Vol. 3, Springer publishing. ISBN 3540430237.

A physicist view to tire traction

J.-M. Vacherand

F32, Centre de recherches de Ladoux, Manufacture francaise des pneumatiques Michelin, 23, Place des Carmes, F-63040 Clermont-Ferrand cedex 09, France,
vacherand@fr.michelin.com

Abstract. Friction of tyres is a major contributor to automotive safety. Usually friction if most demanded on wet roads. Several theories have been proposed about the physical origins of dry and wet friction. They tend to evaluate the respective contributions of adhesion forces and dissipative mechanisms at the relevant scales. We discuss their achievements and limitations. The relevance of taking into account these physical mechanisms to improve the simulation of tire friction problems is addressed.

1 Traction on the road

A large variety of roads and driving conditions can be met by drivers. Putting aside the special case of snow-, ice- and mud- covered roads, the most demanding conditions are met at (rather) low speeds, on city roads where the roughness has been worn away by heavy traffic, and at high speed on flooded roads [1,2].

Part of the friction comes from the deformation of the viscoelastic rubber by the road asperities [3] whose estimation has recently drawn a renewed interest[4-7]; coulomb-like behaviour then arises as a consequence of the statistics of rough surfaces [8,9]. Thermal boundary conditions also are important especially in the dry case, and when large slip is expected. Since most of the dissipation arises near the surface, whose temperature remains that of the road, road renewal by 'convection' also becomes critical. Because of low rubber heat conduction, several μm of rubber can be abraded away by thermal degradation in blocked wheel braking. This effect has been related to the dry friction dependence to tire contact patch length.

Another part of friction comes with local forces, either of dispersive and specific origin. The latter is for instance the energy dissipated by the chains due to a deformation cycle where the chains adsorb on the ground, then are extracted from this place by the sliding motion [10-12], and a dry friction molecular-roughness-induced 'trivial' part.

Dispersive Van der Waals adhesive forces [13-15] between rubber and track may supplement hertzian pressure at the smaller scales; because of nanoscale roughness of surfaces that appear smooth at the macroscale, their contribution has for long obscured the interpretation of 'smooth' friction results as well as the measurement of surface energies.

2 Humid conditions

Let us first remark that, because of the very low viscosity of water, the thickness of a water film able to sustain the typical shearing forces in the contact patch is just about a few nanometres. So this film has either to be eliminated, or its thickness rendered so low as to nullify its effect.

The merit of the hysteretic part of friction is that it remains active through the water films, when their thickness, as is usually the case, is much lower than the deformation due to the asperities [3].

An order of magnitude of the penetration of the track asperities into rubber is several tenth of millimetres. On flooded roads, the internal pressure can squeeze out the water until it is equilibrated by the hydrodynamic pressure of water, which eventually gives rise to hydroplaning [16]. We may thus assume that hydroplaning ceases where film thickness becomes less than a few tenth of millimetres; since the tire's structural deformation due to water dynamic pressure is about the same order of magnitude, the structural reaction couples the water motion to the tire inner structure.

To develop a larger friction coefficient, it is necessary to accommodate lower scales of asperities than these macro-scales.

A film thickness larger than gravel penetration in the rubber *and* a macro-roughness double this value (this rough estimate is supported by experiment) define a point at which a continuous film is able to uniformly support the load and has to be drained at the scale of tread blocks rather than that of the track asperities.

This condition is viscoplaning, and is weakly sensitive to pressure and rate, but very much to tread-block length [17]. Final thicknesses of films corresponding to the time of residence in the contact ($\approx 10^{-2}$s) would be micronic at the centimetre scale, but 0.1 mm at the 10cm scale, if the tread-blocks were undeformable. Viscoplaning is then much more likely on the massive tread-blocks of truck tires; it is controlled by siping.

Deformability of the rubber could give rise to a similar water entrapment at the gravel scale also, with order of magnitude 5 microns. But usually this water entrapment is no obstacle to friction because the penetration of largest scale of roughness sitting on the top of gravels is ten times as much (Fig. 1); a known exception to this situation is signs painted on the road, which can be treacherous for motorcyclists even at very low speeds when mildly humid, because of very low microroughness.

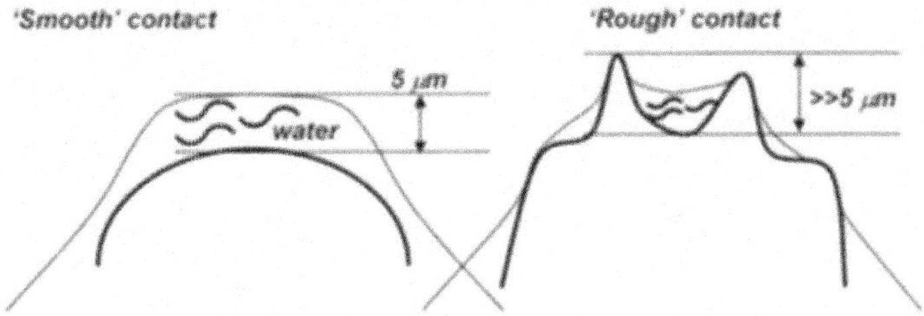

Fig. 1. Water trapping in a contact.

At realistic sliding rates of order m/s, the entrapped films remain thin versus penetration at whatever scale, but nevertheless could prevent adhesion to occur. To avoid this, siping wedges will help build starved lubrication conditions.

3 Adhesive forces

The effect of adhesive Van der Waals forces, yet small at large scales, should dominate strain-induced stresses at small scales ($\approx \mu$m). At this scale the rubber behaves essentially like a wetting liquid. But the rather large contribution (τ ~100 bars) they could bring to sliding friction is usually hampered by a rather low value of intimate contact surface [19].

The same is true for the contribution of molecular-scale dissipation by connector adhesion and molecular-roughness-linked dry friction, whose contribution also would depend much on the chemical nature of the sliding surfaces and pollutants.

Sliding is initiated by nucleation and growth of sliding regions; Stick-slip instabilities at low rates appear connected to either adhesive peeling [20] or stick-slip of connectors [21].

4 Dewetting

Very thin water films are not stable because of their internal cohesive forces. Usually, they break by nucleation and growth of a dry patch. The physics of this growth has been described elsewhere [21]. The water that is evacuated from the contact is collected in a rim. The rate of growth depends on equilibrium between adhesive, elastic and viscous forces, the latter either in the water rim or rubber wedge. Pressure gradients may partially supplement the adhesive forces in real-world situations [22]; Persson has given an estimate [23] which appears to me excessive because care was not taken at the fact that the pressure gradient should only act at the scale of the dewetting rim instead of the whole contact (Fig. 2). [24].

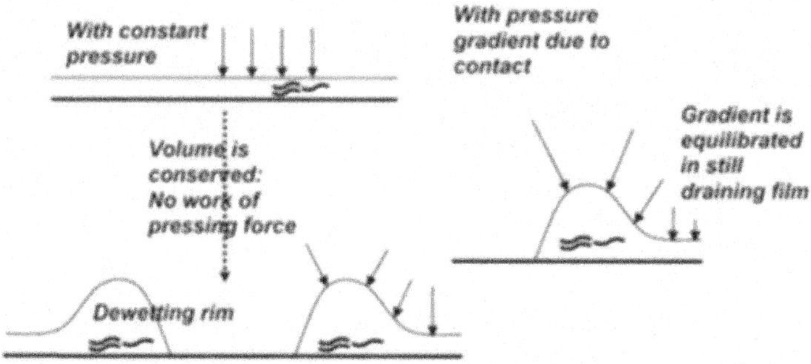

Fig. 2. Effect of pressure gradient on dewetting. A dewetting rim collects the evacuated water. Only archimedes pull may put dewetting rim in motion.

5 Friction scaling

According to the Greenwood-Williamson treatment, Coulomb-like pressure-insensitive friction is the result of a contact surface (and the number of active contacts) varying much faster than the individual contact surface. But individual contacts still grow a little with applied pressure.

Although other statistical distributions have been proposed [9], it is practical to assume that the density of heights of contacts contains a 'tail' represented by a α ~3-6 power law [Moore, 17], and a 'body' represented by a α ~1/2 power law. Notice that for multi-scale roughness, the tail of the macroscale contains the body of the smaller scales.

Assuming that the Greenwood-Williamson approximations hold, we find that contact rate scales slightly slower than pressure, with power 1-1/(5+2α). But local mean pressure on contacts points still scales like the radical of penetration, the same which is found when there is no height fluctuation (formally at least, no fluctuation comes in with α=-1); thus local pressure sensitivity to global pressure is brought in by the small extra term 1/(5+2α) in the exponent.

Applying the same reasoning to two scales [4] brings a hysteretic friction coefficient that scales like pressure to a very small, yet positive, power. When more scales of roughness are taken into account, the qualitative behaviour remains, with an increasingly small exponent. Collecting the terms corresponding to the different scales of roughness, we conclude that the friction then tends to zero for zero pressure (and remains continuous at the loss of contact), albeit extremely slowly.

A logarithmic evolution is expected when less and less scales are considered because of decreasing contact, until scales under which there is no rugosity left 'downwards', where the aforementioned power-law holds. At the adhesive scale, 'snap-on, snap-off' effects [25] could be met. But at least the 'purely hysteretic' part of friction should be continuous with respect to load down to this scale.

For the vertical *rigidity* of contact on a single roughness scale, no divergence at all is obtained, since the load q varies with a high (3 with α=1/2) power of the penetration e, the contact *rigidity* remains finite (with power 2/3 of load, it actually tends to zero) at zero load. With a similar argument as before, the relation between local and global pressure at the different scales allow to conclude that this qualitative behaviour holds under the GW approximation.

According to a similar argument using rate instead of displacement, the sliding rate-friction curve should be regular at very low rates, which has indeed been observed [26]. But I suggest here that this argument does not hold for sticking-sliding transitions. Starting from zero sliding rate, the variation of the sliding rate is so sudden that distribution of the viscoelastic stresses in the contact should never assume a stationary shape. A large part of the deformation energy should be 'suddenly' released during the initial slip of an ($\sim$?adhesive scale) contact length. The maximum friction coefficient at incipient slip could then be much higher than sliding friction coefficient, especially since the relaxation of the contact could occur for a much longer time ($\sim 10^{-2}$ s in a typical tyre contact patch) than the sliding contact time on micronic asperities ($\sim 10^{-6}$ s) which is relevant for steady-state friction; this reasoning can be continued downwards to the adhesive scale.

It is concluded that memory effects similar to those that have been observed by Baumberger's team on a large class of other materials [27-29], should manifest through a memory length of adhesive scale ($\sim \mu$m) [15] order of magnitude.

In the case of wet friction, other memory effects could come in by relaxation of the film thickness or dewetting effects [21]; they let us expect that stick-slip-like motion is much more common in wet conditions, which is indeed well known. Yet a simple rate decreasing friction curve already was able [30] to locate the unstable regions without resorting to any dynamic description of the film thickness, which is not surprising since the stationary picture is at least able to capture the thickness effect when the sliding distance is 'macroscopic'.

Another important issue as far as numerical approach to friction is concerned is the magnitude of the shear interfacial rigidity. Since the characteristic shearing length is of order contact length, this rigidity would be only 4-5 times as much as treadblock rigidity in typical (tourism) tire situations.

We note that, for a single scale of rugosity with no fluctuations of height, the interfacial shear rigidity scales like the pressure to the power 1/3, therefore tends to zero at zero pressure. Introducing fluctuations of height gives rise to a similar scaling as that of the vertical rigidity (power 2/3) if an assumption, maybe realistic when estimating the tangent interfacial rigidity of an established contact, of a similar shearing deflection for every contact point is assumed.

An even more regular contact behaviour (power 4/3) is obtained when contact shearing stresses vanish at each individual contact point when the contact there is established; this description maybe more relevant for secant interfacial rigidity estimation.

Extension to multiscale proceeds along similar lines as those suggested before for vertical interface rigidity, with similar conclusions.

6 Conclusion

I hope to have convinced you that friction physics is much more complicated, and deserves much more interest, than the simple coulomb picture; especially when memory effects are considered.

I also suggest that local analysis could give interesting clues about how to regularize contact problems without losing the physical sense. Also notice that the results for shear rigidity depend on an assumption about loading path.

Acknowledgements

Many thanks are due to my colleagues and academic contacts for stimulating discussions and their help to sharpen my arguments. Publication permission by the Michelin company is also greatly acknowledged.

References

1. Williams, H., Tire science and technology, 1975, vol. 3, p. 267
2. Veith, A.G., Frictional interaction of tire and pavement, ASTM Symposium and ASTM special technical publication 793, paper 1
3. Grosch, K. A., Proceedings of the Royal Society of London Series A, 1963, vol. 274, p. 21–39
4. Yandell, W.O., A new theory of hysteretic sliding friction, Wear, 1971, vol. 17, p. 229
5. Klüppel, M., Heinrich, G., Rubber friction on self-affine road tracks, Rubber chemistry and technology, 2000, vol. 73, p. 578
6. Persson, B.N.J., Theory of rubber friction and contact mechanics, Journal of chemical physics, 2001, vol. 115, n. 8, p. 3840
7. Gabler, A., Straube, E., Heinrich, G., Korrelation des nassrutschverhaltens russgefüllter vulkanisate mit ihren viskoelastischen eigenschaften, Kautschuk Gummi und Kunststoffe, 1993, vol. 46, p.941
8. Archard, J.F., Elastic deformation and the laws of friction, Proceedings of the Royal Society of London Series A, 1957, vol. 243, p. 190
9. Greenwood, J.A., Williamson, J.B.P., Contact of nominally flat surfaces, Proceedings of the Royal Society Series A, 1966, vol. 295, p. 300

10. Deryagin, B.V., Ratner, S.B., On the role of mechanical and molecular roughness in surface friction, DSIR translation, Doklady Akademii Nauk SSSR, 1955, vol. 103, no. 6, p. 1021
11. Bartenev, G.M., Elkin A.I., Friction properties of highly elastic materials, Wear, 1965, vol. 8, p. 8
12. Garrivier, D., Décavé, E., Bréchet, Y., Bruckbert, F., Fourcade, B., European Physics Journal E, 2002, vol. 8, p. 79
13. Johnson, K.L., Kendall, K., Roberts, A.D., Proceedings of the Royal Society of London Series A, 1971, vol. 324, p. 301
14. Barquins, M., Maugis, D., Courtel, R., Rôle de l 'adhésion dans le frottement du caoutchouc; nouvelle définition du coefficient de frottement, Comptes rendus de l'académie des sciences Paris, 1977, vol. 284, p. 127
15. Basire, C., Contact, adhésion et frottement à l'échelle submicrométrique. Etude expérimentale sur des matériaux viscoélastiques. PhD dissertation, Paris: Université Pierre et Marie Curie/Paris VI, 1998
16. Boness, R.J., Journal of automotive engineering, 1968, vol. 58, p. 260
17. Moore, D.F., The friction of pneumatic tires, Elsevier scientific, 1975
18. Persson, B.N.J., Albohr, O., Tartaglino, U., Volokitin, A.I., Tosatti, E., J. Phys : Condens. Matter, 2005, vol. 17, p.R1
19. Ludema, K.C., Tabor, D., The friction and viscoelastic properties of polymeric solids, Wear, 1966, vol. 9, p. 329
20. Barquins, M., Friction and wear of rubber-like materials, Wear, 1993, vol. 160, p. 1
21. Clain-Burckbuchler, J., Friction sèche et mouillée, PhD Dissertation, Paris: Université Pierre et Marie Curie/Paris VI, 2004
22. Roberts, A.D., Tabor, D., Proceedings of the Royal Society of London Series A, 1971, vol. 325, p. 323
23. Persson, B.N.J., Volokitin, A.I., Tosatti, E., Role of the External Pressure on the Dewetting of Soft Interfaces, European Physics Journal E, 2003, vol. 11, p. 409
24. Vacherand, J.-M., to appear in Vysokomol. Soedin.
25. Persson, B.N.J., Tossati, E., Journal of Chemical Physics, 2001, vol. 115 n. 12, p. 5597
26. Persson, B.N.J., Albohr, O., Mancosu, F., Peveri, V., Samoilov, V.N., Sivebaek, I.M., On the nature of the static friction, kinetic friction and creep, Wear, 2003, vol. 254, p. 835
27. Ronsin, Modélisation physique du multicontact, in modélisation numérique et mathématique des problèmes de contact et frottement, problèmes non linéaires appliqués, école CEA-EDF-INRIA, Inria Rocquencourt, 1999
28. Baumberger T., Berthoud P., Caroli C., Physical analysis of the state- and rate-dependent friction law: dynamic friction, Physical Review B, 1999, vol. 60, p. 3928
29. Ronsin, O., Labastie-Coeyrehourcq, K., State, rate and temperature-dependent sliding friction of elastomers, Proceedings of the Royal society of London Series A, 2001, vol. 457, n. 2010, p. 1277
30. Rorrer, R.A.L., Rubber Chemistry and Technology 1955, vol. 73, p. 488

Friction coefficient prognosis for the Grosch-wheel

N. Bouzid, B. Heimann, and H. Blume

Hannover Center of Mechatronics, bouzid@mzh.uni-hannover.de

Abstract. In this paper an approach for friction coefficient prognosis for the Grosch-wheel is presented. Since the solid rubber wheel (80 mm diameter) represents a simplified tyre model this contribution helps to investigate the rolling friction process and especially the feasibility and limitations of the automotive friction prognosis. The grip information is derived from influence parameters like road state, velocity, temperature etc. The methodical innovation of the approach proposed consists in the use of an artificial neural network as data based model that performs multi dimensional interpolation in the learning stage. The operating stage consists of common mathematical operations with low computational effort.

1 Introduction

The optimization of automotive active safety systems like ABS, ESP and ASR is achievable by means of a prognosis of the available grip (maximum friction coefficient) between tyre and road. Furthermore, many dangerous driving situations can be avoided by a warning system that detects a sudden decrease in the grip level prior to the occurrence [6, 15]. Because of the big significance of the application, a variety of approaches for friction coefficient prognosis in measuring cars and trailers have been proposed. Those systems allow to simultaneously measure the influence parameters and grip between tyre and road.

There are two known methods to measure the grip, representing the maximum of the μ-slip-curve: The first one captures the whole characteristic by an artificial variation of the wheel rotational velocity [3]. In that case additional important information like the sliding friction coefficient is provided. The second approach involves measurements during ABS-braking, since it works in the range of maximum friction [15]. In these systems the friction force is either measured by a force/torque sensor or estimated from the deceleration of the car.

To design a prognosis system, the friction process must be first understood. There is a variety of influence parameters on tire road interaction. Observing only dry road conditions the main parameters are generally classified into three groups, see Tab. 1. The first one contains the tire properties and represents the major group. The second one contains the road properties roughness and temperature. The road texture is one of the most important quantities to be measured and can be described by a 2-dimensional profile of the surface after optical measurement based on the triangulation principle. The road temperature which affects the friction process by altering the rubber properties, can also be measured with no contact. The third parameter group contains the vehicle states which are already used by active safety systems.

Table 1. Main influence parameters on the friction process between tyre and dry road track during straight driving, Bachmann [4].

tyre	road	car
tread rubber	micro roughness	velocity
tread pattern	macro roughness	load
tread depth	temperature	slip
tyre pressure		
tyre width		
tyre type (radial/diagonal)		

Here, the Grosch-wheel as a simplified tire model is used for the investigation. A mobile friction coefficient measurement platform using a solid rubber wheel was built [5]. The solid rubber wheel with no tread pattern is adopted from the Grosch rubber wear examinations [1, 2]. It reduces the influence parameters to the rubber compound. The wheel is pressed onto the ground by closed loop force control. The vehicle velocity is kept constant by controlled servomotors. The platform is equipped with sensors for temperature of road and wheel surface, platform velocity, force/torque for friction measurement and a triangulation sensor for the road texture.

The approach for friction prognosis proposed here uses a data based estimation technique, consisting of learning stage in which an artificial neural network is trained with a consistent set of influence parameters and measurement data and an operational stage in which the friction measurement is estimated based on the influence parameters only. A more detailed description of the prognosis system is shown in section 2. In section 3 the data pre-processing with both parts roughness description and mathematical model is presented. Basics for data based modelling and artificial neural networks are given in section 4, followed by experimental results of the systems proposed.

2 System architecture

In the operational stage of the friction prognosis system the grip is estimated on the basis of online measured or unvarying and known parameters. This is possible by means of the data based model that is built in the learning stage (Fig. 1). An example for an apriori known factor is the rubber compound, hence no online processing is needed. In contrast to that, the road roughness is sampled at high rates and pre-processed since the measured profile does not provide any direct information about the roughness contribution to the friction process. This task, denoted by pre-processing 2, is part of both the operational and the learning stage. The influence parameters temperature, load and velocity are captured at a lower rate and are used directly without any further munging. On the other side the μ-slip-characteristic as friction process output is measured and pre-processed for usage in the learning stage. This task consists of determining the effective radius, calculating slip and μ values and fitting to a mathematical model.

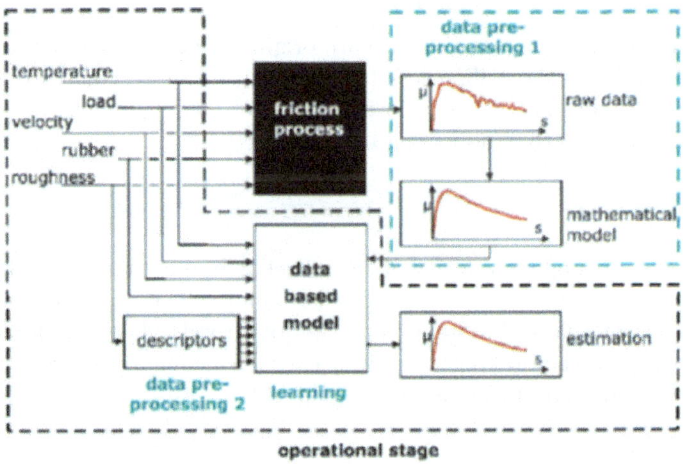

Fig. 1. Principle of the friction prognosis for the Grosch-wheel.

3 Data pre-processing

The data pre-processing consists of computing relevant values for the inputs as well as for the outputs of the friction process. In the first part of the data pre-processing, the mathematical Pacejka-model is considered. The μ-slip-characteristic is described by only five parameters (Eq. 1, 2)

$$\mu(x) = C_3 \cdot \sin[C_2 \cdot \arctan(1 - C_4 \cdot x + C_4 \cdot \arctan(C_1 \cdot x))] \tag{1}$$

$$x = s + offset. \tag{2}$$

In the second part, fractal and bearing ratio curve descriptors are considered to describe the roughness contribution to the friction process.

3.1 Mathematical model

The Pacejka-model is frequently used in tyre modelling. Its convenience is due to the property that each parameter can be assigned to a feature of the μ-slip-characteristic [10]. For a typical μ-slip-characteristic with an ascending and a following descending progression C_3 describes the maximúm friction coefficient μ_{max}, the corresponding slip value is given by C_4, C_1 corresponds to the slope at low slip values and C_2 specifies the curve form. Since the slip is defined as

$$s = \frac{v - \omega \cdot r_{eff}}{v} \cdot 100\% \quad if \quad s > 0 \tag{3}$$

$$s = \frac{v - \omega \cdot r_{eff}}{\omega \cdot r_{eff}} \cdot 100\% \quad if \quad s < 0 \tag{4}$$

the curve does not cross the coordinate origin. In fact the friction coefficient at 0% slip reflects the rolling loss factor because the braking torque crosses zero at this point. Therefore, a horizontal offset has been included according to Eq. 2. The effective radius can be calculated as

$$r_{eff} = \frac{v}{w^*} \tag{5}$$

where ω^* is the angular velocity of the wheel at braking torque zero crossing. v and ω denote the vehicle and the wheel angular velocities, respectively. If we examine the mechanical system consisting of Grosch-wheel and force/torque sensor in Fig. 3 the braking torque T_B can be calculated from the measured entities F_x and T_y and the distance h (Eq. 6)

$$T_B = T_y - F_x \cdot h. \tag{6}$$

The experimental procedure consists of keeping F_N and v constant, step wise increasing the angular wheel velocity so that the desired slip range is covered. A linear function is then fitted to the calculated braking torque and the zero crossing is found. After that the effective radius and the slip are calculated. Using the definition

$$\mu = -\frac{F_F}{F_N} \tag{7}$$

the friction coefficient is chosen to be positive for braking. The data couples (μ,s) are then fitted to the Pacejka-model with a nonlinear least square optimization. Fig. 4 shows an example of resulting data couples and the fitted characteristic.

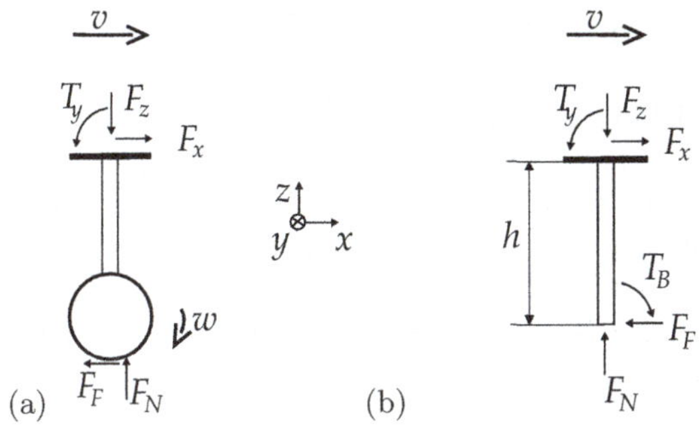

Fig. 2. Definitions: Mechanical system with (a) and without Grosch-wheel (b).

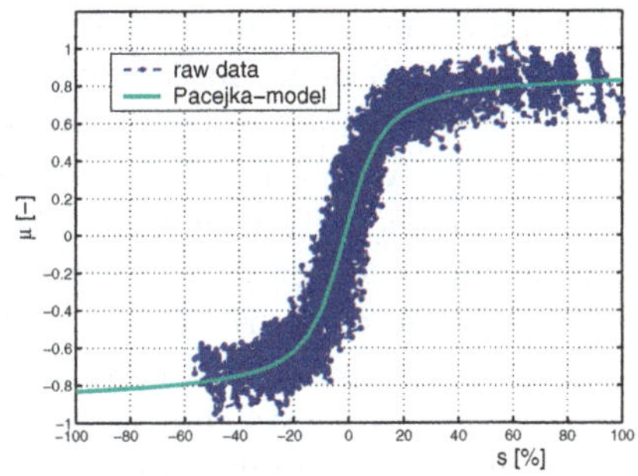

Fig. 3. μ-slip-characteristic after measuring and fitting.

3.2 Roughness descriptors

In literature many roughness descriptors have been proposed. Parallel to the standard height parameters (roughness average Ra, Rp, Rz, etc.) the most descriptors used are the amplitude distribution function ADF parameters (kurtosis, skewness), Schulze-surface-index and Eichhorn-parameters [7]. An investigation in [8] showed that the first group does not have a good correlation with the friction coefficient. Schulze [12] and Eichhorn [6] describe in different ways the capability of road surfaces to establish contact to rubber. In the present work ADF based parameters, the Bearing Ratio Curve descriptors, are introduced to this application field. They are known in the field of metal manufacturing and are used to evaluate the quality of surfaces. This approach (ISO 13565-1996) divides the cumulative ADF into three sections: a central one called main plateau, a section of peaks and a section of valleys. The used BRC-descriptors are main roughness R_k, peak roughness R_{pk}, valley roughness R_{vk}, fraction of peaks M_{R1} and the fraction of valleys M_{R2}. The fractal descriptors build up the second group. Their is due to the ability of giving the power content of the measured roughness profile in dependency of the wave length. This is valid with the assumption of self affinity which is correct for a wide range of road surfaces [9].

4 Data based model

Data based modelling is to find relevant dependencies and interactions between the parameters and the friction coefficient. If valid the model can be used to estimate the grip for new situations. In this application field several approaches have been proposed. The most known are the two procedures in [6] which consist of linear and nonlinear identification methods. In the first one the model output is a linear combination of the input parameters. While in the second method additionally bi-linear and broken terms of the inputs are included.

In this paper the data based modelling of the friction process is done on the basis of Artificial Neural Networks. Inspired by biological neural cells ANN or NN are massively parallel computing systems consisting of a number of simple processors (neurons) with many interconnections. The application areas of ANN are pattern classification, clustering, prediction, optimization, function approximation etc [11]. A neuron weights the inputs and adds them to a total that is transmitted to the output neurons when reaching a specific limit. For the present problem arises the advantage of NN as able to deal with interactions between the different parameters and to reproduce nonlinear contexts without the need to predestine a certain dependency (bi-linear etc.). In the learning stage the neurons weights and bias are learnt using an iterativ optimization procedure called gradient descent with momentum back propagation [16].

An ANN is made of neurons that are arranged in layers. The inputs of the NN are connected to the input layer (IL) neurons which are themselves matched to the next layer. The outputs of the last layer (output layer OL) are the NN-outputs. The number of neurons n_I in IL and $n_i (i = 1..n_{HL})$ in the n_{HL} intercalated layers (hidden layers) are application specific. Generally with increasing n_I, n_{HL} and $n_i (i = 1..n_{HL})$ the interpolation performance of the trained data gets better. However the generality can be lost. As will be shown in section 4 the evaluation of the data based model performance has to be done for both training and validation data. The best result is reached without any hidden layer.

To visualize the multi dimensional interpolation capability of NN (Fig. 5) a simplified model with only 2 inputs $\xi_{\perp}$ and temperature has been taken, as visualization is not possible for higher dimensions. The z-axis represents maximum friction coefficients that are calculated from the pre-processed (circles) and data based model (surface) characteristics. The perfect interpolation in Fig. 5 was achieved with $n_I = 13$, $n_{HL} = 1$ and $n_1 = 13$.

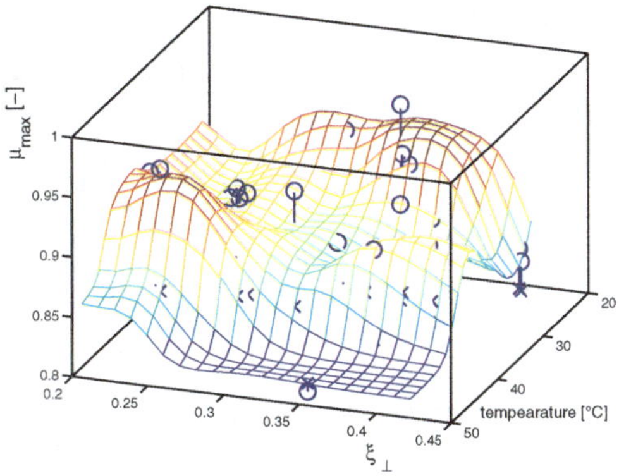

Fig. 4. Interpolation capability of ANN.

5 Experimental results

It is evident that in addition to the mentioned requirements for a wide valid model enough measurements have to be provided during learning. That means measurements on various road surfaces, at different road temperatures, loads, velocities and with different rubber compounds have to be included. Therefore, the measurement planning and executing tasks become time consuming because of the road temperature dependency on the meteorological conditions. To meet all requirements, velocity, load and rubber compound currently have not been included into the data based model. In Total 38 measurements on 8 tarmac surfaces with a temperature range of 20°C-40°C have been considered. The remaining parameters are held constant with $F_N = 100N$, $v = 40mm/s$ and a silicic acid filled EPDM (Ethylene Propylene Diene Monomer) rubber compound. A ANN ($n_I = 10$, $n_{HL} = 0$) has then been trained with half of the data. The second half is used for validation. In Fig. 3 (a) the comparison is done referring (as in the precedent section) to μ_{max}, in (b) to $\mu_{sliding}$ and in (c) to the rolling loss factor. For a better visualization the measurements are arranged in increasing order of μ_{max} and the points representing the data based model are connected with a dashed line. Generally the NN fits the measurements well with the exception of some single failures.

6 Conclusion

In this paper a friction prognosis approach using a data based model for the system Grosch-wheel/road has been presented. The highlights are the pre-processing methods of the measurement data and the use of artificial neural networks for learning. The validity of a data based model, including the parameters roughness and temperature, have been shown. Next work steps will be the extension of the data based model towards varying load, velocity

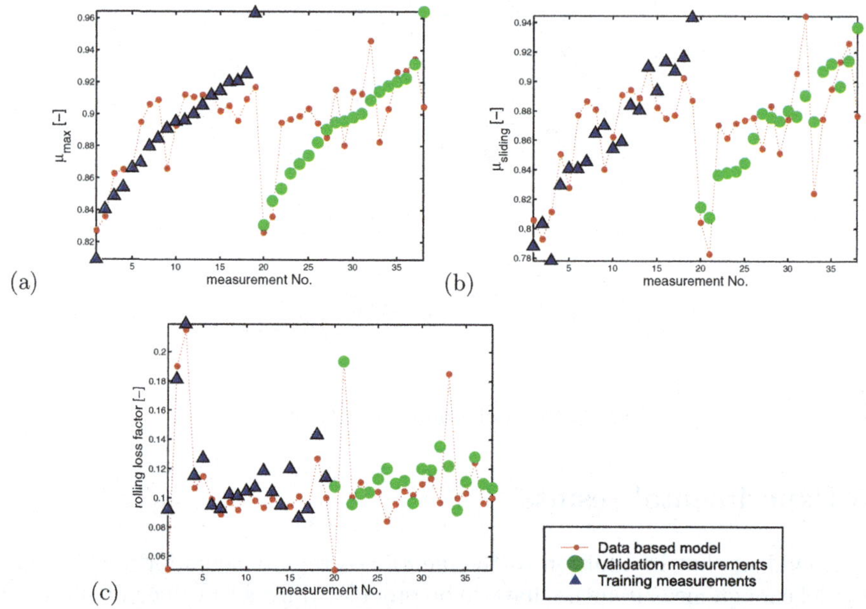

Fig. 5. Validation vor μ_{max}, $\mu_{sliding}$ and rolling loss factor.

and rubber compound. The friction prognosis will be accomplished by the online implementation of the data based model. Furthermore, an extension towards wet road conditions is planned.

References

1. Grosch K A (1996) Kautschuk Gummi Kunststoffe 6:432-441
2. Grosch K A (1997) Kautschuk Gummi Kunststoffe 12:841–851
3. Bachmann T (1996) Berührungslose Kraftschlusserkennung. Finish report, Technische Hochschule Darmstadt, Germany
4. Bachmann T (1998) Wechselwirkungen im Prozess der Reibung zwischen Reifen und Fahrbahn. PhD Thesis, Technische Hochschule Darmstadt, Germany
5. Blume H, Heimann B, Lindner M, Volk H (2003) Friction Measurements on Road Surfaces, Kautschuk Gummi Kunststoffe 56:677–681
6. Eichhorn U (1994) Reibwert zwischen Reifen und Fahrbahn: Einflussgrößen und Erkennung. VDI-Fortschritt-Berichte, Series 12, Band 222
7. Fischlein H (2000) Untersuchung des Fahrbahnoberflächeneinflusses auf das Kraftschlussverhalten von Pkw-Reifen. PhD Thesis, Universität Karlsruhe, VDI-Fortschritt-Berichte, Series 12, Band 414
8. Heinrich G, Schramm J, Müller A, Klüppel M, Kendziorra N, Kelbch S (2002) Zum Einfluss der Straßenoberflächen auf das Bremsverhalten von Pkw-Reifen beim ABS-nass und ABS-trocken Bremsvorgang. In: Proc. of Darmstädter Reifenkolloquium

9. Klüppel M, Mülller A, Le Gal A (2003) Dynamic Contact of Tires with Road Tracks. American Chemical Society, San Francisco
10. Mitschke M(1982) Dynamik der Kraftfahrzeuge. Springer, Berlin
11. Otto P (1995) Identifikation nichtlinearer Systeme mit Künstlichen Neuronalen Netzen
12. Schulze K H (1969) Zur quantitativen Bewertung der Rauheit von Straßenoberflächen in Beziehung zum Reibungswiderstand bei Nässe. PhD Thesis, Technische Universität Berlin
13. Gnadler R, Marwitz H (2004) Neues System zur Ermittlung des Kraftschlusspotentials im Fahrbetrieb, ATZ 5:458–467
14. Trabelsi A, Heimann B, Harting H J, Kendziorra N (2004) Surface State Estimation for the Support of Automotive Systems. In: Proc. of the 8th Int. Conference on Mechatronics Technology, ICMT 2004, Hanoi, Vietnam, 173–178
15. Trabelsi A (2005) Automotive Reibwertprognose zwischen Reifen und Fahrbahn. PhD Thesis, Hannover Center of Mechatronics, University of Hannover, not published yet
16. Hagan M T, Demuth H B, Beale M H (1996) Neural Network Design. MA: PWS Publishing, Boston

Numerical modelling of reinforced geomaterials by wires using the non smooth contact dynamics

R. Laniel[1], O. Mouraille[2], S. Pagano[1], F. Dubois[1], and P. Alart[1]

[1] LMGC, Université de Montpellier II,
romain.laniel@lmgc.univ-montp2.fr
[2] DCT, Delft University of Technology, o.mouraille@tnw.tudelft.nl

Abstract. In Civil Engineering soils may be reinforced by different structures. Wires will interest us. Mixed sand and wire, known as *TexSol*, may be modelled as a continuous medium with classical behaviour laws [6] or with more sophisticated ones taking into account remote interactions [1].

Our approach consists of a discrete model based on the Non Smooth Contact Dynamics. Different choices have been tested on some numerical examples to exhibit at the macroscopic scale the influence of the local models of interaction [5].

First of all we make some numerical tests to compare the mechanical behaviour of a *TexSol* and a sand sample. Then, we compute in both samples the stress tensors of the wires and the sand in order to understand the role of each component.

Our final goal is to define a micro-macro approach and a homogenized realistic behaviour law; if this study is only a first step, it is essential.

1 Motivations

The civil pieces of work needs planed stable floor. The environment configuration often forces civil engineers to raise huge embankments. Moreover, it can be interesting to reinforce them in order to assure a better embankment mechanical behaviour. A lot of different solutions can be used to reinforce soil but one interests us : the *TexSol* process.

Leflaive, Khay and Blivet from LCPC[3], have created the *TexSol* in 1984 [4]. The *TexSol* is a heterogeneous material composed by mixed sand and wires network. This particularity gives to this material a better mechanical resistance than the sand without wires. Of course, the *TexSol* behaviour depends on sand and wire parameters and its frictional angle can be larger than sand one from $0°$ to $10°$ [3]. The wire is described by its linear density with a

[3] Laboratoire Central des Ponts et Chaussées

dtex unit (1 dtex $= 0,1$ g.km^{-1}), its ponderal content, and its stiffness. Classically, the wire density in a *TexSol* sample is included between 100 km.m^{-3} and 200 km.m^{-3}.

To make a *TexSol* bank, a machine named *"Texsoleuse"* is used. It works on throwing sand and, in the same time, injecting wire. The wire is deposed on the free plane of the sand with a random orientation. This machine carries out several passes to raise the bank. The Figure 1 is the *TexSol* microstructure representation. We find, in the literature, a lot of different continuous

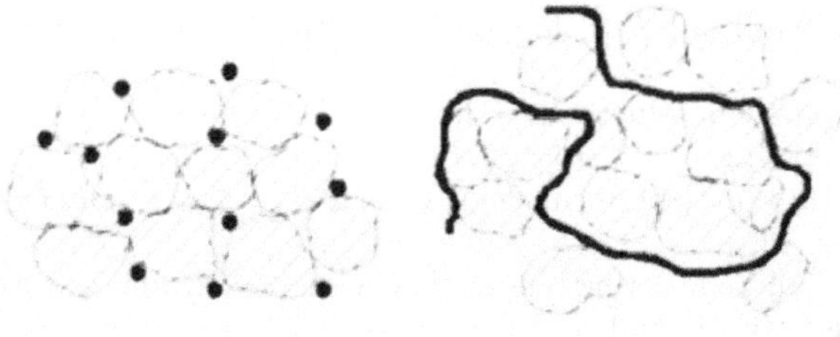

Fig. 1. Schematic *TexSol* sections.

models. The model suggested in [1] is non local and includes remote interactions (corresponding to the wires effects) but also needs an identification of their parameters with macroscopic experiments. Villard proposes a simpler local model in [6]. This one couples a standard model of sand and an equivalent unilateral elastic stiffness contribution corresponding to the wire network. This last contribution is activated only on the traction directions because of the unilateral behaviour of wires. Our first work (exit from the scope of this paper) was to clearly define thermo-dynamic potentials of the Villard local model with both stress and strain formulations in order to identify the best-adapted one. But which micro-mechanisms are working? No continuous theory could ever answer this question.

We thus explore currently a new track using the distinct elements approach. Indeed, thanks to the computation power we have our days, it is possible to carry out some numerical experiments using only microstructural contact laws. Those contact laws must be able to account for the grain/grain, grain/wire and wire/wire interactions.

2 A Numerical discrete model for experiments

We use as a numerical simulation tool, the computer code LMGC90 which uses the Non Smooth Contact Dynamics [2].

2.1 Model of sand

The NSCD method is able to solve multi-body multi-contact problems with rigid and/or deformable bodies. However, our sand sample corresponds to a poly-disperse rigid spheres collection. This kind of problem can thus be computed by LMGC90. On a single contact problem, the principle is to evaluate

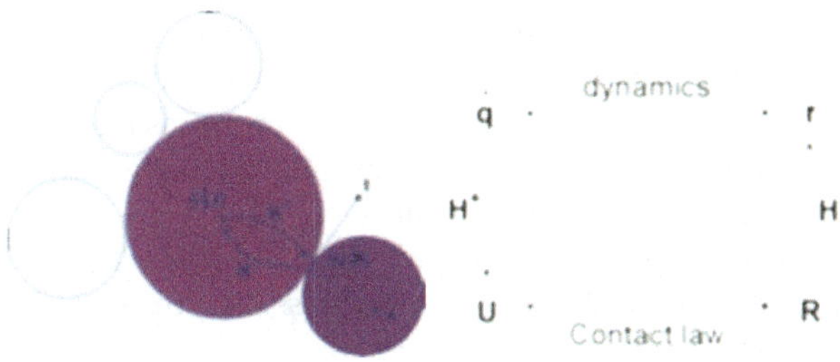

Fig. 2. Local/Global transformations.

external forces dynamic effects on the contactor point. To make such a transformation, we use **H** and $\mathbf{H}^*$ to move variables from the local contact frame to the global body one (cf. Figure 2) and vice-versa. In that way, the PFD[4] is expressed in the local contact frame.

We thus consider $\underline{q}$, $\underline{r}$ respectively the Lagrange coordinates vector of the bodies and the contact reactions/torques vector[5] and $\underline{U}$, $\underline{R}$ the relative velocity and the contact reactions in the contact local frame ($\underline{U} = \mathbf{H}^*\underline{q}$ and $\underline{r} = \mathbf{H}\underline{R}$). $\underline{F}$ are the external forces, **M** the mass matrix and h the time step :

$$\text{Global PFD} : \mathbf{M}d\dot{\underline{q}} = (\underline{F} + \underline{r})\, dt$$
$$\text{Discrete Local PFD} : \underline{U}_{i+1} = \underline{U}_{free} + h\mathbf{W}\underline{R}_{i+1} \tag{1}$$

Smooth dynamic effects are included in the expression of the relative free (of contact) velocity $\underline{U}_{free} = \underline{U}_i + h\mathbf{H}^*\mathbf{M}^{-1}\underline{F}$. $\mathbf{W} = \mathbf{H}^*\mathbf{M}^{-1}\mathbf{H}$ is called the Delassus matrix. This local expression of PFD can "intersect" a normal contact condition (equation 2) modelling an inelastic shock.

$$0 \le U_n \perp R_n \ge 0 \qquad \Leftrightarrow \qquad U_n \ge 0 \quad R_n \ge 0 \quad U_n R_n = 0 \tag{2}$$

Tangential reactions are computed with a frictional condition (Coulomb for example). One Gauss – Seidel loop computes all contact reactions, a convergence criterion (quadratic, maximum et cætera) decides or not to re-execute the loop until a good convergence.

[4] The Principle of Fundamental Dynamics
[5] Reactions imposed on a candidate particle by the neighbour ones

2.2 How making wires with discrete elements

The main difficulty is to model a continuous object with discrete elements. The "discrete wire" is described as a chain of beads [5]. Those are connected with a particular interaction law representing the wire resistance on the normal axis ($R_t = 0$ preserves the wire flexion). We have selected four laws, which can describe the wire behaviour.

- ***elastic wire*** and ***elastic rod*** : respectively unilateral and bilateral elastic laws[6]. Those laws are the most pertinent for small deformations of the sample without large sliding between grains or grains and wire but lead to numerical instabilities for large deformations. Such laws may be only used to validate numerically the local continuous model [6].
- ***rigid rod*** : the simplest law we can use. It imposes $U_n = 0$. Of course, this kind of law may produces some compression stresses in the wire which are not realistic.
- ***rigid wire*** : This law makes possible to free from disadvantages of the preceding one while keeping its advantages. Moreover, no compression component disturbs tensile stresses in the wire.

Unilateral laws must define a reference gap[7] g_{ref}. This last one is a maximum length of the wire between two beads. Beyond this limit, the tensile stress is activated. Let us define a contact candidate particle. We try to solve the α contact (β are neighbour ones) without friction ($R_t^\alpha = 0$), t^-, t^+ the initial and final instants. Let us write a comparative study between a normal Signorini spheres contact law and a *rigid wire* law on a "quasi-inelastic shock" formalism.

Spheres contact	***Rigid wire interaction***
Predicted gap computation : $g^\alpha_{pred} = g^\alpha(t^-) + hU^\alpha_{n\,free}$	
• Case $g^\alpha_{pred} > 0 \quad \Rightarrow \quad R^\alpha_n = 0$	• Case $g^\alpha_{pred} < g^\alpha_{ref} \quad \Rightarrow \quad R^\alpha_n = 0$
• Case $g^\alpha_{pred} \leq 0 \quad \Rightarrow$ (*)	• Case $g^\alpha_{pred} \geq g^\alpha_{ref} \quad \Rightarrow$ (*)

(*) : Modified Inelastic Shock

$U^\alpha_{n\,cont} = -\frac{g^\alpha(t^-)}{h}$	$U^\alpha_{n\,cont} = \frac{g^\alpha_{ref} - g^\alpha(t^-)}{h}$
$\widetilde{U}^\alpha_n = U^\alpha_n(t^+) - U^\alpha_{n\,cont}$	$\widetilde{U}^\alpha_n = U^\alpha_{n\,cont} - U^\alpha_n(t^+)$
$\widetilde{R}^\alpha_n = R^\alpha_n$	$\widetilde{R}^\alpha_n = -R^\alpha_n$

$$
\begin{gathered}
hW^{\alpha\alpha}\widetilde{R}^\alpha_n - \widetilde{U}^\alpha_n = -U^\alpha_{n\,free} - \sum_{\alpha \neq \beta} hW^{\alpha\beta}\widetilde{R}^\beta_n + U^\alpha_{n\,cont} \\
0 \leq \widetilde{U}^\alpha_n \perp \widetilde{R}^\alpha_n \geq 0
\end{gathered} \tag{3}
$$

$U^\alpha_{n\,cont}$ is non null when a contact have to be established during the time step and is the contribution of the velocity to establish this contact. The

[6] The tension is proportional to the gap

[7] It is the minimum distance between two particles

inelastic shock in the second part of the time step leads to define new variables $(\widetilde{U}_n^\alpha, \widetilde{R}_n^\alpha)$ on which are applied the Signorini conditions. Those two problems thus resume to a classical LCP[8] thanks to adapted variable changes.

Let us notice that the wire and sand reactions are computed with two different interaction laws. We can thus separate those contributions in order to analyse how do the wire and the sand work independantly (cf. section 3).

3 A 2D numerical study

This study is a qualitative comparison between sand and *TexSol* [5]. Its aim is to understand the wire contribution towards mechanical solicitations. The two last rigid interaction laws will be used to model the wire but in a first step we start with the *TexSol* sample preparation.

Its wire network must be in a random orientation state (cf. section 1) and the 2D membrane effect must be minimized. Consequently, we define a wire bead diameter close to that of sand particles and the reference gap must be large enough to let pass the coarsest sand grain.

Once the wire generated, two solutions exist to add sand grains. The first superposes a grid of poly-disperse particles and let them deposit by LMGC90. The main problem with this solution is the computing time. Indeed, the NSCD method convergence is slow with weak contact reactions (characteristic of a deposit test). The second uses the Taboada 2D pre-processor. This one makes a geometrical deposit sample of sand with poly-disperse grains. A little LMGC90 deposit relaxes the sample and tightens grains around the wire.

Let us make a biaxial compression test on the final sample in order to compare interaction laws. We consider a 2000 particles *TexSol* sample with 300 for the wire. We carry out three simulations. One with a *rigid wire* interaction law between wire particles, another with a *rigid rod* interaction law and the last one without interaction law (sand). The Figure 3 represents the graph of support reaction according to the crushing percentage. Let us notice that the unilateral or bilateral *TexSol* behaviour is stiffer than sand one. But an accident happens to the bilateral law at the middle of the simulation. A brutal increase of the support reaction shows that a wire compression column has been formed. Sand particles hold it on and it returns a jump of vertical reaction. This bilateral law can makes us a mistake so we would rather use the *rigid wire* law.

The Figure 4 is a deformation state comparison between the *TexSol* and the sand at the same level of an upper side force. It also displays contact reaction chains of spheres contacts (red/grey one) and *rigid wire* interactions (blue/dark one). In the sand sample, reaction chains develop everywhere in every directions. In the *TexSol* sample, *rigid wire* tensile stresses concentrate sphere contact reactions in the centre of the sample. They work on the horizontal direction to prevent the sample from widening.

[8] Linear Complementarity Problem

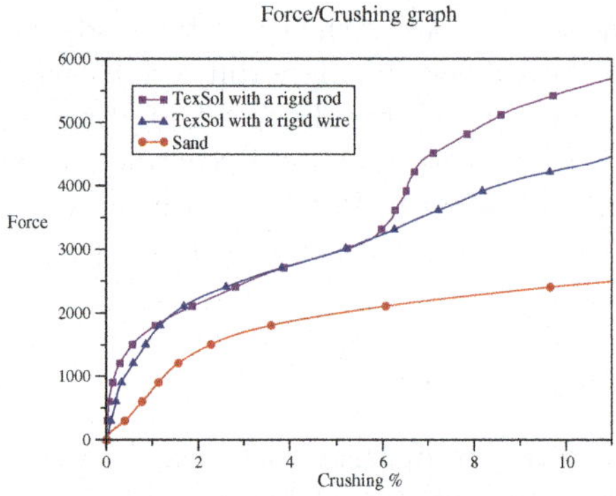

Fig. 3. Different material responses.

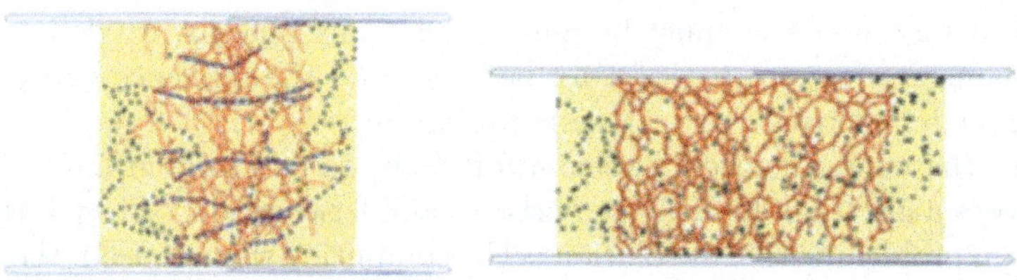

Fig. 4. Reaction chains in *TexSol* (left) and Sand (right).

Thus we interest in the *TexSol* stress tensor. A discrete material stress tensor does not express like continuous material one. We choose the Weber's definition (detailed in [5]). We thus introduce two complementary parts of the *TexSol* stress tensor, one on the wire and the other on sand. The unilaterality of the wire and *TexSol* is highlighted on the Figure 5 graph. Indeed, for the wire component, only one principle stress is positive ; the other one is close

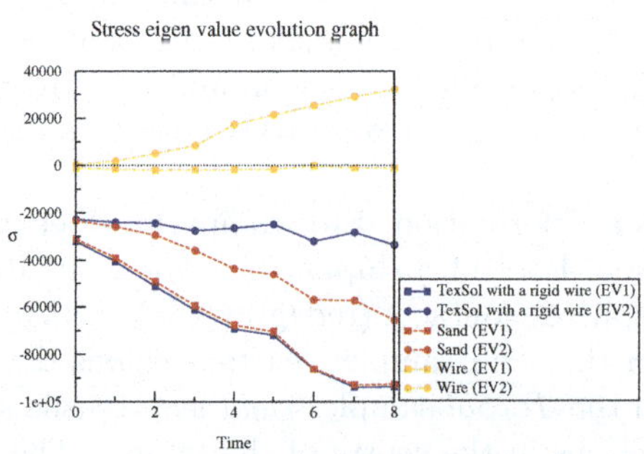

Fig. 5. Evolution of stress tensors in the *TexSol*.

to zero. That is reflected in the global *TexSol* behaviour where one principle stress is equal to one of the sand alone and the other one is reduced by the wire component.

4 A first attempt in 3D modelling

The *TexSol* problem is naturally 3D and the preceding model has some drawbacks. Indeed membrane effects disturbing the material behaviour does not reveal real mechanisms of the wire. We thus have to carry out some 3D numerical study on samples carefully generated. The section 1 tells about the industrial process to raise a *TexSol* bank. Such a process cannot provide an isotropic material. Indeed, the wire is deposed layer by layer and is arranged on parallel planes. The equivalent elastic tensor does not have stiffness on the normal planes direction. It becomes an anisotropic tensor. A 3D pre-processor has been written to define the chain of beads with several rules.

- Each bead is defined with a constant interstice with respect to the previous one.
- The direction of a bead n in the $(O, \underline{x}, \underline{y})$ plane is given by the angle $\theta_n = \theta_{n-1} + \theta_{random}$ with $\theta_{random} \in [-\theta_{max}; \theta_{max}]$ and $\theta_0 = 0$.
- The direction of a bead n in the $\underline{z}$ axis is determinate by the angle $\varphi_n = \varphi_{up} + \varphi_{random}$ with $\varphi_{random} \in [-\varphi_{max}; \varphi_{max}]$ and $\varphi_{up} = cst.$
- **Skirting** of the chain : if the bead n intersects the chain then $\varphi_{random} \in \left[-\frac{2\pi}{3}; \frac{2\pi}{3}\right]$.
- **Rebound** of the chain : if the bead n gets out from lateral box limits then $\theta_{random} \in \left[-\frac{2\pi}{3}; \frac{2\pi}{3}\right]$.
- **Switch** of the chain : if the bead n gets out from vertical box limits then $\varphi_{up} = -\varphi_{up}$.

The raise angle is generally calculated to put the last bead at the top of the box. Sometimes it is impossible to define a bead position, so the wire is cut.

This wire model must have beads diameter around 20% than the smallest sand particle diameter. Indeed, a too large bead would make a rough wire

Fig. 6. 3D wire disposition : isometric (left) and lateral view (right).

and the friction parameter would be more difficult to control. But such a condition increases the number of particle to make a realistic sample and the LMGC90 deposit problem still exists in 3D. In fact, good sample would represent around 10000 particles. We are currently working on a 3D extension of the pre-processor which generates dense sample.

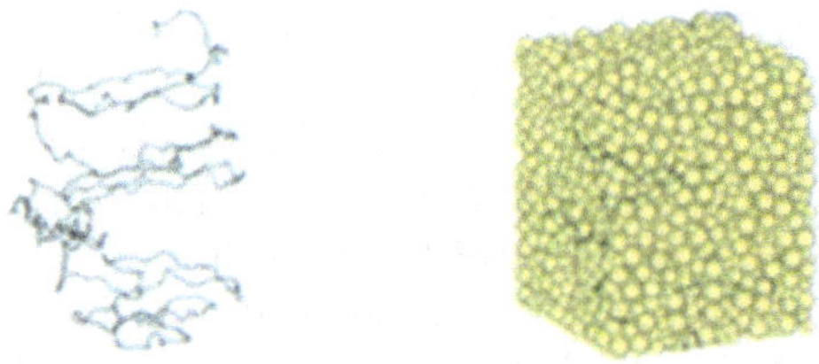

Fig. 7. Wire disposition (left) and dense *TexSol* (right).

Conclusion and perspectives

The distinct elements give us a new approach of the *TexSol* problem. They are able to show us which are the wire deformation mechanisms. Those 3D investigations will be soon compared with the continuous local model in order to determinate if it is the best adapted.

First of all, we have to optimise our simulation tools on several areas : samples preparation, deposit, compaction and mechanical test computation. Then we will be able to analyse 3D wires mechanical influence on the *TexSol*.

References

1. Fremond M. (2002) Non-Smooth Thermo-mechanics. Springer-Verlag Berlin Heidelberg New York
2. Jean M. (1999) The non smooth contact dynamics method. Computer Methods in Applied Mechanic and Engineering 177 (Special issue):235–257
3. Khay M., Gigan J-P. (1990) TEXSOL - Ouvrage de soutènement. LCPC
4. Leflaive E., Khay M., Blivet J-C. (1985) Un nouveau matériau : le TEXSOL. Travaux 602:1–3
5. Mouraille O. (2004) Etude sur le comportement d'un matériau à longueur interne: le *TexSol*. DEA, Université de Montpellier II
6. Villard P. (1988) Etude du renforcement des sables par des fils continus. PhD Thesis, Université de Nantes, ENSM

A unified interface constitutive law for the study of fracture and contact problems in heterogeneous materials

M. Paggi, A. Carpinteri, and G. Zavarise

Department of Structural and Geotechnical Engineering, Politecnico di Torino, C.so Duca degli Abruzzi 24, 10129 Torino, Italy,
marco.paggi@polito.it

Abstract. A unified interface constitutive law for the description of contact and decohesion at bi-material interfaces is proposed. To this aim, a synthesis of the nonlinear models pertaining to Fracture and Contact Mechanics is presented. The issues pertinent to the implementation within the FE discretization framework are also discussed in detail. Finally, a numerical example of fatigue modeling at the mesoscopical level in a fiber-reinforced composite is provided.

1 Introduction

Interfaces between different materials in composites exert an important, and sometimes controlling, influence on the overall mechanical performance. With this respect, it is evident that a proper modeling of the mechanical behavior of interfaces at different length scales is an outstanding issue.

In this framework, nonlinear theories pertaining to Fracture Mechanics and Contact Mechanics can be profitably used for the description of uncoupled decohesion and contact problems at bi-material interfaces. However, since in many engineering applications the interfaces may experience both contact and decohesion during the loading history, the aim is to overcome the classical separation between Fracture and Contact theories. Hence, the present work is undertaken to gain some insight into the possibility to define a unified interface constitutive law able to fully describe the mechanical behavior of the interface. Moreover, the issues pertinent to the implementation in the FE discretization are also deeply discussed.

To demonstrate the effectiveness of this numerical tool, which represents a step forward in modeling the interface mechanical problems at the mesoscopical level, an illustrative example concerning fatigue in fiber reinforced metal matrix composites is proposed.

2 Unified interface constitutive law

Depending on the status of the interface, e.g. bonded or debonded, the interfacial behavior can be described from the mechanical point of view according to constitutive laws traditionally belonging to either Contact Mechanics or Fracture Mechanics. These theories are briefly recalled and summarized in this section. In the sequel the normal and the tangential behaviors of the interface will be separately analyzed.

2.1 Interface normal behavior

When the constituent materials are bonded together, the mechanical behavior of the interface in compression under monotonic and cyclic loading can be modeled according to a constraint formulation on the basis of the Signorini condition of unilateral contact [1]. The mathematical condition for non-penetration can be stated as $g_N \geq 0$, where g_N denotes the normal gap (see curve (a) in Fig. 2.a).

On the other hand, due to the finite interface strength, the mechanical behavior in tension is usually ruled by a cohesive formulation. Decohesion is depicted as a phenomenon of progressive separation across a cohesive zone that is resisted by cohesive tractions [2].

Recently, a cohesive zone model for the study of decohesion between fibers and matrix in metal matrix composites has been proposed by Tvergaard [3]. In this formulation, a measure of the interface opening, λ, is computed from the normal and the tangential gaps, g_N and g_T. Then, normal and tangential cohesive forces are given as functions of interface opening in the process zone (see curve (b) in Fig. 2.a).

In case of unloading with $g_N \geq 0$, two different unloading paths can be considered [4, 5]. The former is based on the analogy with an elastic-damaged solid where unloading is prescribed to occur at the origin of the traction-separation space. The latter employs a condition based on an analogy with

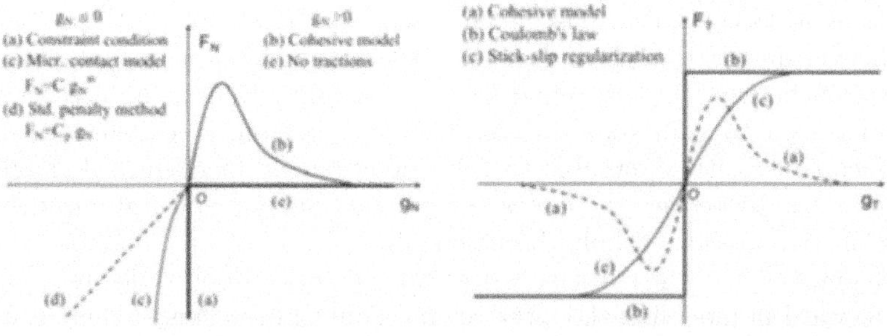

a. b.

Fig. 1. Interface constitutive law: (a) normal behavior and (b) tangential behavior.

an elastic-plastic material where linear unloading occurs with the stiffness of the cohesive law at zero separation. A subsequent reloading takes place on the same unloading paths above described, up to the maximum interface separation experienced during the previous loading history.

In case of cyclic loading, in order to capture finite life effects, a dependence of the softening law on the previous loading history has been recently introduced by Roe and Siegmund [6]. In their model a description of the damage evolution is provided and a degradation of the cohesive formulation according to damage mechanics is proposed.

When complete debonding takes place, physical phenomena at the interface significantly change from the bonded status and different models have to be invoked. Under both monotonic and cyclic compression, the surface roughness plays a fundamental role which has to be carefully taken into account. In this context, microscopical contact models based on a statistical description of the contact geometry can be used (see Ref. [14] for a detailed survey). However, more recent theories pointed out the fact that the statistical parameters, and then the contact predictions of these models, are scale-dependent, being functions of the profilometer resolution. Hence, to overcome this main shortcoming, contact models based on fractal geometry can be adopted [8, 12].

These approaches provide strong nonlinear relationships between the normal force and the corresponding normal gap (see curve (c) in Fig. 2.a):

$$F_N = C g_N^m \tag{1}$$

where the parameter m depends on the deformation assumptions of the profiles in contact and on the probability distribution of asperity heights. Experimentally, m for metals is usually in the range $2 \leq m \leq 3.33$ [10]. Thus the simplest possibility with $m = 1$, which would coincide with a standard penalty method, is excluded (see curve (d) in Fig. 2.a).

While a mutual pressure is exerted at the interface between two bodies in contact, no tensile tractions can be sustained when the interface is debonded (see curve (e) in Fig. 2.a). This assumption agrees with the common experience that, in absence of a specific adhesive, the contact area of rough interfaces falls to zero when the load is removed and that no load is required to separate them.

The normal behavior of a zero thickness interface is summarized in Tab. 1 along with the different mechanical models.

2.2 Interface tangential behavior

Similarly to the normal problem, the tangential behavior of interfaces is significantly influenced by the bonding conditions. When interfaces are bonded together, cohesive formulations are usually employed. Since Mode I and Mode II deformations very often occur at the same time, uniaxial cohesive models have to be generalized to a fully consistent Mixed-Mode formulation, as in

Table 1. Normal behavior of a zero thickness interface.

Interface status	Loading type	Loading condition	Mechanical modeling	Ref.
Bonded	Monotonic	Compression	Constraint formulation	[1]
		Tension	Decohesion	[5, 11]
	Cyclic	Compression	Constraint formulation	[1]
		Tension	Decohesion with damage	[6]
Debonded	Monotonic	Compression	Micromechanical contact models	[14]
		Tension	Adhesion (low roughness)	[12]
			No tensile tractions (high roughness)	[12]
	Cyclic	Compression	Micromechanical contact models	[14]
		Tension	No tractions	[12]

Ref. [3] (see curve (a) in Fig. 2.b). As far as the cyclic tangential loading condition is concerned, a dependence of the softening function on the previous loading history has to be taken into account [6].

On the other hand, when debonding occurs, roughness plays a key role in the mechanical description of the contact phenomenon. Under a constant applied normal force, F_N, and increasing the tangential force, F_T, the contact area progressively changes from the full-stick to the full-slip condition and a small relative motion between the bodies takes place over part of the interface. This complex phenomenon is referred to as *stick-slip* [12]. For spherical bodies, this problem was mathematically solved by Cattaneo [2] and, independently, by Mindlin [3]. More recently, the Cattaneo-Mindlin solution was generalized by Jäger [11] and by Ciavarella [9] to general surfaces in contact and, more specifically, to rough fractal surfaces [17]. According to these models the solution of the tangential contact problem can be directly gained from the solution of the normal contact one. This approach is restricted to bodies of equal material, although Hanson and Keer [18] have shown that, as a first approximation, it can also be applied to dissimilar bodies. Moreover, following the theory by Jäger [11], any general loading scenario can be modeled as a superposition of oblique loading increments using a recursive formula.

According to this approach, for given relative normal and tangential displacements at the interface, the tangential force can be computed as a function of the normal loading history. This method can be used as an alternative to the classical approaches based on the theory of plasticity and provides a regularization of the Coulomb's friction law which is consistent with the stick-slip transition phenomenon observed in rough interfaces (see curves (b) and (c) in Fig. 2.b).

The tangential behavior of the interface and the mechanical modeling as a function of the interface status is summarized in Tab. 2.

Table 2. Tangential behavior of a zero thickness interface.

Interface status	Loading type	Mechanical modeling	Ref.
Bonded	Monotonic	Decohesion	[5, 19]
	Cyclic	Decohesion with damage	[6]
Debonded	Monotonic	Stick-slip	[11, 9, 17]
	Cyclic	Stick-slip with general loading history	[11]

3 FE algorithms

The above described interface constitutive laws can be implemented in a FE code on the basis of the node-to-segment contact strategy [20, 21]. According to this approach, for each node along the master contact surface it is possible to compute the normal and the tangential gaps, that are a measure of the relative displacements of the two bodies along the interface. Then, the interface status can be assessed and an automatic switching procedure can be adopted in order to choose between the different mechanical formulations. More specifically, concerning the contact constraints, a modified penalty method based on micromechanical contact laws can be used [22]. The damage variable can also be computed at each time step.

Contributions of cohesive and contact forces are then added to the global virtual work equation:

$$\delta W = \mathrm{A}\ (F_N \delta g_N + F_T \delta g_T) \tag{2}$$

where the symbol A denotes an assembly operator for all the interface nodes either inside the process zone or in contact. A main difficulty with the analysis, stemming from the contact constraints, is that the debonded surfaces are unknown a priori, and the corresponding boundary value problem must be solved with an iterative method. The Newton-Raphson solution procedures commonly used for solving nonlinear problems require the determination of the tangent stiffness matrix (see Ref. [21] for more details). Then consistent linearization of the equation set (1) has to be performed, as discussed in Ref. [22]. Finally, the normal and tangential force contributions to the virtual work equation can be rewritten in terms of the displacement unknowns, within the classical FE solution scheme. A possible numerical algorithm for the computation of the interface stiffness matrix is presented in Tab. 3.

Table 3. Numerical algorithm for the computation of the interface stiffness matrix.

```
LOOP over iterations: i = 1,..., convergence
        Compute geometrical parameters
        LOOP over all the interfaces Γ_j
                LOOP over all the segments s of Γ_j
                        Step a. Check for the interface status in s
                        Step b. IF g_{N,s} > 0 and 0 ≤ λ ≤ 1 ⇒ cohesive model
                        Step c. IF g_{N,s} ≤ 0 ⇒ contact model
                        Step d. IF g_{N,s} > 0 and λ > 1 ⇒ no tractions
                END
        END
        Solve FE equation system
        Check for convergence
END
Update damage variable for the cohesive formulation
```

Finally, concerning the problem of interface discretization, the virtual node technique [23] has proven to be very effective when problems with multiple heterogeneities are analyzed. The basic idea underling this method consists in changing the integration scheme usually adopted in node-to-segment contact elements. The cohesive/contact contribution to the stiffness matrix and the internal force vector are in fact integrated on the contact element length through a n-point Gauss integration scheme instead of a simpler 2-point Newton-Cotes integration formula. In this way, an arbitrary number of Gauss points can be specified inside each contact element placed along the interface, regardless of the discretization used for the continuum. As a result, this method permits to achieve a fine discretization of the interface without refining the discretization of the continuum.

4 Numerical example

The fatigue response of a fiber-reinforced composite under horizontal displacement cyclic loading in the direction perpendicular to the fibers is investigated. As a reference material we consider an Al-SiC metal matrix composite, whose mechanical parameters are reported in [24]. The Tvergaard cohesive model with the damage formulation by Roe and Siegmund and the Majumdar-Bushan fractal contact model are considered (see Refs. [3, 6] for a description of the input parameters).

The deformed mesh of the composite with 8x8 fibers is shown in Fig. 2.a with a superimposed contour plot of the horizontal stress field. The nondimensional horizontal macroscopic stress is depicted as a function of the horizontal deformation in Fig. 2.b. A magnification of the beginning of the unloading paths is also depicted in Fig. 2.c. From these diagrams the elastic modulus during unloading can be computed. Such parameter, normalized to its initial value E_0, is depicted in Fig. 2.d vs. the nondimensional cycles number. Parameter $N_{\max}$ denotes the cycles number corresponding to the complete debonding of the reinforcement.

A significative reduction of the elastic modulus takes place at the beginning of the test. This is due to the progressive debonding of the interface and to the nonlinear cohesive law adopted for the description of the mechanical behavior of the interface in tension. In this stage, a reduction of 40% in the elastic modulus is achieved when $N/N_{\max}$ is equal to 0.4. Afterwards, the interface debonding phenomenon develops slowly and the elastic modulus remains almost constant.

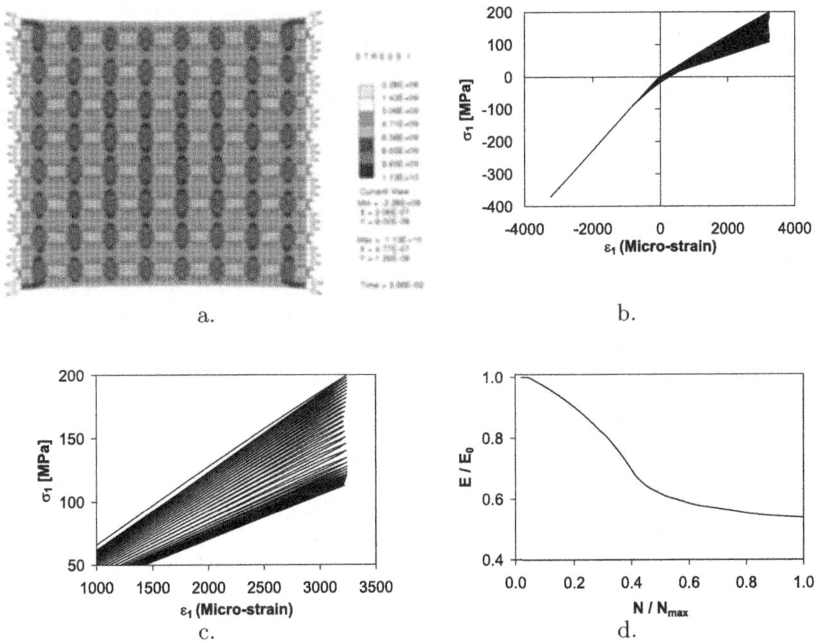

Fig. 2. (a) Deformed mesh. (b) Nondimensional horizontal macroscopic stress vs. horizontal deformation. (c) Magnification of Fig. 2.b. (d) Reduction of the elastic modulus vs. the nondimensional cycles number.

5 Conclusions

A unified interface constitutive law for the study of decohesion and contact problems in heterogeneous materials has been proposed. This formulation can be profitably used for solving mechanical problems with multiple heterogeneities. An application to the study of fatigue damage evolution in fiber-reinforced composites has been presented. This illustrative example shows the effectiveness of the proposed numerical scheme. In fact, due to horizontal displacement cyclic loading, both decohesion and contact problems at bi-material interfaces has been solved during the loading history and a unified interface constitutive law has been adopted.

Acknowledgements

Support of the Italian Ministry of Education, University and Scientific Research (MIUR) and of the European Union (EU) is gratefully acknowledged.

References

1. Signorini A (1933) Atti della Società Italiana per il Progresso delle Scienze
2. Carpinteri A (1989) J. Mech. Phys. Sol., 37:567–582
3. Tvergaard V (1990) Mat. Sci. Engng. A, 107:23–40
4. Chandra N, Li H, Shet C and Ghonem H (2002) *Int. J. Sol. Struct.*, 39:2827–2854
5. de Borst R (2003) Engng. Fract. Mech., 70:1743–1757
6. Roe KL and Siegmund T (2003) Engng. Fract. Mech., 70:209–232
7. Zavarise G, Borri-Brunetto M and Paggi M (2004) Wear, 257:229–245
8. Majumdar A and Bhushan B (1990) ASME J. Tribology, 112:205–216
9. Borri-Brunetto M, Carpinteri A and Chiaia B (1999) Int. J. Fract., 95:221–238
10. Kragelsky IV, Dobychin MN and Kombalov VS (1982) Pergamon Press, (Translated from the Russian by N. Standen)
11. Carpinteri A, editor (1999) Nonlinear Crack Models for Nonmetallic Materials, Kluwer Academic Publisher, Dordrecht
12. Johnson KL (1985) Contact mechanics. Cambridge University Press, Cambridge
13. Cattaneo C (1938) Rendiconti dell'Accademia Nazionale dei Lincei, Serie 6, 27:342–348, 434–436, 474–478
14. Mindlin RD (1949) ASME J. Appl. Mech., 16:259–268
15. Jäger J (1998) ASME J. Tribology, 120:677–684
16. Ciavarella M (1998) ASME J. Appl. Mech., 65:998–1003
17. Borri-Brunetto M, Chiaia B and Ciavarella M (2001) Comp. Meth. Appl. Mech. Engng., 190:6053–6073
18. Hanson MT and Keer LM (1989) Int. J. Mech. Sci., 31:107–119
19. Carpinteri A, Valente S, Ferrara G and Melchiorri G (1995) Computers & Structures, 48:397–413
20. Wriggers P (2002) Computational Contact Mechanics. John Wiley & Sons, Ltd., The Atrium, Southern Gate, Chichester, West Sussex, PO19 8SQ, England
21. Wriggers P, Zavarise G and Zohdi TI (1998) Comp. Mat. Sci., 12:29–56
22. Zavarise G (1991) Problemi termomeccanici di contatto - aspetti fisici e computazionali. PhD thesis, Istituto di Scienza e Tecnica delle Costruzioni, University of Padua, Italy
23. Zavarise G, Boso D and Schrefler BA (2001) A contact formulation for electrical and mechanical resistance. In Proceedings of CMIS, III Contact Mechanics International Symposium, Praja de Consolacao, Portugal, 211–218
24. Peng Y and Chaki TK (1995) Mat. Sci. Engng. A, 200:38–46

Elasto-plastic contact of fractal surfaces

K. Willner

Lehrstuhl für Technische Mechanik, Universität Erlangen-Nürnberg,
Egerlandstraße 5, 91026 Erlangen, Germany,
willner@ltm.uni-erlangen.de

Abstract. The elasto-plastic normal contact of fractal surfaces is numerically investigated using a halfspace model. Artificial surface data are generated using the structure function, to study the influence of different surface parameters with respect to the load-area relationship and the load-gap relationship. The simulations show that for realistic surface parameters the deformation is always in the plastic range.

1 Introduction

The experimentally well confirmed linear relationship between normal and friction force, known as Amonton's law, can only be explained if a linear relationship between normal load and real area of contact can be established. Since contact will occur only at the asperities and the real area of contact A_r will be only a very small fraction of the apparent area of contact A_0, Bowden and Tabor [4] argued that even for moderate normal loads very high local stresses develop, which lead always to plastic deformation. The load-area relationship is then given by simple equilibrium considerations as

$$\frac{p}{H} = \frac{A_r}{A_0} \,, \tag{1}$$

where p denotes the average pressure and H the Hardness of the softer contact partner and thus establishes the required linear relationship. However, for an elastic contact simple Hertzian theory would predict a nonlinear relationship, contradicting the experimental findings, but Archard [1] showed that even for purely elastic contact a linear relationship between load and contact area can be established, if a hierarchy of Hertzian contacts is assumed, thus introducing the concept of fractality well before the invention of the term.

A discussion of Archard's results in view of modern fractal theory was given by Borri-Brunetto et al. [3], who showed that for purely elastic contact of fractal surfaces the real area of contact has fractal dimension $D < 2$ and will consist of an infinite number of infinitely small contact patches with a total area of zero. They point out that this is an unrealistic result since a vanishing area of contact will lead to very high local stresses and thus to plastic deformations. The aim of this paper is therefore to investigate the behavior of fractal surfaces under elasto-plastic conditions.

2 Fractal surfaces

In the following the discussion is restricted to isotropic surfaces, such that a single profile along any direction can be taken as representative. The relation between the fractal dimension of this profile D and the dimension of the corresponding surface D_s is then just $D_s = D + 1$. We seek now parameters that characterize such a profile.

A convenient approach is the use of the structure function defined by

$$S(x_k) = \frac{1}{N-k} \sum_{i=1}^{N-k} (z_{i+k} - z_i)^2 \qquad k = 1, \ldots, N-1 \,. \tag{2}$$

Simple algebra [7] shows that

$$S(x_k) = 2\left(\sigma_z^2 - \mathrm{ACF}(x_k)\right) , \tag{3}$$

where $\mathrm{ACF}(x_k)$ denotes the auto-correlation function. Since for a stationary, stochastic profile

$$\lim_{x_k \to \infty} \mathrm{ACF}(x_k) = 0 , \tag{4}$$

we have

$$\lim_{x_k \to \infty} S(x_k) = 2\sigma_z^2 . \tag{5}$$

For a fractal the structure function becomes

$$S(x_k) = C x_k^{4-2D} , \tag{6}$$

where D denotes the fractal dimension of the profile, which is in the range $1 \le D < 2$, and C is some constant. Thus the usual test for fractality is to calculate the structure function by use of (2) and plot it in a log-log plot. A fractal profile should come out as a straight line, where the fractal dimension D can be easily determined from the slope.

Figure 1 shows some structure functions obtained from profile measurements of typical technical surfaces. As the figure shows, real profiles show a mixed behavior. For small length scales they behave like fractals and the fractal dimension can be determined by the corresponding slope, while for larger length scales they become simple stationary processes and the structure function tends towards the constant value $2\sigma_z$. Berry and Blackwell [2] proposed

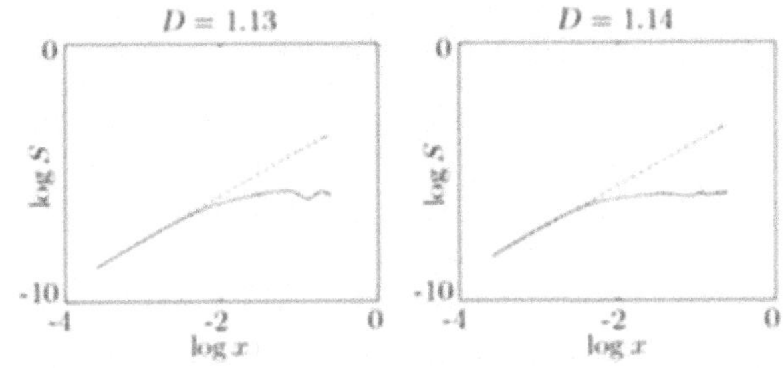

Fig. 1. Structure functions for measured surfaces.

the following function as an approximation to the structure function for this particular case,

$$S(x_k) = 2\sigma_z^2 \left\{ 1 - \exp\left[-\left(\frac{x_k}{x_{\mathrm{T}}} \right)^{4-2D} \right] \right\} , \tag{7}$$

where the parameter x_{T}, which can be interpreted as the transition length between the fractal and the stationary regions, respectively. The parameters σ_z, D and x_{T} are independent of the measurement resolution and are therefore considered intrinsic parameters of the profile.

3 Contact simulation

In order to study the influence of different parameters it is necessary to numerically generate surfaces with specified properties. This can be achieved using the relation (3) and the Wiener-Khinchin theorem to obtain the Fourier transform of height data as shown in [7].

The numerically generated surfaces are then used in a contact simulation based on elastic halfspace theory. The simulation uses a technique developed by Tian and Bhushan [6]. Discretization of the contact area into M regular patches leads to a linear system

$$\mathbf{C}\mathbf{p} = \bar{\mathbf{u}} , \tag{8}$$

where $\mathbf{u}$ denotes the prescribed displacements

$$\bar{u} = \delta - [z_1(x, y) - z_2(x, y)] , \tag{9}$$

where z_i are the height data of the contacting surfaces and δ is the rigid body approach of the contact partners. The flexibility matrix $\mathbf{C}$ is calculated using the Boussinesq solution assuming a constant pressure in each contact patch

$$u_z(x,y) = \frac{1}{\pi E^*} \int\limits_{\Gamma} \frac{p(x',y')}{\sqrt{(x-x')^2 + (y-y')^2}} D\Gamma \,, \tag{10}$$

where E^* is the joint Hertzian modulus of the contacting partners

$$\frac{1}{E^*} = \frac{1-\nu_1^2}{E_1} + \frac{1-\nu_2^2}{E_2} \tag{11}$$

The sytem is solved by a Gauss-Seidel iteration for the unknown pressures $\mathbf{p}$, where the additional restriction

$$p_k \geq 0 \,, \qquad k = 1, \ldots, M \tag{12}$$

has to be observed, since no tension is allowed at the interface.

The elastic algorithm described above can be modified to model elasto-plastic contact. The most simple approach is to limit the local pressure to a maximum value given by the surface hardness of the material. This approach is rigorous if one is only interested in the pressure distribution in the initial loading case, but will not account for the plastic deformations of the asperities. To calculate the deformed surface, which may be of interest in an unloading calculation, Tian and Bhushan [6] propose a modified procedure where local plastic deformations are taken into account at least in an approximate way. However, since we are here only interested in the pressure distribution and not in the deformed surface, we will use the simple limit method for simplicity and require in the elasto-plastic case just the additional restriction

$$p_k \leq p_{\max} \,, \qquad k = 1, \ldots, M \tag{13}$$

which is imposed during the Gauss-Seidel iteration in those contact regions where the calculated local elastic pressure exceeds the limit pressure $p_{\max}$.

The limit pressure $p_{\max}$ can be determined by a micro- or nano-indentation test. Since this value exhibits a considerable size effect at least on the nanoscale, the measurements should be taken if possible at a scale corresponding to the highest resolution used in the calculations. However, in the following always the same value has been used to avoid obscuring the influence of the geometric parameters.

4 Results

While the algorithm described in the last section is capable of treating two rough, deformable surfaces in contact, the following simulations were performed for the contact of a generated, rough fractal surface against a rigid, perfectly flat surface in order to study the influence of the fractal parameters without considering the joint influence of two rough surfaces. The rigid, flat surface is simply obtained by setting the corresponding Young's modulus in equation (11) to infinity and the corresponding height values in equation (9) to zero. For the fractal surface the parameters shown in Tab. 1 were used. While these basic parameters were kept constant, different fractal dimensions D_S, different transition length x_T, and resolutions of the fractals were used by varying the limit wavelength λ_{lim} at the generation of the surfaces.

Table 1. Values used for the contact simulation.

Variance of heights	$\sigma_z = 10^{-6}$ m
Length of patch	$L = 10^{-3}$ m
Number of points	$N = 2^9$
Young's modulus	$E = 2.1 \cdot 10^{11}$ N/m^2
Poisson ratio	$\nu = 0.3$
Indentation hardness	$p_{\mathrm{max}} = 2.5 \cdot 10^9$ N/m^2

Figures 2 and 3 show the calculated apparent pressure vs. gap curves and contact area vs. pressure curves for the elastic and plastic contact, respectively.

In the elastic case the pressure-distance curves are shifted to the right with increasing resolution, i.e., decreasing λ_{lim}. This can be explained by the added roughness features which will lead to higher maximum values of the asperity heights and thus lead to a shift of the initial contact to higher gap values. However, the curves converge once the limit wavelength becomes significantly smaller than the transition length x_T. This shows that the pressure-gap law is dominated by the stationary part of the structure function while the fractal behavior in the small wavelength limit has nearly no influence. As discussed in detail by Borri-Brunetto et al. [3], the real area of contact becomes a *lacunar* fractal surface with a dimension $D_C < 2$. This dimension decreases with increasing fractal dimension D of the contacting surface. This numerically obtained result is confirmed by an analytical analysis of a similar two-dimensional problem by Ciavarella and Demelio [5]. They conclude that the fractal dimension of the contact region is given by $D_C = 4 - D_S$ for the three-dimensional case, which is close to the results obtained by Borri-Brunetto et al. However, the decrease and the fragmentation of the real area of contact will lead to very high local pressures and thus to plastic deformation. If plastic deformation is taken into account the picture changes, especially for the area-pressure relationship in Fig. 3. While for the coarse resolution of $\lambda_{\mathrm{lim}} = 125\ \mu$m the surface behaves still elastically, we have a decidedly different behavior for higher resolutions. The additional smaller features deform plastically and thus lead in the limit to the simple relationship (1).

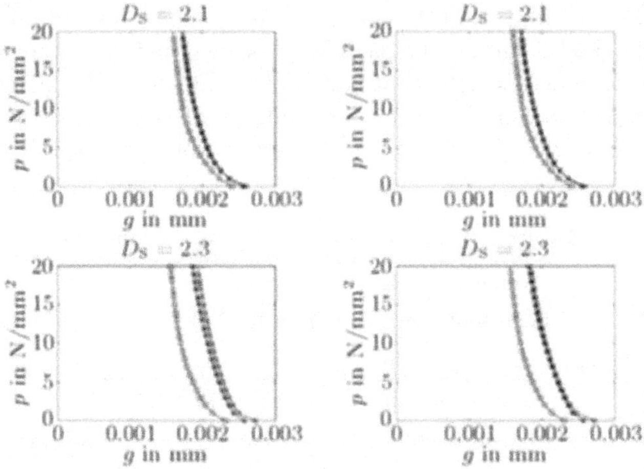

Fig. 2. Pressure-gap relationship for elastic (left) and plastic (right) contact. Variation of fractal dimension D_S, transition length $x_T = 100\ \mu$m. ($\times : \lambda_{\text{lim}} = 7.81\ \mu$m, $\star : \lambda_{\text{lim}} = 31.25\ \mu$m, $\Box : \lambda_{\text{lim}} = 125.0\ \mu$m) .

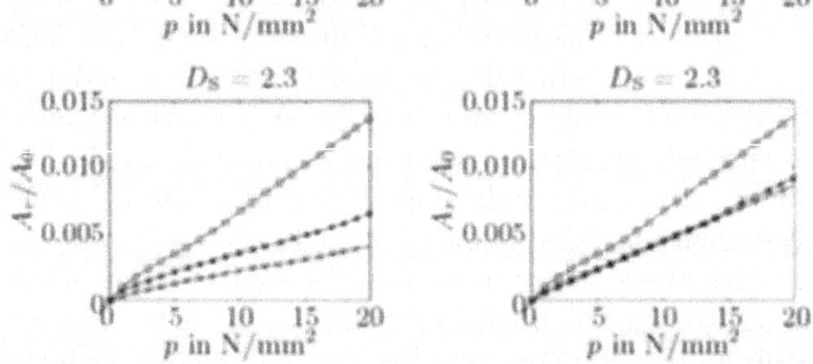

Fig. 3. Area-pressure relationship for elastic (left) and plastic (right) contact. Variation of fractal dimension D_S, transition length $x_T = 100\ \mu$m. ($\times : \lambda_{\text{lim}} = 7.81\ \mu$m, $\star : \lambda_{\text{lim}} = 31.25\ \mu$m, $\Box : \lambda_{\text{lim}} = 125.0\ \mu$m).

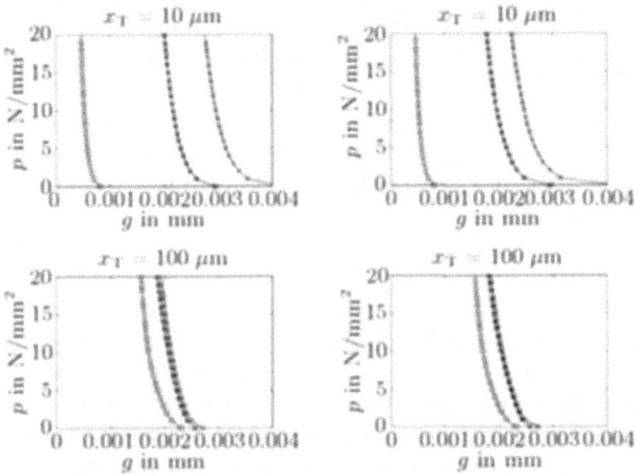

Fig. 4. Pressure-gap relationship for elastic (left) and plastic (right) contact. Variation of transition length x_T, fractal dimension $D_S = 2.3$. ($\times : \lambda_{\text{lim}} = 7.81\ \mu\text{m}$, $\star : \lambda_{\text{lim}} = 31.25\ \mu\text{m}$, $\Box : \lambda_{\text{lim}} = 125.0\ \mu\text{m}$).

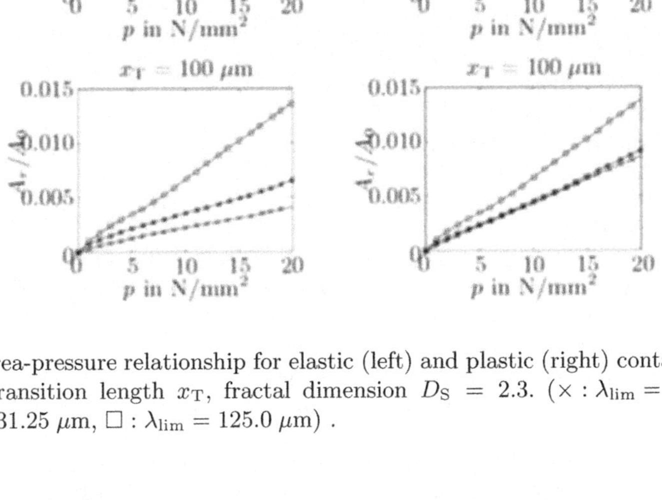

Fig. 5. Area-pressure relationship for elastic (left) and plastic (right) contact. Variation of transition length x_T, fractal dimension $D_S = 2.3$. ($\times : \lambda_{\text{lim}} = 7.81\ \mu\text{m}$, $\star : \lambda_{\text{lim}} = 31.25\ \mu\text{m}$, $\Box : \lambda_{\text{lim}} = 125.0\ \mu\text{m}$) .

Figures 4 and 5 show the influence of the resolution, i.e., the limit wavelength λ_{lim}, on the pressure-gap and the area-pressure relations for a fixed fractal dimension $D = 2.3$. As Fig. 4 shows, the pressure-gap relationship converges in the elastic as well as in the plastic case, if the limit wavelength λ_{lim} is shorter than the transition length x_{T}. However, Fig. 5 shows again that in the elastic case no convergence of the contact area can be obtained.

5 Conclusions

In contrast to the findings of Borri-Brunetto et al. [3] who stated that plasticity has no significant influence since only a few points undergo high stresses, the present paper shows a clear influence of plastic deformations. Actually, it seems as if *all* fractal surfaces will undergo plastic deformations, since there will be always very high local stresses. While no convergence of the area-pressure relation can be observed in the elastic case in accordance with [3], there is a definitive limit, if plastic deformations are taken into account.

The pressure-gap relationship converges both in the purely elastic as well in the plastic case once the wavelength resolution is fine enough to be in the fractal range. Therefore we can conclude that the wavelength limit λ_{lim} has to be chosen just smaller than the transition length x_{T}, which can be determined from a plot of the structure function. A reasonable wavelength limit seems to be $\lambda_{\mathrm{lim}} \approx x_{\mathrm{T}}/10 \ldots x_{\mathrm{T}}/5$. The contact laws are mainly influenced by the long wavelength limit which is completely taken into account by the structure function (7).

References

1. Archard, J. F.: Elastic Deformation and the Laws of Friction. In: *Proceedings of the Royal Society of London, Series A* 243 (1957), S. 190–205.
2. Berry, M. V.; Blackwell, T. M.: Diffractal Echoes. In: *Journal of Physics A* 14 (1981), S. 3101–3110.
3. Borri-Brunetto, M.; Carpinteri, A.; Chiaia, B.: Lacunarity of the Contact Domain Betweeen Elastic Bodies with Rough Boundaries. In: Frantziskonis, G. (Hrsg.): *PROBAMAT – 21st Century: Probabilities and Materials.* S. 45–66. Dordrecht: Kluwer, 1998.
4. Bowden, F. P.: A Review of the Friction of Solids. In: *Wear* 1 (1957), S. 333–346.
5. Ciavarella, M.; Demelio, G.: Elastic Multiscale Contact of Rough Surfaces: Archard's Model Revisited and Comparison with Modern Fractal Models. In: *Journal of Applied Mechanics* 68 (2001), S. 496–498.
6. Tian, X.; Bhushan, B.: A Numerical Three-Dimensional Model for the Contact of Rough Surfaces by Variational Principle. In: *Journal of Tribology* 118 (1996), S. 33–42.
7. Willner, K.: Elasto-Plastic Normal Contact of Three-Dimensional Fractal Surfaces Using Halfspace Theory. In: *Journal of Tribology* 126 (2004), S. 28–33.

A 3D study of the contact interface behavior using elastic-plastic constitutive equations

A.A. Bandeira, P.M. Pimenta, and P. Wriggers

[1] Department of Civil Engineering at Presbyterian Mackenzie University, São Paulo, Brazil,
alex_bandeira@terra.com.br
[2] Department of Structural Engineering, Polytechnic School at University of São Paulo, São Paulo, Brazil,
ppimenta@usp.br
[3] Institute for Mechanics and Computational Mechanics, University of Hannover, Hannover, Germany,
wriggers@ibnm.uni-hannover.de

Abstract. In this work a homogenization method presented by Bandeira et al [2,3,4] is enhanced in order to obtain by numerical simulation the interface law for the normal contact pressure based on statistical surface models. For this purpose elastic-plastic behavior of the asperities is considered. Statistical evaluations of numerical simulations lead to a constitutive law for the contact pressure. The resulting law compared with other laws stemming from analytical investigations, like those presented by Greenwood Williamson [11] and Yovanovich [19, 32]. The non-penetration condition and the interface model for contact that takes into account the surface microstructure are investigated in detail.

This paper can be regarded as a complementary study to that presented by Bandeira et al [2]. Here the plasticity of the asperities is taken into account by assuming a constitutive equation based on an associated von Mises yield function formulated in principal axes, as shown by Pimenta [22]. The basic aim of this paper is to derive constitutive contact laws for a rough surface by using the finite element method. For this purpose one has to model and discretize the rough surface and then, by homogenization procedures develop an interface law for contact. The interface law is obtained from numerical simulation using a model that consists of two deformable bodies in contact. The contact surfaces of both bodies are rough. The law obtained by numerical simulations and statistical evaluation of the numerical results is compared with analytically derived laws that regard plasticity.

An augmented Lagrangian method is applied to solve the contact problem, see Bertsekas [5,6], Fletcher [10], Luenberger [18], Laursen, Maker [15], Laursen, Simo [16, 17], Wriggers [27], Wriggers, Simo [28], Wriggers et al. [29], Wriggers, Zavarise [31] and Heegaard, Curnier [12]. The technique used to solve three-dimensional contact problems with friction, see Curnier [7] and Tabor [26], in finite deformations was already developed and described in Bandeira et al [2,3,4], Simo, Laursen [24], Wriggers [27, 28, 29, 30, 31], Alart, Curnier [1] and Oden, Pires [20].

Numerical examples are selected to show the ability of the algorithm to represent interface law for rough surfaces considering elastoplastic behaviour of the asperities.

1 Introduction

Several formulations concerning the treatment of the contact interface have been presented in the literature, especially when the contact interface of two contacting bodies is rough. Two main lines can be followed to consider the contact conditions in normal direction. One regards the formulation of the non-penetration condition as a purely geometrical constraint and the other develops the constitutive laws with a micromechanical approach within the contact area, which yields a response function for the normal stresses in terms of parameters such as the mean real area of contact or the current mean plane distance. For the tangential stresses, the same situation arises for sticking contact interface, where either a geometrical constraint equation or a constitutive law for the tangential micro displacement can be formulated. For sliding between the bodies in contact, a constitutive equation has to be formulated in form of an evolution equation.

Constitutive equations for the normal contact have been developed by investigating micromechanical behavior within the contact surface. Associated models have been developed based on experiments, like Greenwood, Williamson [11], Evseev, Medvedev, Grigoriyan [8], Zavarise, Schrefler, Wriggers [33] and Kragelsky, Dobychin, Kombalov [14]. In general, the micromechanical behavior depends on material parameters like hardness and on geometrical parameters like surface roughness. It should be noted that the real micromechanical phenomena are extremely complex due to extremely high local pressure at the asperities. The model used in this paper, attempts only to capture the most important phenomena and assume either elastic or plastic deformation of the asperities having real contact in the interface.

This article concentrates on the behavior of the contact interface. The idea is to study the interface behavior by modeling the contact surfaces using a finite element discretization to take into account the geometrical properties of the microstructure. The probabilistic theory is applied based on a statistical model of the micro-geometry like in the microscopic contact mechanics developed by Greenwood, Williamson [11] and Wriggers, Vu Van, Stein [30]. Finally a simple homogenization leads to a contact interface law. The response is then compared to known laws derived, see e.g. [11], [14].

A three-dimensional eight-node brick element, see Kardestuncer, Norrie [13] and Parisch [21], is used for the treatment of finite elastic-plastic deformation of the contacting surfaces. An augmented Lagrangian method is applied to solve the frictional contact problems because high-pressures occur which cannot be treated adequately by standard penalty procedures.

The finite element program is based on a C++ code developed by Bandeira [2]. All numerical examples given are based on three-dimensional calculations. In the numerical examples high-density meshes are used to represent the geometrical irregularity on the surfaces more precisely.

2 Constitutive equation for contact interface

Different approaches have been proposed to represent microscopic contact mechanisms. The available formulations are based either on curve fitting of experimental results or on theoretical analyses of microscopically rough surface, like Greenwood, Williamson [11], Evseev, Medvedev, Grigoriyan [8], Zavarise, Schrefler, Wriggers [33] and Kragelsky, Dobychin, Kombalov [14].

The current normal approach g_N can be defined by

$$g_N = \xi - d \tag{1}$$

where ξ denotes the maximum initial asperities height and d the current mean plane distance. The graphical interpretation of equation (1) is illustrated in Figure 2.

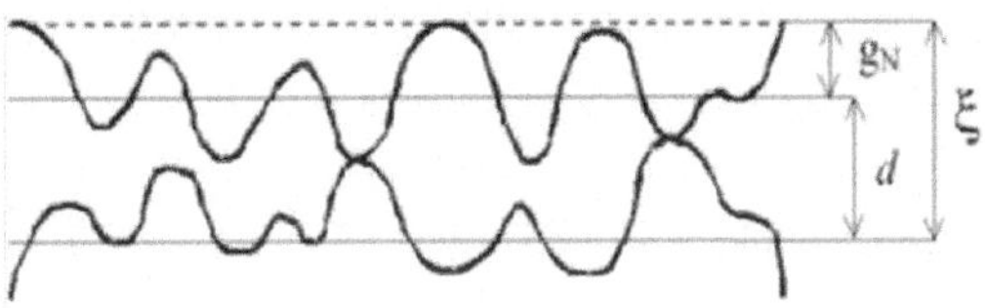

Fig. 1. Physical approach on the contact interface (cross section)

The microscopic contact law proposed in Zavarise, Schrefler, Wriggers [33] is used to compare the numerical results. This relationship that correlates the current mean plane distance d with the apparent mechanical pressure p is defined by

$$d = 1.363\sigma\sqrt{-\ln\left(5.589\frac{p}{H_e}\right)}, \tag{2}$$

where σ is the RMS surface roughness (mean equivalent roughness) and H_e is current yield limit of the asperities. Replacing the mean plane distance d from equation (1) the following equation is obtained

$$\frac{p}{H_e} = \frac{\exp\left[-\left(\frac{\xi - g_N}{1.363 \cdot \sigma}\right)^2\right]}{5.589}. \tag{3}$$

Considering the nonlinear dependence of H_e from the mean planes approach, the following equation was obtained in Song, Yovanovich [25]

$$\frac{p}{H_e} = \left[\frac{p}{c_1}\left(1617646.152\frac{\sigma}{m}\right)^{-c_2}\right]^{\frac{1}{1+0.0711c_2}}, \tag{4}$$

where c_1 and c_2 are experimental constants related to hardness nonlinear distribution (hardness coefficients) and $m$is the mean absolute asperity slope (mean equivalent slope). Combining equations (3) and (4) a relationship expressing the dependence of apparent contact pressure from the surface approach is obtained

$$\frac{p}{c_1}\left(1617646.152\frac{\sigma}{m}\right)^{-c_2} = \left\{\exp\left[-\left(\frac{\xi - g_N}{1.363\cdot\sigma}\right)^2 \frac{1}{5.589}\right]\right\}^{1+0.0711c_2} . \tag{5}$$

The normal contact force F_N is finally obtained as a product of the apparent pressure by the apparent contact area A, as follows

$$F_N = p_N A = c_1\left(1617646.152\frac{\sigma}{m}\right)^{c_2} \frac{\exp\left[-(\xi - g_N)^2 \frac{1+0.0711c_2}{(1.363\sigma)^2}\right]}{5.589^{1+0.0711c_2}} A. \tag{6}$$

Following relationship correlates the current normal approach $g_N$with the apparent mechanical pressure p_N

$$p_N = \frac{c_1\left(1617646,152\frac{\sigma}{m}\right)^{c_2}}{5,589^{1+0,0711c_2}} \exp\left[\left(-\frac{1+0,0711c_2}{(1,363\sigma)^2}\right)(\xi - g_N)^2\right] . \tag{7}$$

The mechanical constants mentioned above depend on the material and the micromechanics of the surface. The constants assumed in this work were presented in Zavarise, Schrefler, Wriggers [33].

3 A simple homogenization method for contact interface

The basic aim of this paper is to derive constitutive contact laws as stated in section 2 for a rough surface by using the finite element method. For this purpose one has to model and discretize the rough surface and then, by homogenization procedures develop an interface law for contact. This section summarizes the homogenization method leading to the contact interface laws. The interface law is obtained from numerical simulation using a model that consists of two deformable bodies in contact, see Figure 2. The contact surfaces of both bodies are rough. A representative surface at the contact interface is represented by the Figure 3.

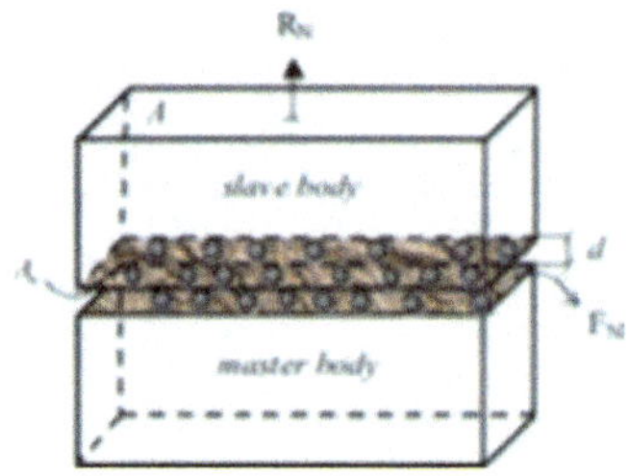

Fig. 2. Model to obtain an interface law.

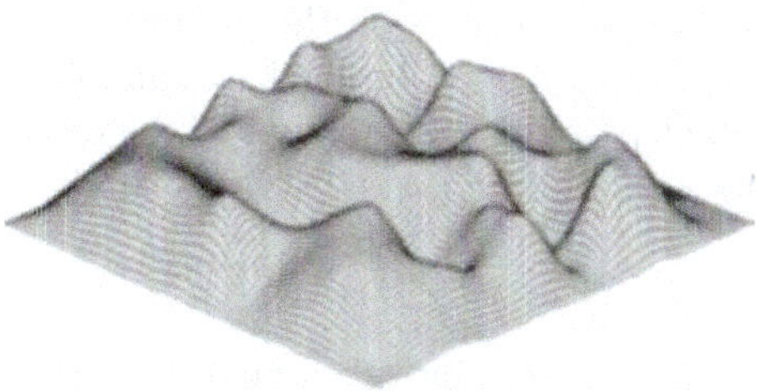

Fig. 3. Contact surface with macroscopic asperities.

The procedure to find the associated interface law is performed in several steps. The bodies are discretized using standard hexahedral finite elements. One body is placed above another with initial distance to separate them. The inferior body is kept fixed in position and superior body is moved towards the inferior body by a displacement imposed on the top surface A. The fixed block is defined as master body and one that is in motion is defined as slave body. The prescribed displacement is applied in several increments. In each increment, the resultant force R_N at the top of superior body is calculated by summing up the reaction forces R_{Nk} on the surface A related to each node k. Then,

$$R_N = \sum_k R_{Nk}. \tag{8}$$

The existence of a reaction force R_N indicates the first contact between the bodies. The maximum initial asperities height ξ between the two middle planes, which are in contact, is determined at this step, see Figure 1 for the geometrical relations.

The normal contact force F_N at the contact interface is obtained by taking into account the real contact area A_r. The actual contact occurs at n-discrete areas A_n on the discrete boundary Γ_c^n, as shown in Figure 4. This yields $A_r = \sum_{n=1}^{n_c} A_n$, when n_c asperities are in contact.

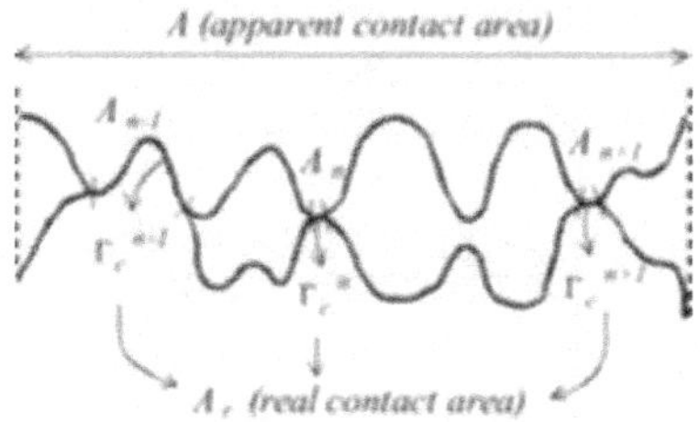

Fig. 4. Contact interface (cross section).

The real contact pressure t_{Nn} occurs on the discrete contact surface Γ_c^n. The actual normal contact force F_N is obtained by summing up the integrals of the real normal contact pressures t_{N_n} related to the discrete area A_n at each boundary Γ_c^n. Then:

$$F_N = \sum_n \int_{\Gamma_c^n} t_{Nn} dA_n. \tag{9}$$

In general it is difficult to determine the actual discrete contact areas and to compute F_N from (9). Here a different procedure is followed. Considering the equilibrium of the bodies, it is clear that the normal contact force developed at the interface is equal to the normal reactions developed at the top of superior body. Therefore, within the finite element treatment it is sufficient, instead of computing (9), to calculate the normal reaction force R_N. Hence

$$F_N = R_N. \tag{10}$$

The total contact force F_N at the interface can be distributed on the apparent contact area A to yield a uniform apparent contact pressure p_N. Therefore, with equation (10),

$$p_N = \frac{R_N}{A}. \tag{11}$$

Since R_N depends on the current mean plane distance d, the penetration law is displayed as

$$p_N = p_N\left(d\right). \tag{12}$$

At the end of each step the current mean plane distance d, the total reaction force R_N and the apparent contact pressure p_N are calculated. The analysis ends when the current mean plane distance d goes to zero. The numerical results after homogenization yield a microscopic contact law analogous to the theoretical law presented in section 2. The homogenization is computed in terms of the average normal pressure p_N. The plotted curve describes the penetration law relating the apparent contact pressure p_N to the current mean plane distance d, see equation (12).

The generation of the smoothen contact surfaces with Bézier interpolations (see Farin [9]), the procedures to obtain a statistical law and the techniques to obtain the maximum initial asperities height ξ for a generated surface were presented in detail in Bandeira et al. [2].

The von Mises elastic-plastic constitutive law is based on the following multiplicative decomposition of the deformation gradient

$$\mathbf{F} = \mathbf{F}^e \mathbf{F}^p \tag{13}$$

where the superscript e and p describe the elastic and the plastic part, respectively. The elastic logarithmic strain tensor is given by

$$\mathbf{E}^e = \ln \mathbf{V}^e \ , \quad \text{where} \quad \mathbf{V}^e = \left(\mathbf{F}^e \mathbf{F}^{e^T} \right)^{\frac{1}{2}} \tag{14}$$

is the elastic left stretching tensor. The elasticity is described by the following strain energy function

$$\psi \left(\mathbf{E}^e \right) = \frac{1}{2} \kappa \vartheta^2 + \mu \left(\bar{\mathbf{E}}^e : \bar{\mathbf{E}}^e \right) \ , \tag{15}$$

where

$$\vartheta = \mathrm{tr} \mathbf{E}^e \quad \text{and} \quad \bar{\mathbf{E}}^e = \mathrm{Dev} \mathbf{E}^e \ . \tag{16}$$

that leads to the following Kirchhoff-Trefftz stress tensor

$$\Sigma = \mathbb{D}^e \mathbf{E}^e \ , \quad \text{where} \quad \mathbb{D}^e = \kappa \mathbf{I} \otimes \mathbf{I} + 2\mu \left(\mathbb{I} - \frac{1}{3} \mathbf{I} \otimes \mathbf{I} \right) \ . \tag{17}$$

The logarithmic isotropic linear elastic material simplifies the volumetric-isochoric splitting. Note also that (17) is similar to the expression of the small strain Hooke's Law. The fourth-order tangent tensor is obtained from (17) as shown in detail in Pimenta [22].

For computational purposes the classical radial return algorithm along with the von Mises plasticity with linear isotropic hardening is summarized below.

1) Trial step:

1. $\mathbf{F}_t^e = \mathbf{F}_{i+1} \mathbf{U}_i^{p-1}$; $\quad \mathbf{C}_t^e = \mathbf{F}_t^{eT} \mathbf{F}_t^e$;
2. $\mathbf{U}_t^e = \left(\mathbf{C}_t^e \right)^{1/2} = \lambda_i^e \, \mathbf{c}_i^e \otimes \mathbf{c}_i^e$; $\quad \varepsilon_i = \ln \lambda_i^e$; $\quad \mathbf{E}_t = \ln \mathbf{U}_t^e$;
3. $\mathbf{R}_t^e = \mathbf{F}_t^e \mathbf{U}_t^{e^{-1}}$; $\quad \mathbf{b}_i^e = \mathbf{R}_t^e \mathbf{c}_i^e$;
4. $\vartheta_t = tr \left(\mathbf{E}_t \right)$; $\quad \bar{\mathbf{E}}_t = \mathrm{Dev} \left(\mathbf{E}_t \right)$;
5. $\bar{\Sigma}_t = 2\mu \bar{\mathbf{E}}_t$; $\quad \bar{\sigma}_t = \sqrt{\frac{3}{2} \bar{\Sigma}_t : \bar{\Sigma}_t}$; $\quad F_t = \bar{\sigma}_t - \sigma_y \left(\alpha_i \right)$.

2) Radial return algorithm

if $(F_t < 0)$ then

elastic step:

1. $\alpha_{i+1} = \alpha_i$;
2. $\mathbf{U}_{i+1}^p = \mathbf{U}_i^p$;
3. $\bar{\Sigma}_{i+1} = \bar{\Sigma}_t$;

4. $\bar{\mathbb{D}}_{i+1} = 2\mu\left(\mathbb{I} - \frac{1}{3}\mathbf{I}\otimes\mathbf{I}\right)$;

else if $(F_t \geqslant 0)$ then

elastic - plastic step:

1. $\Delta\alpha = \frac{\bar{\sigma}_t - \sigma_y(\alpha_i)}{3\mu+h}$; $\quad \alpha_{i+1} = \alpha_i + \Delta\alpha$;
$\sigma_y(\alpha_{i+1}) = \sigma_y(\alpha_i) + h\Delta\alpha$;
2. $\Delta\mathbf{E}^p = \Delta\alpha\frac{3}{2\bar{\sigma}_t}\bar{\Sigma}_t$; $\quad \Delta\mathbf{U}^p = e^{\Delta\mathbf{E}^p}$;
$\mathbf{F}^p_{i+1} = \Delta\mathbf{U}^p\,\mathbf{U}^p_i$; $\quad \mathbf{U}^p_{i+1} = \left(\mathbf{F}^{pT}_{i+1}\mathbf{F}^p_{i+1}\right)^{\frac{1}{2}}$;
3. $\bar{\Sigma}_{i+1} = \frac{\sigma_{y_{i+1}}}{\bar{\sigma}_t}\bar{\Sigma}_t$;
4. $\bar{\mathbb{D}}_{i+1} = 2\mu\frac{\sigma_y(\alpha_{i+1})}{\bar{\sigma}_t}\left(\mathbb{I} - \frac{1}{3}\mathbf{I}\otimes\mathbf{I}\right) +$
$\frac{3\mu}{\bar{\sigma}_t^2}\left(\frac{h}{3\mu+h} - \frac{\sigma_y(\alpha_{i+1})}{\bar{\sigma}_t}\right)\bar{\Sigma}_t\otimes\bar{\Sigma}_t$;

end if;

3) add volumetric part:

1. $\Sigma_{i+1} = \kappa\vartheta_t\mathbf{I} + \bar{\Sigma}_{i+1}$;
2. $\mathbb{D}_{i+1} = \kappa\mathbf{I}\otimes\mathbf{I} + \bar{\mathbb{D}}_{i+1}$.

The assumed linear isotropic hardening used in this paper is presented in the Figure 5. Its function behavior is defined as following

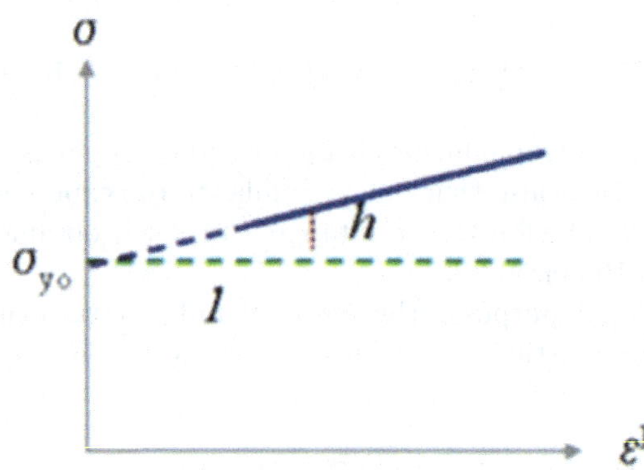

Fig. 5. Constitutive equation for steel material - Linear.

$$\sigma_y(\alpha) = \sigma_{yo} + h\varepsilon^p. \tag{18}$$

where σ_{yo} is the initial yield stress, $\alpha = \varepsilon^p$ is the equivalent plastic strain and h is the linear hardening parameter.

4 Numerical simulation

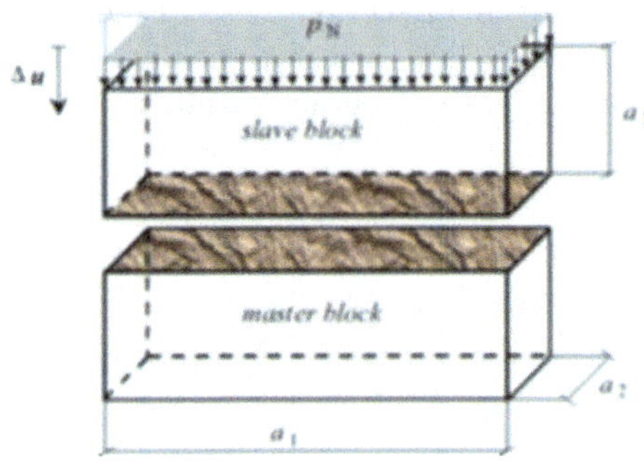

Fig. 6. Contact with two deformable bodies.

In this section, a numerical example is presented to obtain an interface law for rough surfaces numerically. This is done for three-dimensional bodies in contact. The homogenization method used was presented in section 3. Here two blocks with the same geometrical and material properties are considered as shown in Figure 6.

Each block is discretized with a high-density finite element mesh. The total number of unknowns is about 100.000. The number of master and slave surfaces is around 4000 elements and the complete mesh is around 25000 bricks elements. The Preconditioned Bi-Conjugate Gradient Method (PBCG) is used to solve the linear equations system (see Press [23]). The average number of PBCG iterations per load increment is 250.

The base of the master block is fixed and lateral displacements of both blocks are released. The elastic-plastic material law presented in section 4 is used in all examples.

In the numerical examples two steel blocks with Lamé constants $\Lambda = 115384.62$ MPa and $\mu = 76923.08$ MPa are considered. The adopted initial yield stress is $\sigma_{\text{YO}} = 400$ MPa. Two hardening parameters are used, namely $h = E/10 = 20,000$ MPa and $h = E/10000 = 20$ MPa.

The geometrical parameters of the blocks are $a_1 = 2.50$ mm,$a_2 = 5.00$ mm and$a_3 = 1.25$ mm. The contact surfaces are modified according to the theory presented Bandeira et al. [2], such that the maximum initial asperities height ξ is 0.038 mm. A uniform displacement of 0.004 mm is prescribed at the top of the slave block in several increments. Each analysis ends when the current mean plane distance d approaches zero. The material parameters assumed in the theoretical law presented in equation (7) are σ equal 0.011 mm, c_1 equal 627.1 MPa, c_2 equal t0.229 and m equal 0.022. These constants parameters are presented in Zavarise, Schrefler, Wriggers [33].

After all generated surfaces were analyzed; the mean value curve of the normal pressures is depicted in Figure 7, which represents the constitutive interface law for different hardness. The numerical law is compared with the theoretical formulation given in section 2, see Figure 7. It is important to mention that each numerical laws are statistically computed curves resulted by 20 different random generated contact surfaces.

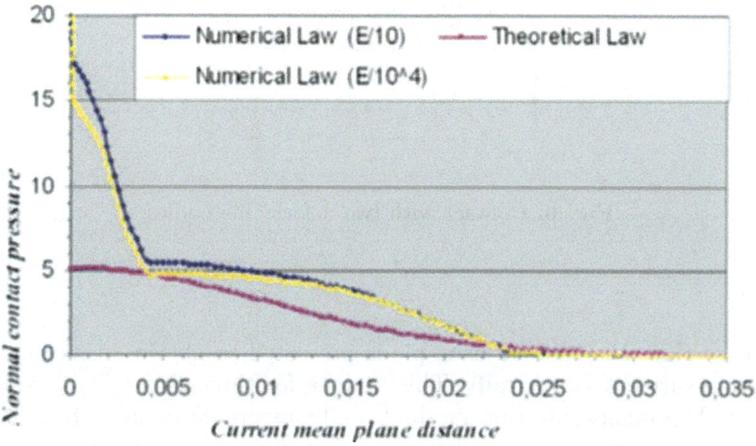

Fig. 7. Penetration behaviour (h = E/10).

5 Conclusions

This work can be regarded as a complementary study of Bandeira et al. [2]. In that paper the elastic Neo-Hookean constitutive equation is assumed to compare the contact interface behavior obtained by numerical simulation with the theoretical one of Kragelsky, Dobychin, Kombalov [14].

Here, the plasticity of the asperities is taken into account by assuming a constitutive equation based on associated von Mises yield function formulated in principal axes. The same procedure was done to obtain the numerical law. The interface law obtained from numerical simulation is in agreement with the theoretical law presented in Zavarise, Schrefler, Wriggers [33] for a low pressure. When the current mean plane distance goes to zero (d $\leqslant$ 0.004 mm) the numerical law tend to infinite.

It is important to mention that the parameters are adjusted for these examples. More experiments are needed to have more precise conclusions. It is verified that the contact parameters presented in equation (2) are very sensitive and can lead to completely different behaviors of the theoretical law presented in section 2.

References

1. Alart, P.; Curnier, A. A Mixed Formulation for Frictional Contact Problems prone to Newton like Solution Methods. *Computational Methods in Applied Mechanics and Engineering*, v. 92, p. 353-375, 1991.
2. Bandeira, A. A., Wriggers, P., Pimenta, P. M. Numerical Derivation of Contact Mechanics Interface Laws using a Finite Element Approach for large 3D Deformation. *International Journal for Numerical Methods in Engineering: IJNME*, 2005 (presented in 3rd contact mechanics international symposium: cmis, peniche, portugal, 2001).
3. Bandeira, A. A., Wriggers, P., Pimenta, P. M. Numerical Simulation of 3D Contact Problems under Finite Elastic-Plastic Deformation. *7th US National Congress on Computational Mechanics*, Albuquerque, New Mexico, USA, 2003.
4. Bandeira, A. A., Wriggers, P., Pimenta, P. M. Computational Analysis of Contact Mechanics Undergoing Large 3D Deformation. *European Conference on Computational Mechanics: ECCM*, Krakow, Poland, 2001.
5. Bertsekas, D. P. *Nonlinear programming.* Belmont, Athena Scientific, 1995.
6. Bertsekas, D. P. *Constrained optimization and Lagrange multiplier methods.* N. Y., Academic Press, 1984.
7. Curnier, A. A Theory of Friction. *International Journal for Solids Structures*, v. 20, p. 637-47, 1984.
8. Evseev, D. G.; Medvedev, B. M.; Grigoriyan, G. G. Modification of the elastic-plastic model for the contact of rough surfaces. *Wear*, v. 150, p. 79-88, 1991.
9. Farin, G. *Curves and Surfaces for Computer Aided Geometric Design: A Practical Guide.* Department of Computer Science, Arizona State University, Arizona.
10. Fletcher, R. *Practical methods of optimization.* Chichester, John Wiley & Sons, 1980. 2v.
11. Greenwood, J. A; Williamson, J. B. P. *Contact of nominally flat surfaces.* Burndy Corporation Research Division, Norwalk, Connecticut, U.S.A, 1966.
12. Heegaard, J. H.; Curnier, A. An Augmented Lagrangian Method for Discrete Large-Slip Contact Problems. *International Journal for Numerical Methods in Engineering*, v. 36, p. 569-93, 1993.
13. Kardestuncer, H; Norrie, D. H. *Finite Element Handbook.* McGRAW-HILL Book Company, 1987.
14. Kragelsky, I.V; Dobychin, M. N.; Kombalov, V. S. *Friction and Wear – Calculation Methods.* (Translated from The Russian by N. Standen). Pergamon Press, 1982.
15. Laursen, T. A.; Maker, B. N. An augmented Lagrangian quasi-Newton solver for constrained nonlinear finite element applications. *International Journal for Numerical Methods in Engineering*, v. 38, p. 3571-90, 1995.
16. Laursen, T. A.; Simo, J. C. A Continuum-Based Finite Element Formulation for the Implicit Solution of Multibody, Large Deformation Frictional Contact Problems. *International Journal for Numerical Methods in Engineering*, v. 36, p. 3451-85, 1993a.
17. Laursen, T. A.; Simo, J. C. Algorithmic Symmetrization of Coulomb Frictional Problems using Augmented Lagrangians. *Computer Methods in Applied Mechanics and Engineering*, v. 108, p. 133-46, 1993b.
18. Luenberger, D. G. *Linear and Nonlinear Programming.* 2. ed. Reading, Addison-Wesley Publishing Company, Massachusetts, 1984.

19. Yovanovich, M. M. Thermal Contact Correlations. *AIAA Paper*, p. 81-1164, 1981.
20. Oden, J. T.; Pires, E. B. Algorithms and Numerical Results for Finite Element Approximations of Contact Problems with Non-Classical Friction Laws. *Computer & Structures*, v. 19, p. 137-47, 1983a.
21. Parisch, H. A Consistent Tangent Stiffness Matrix for Three-Dimensional Non-Linear Contact Analysis. *International Journal for Numerical Methods in Engineering*, v. 28, p. 1803-12, 1989.
22. Pimenta, P. M. Finite deformation soil plasticity on principal axes, in: Proceedings of the *Third International Conference on Computational Plasticity, Fundamentals and applications*, ed. by D.R.J. Owen, E. Oñate & E. Hinton, Pineridge Press Limited, Swansea, U.K., 1992, pp 859-870.
23. Press, W. H. *Numerical recipes in C: the art of scientific computing.* 2 ed. Cambridge, University Press, 1995.
24. Simo, J. C.; Laursen, T. A. An augmented Lagrangian Treatment of Contact Problems Involving Friction. *Computers & Structures*, v. 42, p. 97-116, 1992.
25. Song, S.; Yovanovich, M. M. Explicit relative contact pressure expression: dependence upon surface roughness parameters and Vickers microhardness coefficients. *AIAA Paper*, p. 87-152, 1987.
26. Tabor, D. Friction – The Present State of Our Understanding. *Journal of Lubrication Technology*, v. 103, p. 169-116, 1981.
27. Wriggers, P. Finite Element Algorithms for Contact Problems. *Archives of Computational Methods in Engineering*, v. 2, n.º 4, p. 1-49, 1995.
28. Wriggers, P.; Simo, J. C. A Note on Tangent Stiffness for Fully Nonlinear Contact Problems. *Communications in Applied Numerical Methods*, v. 1, p. 199-203, 1985.
29. Wriggers, P.; Simo, J. C.; Taylor, R. L. Penalty and Augmented Lagrangian Formulations for Contact Problems. *In Proceedings of NUMETA 85 Conference*, eds. J. Middleton & G. N. Pande, Balkema, Rotterdam, 1985.
30. Wriggers, P.; Vu Van, T.; Stein, E. Finite Element Formulation of Large Deformation Impact-Contact Problems with Friction. *Computers & Structures*, v. 37, p. 319-31, 1990.
31. Wriggers, P.; Zavarise, G. On the Application of Augmented Lagrangian Techniques for Nonlinear Constitutive Laws in Contact Interfaces. *Comm. Num. Meth. Engng.*, v. 9, p. 815-24, 1993.
32. Yovanovich, M. M. Thermal Contact Correlations. AIAA Paper, p. 81-1164, 1981.
33. Zavarise, G; Schrefler, B. A; Wriggers, P. Consistent Formulation for Thermomechanical Contact based on Microscopic Interface Law, in Proceedings of COMPLAS III, eds. D.R.J. Owen, E. Hilton, E. E. Oñate, Pineridge Press, 1992.

Micromechanical analysis of deformation and temperature inhomogeneities within rough contact layers

S. Stupkiewicz and P. Sadowski

Institute of Fundamental Technological Research (IPPT), Warsaw, Poland,
sstupkie@ippt.gov.pl, psad@ippt.gov.pl

Abstract. Boundary layers are studied which are induced by micro-periodic boundary conditions as, for instance, in the case of contact of rough bodies. The adopted micromechanical framework is outlined briefly and two applications are provided. The finite element method is applied to study the normal contact compliance in the elasto-plastic regime, and the effect of macroscopic in-plane strain on contact response is analyzed. Secondly, a simple model for prediction of the effective heat transfer coefficient in steady-state conditions is presented.

1 Introduction

In the case of contact of rough bodies, the characteristic dimension of roughness is typically much smaller than that of the contacting bodies, see Fig. 1. Thus two points of view can be adopted. At the *micro-scale*, stress transfer is concentrated at small spots, so-called real contacts, and the distribution of contact traction is highly inhomogeneous. These inhomogeneities govern the interaction and deformation of surface asperities. Furthermore, a thin subsurface layer of inhomogeneous deformation is induced. At the *macro-scale*, it is the slowly-varying average (macroscopic) contact traction that determines the overall deformation of the contacting bodies.

While frictional interactions are governed by the *local* phenomena at the micro-scale, friction laws are usually formulated in terms of the normal and tangential components of the *macroscopic* contact traction vector. Consequently, only the *exterior* part [1] of the stress tensor is involved in the description, and the complete stress and strain state in the vicinity of the contact surface is typically not accounted for. However, in some situations, the *interior* (in-plane) parts of stress or strain may significantly affect friction and other contact phenomena. This is, for example, the case of metal forming processes where surface asperities are flattened more easily in the presence of macroscopic (bulk) plastic deformation [2, 3, 4]. This leads to high real contact area

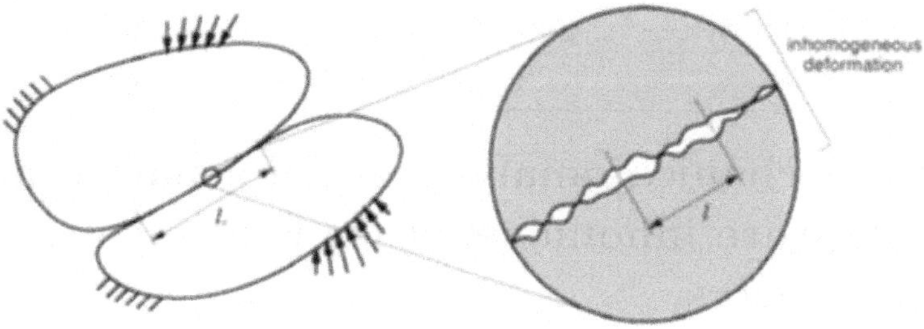

Fig. 1. Contact of rough bodies: macro- and micro-scale.

fractions, even at moderate contact pressures. A closely related effect is also observed in hardness indentation testing, where the in-plane stresses affect the force-penetration response [5, 6].

The importance of surface roughness in the mechanics of contact interactions is, of course, very well recognized [7, 8]. The area of real contact of rough bodies is directly related to asperity interaction, and the related effects at the micro-scale govern the contact phenomena (friction, wear, contact compliance, heat transfer, lubrication, etc.) observed at the macro-scale.

The present paper briefly presents our recent work concerned with micromechanical analysis of boundary layers induced by contact interaction of rough bodies.

2 Contact boundary layers

This section outlines the micromechanical framework that has been developed in [9] for the analysis of contact boundary layers. For brevity, the case of micro-periodic surface traction is only addressed. Discussion of boundary layers associated with frictional contact of rough bodies can be found in [9].

Consider thus a homogeneous body occupying domain Ω with *micro-periodic* surface traction $\mathbf{t}^*$ prescribed on the boundary $\partial_t\Omega$. By micro-periodicity of $\mathbf{t}^*$ we understand that it consists of a slowly varying, average field $\bar{\mathbf{t}}^*$ and its almost periodic fluctuation $\tilde{\mathbf{t}}^*$. The wave-length l of the fluctuation field is assumed small compared to L, the characteristic dimension of Ω, thus $\epsilon = l/L \ll 1$. By the Saint-Venants's principle, inhomogeneity of deformation induced by the micro-inhomogeneous boundary condition is confined to a thin sub-surface layer, the *boundary layer*, along $\partial_t\Omega$, while the thickness of the boundary layer is of the order of l.

The equations of the boundary layer are obtained using the method of asymptotic expansions of the homogenization theory [10]. As a result, in addition to the *macroscopic* b.v.p. in Ω with *micro-homogeneous* boundary conditions $\bar{\mathbf{t}}^*$ on $\partial_t\Omega$, the *microscopic* b.v.p. of the boundary layer is obtained for each point on the boundary $\partial_t\Omega$. The microscopic b.v.p. is found to be a problem of a homogeneous half-space with a periodic traction $\tilde{\mathbf{t}}^*$ prescribed along its boundary.

The strain field $\varepsilon(\mathbf{y})$ within the boundary layer is obtained as a sum of the macroscopic strain $\mathbf{E}$ and strain $\mathbf{e}(\mathbf{w})$ derived from the displacement correction term $\mathbf{w}(\mathbf{y})$

$$\varepsilon(\mathbf{y}) = \mathbf{E} + \mathbf{e}(\mathbf{w}(\mathbf{y})), \tag{1}$$

where $\mathbf{y} = \mathbf{x}/\epsilon$ is the microscopic spatial coordinate and $\mathbf{e}(\,\cdot\,)$ denotes the symmetrized gradient with respect to $\mathbf{y}$. A local coordinate system is adopted such that the y_3-axis is normal to the boundary and directed inwards. The boundary layer correction $\mathbf{w}(\mathbf{y})$ is periodic with respect to y_1 and y_2 (periodic in the tangent direction) and its gradient vanishes far from boundary (i.e. far in the y_3 variable). In order to keep the notation simple, the usual two-scale description is dropped in the present exposition, and thus the dependence on the macro variable $\mathbf{x}$ is not indicated explicitly.

The averaging operation is next defined. For a field $\varphi(\mathbf{y})$ its average $\bar{\varphi}$ at fixed y_3 is given by

$$\bar{\varphi}(y_3) = \langle\varphi\rangle(y_3) \equiv \frac{1}{|S|}\int_S \varphi(\mathbf{y})\,\mathrm{d}y_1\mathrm{d}y_2, \qquad |S| = \int_S \mathrm{d}y_1\mathrm{d}y_2, \tag{2}$$

where S is the periodicity cell within the (y_1, y_2)-plane. The inhomogeneous field $\varphi(\mathbf{y})$ can now be decomposed into its average value $\bar{\varphi}(y_3)$ and fluctuation $\tilde{\varphi}(\mathbf{y})$, so that $\varphi(\mathbf{y}) = \bar{\varphi}(y_3) + \tilde{\varphi}(\mathbf{y})$ and $\langle\tilde{\varphi}\rangle = 0$.

Several properties of stress and strain averages can now be derived. The derivation is omitted here, and three important properties are provided below. Firstly, the following compatibility conditions hold,

$$\Delta\bar{\boldsymbol{\sigma}}_A(y_3) \equiv \bar{\boldsymbol{\sigma}}_A(y_3) - \boldsymbol{\Sigma}_A = \mathbf{0}, \qquad \Delta\bar{\varepsilon}_P(y_3) \equiv \bar{\varepsilon}_P(y_3) - \mathbf{E}_P = \mathbf{0}, \tag{3}$$

where $\boldsymbol{\Sigma}$ is the macroscopic stress, and subscripts A and P denote, respectively, the exterior and the interior parts of a symmetric tensor [1].

Secondly, the average work of a statically admissible stress $\boldsymbol{\sigma}(\mathbf{y})$ on strain $\varepsilon(\mathbf{y})$, derived from a kinematically admissible displacement correction $\mathbf{w}(\mathbf{y})$, cf. (1), is given by

$$\langle\boldsymbol{\sigma}\cdot\varepsilon\rangle = \bar{\boldsymbol{\sigma}}\cdot\bar{\varepsilon} - \frac{\mathrm{d}}{\mathrm{d}y_3}\langle\tilde{\mathbf{w}}\cdot\mathbf{n}\tilde{\boldsymbol{\sigma}}\rangle. \tag{4}$$

Consider now an elastic-plastic body. The local constitutive relation is given by $\boldsymbol{\sigma} = \mathbf{L}(\varepsilon - \varepsilon^p)$ with $\mathbf{L}$ a constant elastic moduli tensor and ε^p an inelastic strain periodic with respect to y_1 and y_2. Accordingly, an analogous relation holds for the averages, $\bar{\boldsymbol{\sigma}} = \mathbf{L} : (\bar{\varepsilon} - \bar{\varepsilon}^p)$, while far from the surface we have $\boldsymbol{\Sigma} = \mathbf{L} : (\mathbf{E} - \mathbf{E}^p)$. In view of compatibility conditions (3), the average stress $\bar{\boldsymbol{\sigma}}(y_3)$ and the average strain $\bar{\varepsilon}(y_3)$ are fully determined if $\Delta\bar{\varepsilon}^p(y_3) = \bar{\varepsilon}^p(y_3) - \mathbf{E}^p$ is known, namely

$$\Delta\bar{\varepsilon}(y_3) = \mathbf{P}^0\mathbf{L}\Delta\bar{\varepsilon}^p(y_3), \qquad \Delta\bar{\boldsymbol{\sigma}}(y_3) = -\mathbf{S}^0\Delta\bar{\varepsilon}^p(y_3), \tag{5}$$

where the operators $\mathbf{P}^0$ and $\mathbf{S}^0$ depend on $\mathbf{L}$ and $\mathbf{n}$. Equations (5) resemble the interfacial relationships which hold locally at a discontinuity surface [1].

3 Applications

3.1 Normal contact compliance of a sand-blasted surface

In this section, the elasto-plastic normal contact compliance of a real three-dimensional rough surface is analyzed using the finite element method. The aim of this example is to study the interaction of local elasto-plastic deformations, associated with flattening of asperities, with the macroscopic deformation field, and, specifically, to investigate the effect of the macroscopic strain on the normal contact compliance. The present direct approach is, in essence, similar to numerous previous studies, e.g. [11, 12]. The original contribution of the present study is that, here, the macroscopic strain and its effect on the contact response are directly accounted for and analyzed.

In brief, the present approach involves the following steps. First, a three-dimensional topography of a real surface is measured. A representative part of the surface is next chosen, and a finite element model of the unit cell of the corresponding boundary layer is generated. A boundary value problem is then solved by considering the frictionless contact of the rough surface with a rigid and smooth counter-surface. In addition to all the microscopic quantities, such as local displacements, stresses, etc., the analysis provides the macroscopic contact pressure and the real contact area fraction as a function of the relative approach of the surfaces.

A steel specimen has been sand-blasted in order to create a severely rough surface. Tree-dimensional roughness topography has then been measured using a 3D stylus scanning profilometer ($S_a = 6.01\,[\mu\text{m}]$, $S_q = 7.98\,[\mu\text{m}]$). The material is a carbon steel C45 (ISO) with the initial yield stress of 400 [MPa] and the ultimate tensile strength of 640 [MPa]. Although the hardening curve of the bulk material is readily available, strain hardening is neglected in the simulations, and a uniform yield stress $\sigma_y = 400\,[\text{MPa}]$ is assumed in the whole volume.

A structured, three-dimensional finite element mesh has been designed such that the mesh is substantially refined in the vicinity of the surface, Fig. 2. The contact surface is divided into a regular mesh of 54×54 quadrilateral elements, which corresponds to the size of the sample equal to 1.08×1.08 [mm^2] for node spacing of 20 [μm].

In order to verify the finite element model, measurements of the normal contact compliance have been performed at the Surface Layer Laboratory of the Institute of Fundamental Technological Problems (IPPT) using the technique developed by Handzel-Powierża et al. [13] (three hard, smooth, and flat punches are pressed into the specimen and the relative approach and the force are measured). Unknown elastic deflections in the experimental setup, which contribute to the measured approach, and the initial gap have been identified by fitting the *unloading* branch of the pressure–approach curve corresponding to a maximum contact pressure of 250 [MPa]. After application of this correction procedure [9] a very good agreement of the predicted and the measured

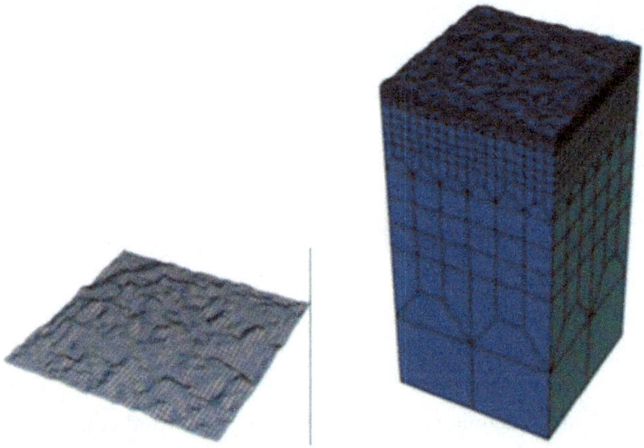

Fig. 2. Sand-blasted surface (roughness topography of a 1.08×1.08 [mm^2] sample) and the finite element mesh of the corresponding boundary layer (unit cell).

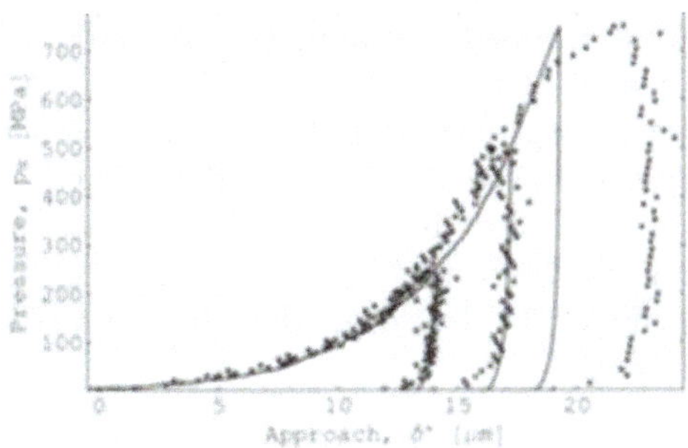

Fig. 3. Normal contact response of the sand-blasted surface for three different maximum contact pressures (solid line—finite element prediction, dots—experiment).

contact compliance has been obtained, cf. Fig. 3. The discrepancy observed for pressures higher than 600 [MPa] is due to macroscopic indentation of the punches into the specimen.

Subsequently, contact response was studied for different macroscopic in-plane strains. In the simulations, the macroscopic strain was enforced before the normal pressure was applied and it was then hold constant during compression. Four cases were considered: (i) $E_{11} = E_{22} = 0$, (ii) $E_{11} = E_{22} = 0.001$, (iii) $E_{11} = E_{22} = -0.001$, (iv) $E_{11} = -E_{22} = 0.001$, where E_{11} and E_{22} denote the in-plane components of the macroscopic strain tensor, and $E_{12} = 0$ was assumed in all cases.

Contact response predicted for cases (i)–(iii) is reported in Fig. 4. Case (iv), i.e. combined tensile-compressive in-plane strain, yielded results which are hardly distinguishable from case (i), and thus the corresponding diagrams are omitted in Fig. 4. The results corresponding to the original roughness are

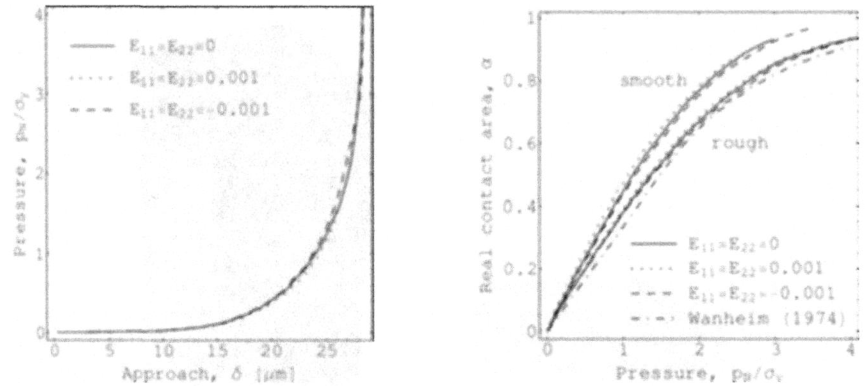

Fig. 4. Effect of macroscopic in-plane strain on (a) normal contact compliance and (b) real contact area.

marked 'rough' in Fig. 4(b), while the results corresponding to an artificial roughness topography obtained by scaling the original roughness by the factor of 0.2 are marked 'smooth'.

The effect of macroscopic strain on the normal contact compliance is noticeable (in-plane tension promotes asperity flattening) though not substantial. A more pronounced effect is observed in the case of real contact area. In Fig. 4(b), the FE results are additionally compared to the predictions of the Wanheim-Bay model [14] based on the slip-line analysis of flattening of wedge-shapes asperities.

3.2 Determination of the effective heat transfer coefficient

The approach outlined in Section 2 can also be directly applied to the case of boundary layers in steady-state heat conduction problems. Due to surface roughness, heat transfer through the contact interface is associated with micro-inhomogeneity of temperature in the boundary layer. A simple model aimed at predicting the effective heat transfer coefficient in conditions typical for metal forming processes is presented below. One of the bodies (the tool) is thus assumed to be smooth, and real contact area fractions approaching unity are considered.

The local contact heat flux q_c is assumed to be governed by

$$q_c = h(T_1 - T_2), \tag{6}$$

where T_1 and T_2 denote the local temperatures of the contacting surfaces, and h is the local heat transfer coefficient. As a particular form of the local law (6), we assume that $h = h_{loc} = \text{const}$ at real asperity contacts and $h = 0$ at the points of separation. The area of real contact is determined by simply assuming that, for a given nominal separation of surfaces, the points at which the surfaces overlap are in contact. This assumption could be released by combining the present approach with that of Section 3.1.

The macroscopic counterpart to (6), constituting the contact boundary condition in the macroscopic problem, can written as

$$\bar{q}_c = h_{eff}(\bar{T}_1 - \bar{T}_2), \tag{7}$$

where $\bar{q}_c = \langle q_c \rangle$ and $\bar{T}_i = \langle T_i \rangle$ are the average (macroscopic) quantities, and $\langle \cdot \rangle$ denotes the averaging operation, c.f. (2). The effective heat transfer coefficient h_{eff} is obtained by solving the microscopic b.v.p. on the periodic unit cell of the boundary layer corresponding to a uniform unit heat flux prescribed far from the boundary.

As the steady-state heat conduction problem is linear, the fluctuations of temperature are proportional to the macroscopic heat flux, thus $\tilde{T}_i = \tilde{\theta}_i \bar{q}_c$, where functions $\tilde{\theta}_i$, defined on the boundary, follow from the microscopic problem. By averaging (6), the effective heat transfer coefficient h_{eff} is found to satisfy the following equation

$$h_{eff} = \frac{\langle h \rangle}{1 - \langle \tilde{h}(\tilde{\theta}_1 - \tilde{\theta}_2) \rangle}, \tag{8}$$

where, in the present case, $\langle h \rangle = \alpha h_{loc}$, and α is the real contact area fraction.

Figure 5 presents the results obtained for the sand-blasted surface considered in Section 3.1. The normalized effective heat transfer coefficient h_{eff}/h_{loc} is shown as a function of the real contact area fraction α and dimensionless parameter $h_{loc}l/k_e$, where l is the characteristic dimension and $k_e = k_1 k_2/(k_1 + k_2)$ is the equivalent thermal conductivity, k_1 and k_2 being the thermal conductivities of the two bodies. The dotted lines denote FE predictions while the solid lines correspond to a simple analytical approximation $h_{eff}/h_{loc} = B\alpha/(1 - \alpha + B\alpha)$, where parameter B $(0 < B < 1)$ is a function of the dimensionless parameter $h_{loc}l/k_e$.

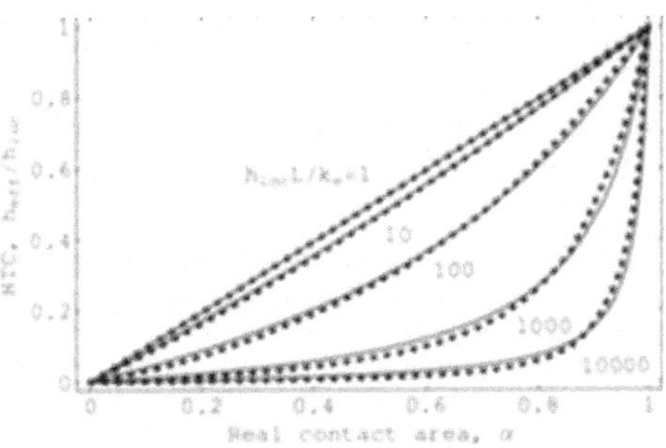

Fig. 5. Effective heat transfer coefficient as a function of real contact area fraction.

Acknowledgements

This work has been supported by the European Commission through a FP5-GROWTH project ENLUB (G1RD-CT-2002-00740).

References

1. R. Hill. Interfacial operators in the mechanics of composite media. *J. Mech. Phys. Solids*, 31(4):347 357, 1983.
2. M.P.F. Sutcli e. Surface asperity deformation in metal forming processes. *Int. J. Mech. Sci.*, 30(11):847 868, 1988.
3. W.R.D. Wilson and S. Sheu. Real area of contact and boundary friction in metal forming. *Int. J. Mech. Sci.*, 30(7):475 489, 1988.
4. S. Stupkiewicz and Z. Mroz. Phenomenological model of real contact area evolution with account for bulk plastic deformation in metal forming. *Int. J. Plast.*, 19(3):323 344, 2003.
5. A.E. Giannakopoulos. The influence of initial elastic surface stresses on instrumented sharp indentation. *Trans. ASME J. Appl. Mech.*, 70:638 643, 2003.
6. Y.-H. Lee and D. Kwon. Estimation of biaxial surface stress by instrumented indentation with sharp indenters. *Acta Mater.*, 52:1555 1563, 2004.
7. F.P. Bowden and D. Tabor. *The Friction and Lubrication of Solids.* Clarendon Press, Oxford, 1953.
8. K.L. Johnson. *Contact Mechanics.* Cambridge University Press, 1985.
9. S. Stupkiewicz. Micromechanics of contact and interphase layers. IFTR Reports 2/2005, Institute of Fundamental Technological Research, Warsaw, 2005.
10. E. Sanchez-Palencia. Boundary layers and edge e ects in composites. In E. Sanchez-Palencia and A. Zaoui, editors, *Homogenization Techniques for Composite Media*, volume 272 of Lecture Notes in Physics, pages 121 192. Springer, Berlin, 1987.
11. A.A. Bandeira, P. Wriggers, and P. de Mattos Pimenta. Numerical derivation of contact mechanics interface laws using a finite element approach for large 3D deformation. *Int. J. Num. Meth. Engng.*, 59:173 195, 2003.
12. K. Varadi, Z. Neder, and K. Friedrich. Evaluation of the real contact areas, pressure distributions and contact temperatures during sliding contact between real metal surfaces. *Wear*, 200:55 62, 1996.
13. Z. Handzel-Powierza, T. Klimczak, and A. Polijaniuk. On the experimental verification of the Greenwood-Williamson model for the contact of rough surfaces. *Wear*, 154:115 124, 1992.
14. T. Wanheim, N. Bay, and A.S. Petersen. A theoretically determined model for friction in metal working processes. *Wear*, 28:251 258, 1974.

Micro-slip of rough surfaces under cyclic tangential loading

M. Borri-Brunetto, A. Carpinteri, S. Invernizzi, and M. Paggi

Department of Structural and Geotechnical Engineering, Politecnico di Torino, C.so Duca degli Abruzzi 24, 10129 Torino, Italy,
marco.paggi@polito.it

Abstract. A numerical model based on the solution of the normal contact between elastic half-spaces and subsequent post-processing according to the Mindlin and Deresiewicz solution for cyclic tangential loading is presented. Thanks to a recent extension of the Cattaneo-Mindlin analogy to the solution of tangential contact between non-convex domains, the proposed approach enables the study of cyclic micro-slip and energy dissipation between elastic bodies with general shapes in contact. In order to make the procedure straightforward and as general as possible, a non-dimensional formulation, based only on the normal contact load displacement curve, is proposed. The cyclic behaviour of the tangential contact of self-affine fractal surfaces, like those generated by fracture of concrete or rock, is described with several examples.

1 Introduction

The first complete analysis of the frictionless normal contact of two elastic solids with quadratic profile was given by Hertz [1]. After his pioneering study, the first contact problem involving tangential forces was solved by Cattaneo [2], who included friction to treat the case of a monotonically increasing tangential force. Similar results were obtained in an independent manner by Mindlin [3]. Cattaneo's results were extended to other loading scenarios by Mindlin and Deresiewicz [4], who considered also the case of an oscillating tangential force acting under a constant normal load.

When an increasing tangential force is applied, small relative displacements occur locally at the contacting surface (micro-slip), before sliding (also referred to as full slip) takes place. During this phenomenon a frictional energy dissipation occurs. Micro-slip regime is important in different fields, e.g. in fretting fatigue [5], geotechnical engineering [6], damping and stiffness of connections [7], and control engineering [8].

Recently, Ciavarella [9, 10] and Jäger [11] proved that the so-called Cattaneo analogy, which provides the solution of the tangential problem by a linear superposition of two normal solutions, applies not only to smooth convex contacting profiles but also to more general cases, including rough surfaces.

In the present paper this generalized approach is adopted, following the Mindlin and Deresiewicz procedure, to study the cyclic behavior of two bodies in contact under an oscillating tangential force. In order to make the procedure straightforward and as general as possible, a non-dimensional formulation, based only on the normal contact load-displacement curve, is proposed. This formulation allows us to apply the code ICARUS (Incremental Contact Analysis of Rough Unmated Surfaces) [12], developed originally to describe the normal contact of surfaces, also to the study of the tangential problem, under monotonically increasing or oscillating shear forces. Moreover, since the normalized tangential load-displacement curve and energy dissipation are expressed as a function of the normalized normal load-displacement relation only, the same method applies directly also to experimental data. The procedure can be applied to rough surfaces with different statistical properties, and the effect of spatial correlation of peaks can be investigated.

2 Generalized Mindlin-Deresiewicz procedure for cyclic loading

In order to describe the load-displacement relation and energy dissipation involved in micro-slips with rough interfaces, a quite general loading condition must be considered. In particular, it can be useful to follow the procedure proposed by Mindlin and Deresiewicz [4]. When rough surfaces are considered, difficulties arise because the stick region is not easily quantifiable, contrarily to problems involving profiles described by closed-form expressions. In this case, it can be useful to introduce the following non-dimensional formulation, in order to generalize the original Mindlin and Deresiewicz procedure. First the normal load P is applied, and, if the elastic solids are similar, no slips occur on the onset of contact. The load-displacement curve is, as well-known, non linear. Figure 1 shows this general relationship where the load is normalized with respect to the maximum achieved load P_0, whereas the normal approach δ_z is normalized analogously, with respect to the maximum δ_{z0}. For our purpose, this curve can be either calculated by applying the code ICARUS, or obtained experimentally. Subsequently, the tangential load Q is increased monotonously. We assume that the load is increased up to the value $\overline{Q}$ lower than or equal to the maximum admissible load (i.e. $Q_0 = fP_0$). Thanks to the Cattaneo analogy [9, 11], it is straightforward to obtain the following nondimensional parametric expression:

$$\begin{cases} \frac{\delta_x^{\mathrm{IL}}}{\delta_{x0}} = 1 - \frac{\delta_z^*}{\delta_{z0}}, \; 0 \le \delta_x^{\mathrm{IL}} \le \overline{\delta}_x, \\ \frac{Q^{\mathrm{IL}}}{Q_0} = 1 - \frac{P^*}{P_0}, \; 0 \le Q^{\mathrm{IL}} \le \overline{Q}, \end{cases} \tag{1}$$

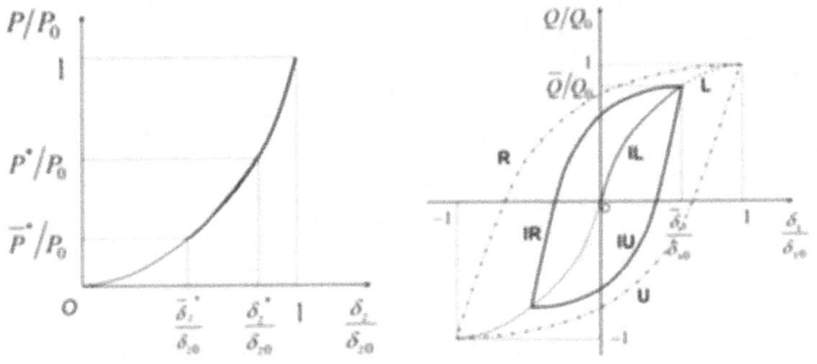

a. b.

Fig. 1. (a) Non-dimensional normal load vs. non-dimensional normal approach diagram. The thick line represents the range of correction solutions. (b) Non-dimensional tangential load vs. displacement diagram. The thick line represents the general load cycle, the dot line curve is the largest load cycle for a given normal load.

where the superscript (IL) refers to an incomplete loading (i.e. which does not overcome the full-slip load), and superscript (*) refers to the elastic correction of the Cattaneo analogy. It is worth noting that the range of corrections for incomplete loading are represented in Fig. 2.a with the thick line.

For some classical problems involving contact between smooth surfaces, the relationship depicted in Fig. 2.a is a power law of the following kind:

$$\frac{P}{P_0} = \left(\frac{\delta_z}{\delta_{z0}}\right)^\alpha , \tag{2}$$

where e.g. $\alpha = 3/2$ for Hertzian spheres in contact. It is therefore possible, in those cases, to give an explicit expression of the load-displacement relation. By substituting Eq. (2) into Eq. (1), and solving with respect to the normalized tangential load, we obtain:

$$\frac{Q^{\mathrm{IL}}}{Q_0} = 1 - \left(1 - \frac{\delta_x}{\delta_{x0}}\right)^\alpha \tag{3}$$

Once the tangential load $\overline{Q}$ is reached, the tangential displacement is reversed. We will refer to this phase as incomplete unloading (IU). Mindlin and Deresiewicz noted that, as soon as the displacement is reversed, two hypotheses could be put forward. First, one can suppose that the contact area is entirely in stick condition. In this case infinite tangential stresses should arise, so this eventuality is not interesting in the present case. The second alternative is that slip must take place, starting from the points of the contact area that are less loaded by normal pressure, i.e. where slip of reverse sign has already taken place. Consequently, the incremental tangential stress must be opposite in sign, and with double magnitude, expressed as follows:

$$\begin{cases} \Delta\delta_x^{\mathrm{IU}} = -2f\left(\delta_{z0} - \delta_z^*\right), \\ \Delta Q^{\mathrm{IU}} = -2f\left(P_{z0} - P_z^*\right). \end{cases} \tag{4}$$

The total amounts become:

$$\begin{cases} \delta_x^{\mathrm{IU}} = \overline{\delta}_x - 2f\left(\delta_{z0} - \delta_z^*\right), \\ Q^{\mathrm{IU}} = \overline{Q} - 2f\left(P_{z0} - P_z^*\right), \end{cases} \tag{5}$$

which provide, given the adequate normalization, the following non-dimensional relationships:

$$\begin{cases} \frac{\delta_x^{\mathrm{IU}}}{\delta_{x0}} = \frac{\overline{\delta}_x}{\delta_{x0}} - 2 + 2\frac{\delta_z^*}{\delta_{z0}}, \\ \frac{Q^{\mathrm{IU}}}{Q_0} = \frac{\overline{Q}}{Q_0} - 2 + 2\frac{P^*}{P_0}. \end{cases} \tag{6}$$

It is worth noting that Eq. (6) consists again in a couple of parametric expressions where the parameter is the generalized size of the stick area (recalled by the (*) superscript). In the particular case when Eq. (2) holds, after some algebraic manipulation, an explicit expression can be obtained as well:

$$\frac{Q^{\mathrm{IU}}}{Q_0} = \frac{\overline{Q}}{Q_0} - 2 + 2\left\{1 + \frac{1}{2}\left[\frac{\delta_x}{\delta_{x0}} - 1 + \left(1 - \frac{\overline{Q}}{Q_0}\right)^{1/\alpha}\right]\right\}^{\alpha} \tag{7}$$

The ratio $\overline{Q}/Q_0$ is the amplitude of the load cycle with respect to the allowable maximum value for a given normal load. When $\overline{Q}/Q_0 = 1$, the load cycle can be called complete. Equation (6) holds until the normalized load $-\overline{Q}/Q_0$ is reached. At that point, the tangential displacement is reversed again, and, consequently, the previous reasoning about the sign and magnitude of differential tangential tractions can be repeated. The normalized parametric expression for the incomplete reloading (IR) is thus the following:

$$\begin{cases} \frac{\delta_x^{\mathrm{IR}}}{\delta_{x0}} = -\frac{\overline{\delta}_x}{\delta_{x0}} + 2 - 2\frac{\delta_z^*}{\delta_{z0}}, \\ \frac{Q^{\mathrm{IR}}}{Q_0} = -\frac{\overline{Q}}{Q_0} + 2 - 2\frac{P^*}{P_0}, \end{cases} \tag{8}$$

which emphasizes that the reloading process is exactly symmetric to the case of unloading. Also in this case, when Eq. (2) is valid, an explicit expression can be obtained after some algebraic manipulation:

$$\frac{Q^{\mathrm{IR}}}{Q_0} = -\frac{\overline{Q}}{Q_0} + 2 - 2\left\{1 - \frac{1}{2}\left[\frac{\delta_x}{\delta_{x0}} + 1 - \left(1 - \frac{\overline{Q}}{Q_0}\right)^{1/\alpha}\right]\right\}^{\alpha} \tag{9}$$

When the normalized tangential load $\overline{Q}/Q_0$ is reached, the cycle is completed. If afterwards the load decreases, another cycle can be performed. On the other hand, if the load is increased beyond the previous maximum value, $\overline{Q}$, any memory of the past load history is lost, and the load displacement curve is described by Eq. (8). Therefore, it can be suggested that, when the normal load is kept constant, the incomplete load curve acts like a sort of virgin-material curve (or hardening curve).

Finally, it is possible to determine the energy dissipated in a loading cycle. To this purpose, it is not convenient to integrate the tangential stress field over the slip domain, but rather to calculate the area enclosed in the diagram of Fig. 2.b. The non-dimensional dissipated energy per cycle becomes:

$$\frac{\Delta E}{Q_0 \delta_{x0}} = \int\limits_{-\frac{\overline{\delta}_x}{\delta_{x0}}}^{\frac{\overline{\delta}_x}{\delta_{x0}}} \left(\frac{Q^{\mathrm{IR}}}{Q_0} - \frac{Q^{\mathrm{IU}}}{Q_0} \right) \mathrm{d}\frac{\delta_x}{\delta_{x0}}, \tag{10}$$

where the expressions for Q^{IU}/Q_0 and Q^{IR}/Q_0 are reported in Eqs. (7) and (9), respectively.

Introducing $\overline{Q} = Q_0$ and $\alpha = 3/2$ into Eq. (10), i.e. by considering spheres in contact and a complete cycle, the non-dimensional energy dissipated is equal to 4/5, consistently with the analytical results by Mindlin and Deresiewicz [4]. The above described procedure has been implemented in the improved version of the ICARUS software.

3 Hysteretic energy dissipation on rough surfaces

Fractal rough surfaces are generated according to the random midpoint displacement technique (RMD) which can produce realistic height-fields at any resolution [13, 14]. In the RMD algorithm a random number generation, whose variance changes with scale, adds the requested details at the gridpoints of a progressively refined square mesh. Few input parameters are needed to generate these complex geometries: the surface fractal dimension, D, and the surface resolution, s. From experimental measures, the former parameter usually ranges between 2.0 and 2.5. The latter parameter is set equal to 1/256 in the computations, considering sides of the square mesh of unitary length and 257 heights per side.

The study of these fractal surfaces with self-affine properties is motivated by the fact that most of rough surfaces originated by rock and concrete fracture show multi-scale features of roughness at different scale lengths, from the length of the sample to the lower cut-off length of the system. Since the lower cut-off length may coincide with the atomic scale, the description of the topography of rough surfaces using fractal concepts seems to be very promising also in nanotribology [15]. Hence, the proposed numerical approach can be considered as an important tool for the mechanical interpretation of the tangential response of fractal rough surfaces.

Aiming at investigating the role played by the surface fractal dimension, three values of D are considered: $D = 2.1$, 2.3 and 2.5. For each dimension D, 20 surfaces are generated according to the RMD technique, with the same maximum heights amplitude. This permits to have enough data for a statistical analysis of the results of the numerical simulations. According to the numerical procedure implemented in the ICARUS code, the height-fields of the fractal surfaces are used as input for the simulations of the normal contact between the rough surfaces and a smooth plane made of the same material. The parameter α of Eq. (2) has been computed from a best-fit procedure on the average numerical results. To be more specific, for $D = 2.1$, 2.3 and 2.5 we have, respectively, $\alpha = 2.55$, 2.41 and 2.64. The corresponding normal load vs. normal approach curves are shown in a non-dimensional form in Fig. 2.a. Hence, the exponent α of these curves seems to be approximately independent of the surface fractal dimension and it is higher than that for smooth spheres.

Nevertheless, it has to be noticed that numerical predictions are characterized by a large scatter which can be quantified in Fig. 2.b, where the envelope of numerical results is shown for the set of fractal surfaces with $D = 2.3$. Similar diagrams are obtained for the other fractal dimensions. This scatter has to be ascribed to the different statistical distribution of the asperity heights for each fractal surface. For instance, a statistical analysis of the asperity heights pertaining to the contact domains of the fractal surfaces with $D = 2.3$ reveals that the exponent α is close to the average value of 2.41 when the height distribution is approximately uniform. On the other hand, higher values of α are observed when the asperity heights distribution is closer to the Gaussian distribution ($\alpha \cong 3.2$). Bi-modal distributions are instead observed for rough contact domains leading to $\alpha \cong 2.0$ (see Fig. 2.b). It has to be noticed that this observation is in good agreement with the predictions of the statistical models proposed by Olofsson and Hagman [16] and by Björklund [17]. In fact, the former model assumes a uniform distribution of asperity heights with $\alpha = 5/2$, whereas the latter considers a Gaussian distribution leading to $\alpha \cong 4.3$.

This result has an important implication for the energy dissipation during a loading cycle. In fact, according to Eq. (10), the higher is the exponent α, the larger is the dissipated energy. To quantify the amount of energy dissipated during cyclic micro-slips of fractal surfaces, the Cattaneo-Mindlin analogy is firstly applied to the normal contact data in order to simulate a contact problem with constant normal load and a monotonically increasing tangential force. Then, the generalized Mindlin-Deresiewicz procedure for cyclic tangential loading is applied by considering different values of the ratios between the tangential displacement and the tangential displacement corresponding to full slip, δ_x/δ_{x0}. Finally, the area of each hysteresis loop is computed, and the non-dimensional energy dissipation averaged over 20 surfaces for each set of fractal surfaces is plotted vs. the non-dimensional tangential displacement in Fig. 3.a. The effect of roughness is particularly evident, since the maximum non-dimensional energy dissipation is approximately equal to 1.8, far higher than 0.8, as observed for smooth spheres.

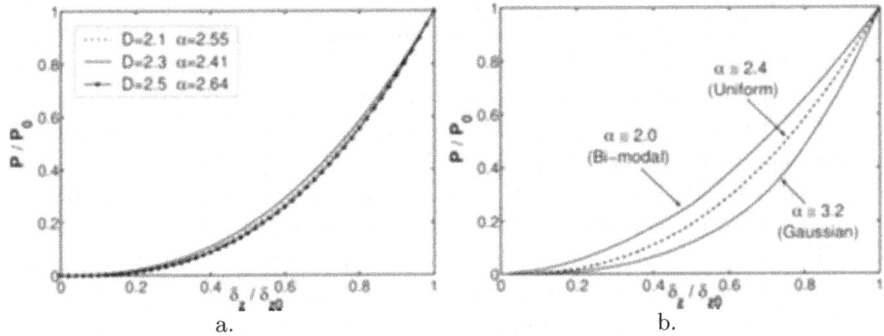

Fig. 2. Non-dimensional normal load vs. non-dimensional normal approach. (a) Average response for fractal surfaces with $D = 2.1$, 2.3 and 2.5. (b) Envelope of numerical results for fractal surfaces with D=2.3. The average response is also reported with dashed line.

The effect of the statistical distribution of asperity heights on the dissipated energy for the fractal surfaces with $D = 2.3$ is also quantified in Fig. 3.b. The larger the exponent α, which is directly related to the height distribution of the asperities in contact, the larger is the energy dissipated.

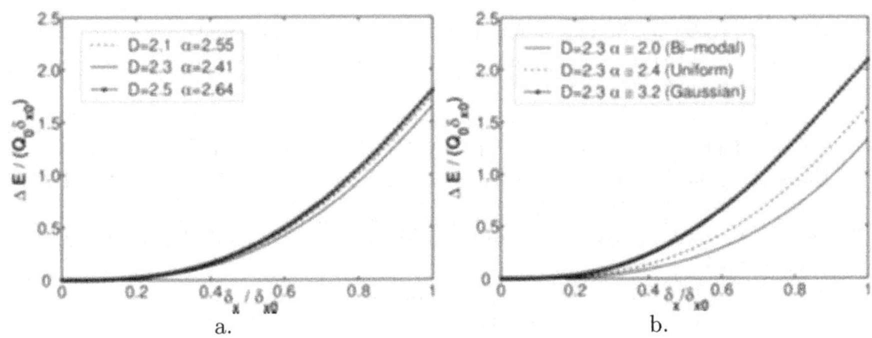

Fig. 3. Non-dimensional energy dissipation vs. non-dimensional tangential displacement diagram. (a) Average response for fractal surfaces with $D = 2.1$, 2.3 and 2.5. (b) Fractal surfaces with $D = 2.3$ and different distribution of asperity heights.

4 Conclusions

A generalized approach has been adopted following the Mindlin and Deresiewicz procedure to study the cyclic behavior of surfaces in contact under a small oscillating tangential force. In order to make the procedure straightforward and as general as possible, a non-dimensional formulation, based only on the normal contact load-displacement curve, has been provided. It turns out that the non-dimensional behavior under normal or tangential loading, as well as the energy dissipation involved in cyclic loading, strongly depends on the exponent α. This exponent can be calculated explicitly in the case of profiles described by mathematical expressions (e.g. cylinders and spheres), or can be determined from best fitting of experimental or numerical data (e.g. from ICARUS simulations). The larger the exponent α, the larger the energy dissipated in small amplitude (partial-slip) tangential cycles.

The calculation of the exponent α allows us to show the influence of the different statistical distribution of surface heights on the hysteretic energy dissipation.

Acknowledgements

Support of the Italian Ministry of Education, University and Scientific Research (MIUR) and of the European Union (EU) is gratefully acknowledged.

References

1. Hertz H (1882) J. für Die Reine und Angewandte Mathematik, 92:156–171
2. Cattaneo C (1938) Rendiconti dell'Accademia Nazionale dei Lincei, Serie 6, 27:342–348, 434–436, 474–478
3. Mindlin RD (1949) ASME J. Appl. Mech., 16:259–268
4. Mindlin RD and Deresiewicz H (1953) ASME J. Appl. Mech., 20:327–344
5. Rooke DP and Edwards PR (1988) Fatigue Fract. Engng. Mat. Struct., 11:447–465
6. Desai CS, Drumm EC and Zaman MM (1985) ASCE J. Geotech. Engng., 111:793–815
7. Kirsonava VN (1967) Machineering Tooling, 38:30–34
8. Harnoy A, Friedland B and Rachoor H (1994) Wear, 172:155–165
9. Ciavarella M (1998) Int. J. Sol. Struct., 35:2349–2378
10. Ciavarella M (1998) ASME J. Appl. Mech., 64:998–1003
11. Jäger J (1998) ASME J. Tribology, 120:677–684
12. Borri-Brunetto M, Carpinteri A and Chiaia B (1999) Int. J. Fract., 95:221–238
13. Curnier A (1984) Int. J. Sol. Struct., 20:637–647
14. Zavarise G, Borri-Brunetto M and Paggi M (2004) Wear, 257:229–245
15. Carpinteri A and Paggi M (2005) Int. J. Sol. Struct., 42:2901–2910
16. Olofsson U and Hagman L (1997) Tribol. Int., 30:599–603
17. Björklund S (1997) ASME J. Tribology, 119:726–732

Part IV

Stability

Stability of discrete systems involving shocks and friction

P. Ballard, A. Léger, and E. Pratt

Laboratoire de Mécanique et d'Acoustique - CNRS, 31, Chemin Joseph Aiguier, 13402 Marseille Cedex20, France,
ballard@lma.cnrs-mrs.fr, leger@lma.cnrs-mrs.fr, elaine@lma.cnrs-mrs.fr

Abstract. The stability of the equilibrium states of a simple mechanical system with unilateral contact and Coulomb friction is explored. When the external force is constant, the equilibrium states are completely determined by the mechanical properties of the system and the stability or instability of each of these states is proved. When the external force varies in time two stability results are given.

1 Introduction

The aim of this work is to determine the stability of equilibrium states for a mechanical system involving unilateral contact with Coulomb friction. Because of the nonsmoothness of such systems due to the presence of frictionnal contact, the usual methods used to determine the stability of equilibrium states are practically inoperate. The analysis is performed here by direct integration of the dynamics with initial data close to equilibrium. The stability or instability results are proved by considering the discrete dynamical system induced by the numerical integration method. The convergence of this method has been proved and used to establish the existence result.

We begin by setting the equations and by giving the existence and uniqueness results. In the case of a constant external force the set of equilibria is determined explicitly by the mechanical parameters of the system, namely its stiffness, friction coefficient and external forces. The stability results concerning all these equilibria are stated and the proof is briefly sketched.

We then consider the case where the external force depends on time. In fact we operate a small time dependant perturbation of the constant external force previously studied. Interesting results concerning the equilibria of the perturbed problem and the evolution in time of solutions to this problem with initial data out of equilibrium are given.

2 The basic equations and the dynamics

We restrict our attention to the mass-spring system, now quite classical since the work of Klarbring [4], represented on Figure 1.

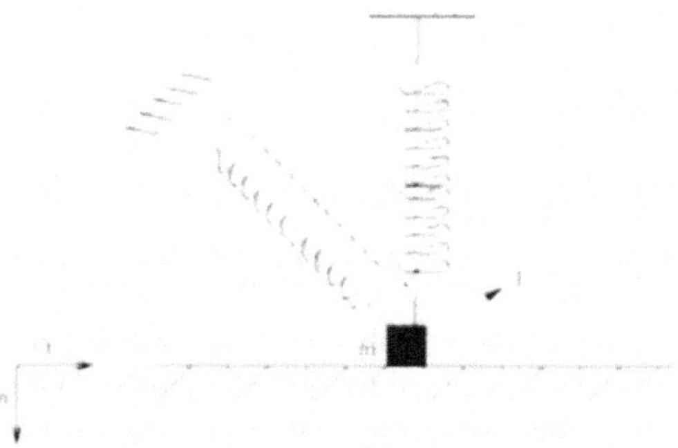

Fig. 1. The mass-spring model.

n and t denote respectively the normal and tangential components of the displacement U and of the reaction R of the mass m. We recall the nonregularized unilateral contact and Coulomb friction laws in which μ denotes the friction coefficient:

$$\begin{cases} U_n \leq 0, \ R_n \leq 0, \ U_n.R_n = 0, \\ \\ |R_t| \leq \mu |R_n|, \\ |R_t| < \mu |R_n| \Longrightarrow \dot{U}_t = 0, \\ |R_t| = \mu |R_n| \Longrightarrow \dot{U}_t = -\lambda R_t, \ \lambda \geq 0, \end{cases} \tag{1}$$

We denote by $MMA([0,T];\mathrm{R}^2)$ (motions with measure acceleration) the vector space of those integrable functions of $[0,T]$ into R^2 whose second derivative in the sense of distributions is a measure. It is nothing but the space of integrals of functions of bounded variation over $[0,T]$. Functions U in MMA are continuous and admit left and right derivatives (in the classical sense) $\dot{U}^-$, $\dot{U}^+$, at any point, both being functions of bounded variation. We recall that a function of bounded variation, being a uniform limit of a sequence of step functions, is universally integrable (integrable with respect to any measure).

The evolution problem, formulated along the lines of MOREAU [7], [8], is the following:

Problem $\mathcal{P}$. Find $U \in MMA([0,T];\mathrm{R}^2)$ and $R \in \mathcal{M}([0,T];\mathrm{R}^2)$ such that:

$$\begin{aligned}
&\bullet \quad U(0) = U_0 \quad ; \quad \dot{U}^+(0) = V_0 \\
&\bullet \quad \ddot{U} + K \cdot U = F + R, \quad in\ [0,T] \\
&\bullet \quad U_n \le 0, \quad R_n \le 0, \quad U_n R_n = 0 \\
&\bullet \quad \int_{[0,T]} \left[R_t \cdot (V - \dot{U}_{\mathrm{T}}^+) - \mu R_n(|V| - |\dot{U}_t^+|) \right] \ge 0, \\
&\qquad \forall V \in C^0([0,T];\mathrm{R}) \\
&\bullet \quad U_n(t) = 0 \quad \Longrightarrow \quad \dot{U}_n^+(t) = -e\dot{U}_n^-(t), \quad in\ ]0,T].
\end{aligned} \tag{2}$$

F denotes the external force, $e \in [0,T]$ a real constant (the so-called restitution coefficient) and (U_0, V_0) some initial condition, assumed to be compatible with the unilateral constraint, that is

$$U_{0n} \le 0 \ \ and \ \ U_{0n} = 0 \quad \Longrightarrow \quad V_{0n} \le 0. \tag{3}$$

In the following we consider completely inelastic shocks, i.e. $e = 0$. In this case, the unilateral contact conditions and the shock law can be replaced in (2) by

$$U_n \le 0 \ \ with \ \begin{cases} U_n < 0 \Longrightarrow R_n = 0, \\ U_n = 0 \Longrightarrow \dot{U}_n^+ \le 0, R_n \le 0, \dot{U}_n^+ R_n = 0. \end{cases} \tag{4}$$

It has been recently established that problem $(\mathcal{P})$ has a solution as soon as the data F is an integrable function of $[0,T]$. The proof of this result uses the convergence of a time discretization, and is heavily inspired by the proof given by Monteiro-Marques in [6]. The main qualitative point of this proof is that the time discretization is the same as that of the numerical software $NSCD$ [3] and will be used in the following as a discrete dynamical system.

It has also been obtained recently [1] that, for given intial data, the trajectory is not generally unique, and that uniqueness holds only if the data F is an analytical function of $[0,T]$.

3 The set of equilibria

The equilibrium states of problem $(\mathcal{P})$, which have been explored in [2], are the set of displacements U and reactions R satisfying, in addition to conditions (1), the following system:

$$\begin{cases} K_t.U_t + W.U_n = F_t + R_t \\ W.U_t + K_n.U_n = F_n + R_n, \end{cases} \tag{5}$$

where (F_t, F_n) denotes the external force and $K = \begin{pmatrix} K_t & W \\ W & K_n \end{pmatrix}$ the stiffness matrix of the system of springs.

Looking first for solutions without contact, immediatly leads to:

$$\begin{cases} U_n = \dfrac{A}{detK}, \; U_t = \dfrac{K_n.F_t - W.F_n}{detK} \\ \\ R_n = R_t = 0, \end{cases} \tag{6}$$

where quantity A is defined as $A = K_t.F_n - W.F_t$. Conditions (1) then imply that (6) will be an equilibrium solution only if $A \leq 0$.

Looking for equilibrium solutions in contact with the obstacle, that is such that $U_n = 0$, we find that system (1), (5) reduces to:

$$\begin{cases} W.U_t = F_n + R_n \\ K_t.U_t = F_t + R_t \\ |R_t| \leq \mu |R_n|, \end{cases} \tag{7}$$

which gives

$$\begin{cases} R_t = \dfrac{K_t}{W} R_n + \dfrac{A}{W} \\ \\ |R_t| \leq \mu |R_n|. \end{cases} \tag{8}$$

The solutions of (8) determine the equilibria in the $\{R_t, R_n\}$ plane. As shown on figure 2, these equilibria belong to the intersection of an affine line which represents the equilibrium equation and of the Coulomb cone.

The dependance of the set of equilibria on the stiffness parameters, friction coefficient and external load is summarized in table 1.

4 Stability analysis

4.1 Main steps of the analysis

Having investigated the equilibrium states under constant data, we analyze the stability of these equilibria. As announced in the introduction, this stability analysis consists in choosing an initial data for problem $(\mathcal{P})$ in a neighborhood of any of the equilibria in a classical phase space, and studying the evolution in time of the distance between the corresponding trajectory and the equilibrium. This analysis uses estimates on the iterates of the time discretization, considered as a dynamical system, that was used to establish the existence result.

Proposition 1 summarizes the stability results in the case of a constant force. We conclude this section by studying the case where the external force varies in time.

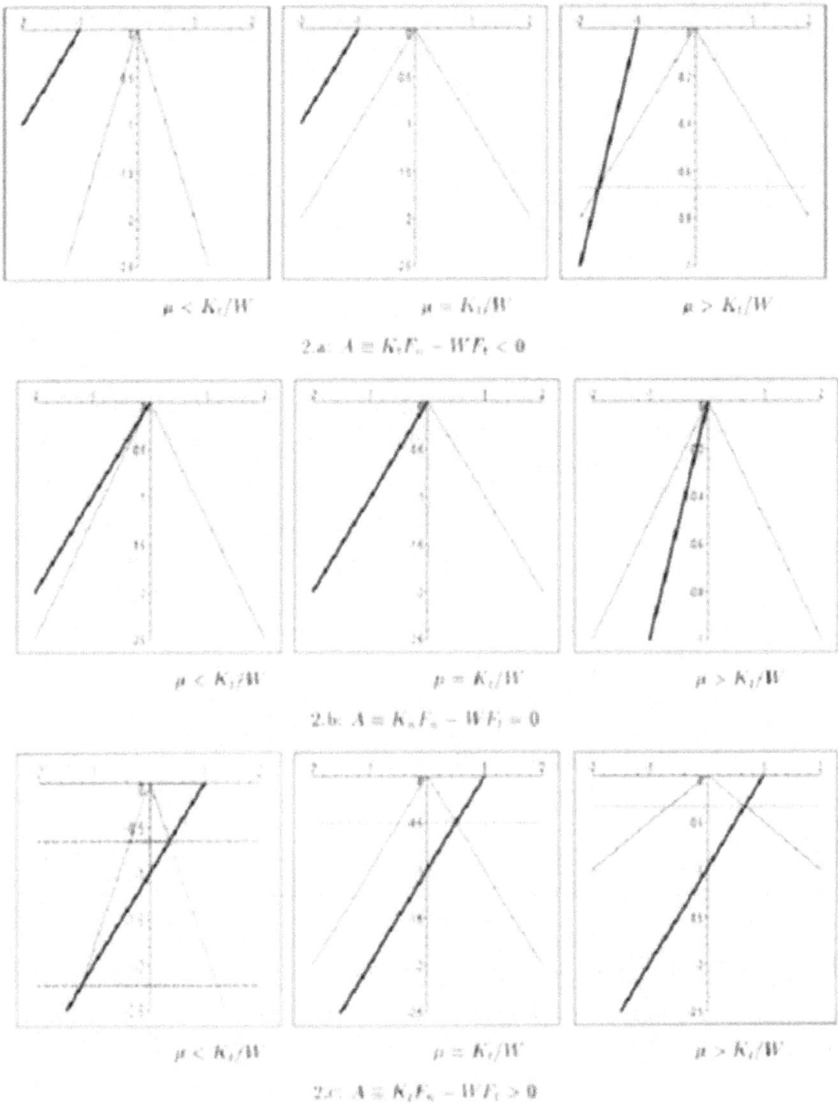

Fig. 2. The sets of equilibria in the $\{R_t, R_n\}$ plane.

Table 1. The equilibrium states with respect to parameters A and μ.

	$A < 0$	$A = 0$	$A > 0$
$\mu < \frac{K_T}{W}$	1 solution without contact	1 solution in grazing contact	2 solutions in impending positive and negative slip + infinitely many solutions in strict sticked contact
$\mu = \frac{K_T}{W}$	1 solution without contact	1 solution in grazing contact + infinitely many solutions in impending positive slip	1 solution in impending negative slip + infinitely many solutions in strict sticked contact
$\mu > \frac{K_T}{W}$	1 solution without contact + 1 solution in impending positive slip	1 solution in grazing contact + infinitely many solutions in strict sticked contact	1 solution in impending negative slip + infinitely many solutions in strict sticked contact

Proposition 1. *Under a constant external force*

- *if there are only two equilibria, the one which is in contact is unstable;*
- *if there are infinitely many equilibrium states in impending sliding, they are all unstable, except the vertex of the cone which is Lyapunov stable;*
- *as soon as there are infinitely many equilibrium states in strictly sticked contact, all the equilibria are Lyapunov stable;*
- *if the parameters are such that the vertex of the cone is the unique equilibrium state, then this equilibrium is asymptotically stable.*

These results are obtained thanks to several technical lemmas, of which we give hereafter three significative examples.

4.2 Some technical lemmas

The first two lemmas are simple and intuitive. Considering an equilibrium state in contact perturbed by a normal or a tangential velocity, then, under some conditions, the trajectory will come into contact again. More precisely:

Lemma 1. *Let* (U^{eq}, R^{eq}) *with* R_n^{eq} *strictly negative be an equilibrium solution, and* V_{n0} *a perturbation of this equilibrium at time* t_0*, then there exists* $\bar{t}$*,* $t_0 < \bar{t} < +\infty$*, such that* $U_n(\bar{t}) = 0$.

Lemma 2. *Let* (U^{eq}, R^{eq}) *be a grazing equilibrium state* ($R_n^{eq} = 0$ *and* $U_n^{eq} = 0$) *and* V_{t0} *a perturbation of this equilibrium at time* t_0*, then there exists* $\tilde{t}$*,* $t_0 < \tilde{t} < +\infty$*, such that* $U_n(\tilde{t}) = 0$.

The last lemma is more technical, and is already very close to a stability result:

Lemma 3. *Let* (U^{eq}, R^{eq}) *be an equilibrium state with* R_n^{eq} *strictly negative. Let the dynamics after any perturbation be such that there exists a time* $t^* < +\infty$ *with* $V_t(t^*) = 0$ *and* $R_n(t^*) < 0$.

Then, if there exists an equilibrium state $(\hat{U}^{eq}, \hat{R}^{eq})$ *such that* $R_n^{eq}(t^*) = \hat{R}_n^{eq}$*,* $V_t(t) = 0 \ \forall t > t^*$.

4.3 Stability under nonconstant forces

The two theorems given below prove a conjecture recently suggested to the authors by Michel Jean.

Let us choose the external force F under the form $F(t) = F_0 + \epsilon\xi(t)$, where $\xi(t)$ is an analytical function and ϵ a positive parameter. $(\mathcal{P}_0)$ (resp. $(\mathcal{P}_\epsilon)$) denotes problem $(\mathcal{P})$ where $\epsilon = 0$ (resp. $\epsilon > 0$). In fact $(\mathcal{P}_0)$ represents the problem previously studied, i.e. with a constant external force and $(\mathcal{P}_\epsilon)$ a perturbation of $(\mathcal{P}_0)$.

A solution (U_ϵ, R_ϵ) of problem $(\mathcal{P}_\epsilon)$ where U_ϵ is constant will be referred to as a *space-equilibrium* solution.

The two following results are established:

Theorem 1. *The following statements i) and ii) are shown to be equivalent:*
i) (U_0, R_0) *is an equilibrium solution of problem* $(\mathcal{P}_0)$ *such that the reaction* (R_{n0}, R_{t0}) *is strictly inside the Coulomb cone.*
ii) $\exists\ \epsilon_0 > 0$ *such that* $\forall \epsilon < \epsilon_0$*, the solution* (U_ϵ, R_ϵ) *of problem* $(\mathcal{P}_\epsilon)$*, obtained with the equilibrium solution of* $(\mathcal{P}_0)$ *as initial data, is a space-equilibrium solution with* $U_\epsilon = U_0$.

Theorem 2. *Assume* ϵ *is such that there exists a space-equilibrium solution to problem* $(\mathcal{P}_\epsilon)$*. Then any solution of* $(\mathcal{P}_\epsilon)$*, with a nongrazing equilibrium solution of problem* $(\mathcal{P}_0)$ *as initial data, leads in finite time to a space-equilibrium solution of* $(\mathcal{P}_\epsilon)$.

References

1. Ballard P. and Basseville S. (2005) Existence and uniqueness for dynamical unilateral contact with Coulomb friction: a model problem. Math. Modelling and Num. Analysis
2. Basseville S., Leger A. and Pratt E. (2003) Investigation of the equilibrium states and their stability for a simple model with unilateral contact and Coulomb friction. Arch. Appl. Mechanics 73: 409-420
3. Jean M. (1999) The nonsmooth contact dynamics method. Cmp. Meth. Appl. Mech. Engn. 177: 235-257
4. Klarbring A. (1990) Examples of nonexistence and nonuniqueness of solutions to quasistatic contact problems with friction. Ing. Arch. 60: 529-541
5. Martins J.A.C., Monteiro Marques M.D.P. and Gastaldi F. (1994) On an example of nonexistence of solution to a quasistatic frictional contact problem, Eur. J. Mech. A/Solids 13: 113-133
6. Monteiro Marquess M.D.P. (1993) Differential Inclusions in Nonsmooth Mechanical Problems. Birkhäuser, Basel Boston Berlin
7. Moreau J.J. (1988) Standard inelastic shocks and the dynamics of unilateral constraints. In: Del Piero G, Marceri F. (eds) Unilateral Problems in Structural Analysis. Springer Verlag, Berlin Heidelberg New York
8. Moreau J.J. (1993) Unilateral contact and dry friction in finite freedom dynamics. In: Moreau J.J., Panagiotopoulos D.D. (eds) Nonsmoooth Mechanics and Applications. CISM Courses and Lectures 302, Springer-Verlag, Berlin Heidelberg New York.

On the stability of quasi-static paths of a linearly elastic system with friction

J.A.C. Martins[1], M.D.P. Monteiro Marques[2], and N.V. Rebrova[1]

[1] Dep. Eng. Civil and ICIST Instituto Superior Técnico Av. Rovisco Pais, 1049-001 Lisboa, Portugal,
jmartins@civil.ist.utl.pt, rena@civil.ist.utl.pt

[2] Centro de Matemática e Aplicacões Fundamentais Faculdade de Ciências da Universidade de Lisboa Av. Prof. Gama Pinto 2, 1649-003 Lisboa, Portugal,
mmarques@ptmat.fc.ul.pt

Abstract. In this paper we discuss the stability of quasi-static paths of a single degree of freedom linearly elastic system with Coulomb friction and known normal force. A common and useful approximation for the equations that govern the slow evolution of many mechanical systems is to neglect inertia effects in the dynamic balance equations, and replace them by static equilibrium equations. Slow evolutions calculated with this approximation are called quasi-static evolutions. The relationship of this issue with the theory of singular perturbations has been established in [1], where the existence of fast (dynamic) and slow (quasi-static) time scales was recognized: a change of variables is performed that replaces the (fast) physical time t by a (slow) loading parameter λ, whose rate of change with respect to time, $\varepsilon = d\lambda/dt$, is decreased to zero. This change of variables leads to a system of dynamic differential equations or inclusions that defines a singular perturbation problem: the small parameter ε multiplies some of the highest order derivatives in the system. The concept of stability of quasi-static paths used here is essentially a continuity property relatively to the size of the initial perturbations (as in Lyapunov stability) and to the smallness of the rate of application of the external forces, ε (as in singular perturbation problems). This study applies for the first time to a non-smooth context, the definition of stability of quasi-static paths, recently proposed by Martins et al. ([2], [3]).

1 Introduction

The study of the stability of frictional contact systems has deserved an increasing attention ([4], [5], [6], [7], [8], [9]) due to its relevance in many engineering applications ([10], [11], [12]) as well as in geophysics ([13], [14]).

The concept of stability that one has in mind in many mechanical situations is the concept of Lyapunov stability. This concept has been used for long to study the stability of dynamic trajectories of mechanical systems, in particular the *stability of the equilibrium configurations under constant applied loads* (dynamic trajectories with zero acceleration). The problems with Coulomb friction have the additional difficulty that the friction law is a multivalued application, which leads to non-smooth dynamic problems with (at least) discontinuous accelerations and friction forces. In [6] one can find a discussion on the attractiveness of equilibrium sets with the application of LaSalle's principle. The works [4] and [7] develop non-smooth Lyapunov functions.

A related but different issue is the *stability of quasi-static paths* of mechanical systems. In general, the concept of Lyapunov stability cannot be applied to quasi-static paths because such paths are not, in general, true solutions of the original governing dynamic equations (Loret et al. [1]). But the "stability of quasi-static paths" can be related to the theory of singular perturbations [1]: the physical time t can be recognized as a fast (dynamic) time scale and a loading parameter λ, whose rate of change with respect to time, $\varepsilon = d\lambda/dt$, is arbitrarily small, can be recognized as a slow (quasi-static) time scale. Changing the independent variable t into λ in the governing system of dynamic differential equations or inclusions, one is led to a system in which some of the highest order derivatives with respect to λ appear multiplied by the small parameter ε. In this manner, following the mathematical definition of stability of quasi-static paths proposed by Martins et al. ([2],[3]) a quasi-static path is stable at some point if, in some subsequent finite interval of the load parameter, any dynamic trajectory does not deviate from the quasi-static one more than some desired amount, provided that the initial conditions for the dynamic evolution are sufficiently close to the quasi-static path, and the loading is applied sufficiently slowly.

After the study of some smooth cases and some problems that have a not very severe non-smoothness (the elastic-plastic problems with linear hardening) [3], this paper applies the same definition to a class of linearly elastic problems with friction that has a more severe non-smoothness: discontinuous acceleration and friction forces, as mentioned earlier.

The structure of the article is the following. In Section 2, the governing dynamic and quasi-static equations and conditions are presented, and the definition of stability of quasi-static paths is recalled. Section 3 contains some auxiliary results on the regularity of the solution of the quasi-static problem, which are then used in Section 4, in the proof of the main result of this paper.

2 Governing dynamic and quasi-static systems. Stability of the quasi-static path

We consider a particle in frictional contact with a flat surface, moving along some direction of that surface, and restrained by a linear elastic spring that acts along the same direction. For simplicity, and without loss of generality, we assume that the mass of the particle, the spring stiffness and the normal contact force are all unitary. For each value of the (time-like) load parameter $\lambda \in R$, $u(\lambda) \in R$, $f(\lambda) \in R$ and $r(\lambda) \in R$ denote the displacement of the particle, the applied force and the frictional reaction, respectively, all along the same direction. The function f is assumed to belong to C^2 and its derivative f' changes sign only a finite number of times. The small parameter $\varepsilon = d\lambda/dt > 0$ is the rate of change of the load parameter λ with respect to the physical time t, and $\mu \geq 0$ is the coefficient of friction.

Denoting each derivative $d()/d\lambda$ by $()'$, the governing dynamic equation of the particle has the form:

$$\varepsilon^2 u'' + u - f(\lambda) = r, \tag{1}$$

In addition, the frictional reaction r satisfies the well-known Coulomb friction law:

$$r \in \mu sign(-u'), \tag{2}$$

where $sign$ is the multivalued application given, for each $x \in R$, by:

$$sign(x) = \begin{cases} +1, \text{ if } x > 0, \\ [-1, +1], \text{ if } x = 0, \\ -1, \text{ if } x < 0. \end{cases} \tag{3}$$

The *dynamic problem* (1)-(2) can be equivalently written as the system of first-order differential inclusions:

$$\begin{cases} \varepsilon u' = v, \\ \varepsilon v' \in -u + f(\lambda) + \mu sign(-u'), \end{cases} \tag{4}$$

with initial conditions $u(\lambda_1)$ and $v(\lambda_1)$ at some value λ_1 of the load parameter.

Letting $\varepsilon = 0$ in (4) we obtain the governing system for the *quasi-static problem*:

$$\begin{cases} 0 = \bar{v}, \\ 0 \in -\bar{u} + f(\lambda) + \mu sign(-\bar{u}'), \end{cases} \tag{5}$$

which has an initial condition only on the displacement variable $\bar{u}(\lambda_1)$, necessarily given in the interval $[f(\lambda_1) - \mu, f(\lambda_1) + \mu]$. The displacement solution $\bar{u}(\lambda)$ of this problem defines a quasi-static path.

Existence and uniqueness of solution for the above dynamic problem in some interval $[\lambda_1, \lambda_2]$ of the time-like parameter λ has been established, for instance, in [16]. We have $u \in H^2(\lambda_1, \lambda_2)$ under the above assumptions.

In what concerns the quasi-static problem, it can be equivalently formulated as a sweeping process

$$-\bar{u}' \in N_{C(\lambda)}(\bar{u}) \tag{6}$$

where $C(\lambda) = [f(\lambda) - \mu, f(\lambda) + \mu]$ is a moving closed interval, and $N_{C(\lambda)}(\bar{u})$ is the outward normal cone to that interval at $\bar{u}$. The solution $\bar{u}$ of this sweeping process is known to be Lipschitz continuous (cf. Moreau [17]).

The purpose of the present paper is to discuss the stability of the quasi-static paths of the above frictional system.

First we recall the definition of stability of a quasi-static path ([2],[3]).

Definition The quasi-static path $\bar{u}(\lambda)$ is said to be stable at λ_1 if there exists $0 < \Delta\lambda \leq \lambda_2 - \lambda_1$, such that, for all $\delta > 0$ there exists $\bar{\rho}(\delta) > 0$ and $\bar{\varepsilon}(\delta) > 0$ with the property: for all initial conditions $(v(\lambda_1), u(\lambda_1))$ and all $\varepsilon > 0$ such that

$$|v(\lambda_1)| + |u(\lambda_1) - \bar{u}(\lambda_1)| < \bar{\rho}(\delta) \quad and \quad \varepsilon < \bar{\varepsilon}(\delta), \tag{7}$$

the solutions $v(\lambda), u(\lambda)$ of the dynamic problem (4) satisfy

$$|v(\lambda)| + |u(\lambda) - \bar{u}(\lambda)| < \delta, \tag{8}$$

for all $\lambda \in [\lambda_1, \lambda_1 + \Delta\lambda]$.

Next we recall that the Coulomb friction law (2) holds if and only if the following variational inequality holds:

$$r(w - v) + \mu(|w| - |v|) \geq 0, \quad \forall w \in R. \tag{9}$$

And in the next section we show that $\bar{u}'$ is a function of bounded variation.

3 Variation of the derivative of the quasi-static path

The solution of the quasi-static problem satisfies the differential inclusion (6). We study the total variation of $\bar{u}'$ assuming that on any bounded interval the derivative $f'(\lambda)$ changes its sign at most a finite number of times. We begin by considering the possible values of the right velocity $\bar{u}'^{+}$ and of the left velocity $\bar{u}'^{-}$.

If $\bar{u}(\bar{\lambda}) \in int\, C(\bar{\lambda})$, then, for λ close to $\bar{\lambda}$, $\bar{u}(\bar{\lambda}) \in int\, C(\lambda)$, so that the solution is locally constant, $\bar{u}(\lambda) \equiv \bar{u}(\bar{\lambda})$, with $\bar{u}'(\lambda) = 0$ and $\bar{u}'^{+}(\bar{\lambda}) = \bar{u}'^{-}(\bar{\lambda}) = 0$.

Now assume that $\bar{u}(\bar{\lambda})$ is on the boundary of $C(\bar{\lambda})$, say, $\bar{u}(\bar{\lambda}) = f(\bar{\lambda}) + \mu$. We fix a $\delta > 0$ such that $\bar{u}(\lambda) > f(\lambda) - \mu$ for $\lambda \in [\bar{\lambda}, \bar{\lambda} + \delta]$. This implies that the lower boundary $(f(\lambda) - \mu)$ does not act on the solution $\bar{u}$ during that interval of the loading parameter. It can be shown that $\bar{u}$ is then also the solution of the sweeping process by the set $\tilde{C}(\lambda) = (-\infty, f(\lambda) + \mu]$ and that

$$\bar{u}(\lambda) = \min\{\bar{u}(\bar{\lambda}), \inf_{\bar{\lambda} \leq \xi \leq \lambda} (f(\xi) + \mu)\} \tag{10}$$

Indeed, as the outward normal cone to to $\tilde{C}(\lambda)$ is either $\{0\}$ or $[0; +\infty)$, $\bar{u}'(\lambda)$ must be less or equal to zero to satisfy the differential inclusion.

Given the assumptions on f, we may restrict δ so that either

1. $f'(\lambda) \geq 0$, for all $\lambda \in (\bar{\lambda}, \bar{\lambda} + \delta]$, or
2. $f'(\lambda) \leq 0$, for all $\lambda \in (\bar{\lambda}, \bar{\lambda} + \delta]$.

In case 1, formula (10) implies that $\bar{u}(\lambda) \equiv f(\bar{\lambda}) + \mu = \bar{u}(\bar{\lambda})$, so that $\bar{u}'^{+}(\bar{\lambda}) = 0$. In case 2, $\bar{u}(\lambda) = f(\lambda) + \mu$ and $\bar{u}'^{+}(\bar{\lambda}) = f'(\bar{\lambda})$.

Now, consider the left velocity at $\bar{\lambda}$. (We assume that $\bar{u}(\bar{\lambda}) = f(\bar{\lambda}) + \mu$). Fix $\rho > 0$ such that either

1. $f'(\lambda) \leq 0$, for all $\lambda \in [\bar{\lambda} - \rho, \bar{\lambda}]$, or
2. $f'(\lambda) > 0$, for all $\lambda \in [\bar{\lambda} - \rho, \bar{\lambda}]$.

In case 1, there are only two possible solutions on the interval $[\bar{\lambda} - \rho, \bar{\lambda}]$:

(a) $\bar{u}(\lambda) \equiv f(\bar{\lambda}) + \mu$ for all $\lambda \in I$, or
(b) $\bar{u}(\lambda) = f(\lambda) + \mu$ for all $\lambda \in I$.

In case (a), the left-velocity satisfies $\bar{u}'^{-}(\bar{\lambda}) = 0$, and then $|\bar{u}'^{-}(\bar{\lambda}) - \bar{u}'^{+}(\bar{\lambda})| = |f'(\bar{\lambda})|$. In case (b), $\bar{u}'^{-}(\bar{\lambda}) = \bar{u}'^{+}(\bar{\lambda}) = f'(\bar{\lambda})$ and $\bar{u}'$ is continuous at $\bar{\lambda}$.

In case 2, the solution has no "previous" history, i.e. it is not possible to come to such a point by a sweeping process.

The case of the left boundary ($\bar{u}(\bar{\lambda}) = f(\bar{\lambda}) - \mu$) can be treated analogously. Therefore, we see that the right and the left velocities can only take values 0 or $f'(\lambda)$. Moreover, there is only a finite number of possible discontinuity points of the velocity (the value of the jump being always equal to $|f'(\bar{\lambda}_i)|$). Indeed, the above study shows that jumps may only occur when

1. $\bar{u}(\bar{\lambda}) = \mu + f(\bar{\lambda})$, $f'(\bar{\lambda}) < 0$ and $\bar{u}'(\lambda) = 0$ to the near left of $\bar{\lambda}$;
2. $\bar{u}(\bar{\lambda}) = \mu - f(\bar{\lambda})$, $f'(\bar{\lambda}) > 0$ and $\bar{u}'(\lambda) = 0$ to the near left of $\bar{\lambda}$.

As $f'(\lambda)$ has to change sign between two such discontinuities of $\bar{u}'(\lambda)$, even when they are of the same type, it is clear that such discontinuities form a finite set. Then, we take the right-velocity, which is a right-continuous function, as a representative of the velocity, and we keep the same notation.

Hence,

$$var(\bar{u}'; \lambda, \lambda + \Delta\lambda) \leq var(f'; \lambda, \lambda + \Delta\lambda) + \sum_i |f'(\bar{\lambda}_i)|, \tag{11}$$

where $\bar{\lambda}_i$ are points of discontinuity of $\bar{u}'$. Given that the sum is finite and f' is C^1, the total variation of the velocity is indeed finite.

4 Stability of the quasi-static path

We consider again the inclusions in the dynamic and the quasi-static governing systems (4) and (5), respectively. Due to (9) we have:

$$\forall w \quad (\varepsilon^2 u'' + u - f(\lambda))(w - u') + \mu(|w| - |u'|) \geq 0,$$

$$\forall w \quad (\bar{u} - f(\lambda))(w - \bar{u}') + \mu(|w| - |\bar{u}'|) \geq 0.$$

Letting $w = \bar{u}'$ in the first of these inequalities and $w = u'$ in the second one, and adding the resulting inequalities, we get

$$(\varepsilon^2 u'' + (u - \bar{u}))(\bar{u}' - u') \geq 0.$$

Then, we rewrite this inequality in the form

$$\varepsilon^2 u'' u' + (u - \bar{u})(u' - \bar{u}') \leq \varepsilon^2 u'' \bar{u}',$$

$$\frac{1}{2}\frac{d}{d\lambda}\left(\varepsilon^2 (u')^2 + (u - \bar{u})^2\right) \leq \varepsilon^2 u'' \bar{u}',$$

and we integrate it:

$$v^2(\lambda) + [u(\lambda) - \bar{u}(\lambda)]^2 \leq v^2(\lambda_1) + [u(\lambda_1) - \bar{u}(\lambda_1)]^2 + 2\int_{\lambda_1}^{\lambda} \varepsilon v'(l)\bar{u}'(l)dl. \quad (12)$$

Since we already know that $\bar{u}'$ is a right-continuous function of bounded variation the differential measure of the product $v\bar{u}'$ can be transformed according to the following formula:

$$d(v\bar{u}') = v d\bar{u}' + \bar{u}' dv = v d\bar{u}' + \bar{u}' v' d\lambda,$$

and integration by parts becomes possible in (12). Denoting $G_1(\lambda_1) = v^2(\lambda_1) + [u(\lambda_1) - \bar{u}(\lambda_1)]^2 - 2\varepsilon v(\lambda_1)\bar{u}'(\lambda_1)$, we obtain

$$v^2(\lambda) + [u(\lambda) - \bar{u}(\lambda)]^2 \leq G_1(\lambda_1) + 2\varepsilon v(\lambda)\bar{u}'(\lambda) - 2\int_{(\lambda_1,\lambda]} \varepsilon v(l) d\bar{u}'(l),$$

And then

$$(1 - 2\varepsilon)\|v(\lambda)\|^2 + \|u(\lambda) - \bar{u}(\lambda)\|^2 \leq G(\lambda_1) + G_2(\varepsilon), \quad (13)$$

where $\|\cdot\|$ is the maximum norm, $G_2(\varepsilon) = \varepsilon\|\bar{u}'(\lambda)\|^2 + \varepsilon var^2(\bar{u}'(l); \lambda_1, \lambda)$.

The left hand side of (13) represents the square of the difference between the dynamic solution and the quasi-static solution. On the other hand, the first term on the right hand side of (13) depends on the initial conditions, namely on the square of the difference between the dynamic and the quasi-static initial conditions. The last term on the right hand side consists of bounded values multiplied by ε. Hence, (13) implies that the quasi-static path is stable.

We summarize this result in the following statement.

Proposition 6 *Suppose that the function $f(\lambda)$ in the dynamic problem (1)-(2) belongs to C^2 and the derivative $f'(\lambda)$ changes its sign a finite number of times on the interval $[\lambda_1, \lambda_2]$. Then the quasi-static path $\bar{u}(\lambda)$ is stable at λ_1.*

5 Final remarks

The present paper shows some results relative to the stability of quasi-static paths of frictional systems with linear elastic behavior and the following simplifying features:

- a single particle, with
- known normal force.

In the situations considered here the non-smoothness of the dynamic problem corresponds to the possible occurrence of discontinuous accelerations and frictional reactions.

The cases with unilateral contact are, of course, much more delicate, not only due to the possible occurrence of velocity discontinuities and impulsive reactions, but also due to the fact that the normal reactions become unknown, and, for sufficiently large friction, the non-associated character of the friction law may lead to loss of stability, bifurcations of the quasi-static paths, and absence of quasi-static solutions for some loading directions.

The present work is currently being generalized in the following directions: (i) arbitrary number of particles in the system; and (ii) a priori unknown normal forces, although, at least during the time interval of our analysis, the region of contact is preserved and the product of the friction coefficient with mass coupling terms is sufficiently small. In these circumstances velocity discontinuities and impulsive reactions cannot arise, neither due to collisions nor due to "frictional catastrophes" [18]. A preliminary work in this direction was the formulation as differential inclusion and the mathematical study of that finite dimensional dynamic problem with friction and persistent contact [19].

References

1. Loret B, Simões FMF, Martins JAC (2000) Flutter instability and ill-posedness in solids and fluid-saturated porous media. In: Petryk H (eds) Material Instabilities in Elastic and Plastic Solids. International Centre for Mechanical Sciences, Courses and Lectures No. 414. Springer-Verlag, Wien New York.
2. Simões FMF, Pinto da Costa A, Martins JAC, Coelho I (2004) Three examples on "(in)stability of quasi-static paths" (submitted).
3. Martins JAC, Monteiro Marques MDP, Petrov A, Rebrova NV, Sobolev VA, Coelho I (2005) (In)stability of quasi-static paths of some finite dimensional smooth or elastic-plastic systems. J Phys Conf–Inst of Physics, to appear.
4. Shevitz D, Paden B (1994) Lyapunov stability theory of nonsmooth systems, IEEE Transactions on Automatic Vontrol 39 (9): 1910–1914.
5. Adly S, Goeleven D (2004) A stability theory for second-order nonsmooth dynamical systems with application to friction problems, Journal de Mathématiques Pures et Appliquées 83 (1): 17–51.
6. Van de Wouw N, Leine RI (2004) Attractivity of equilibrium sets of systems with dry friction, Nonlinear Dynamics 35 (1): 19–39.

7. Brogliato B (2004) Absolute stability and the Lagrange-Dirichlet theorem with monotone multivalued mappings, Systems and Control Letters 51: 343–353.
8. Sinou JJ, Thouverez F, Jezequel L (2003) Center manifold and multivariable approximants applied to non-linear stability analysis, International Journal of Nonlinear Mechanics 38 (9): 1421–1442.
9. Duffour P, Woodhouse J (2004) Instability of systems with a frictional point contact. Part 1: basic modelling. Part 2: model extensions, Journal of Sound and Vibration 271 (1-2): 365–410.
10. Ibrahim RA (1994) Friction-induced vibration, chatter, squeal, and chaos, Part I: Mechanics of contact and friction, Part II: Dynamic and modelling, ASME Applied Mechanics Reviews 47: 209–253.
11. Kinkaid NM, O'Reilly OM, Papaclopoulos P (2003) Automotive disc brake squeal, Journal of Sound and Vibration 267 (1): 105–166.
12. Sinou JJ, Thouverez F, Jezequel L (2004) Application of a nonlinear modal instability approach to brake systems, Journal of Vibration and Acoustics-Transactions of the ASME, 126 (1): 101–107.
13. Gu JC, Rice JR, Ruina AL, Tse ST (1984) Slip motion and stability of a single degree of freedom elastic system with rate and state dependent friction, J Mech Phys Solids 32: 167–196.
14. Scholz CH (1998) Earthquakes and friction Laws, Nature 391: 37–42.
15. Monteiro Marques MDP (1993) Differential Inclusions in Nonsmooth Mechanical Problems. Shocks and Dry Friction. Birkhauser, Basel Boston Berlin.
16. Monteiro Marques MDP (1994) An existence, uniqueness, and regularity study of the dynamics of systems with one-dimensional friction, Eur J Mech A/Solids 13:273–306.
17. Moreau J-J (1977) Evolution problem associated with a moving convex set in a Hilbert space, J Differ Equa 26:347–374.
18. Moreau J-J, Panagiotopoulos PD (1988) Nonsmooth mechanics and applications, CISM Courses and Lectures 302. Springer-Verlag, Vienna.
19. Martins JAC, Monteiro Marques MDP, Petrov A (2004) Dynamics with friction and persistent contact, ZAMM (accepted).

The T.G.V. disk brake squeal

X. Lorang[1], Q.S. Nguyen[1], F. Margiocchi[2], and P.E. Gautier[2]

[1] LMS, UMR 7649, Ecole Polytechnique, F-91128 Palaiseau cedex, France, lorang@lms.polytechnique.fr
[2] SNCF, Direction de la Recherche, SFC, 45 rue de Londres 75379 Paris cedex 08, florence.margiocchi@sncf.fr

Abstract. Discomfort problems due to the noise emittence of braking systems in trains have suggested recently several mechanical analyses. This paper gives some of our results obtained from the numerical modelling of TGV brakes in relation with some experimental data. The numerical discussion is based upon the Coulomb's law of contact with a constant coefficient of friction. A dynamic stability analysis enables us to show the loss of stability by flutter of the steady sliding response of the pad on the brake disks. We will present the numerical calculation of the unstable modes for the entire brake system which is a large structure. From an experimental point of view, the modal parameters have been measured for the brake system. We will present also the measures done at the train station giving the close-field acoustic. A comparison between these data and the numerical calculation help us to understand what happen during squeal.

1 Introduction

The brake squeals generated by TGV disk brakes are a source of discomfort for the passengers in or outside the train. This explains why a refined mechanical modelling of the phenomenon in order to understand the mechanism of the squeal generation is searched for. The squeal phenomenon is due principally to a self-excited vibration of the brake components, of high frequency (from 4000 to 17000 Hz), and high intensity (up to 130dB). It occurs only at small velocity, thus principally when the train arrives at the station. A theoretical and numerical discussion is given here and compared to the experimental data in order to understand the principal mechanisms of squeal generation. The System of TGV disk brake is first described briefly. Some experimental data are then reported and describe the measures done at the train station by microphones giving the close-field acoustic and on the brake components of a train at rest. A numerical simulation by the finite element method is undertaken, the steady sliding solution is computed. A stability analysis is then performed in order to analyze the flutter instability of the steady sliding response of the pads on the disks, under the assumption of a Coulomb's friction with constant friction coefficient.

2 The brake system

The disk-brake system of TGV trains is mounted on a bogie as shown in (Fig. 1). There are four disks on each axle. A disk brake is composed of two symmetric plates of lining with cylindrical pads (Fig. 3) which compress the disk. The disks are fixed on the axle thanks to a bell which is a very thin structure clamped on the axle. The compression force is about 19kN at maximum.

Fig. 1. Bogie.

Fig. 2. Linings.

3 Experimental data

The experimental measures at the station furnish the spectrum of the brake squeal. For this, a microphone is mounted (Fig. 3) near a disk brake. The obtained spectrum (Fig. 5) shows clearly the existence of 7 frequencies which merge from the background noise. They are included in the frequency interval 6 kHz to 17 kHz.

An experimental modal analysis to determine the modal vibration has been also performed for a bogie at rest. Figure 6 shows the superposed spectra which show clearly that some vibration frequencies of the disk coincide also with some measured noise frequencies. The first frequency response function comes from excited bending modes (the captor measure axial accelerations) whereas the second one is related to in-plane modes.

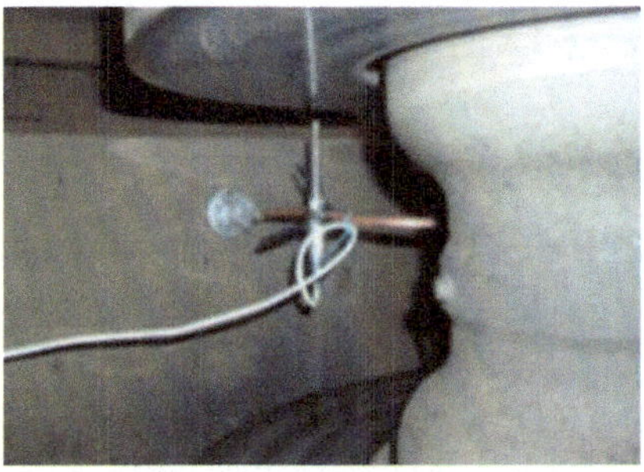

Fig. 3. Microphone.

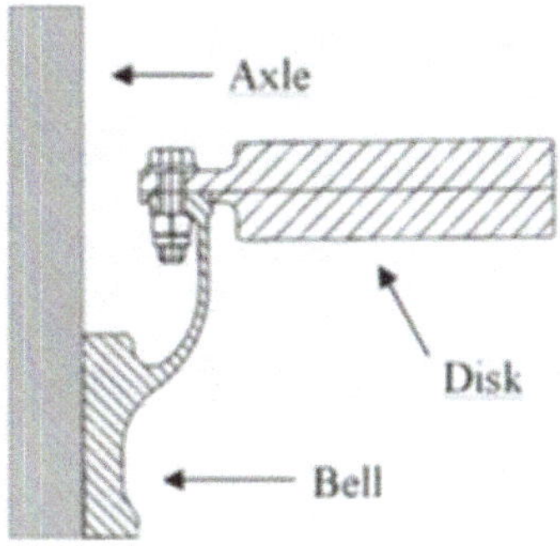

Fig. 4. Sketch.

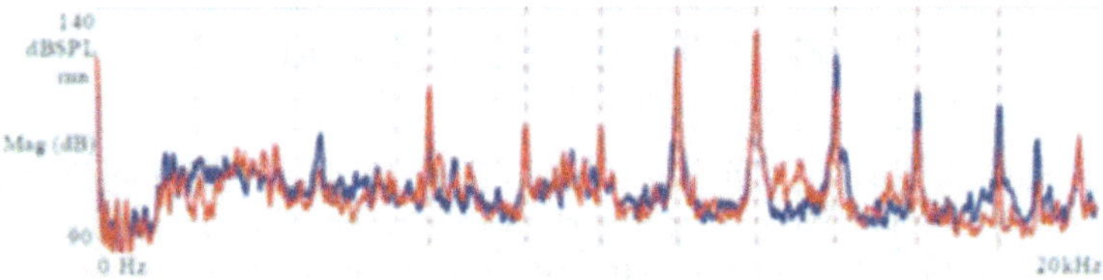

Fig. 5. A squeal spectrum.

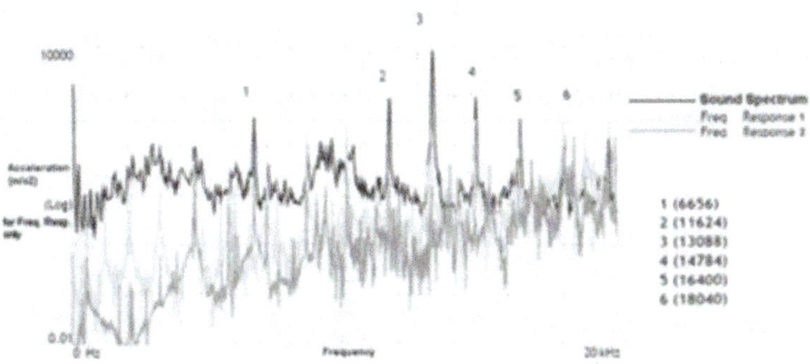

Fig. 6. The superposed spectra.

4 A mechanical modeling

The disk and the linings are assumed to be linearly elastic, isotropic and homogeneous solids in small deformation. The contact is unilateral with Coulomb friction and the friction coefficient is assumed constant. A discretization by finite elements (Fig. 7) is adopted with the same nodes in contact.

The first considered model was a clamped disk. Different studies have shown that the bell did not add so much stiffness to the disk which behaves like a free-free disk. From a numerical point of view we have shown that it is necessary to use quadratic elements for the disk and linings to correctly approximate the bending modes. As a consequence, the model is composed

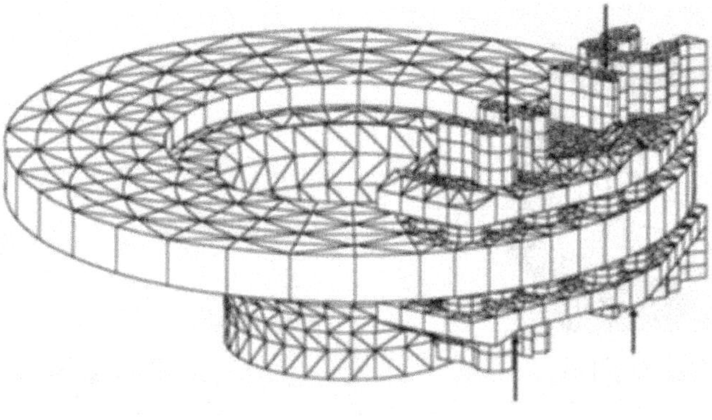

Fig. 7. The Finite Element model.

of shell elements for the bell and composed of quadratic volume elements for the disk and linings. We can see in the table 1 the comparison between the numerical free vibration of the bell/disk system versus the experimental one.

Table 1.

Modes with nodal diameters	Experimental results (frequencies Hz)	Finite element model (frequencies Hz)	Relative difference %
2	520	586	12.7
3	1200	1224	2.0
4	2120	2120	0.0
5	3208	3200	-0.25
6	4424	4422	-0.05
7	5736	5760	0.42
8	7128	7190	0.87
9	8575	8710	1.57
10	10080	10300	2.18
11	11624	11970	2.98
12	13088	13700	4.68
13	14784	15490	4.78

5 The governing equations

The system is governed by the virtual work equation with unilateral and frictional contact conditions. If $\mathbf{u}$ denotes the displacement of the bell/disk D, and $\mathbf{u^L}$ the displacement of the pads/linings L, in contact on , then:

$$\begin{array}{l}\int_D \sigma(\mathbf{u}) : \varepsilon(\mathbf{u}^*).d\Omega + \int_D \rho.\gamma \cdot \mathbf{u}^*.d\Omega = \int_{\partial\Omega_C} N.u_n^* + \mathbf{T}\cdot\mathbf{u}_t^* d\Omega \\ \forall \mathbf{u}^* \in U_{ad_0} \text{ with } U_{ad_0} = \left\{\mathbf{u}/\mathbf{u} = 0 \text{ on } \partial\Omega_U\right\}, \mathbf{u} = \mathbf{U_g} \text{ on } \partial\Omega_U\end{array} \quad (1)$$

$$\begin{cases} u_n = \mathbf{u}\cdot\mathbf{n} \\ \mathbf{u_t} = \mathbf{u} - u_n.\mathbf{n}\end{cases} \text{and} \begin{cases} N = (\sigma\cdot\mathbf{n})\cdot\mathbf{n} \\ \mathbf{T} = \sigma\cdot\mathbf{n} - N.\mathbf{n}\end{cases}$$

the constitutive equations are :

$$\sigma(\mathbf{u}) = \mathbf{E} : \varepsilon(\mathbf{u}) \text{ in } D$$

$$\begin{cases} N \leq 0 \\ \mathbf{u}\cdot\mathbf{n} + \mathbf{u^L}\cdot\mathbf{n^L} \leq 0 \\ N.\left[\mathbf{u}\cdot\mathbf{n} + \mathbf{u^L}\cdot\mathbf{n^L}\right] = 0\end{cases} \text{and} \begin{cases} |\mathbf{T}| \leq -f.N \\ \mathbf{T}\cdot\mathbf{w} - f.N\,|\mathbf{w}| = 0\end{cases} \text{on } \partial\Omega_C$$

$\mathbf{E}$ is the Hooke fourth order tensor, $\mathbf{n}$ is the outward normal to the planar surface of the disk and $\mathbf{n^L}$ is the normal to the pads. On the contact surface $\partial\Omega_C$, $\mathbf{w}$ is the sliding velocity:

$$\mathbf{w} = \mathbf{v} + \dot{\mathbf{u}} - \dot{\mathbf{u}}^{\mathbf{L}}$$

with $\mathbf{v}$ the velocity due to the rotation of the disk. The coefficient of friction is f. In a same spirit $\mathbf{u^L}$ must satisfy similar equation as (1).

Under the assumption of small speed of rotation, the approximation $\gamma = \mathbf{u}_{,tt}$ holds, cf. to [1] for a more complete expression of γ .

6 The steady sliding response

At the steady sliding equilibrium, $\mathbf{w} = \mathbf{v}$ and $\gamma = 0$. With these conditions, the displacement at equilibrium $\mathbf{u}$ must satisfy from (1).

$$\begin{array}{l}\int_D \varepsilon(\mathbf{u}) : \mathbf{E} : \varepsilon(\mathbf{u}^*).d\Omega = \int_{\partial\Omega_C} N.u_n^* + \mathbf{T}\cdot\mathbf{u}_t^* d\Omega \\ \forall \mathbf{u}^* \in U_{ad_0} \text{ with } U_{ad_0} = \left\{\mathbf{u}/\mathbf{u} = \mathbf{0} \text{ on } \partial\Omega_U\right\}, \mathbf{u} = \mathbf{U_d} \text{ on } \partial\Omega_U\end{array}$$

$$\text{with} \begin{cases} N \leq \mathbf{0} \\ \mathbf{u}\cdot\mathbf{n} + \mathbf{u^L}\cdot\mathbf{n^L} \leq \mathbf{0} \\ N.\left[\mathbf{u}\cdot\mathbf{n} + \mathbf{u^L}\cdot\mathbf{n^L}\right] = \mathbf{0}\end{cases} \text{and } \mathbf{T} = f.N.\frac{\mathbf{v}}{|\mathbf{v}|} = f.N.\mathbf{t} \text{ on } \partial\Omega_C$$

After the discretization of the previous equations, the steady sliding equilibrium is computed with the status method.

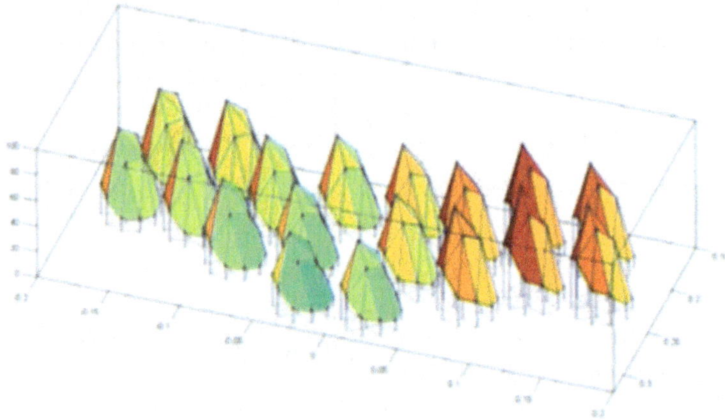

Fig. 8. The local normal contact force N_e (Newton) over $\partial\Omega_C$.

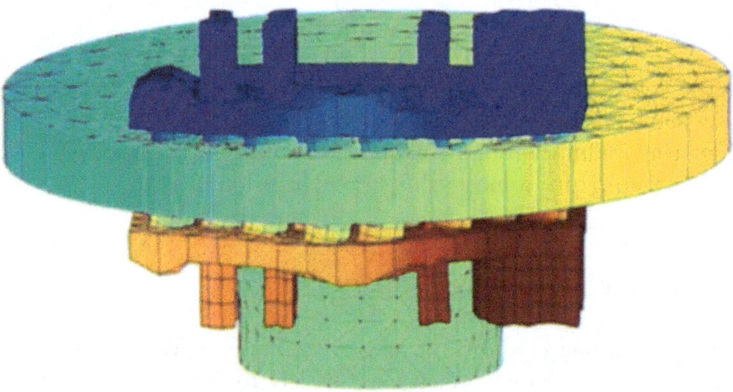

Fig. 9. The sliding equilibrium of the brake.

7 The flutter instability of the steady sliding equilibrium

A small perturbed motion near the equilibrium is now considered. However, the governing equations are not differentiable when there is separation, and thus only sliding motions are considered on the present contact zone. Such a perturbed motion is then governed by the linearized equations. From the expressions

$$\mathbf{u} = \mathbf{u_e} + \hat{\mathbf{u}}, \;\; \hat{\mathbf{u}} \cdot \mathbf{n} + \hat{\mathbf{u}}^{\mathbf{L}} \cdot \mathbf{n}^{\mathbf{L}} = \mathbf{0} \text{ on } \partial\Omega_C^e$$

The contact conditions:

$$\mathbf{T} = f.N.\frac{\mathbf{w}}{|\mathbf{w}|}$$

gives after linearization

$$\hat{\mathbf{T}} = f.\hat{N}.\frac{\mathbf{v}}{|\mathbf{v}|} + \frac{f.N_e}{|\mathbf{v}|}\left(\hat{\mathbf{w}} - \frac{\mathbf{v}\cdot\hat{\mathbf{w}}}{|\mathbf{v}|^{\mathbf{2}}}\mathbf{v}\right) = f.\hat{N}.\mathbf{t} + \frac{f.N_e}{|\mathbf{v}|}\left(\mathbf{1} - \mathbf{t}\otimes\mathbf{t}\right).\hat{\mathbf{w}}$$

The second term of $\hat{\mathbf{T}}$ is due to the planar nature of the friction force. The perturbed displacement $\hat{\mathbf{u}}$ is thus governed by

$$\int_D \rho.\hat{\mathbf{u}}_{,\mathbf{tt}} \cdot \mathbf{u}^* d\Omega + \int_D \varepsilon(\hat{\mathbf{u}}) : \mathbf{E} : \varepsilon(\mathbf{u}^*).d\Omega = \int_{\partial\Omega_C} \hat{N}.u_n^* + \hat{\mathbf{T}}\cdot\mathbf{u}_{\mathbf{t}}^* d\Omega$$

After discretization, the discrete perturbed displacement $\mathbf{U}$ verify:

$$\mathbf{M(f)\ddot{U} + C(f)\dot{U} + K(f)U = 0}$$

It is important to notice that $\mathbf{M}$ and $\mathbf{K}$ are unsymmetric matrices. However, the damping matrix $\mathbf{C(f)}$ is symmetric. It comes from the discretization of the second term of $\hat{\mathbf{T}}$.

Assuming that $\mathbf{U} = \mathbf{X}\mathbf{e}^{\lambda.\mathbf{t}}$ leads to the following eigenvalue problem:

$$(\lambda^2\mathbf{M} + \lambda\mathbf{C} + \mathbf{K})\mathbf{X} = \mathbf{0}$$

The lack of symmetry of these matrices leads to complex modes and to complex eigenvalues. A mode is unstable if $Re(\lambda) > 0$. For an unstable mode, there is flutter instability with a growing displacement in the form:

$$\mathbf{U} = e^{\mathrm{Re}(\lambda)t}\left[\mathrm{Re}(\mathbf{X})\cos(\mathrm{Im}(\lambda)t) - \mathrm{Im}(\mathbf{X})\sin(\mathrm{Im}(\lambda)t)\right]$$

On the figure 10, are presented the results of the eigenvalue analysis. We decide to plot $\xi = \mathrm{Re}(\lambda)/|\lambda|$ as a function of the frequency (we can compare ξ with the modal damping ratio of the material). On the figures, the straight lines are disk modes whereas dashed-dot lines are linings modes. The noisy modes identified experimentally are represented with dashed lines.

8 Conclusion and future work

This discussion presents some interesting directions of research to understand the mechanism of brake squeal. The numerical analysis gives the flutter modes. For some modes, there is a good agreement between the numerical analysis and the experiments. But the phenomenon is not yet completely explained. This is the first step of the analysis. A future work to be considered consists of computing the cyclic response resulting from this dynamic bifurcation in the sense of Hopf.

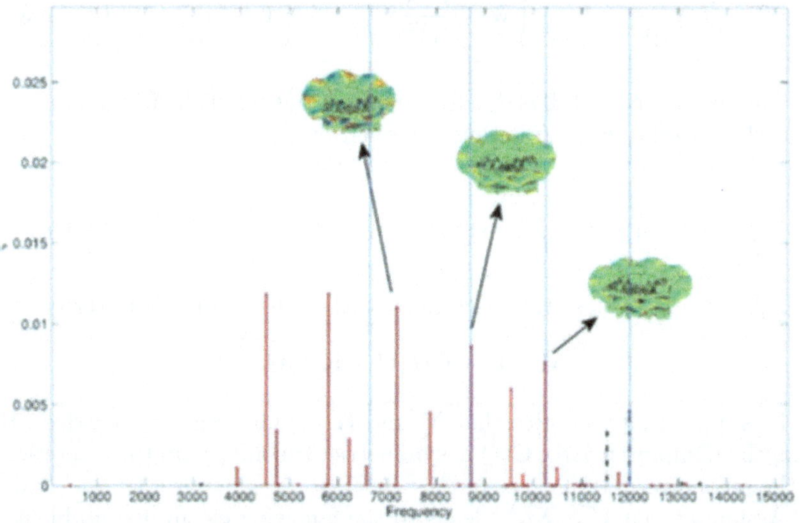

Fig. 10. Eigenvalues in the complex plane.

References

1. F. Moirot, Q.S. Nguyen, Brake squeal: a problem of flutter instability of the steady sliding solution ? Arch. Mech., 52, 645-662 (2000).
2. F. Moirot, Q.S. Nguyen, An example of stick-slip waves. C. R. Mécanique, 328, 663-669 (2000).
3. J.A.C. Martins, J. Guimaraes, L.O. Faria, Dynamic surface solutions in linear elasticity and viscoelasticity with frictional boundary conditions. J. Vibration and Acoustics, 117, 445-451 (1995).
4. D.A. Crolla, A.M. Lang, Brake noise and vibration: state of art. Vehicle Tribologie, 18, 165-174 (1991).
5. M.R. North, Disc brake squeal, in: Braking of Road Vehicles, Automobile Division of the Institution of Mechanical Engineers, Mechanical Engineering Publications Limited, London, England, 1976, pp. 169-176.
6. F. Moirot, Étude de la stabilité d'un équilibre en présence du frottement de Coulomb. Application au crissement des freins à disque. Thèse, École Polytechnique, Paris (1998).
7. H. Murakami, N. Tsunada, T. Kitamura, A study concerned with a mechanism of disc-brake squeal, Technical Report 841233, SAE, Warrendale, PA, 1984.
8. N.M. Kinkaid, O.M. O'reilly, P. Papadopoulos, Automotive disc brake squeal. Journal of Sound and Vibration 267 (2003) 105-166.

Part V

Poster session

Sliding path curvature dependent friction and wear

A. Zmitrowicz

Institute of Fluid-Flow Machinery, Polish Academy of Sciences, ul.J.Fiszera 14, PL-80-952 Gdańsk, Poland,
azmit@imp.gda.pl

1 Introduction

Kinematics of sliding can initiate microstructural and frictional changes in the surface and near-surface material of some polymers [1] and in layer-lattice materials as graphite and molybdenum disulphide [4]. In these cases, microstructures of the sliding surfaces reorient themselves, e.g. in the direction of sliding. It may reduce (or increase) friction and wear of the materials. From the mathematical point of view, the induced friction and wear depend on the sliding direction and on a shape of the sliding path, i.e. they depend on a sliding path curvature. First- second- and higher-order equations describe evolutions of friction and wear induced by the sliding path curvature, see [2, 3, 4].

2 Models of friction and wear

The friction equation is a relation between a friction force vector $\mathbf{p}_t$ and a normal pressure p_n, a sliding velocity unit vector $\mathbf{v}$ and its derivative $\partial \mathbf{v}/\partial s$ with respect to a one-dimensional parameterization s of the sliding path. A first-order friction force equation is defined by a sum of two single-term polynomials as follows

$$\mathbf{p}_t = -p_n \left(\mathsf{C}_1 \mathbf{v} + \mathsf{E}_1 \mathbf{n} \frac{1}{r} \right) , \tag{1}$$

where, according to the Frenet-Serret first formula $\partial \mathbf{v}/\partial s = \mathbf{n}/r$, $\mathbf{n}$ is a unit vector normal to the sliding path, r is the sliding path radius [2]. Two second-order tensors C_1 and E_1 describe frictional anisotropy and heterogeneity and effects associated with the sliding kinematics.
Constitutive equations governing the phenomenon of kinematics dependent wear are created in the frame of Archard law. A wear intensity coefficient is assumed to be a similar function of the curvature as $\mu_\alpha^{||} = f(1/r)$, where $\mu_\alpha^{||}$ is a coefficient of the friction force component tangent to the trajectory [4].

Fig. 1. Various radii of circular sliding trajectories in pin-on-disc tests [1].

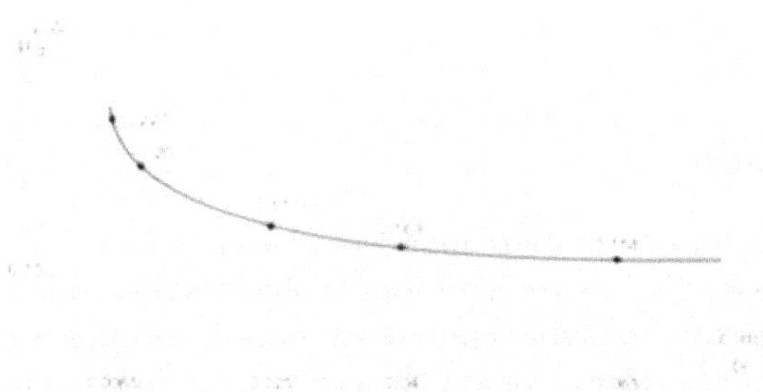

Fig. 2. The friction coefficient in dependence on the radii of the trajectories.

3 Results

The friction model (1) is illustrated by an example of circular trajectories of a polymeric pin sliding with respect to a rotating steel disc, see Fig. 1. The parameters of the model were estimated with the aid of Briscoe and Stolarski test data [1]. The sliding path curvature generates: (a) additional resistance to sliding (dissipative type forces), (b) constraint forces normal to the sliding path (gyroscopic type forces). It can induce positive or negative additional friction (see Fig. 3), and it can change essentially the sliding trajectory [2]. There is very good agreement between the experiment [1] and the model (1)[1]

References

1. Briscoe BJ, Stolarski TA (1991) Wear 149: 233-240
2. Zmitrowicz A (1999) Int. J. Solids Structures 36:2825 V2848, 2849 V2863
3. Zmitrowicz A (2004) Evolutions of friction anisotropy and heterogeneity. In: Gutkowski W, Kowalewski TA (eds) Proc. of the 21st Int. Congress of Theoretical and Applied Mechanics. Warsaw, http://ictam04.ippt.gov.pl
4. Zmitrowicz A (2005) Int. J. Solids Structures (in press)

[1] The author thanks the Alexander von Humboldt Foundation for a financial help.

Composition duality methods in contact mechanics

G. Alduncin

Institute of Geophysics, UNAM, Coyoacan 04510 D.F., Mexico,
alduncin@geofisica.unam.mx

Composition duality methods in the context of optimization problems in mechanics have been the basis for analysis and approximation of minimization and related minimax problems, as studied by Ekeland and Temam [1]. Abstract convex functionals of the type $J(v) = F(v) + G(\Lambda v)$, $v \in V$, have been considered on a reflexive Banach space V. Associated two-field Lagrangians have been of the type $L(v, q^*) = F(v) - G^*(q^*) + < q^*, \Lambda v >_Y$, $(v, q^*) \in V \times Y^*$, Y^* denoting the topological dual of another reflexive Banach space Y. Then, corresponding duality principles establish conditions for the solvability equivalence of the primal and mixed optimality condition problems. The natural compatibility condition handled by the classical composition duality principles has required the existence of a point Λv at which functional G is finite and continuous; i.e, $(\mathbf{C}_C)$ $\mathrm{int}\mathcal{D}(G) \cap \mathcal{R}(\Lambda) \neq \emptyset$.

In the case of nonpotential mixed problems from mechanics, the classical composition duality principle was extended by Gabay [2] in a Hilbert framework to mixed problems with a primal operator $\mathcal{A} : V \to 2^{V^*}$, maximal monotone and not necessarily potential. Also, the important augmented or exactly penalized composition duality theory for two- and three-field Lagrangians treated in [1] by perturbations, was extended in [2] through resolvent characterizations, which interpreted as proximation variational versions of mixed problems further lead naturally to proximal-point resolution algorithms. In a general framework of multivalued operators, duality principles have also been established recently by Attouch and Théra [3] and, with a composition component, by Robinson [4] where the compatibility condition is given by $(\mathbf{C}_R)$ $\mathcal{R}(\Lambda)$ is closed and $\partial G + \partial I_{\mathcal{R}(\Lambda)}$ is maximal monotone, guaranteeing the maximal monotonicity of the composition $\Lambda^T \partial G \circ \Lambda$.

The purpose of this study is to develop composition duality methods for contact problems on a functional framework of reflexive Banach spaces, without utilizing the classical compatibility condition $(\mathbf{C}_C)$, nor Robinson's condition $(\mathbf{C}_R)$. Motivated by the analysis and finite element approximation of hybridized and mixed variational models of boundary value mechanical problems, we demonstrated in [5–7] and applied in [8], that an alternative and more appropriate compatibility condition should be: (**C**) $\Lambda \in \mathcal{L}(V,Y)$ is surjective. This Λ-surjectivity condition, independent of the interior of the effective domain $\mathcal{D}(G)$, and implying Robinson's condition, is in fact an operator version of the general Ladysenskaja-Babuska-Brezzi inf-sup condition, which is in turn equivalent to the lower boundedness of the transpose Λ^T.

Thereby, under the alternative compatibility condition (**C**), we consider composition duality principles for the solvability analysis of mixed variational inclusions in contact mechanics. For computational purposes, we introduce regularizations via mass-preconditioned augmented or exactly penalized formulations. Further, three-field and macro-hybrid variational versions are treated. Three-field formulations introduce an additional intermediate field that could be of technological interest, and provide of a linear dual mixed structure that may facilitate numerical analysis and proximation realization of penalty-duality algorithms. Also, macro-hybrid localizing formulations, based on nonoverlapping domain decompositions, have proved to be very efficient for numerical models with a large scale and nonsmooth geometry, discontinuous parameters and multiscale and multiphysics behavior, as well as for multialgorithmia and parallel computing.

At a finite dimensional level, we also elaborate on corresponding discrete mixed and macro-hybrid internal approximations, the latter having the important property of being implementable via local conforming finite element discretizations without matching constraints at the subdomain interfaces. Moreover, we discuss proximal-point iterative algorithms which correspond to numerical realizations of mass-preconditioned augmented discrete versions.

References

1. Ekeland I, Temam R (1974) Analyse convexe et problèmes variationnels. Dunod Gauthier Villars, Paris
2. Gabay D (1982) Application de la méthode des multiplicateurs aux inéquations variationnelles. In: Fortin M, Glowinski R (eds) Méthodes de Lagrangien Augmenté. Dunod-Bordas, Paris, pp 279-307
3. Attouch H, Thera M (1996) A general duality principle for the sum of two operators. J Convex Anal 3: 1–24
4. Robinson S M (1999) Composition duality and maximal monotonicity. Math Program 85: 1–13
5. Alduncin G (2005) Composition duality principles for mixed variational inequalities. Math Comp Modelling 41: 639–654
6. Alduncin G Macro-hybrid variational formulations of constrained boundary value problems. Numer Funct Anal Optimiz (to appear)
7. Alduncin G Composition duality methods for mixed variational inclusions. Appl Math Optim (to appear)
8. Alduncin G (2004) Augmented domain decomposition methods in contact mechanics. In: Neittaanmäki P, Rossi T, Korotov S, Oñate E, Périaux J, Knörzer D (eds) ECCOMAS 2004. University of Jyväskylä, Jyväskylä, pp 1–20

Modeling of behaviour of the flat friction lining under the external

A.S. Khachikyan, K.A. Karapetyan, R.A. Shirinyan, and V.V. Hakobyan

Institute of Mechanics of NAS RA

The model problem is considered for a contact interaction between plate and massive body under adhesive joint, which simulates the work of the friction lining and its metal base. The conditions are sought by way of the choice of favorable angles of the joint geometry depending on the mechanical properties of joining materials, which promote development of the optimal structure of the friction lining joint with a metal base under external thermal and mechanical action.

The main idea consists in the assumption, that serviceability of friction lining would be higher if the order of stress concentration would be less and the stress distribution would be more close to the uniformity.

The non-uniformity of the stress state of the friction lining can be caused due to influence of several factors. First of all, the regions of strong stress concentration arise near the boundary of contact surface due to difference of elasticity module magnitudes of joint parts. The stress non-uniformity arise also due to non-uniformity of the distribution of temperature and difference of coefficients of linear thermal expansion of joint parts. The stress distribution also depends both on ways of mechanical forces application to a joint and geometrical forms of joint as well as precision of manufacturing parts of the device. These mentioned causes are common for many devices.

An integrated approach to the solution of formulated problem is applied. The geometrical parameters are obtained theoretically for friction lining surface near the boundary line of contact zone, which provide the low-stress state in dangerous points of this line. Here, the well known method of the elasticity theory local solutions is applied.

The corresponding photoelastic models are developed and the stress distribution is investigated in the loaded specimens. The external mechanical action is simulated by the normal compressive force (700 N) and lateral tension force (280 N), the last one is applied through intermediate rubber layers substitute the action of tangential stresses. The thermal action is simulated by the uniform increasing of temperature as well as by the testing of specimens with the prestressed joint.

Three types of specimens are tested (fig. 1). The first type of specimens corresponds to the most widely used kind of joints, when the edge surface of composite is perpendicular to the facial surface of metal base, i.e. opening angle of composite material equal 90 and opening angle of metal base equal 180 .

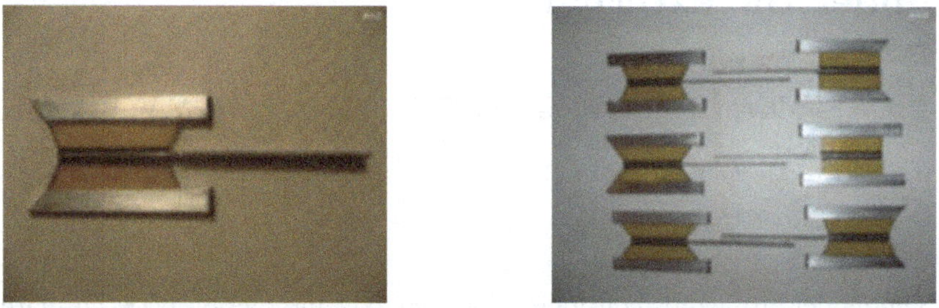

In accordance with the theoretical recommendations the second type of specimens were made so, that the opening angle of composite material equal 70 and 50 and opening angle of metal base equal 180 .

The third type of specimens were made with opening angle of composite material equal to 90 , 70 and 50 and opening angle of metal base equal 130 .

The analysis of stress state of photoelastic models (fig. 2) shows that favourable influence of the decreasing of opening angles of composite material on the order of stress concentration and low effectiveness of the variation of opening angle of metal base. The experiments also show the sensitivity of the stress distribution from the precision of joint-parts preparation, stability of direction of acting forces and rigidity of metal base. The less rigid base leads to increasing of non-uniformity of stress distribution.

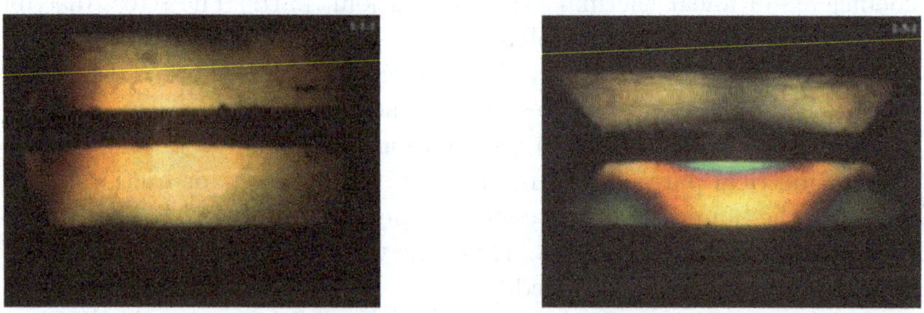

Thus, the main condition of achievement of uniformity of stress distribution is the decrease of opening angles of composite material near the boundary of contact surface. The exact values of optimal angles must be determined for any concrete materials and devices. The sufficient rigidity of metal base and the provision of precision of details sizes are also important.

The non smooth contact dynamic method: recent LMGC90 software developments and application

F. Dubois[1] and M. Jean[2]

[1] Laboratoire de Mécanique et Génie Civil, CNRS, Université Montpellier II, France, dubois@lmgc.univ-montp2.fr
[2] Laboratoire de Mécanique et d'Acoustique, UPR CNRS, Marseille, France, mjean@imtumn.imt-mrs.fr

The *Non Smooth Contact Dynamics* method (NSCD) deals with frictional unilateral contact between rigid or deformable bodies. It was originated around 1984 by J.J. Moreau as the *Contact Dynamics* method (*CD*). It was extended as the (*NSCD*) method by M. Jean to deal with more general applications, such as nite element modelling. Numerical applications of the *CD* method for granular materials are currently customized in C language by J.J. Moreau. The *NSCD* method has been implemented by M. Jean in a now obsolete fortran77 general purpose software. F. Dubois is the author of a completely remodelled new version *LMGC90* written in fortran90, open to research scientists for further developments and applications. This is an open source software governed by a *Cecill* license (i.e. GPL). In the NSCD method the basic laws such as Coulomb's law and the inelastic shock law are described as non smooth laws in terms of multimappings. The dynamical equation is discretized according to a low order implicit algorithm. The main unknowns are the relative velocities between contactors at some overlapping moments with the time steps (leap frog technique) and the mean reaction impulses during the time step. Assuming provisional values for contacts neighbouring of a given contact, values of the reactions for this given contact are obtained discussing the intersection of the graphs of affine mappings. Values of the reactions are updated, and all contacts are processed successively as long as necessary to obtain a satisfactory convergence. This may be described as a non linear block Gauss Seidel method. The *LMGC90* software is dedicated to applications with a large number of 2D or 3D contactors. It is a modular software written in fortran90 with an object oriented organisation:

- Mechanical bodies moduli are listing bulk geometric and physical properties, degrees of freedom, descriptors of contactors (or contacting boundaries) attached to the bodies, and the methods to read, write and precompute data.

- Contact moduli are dedicated to classes of contactors, pairs of a candidate and antagonist contactors. They are listing all local properties, relative velocities, gaps, reaction impulses, interaction laws, detection methods, methods allowing to compute relative velocities from bodies velocities, to compute generalized impulses from reaction impulses.
- A core solver uses the above methods to transfer data to the dynamical algorithm relating bodies velocities and generalized impulses, to the interaction law relating relative velocities and reaction impulses.

Recent developments are concerning core solvers. The non linear block Gauss Seidel method is now available as a shared memory process; a generalized Newton algorithm is implemented. Several interaction laws, so called "Signorini Coulomb derived laws" are implemented, including elastic repell, adhesion and wear laws. The software allows also coupling with thermo elastic problems. Contactors are of the kind: disks, polygons, spheres, polyhedrons, planes, wires, rods, blocks, and a wide variety of 3 or 4 nodes element contactors de ned by nite elements. Applications are concerning: Granular materials, for research purposes such as stability of slopes, segregation, compaction behaviour, or industrial purposes such as railway ballast behaviour. Behaviour of buildings made of blocks, walls, arches, bridges. Dynamic fracture of deformable bodies. Delamination. Biomechanics, micro brilla models, tensegrity cell models.

Fig. 1. Masonry: Pont du Gard, 35000 blocks, 250000 contacts, author Brahim Chetouane.

Fig. 2. Masonry: Pont du Gard, 35000 blocks, 250000 contacts, author Brahim Chetouane.

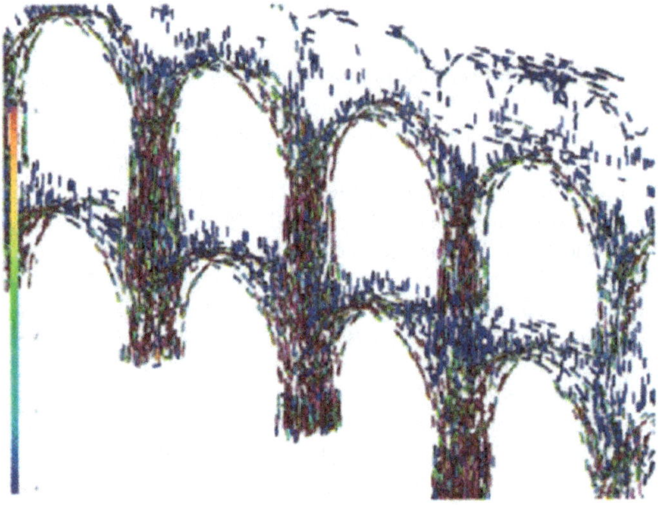

Fig. 3. Pont Du Gard: main stress chains.

Fig. 4. Pont Du Gard: pressure colored blocks, high pressure pale blue, low pressure dark blue.

References

1. J.J. Moreau. *Unilateral contact and dry friction in nite freedom dynamics, volume 302 of International Centre for Mechanical Sciences, Courses and Lectures.* pp. 1-82, J.J. Moreau, P.D. Panagiotopoulos, Springer, Vienna, 1988.
2. J.J. Moreau. *Some basics of unilateral dynamics*, in F. Pfeiffer, C. Glocker (ed), Unilateral Multibody Dynamics, Kluwer, Dordrect, 1999.
3. M. Jean, J.J. Moreau. Unilaterality and dry friction in the dynamics of rigid bodies collections. In A. Curnier, (ed), *Proc. of Contact Mech. Int. Symp.*, pp. 31-48, 1992.
4. M. Jean. *Frictional contact in rigid or deformable bodies: numerical simulation of geomaterials*, pp. 463-486, A.P.S. Salvadurai J.M. Boulon, Elsevier Science Publisher, Amsterdam, 1995.
5. M. Jean. The Non Smooth Contact Dynamics Method, *Computer Methods in Applied Mechanics and Engineering*, special issue on computational modeling of contact and friction, J.A.C. Martins, A. Klarbring (ed), 177, pp. 235-257, 1999.
6. B. Cambou, M. Jean. Micro Mécanique des Matériaux Granulaires. Hermes, Paris, 2001.

Frictional contact of elastomer materials on rough rigid surfaces

J. Nettingsmeier and P. Wriggers

Institute of Computational Mechanics, University of Hannover, Appelstrasse 9a, D-30167 Hannover,
netting@ibnm.uni-hannover.de

Abstract. The frictional behaviour of elastomer materials still raises a lot of questions. The sliding of rubber on a rough surface cannot be described by using a constant friction coefficient. An appropriate constitutive law has to represent the complex dependancies on sliding velocity or normal stress. As rough tracks have a fractal character it is necessary to use a staggered procedure over the relevant length scales.

1 Rubber friction

Rubber friction consists of two main uncoupled effects - adhesion and hysteresis. The adhesional part is based on the intermolecular boundings between two surfaces, in contrast hysteretic friction results from the energy losses inside th e material due to cyclic loading wich appears while the rubber is sliding over the surface asperities.

2 Surface description and scale transition

Usual road tracks are rough on different length scales, which complicates the numerical treatment. A staggered procedure will be necessary, therefore the fractal surface is converted into harmonic functions, which are calculated seperately. The observed friction coefficient as a function of sliding velocity and normal stress will then be added locally on the larger length scales.

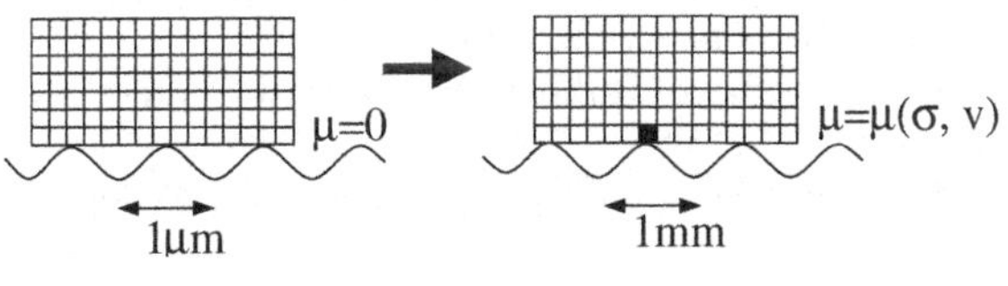

3 Contact Treatment

The stiffness of an asphalt road track is much higher than the rubber stiffness. It is therefore acceptable to simplify the rough surface as rigid and fixed. The resulting friction coefficient will in the following be calculated as the relation of the averaged horizontal ($t_{T\alpha}$) to the applied vertical stresses (t_N) in the contact zone.

4 Numerical results

The derivation of a friction law on the microscale is done on a sine wave with a wavelength of 1 μm. The rubber sample is pushed onto the surface vertically and afterwards pulled horizontally with a constant sliding velocity.
The strong dependance on the loading frequency can be detected from the progress of the friction coefficient against the logarithmic sliding velocity for different normal stresses.

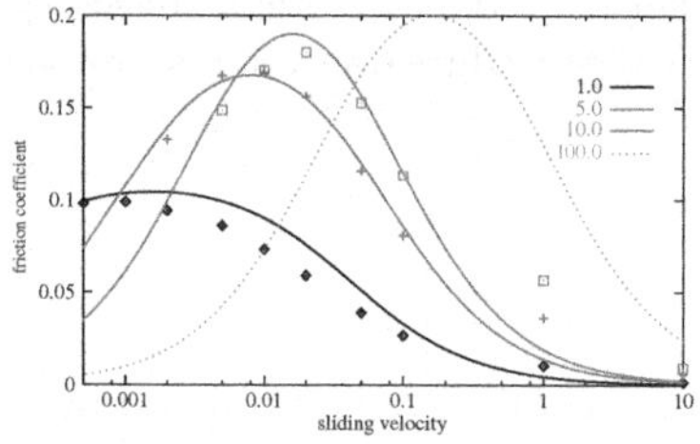

$$\mu(v) = \frac{a\tilde{v}}{(\tilde{v}+b)^2+c} \tag{1}$$

$$a, b, c = f(\sigma_N) \tag{2}$$

On the next scale the micro roughnesses will be smoothed and instead the calculated friction value will be put onto the contact zone. Thus the maximum of the friction curve broadens and raises due to the superposition of the effects on different length scales. The dotted line is depicted for comparison.

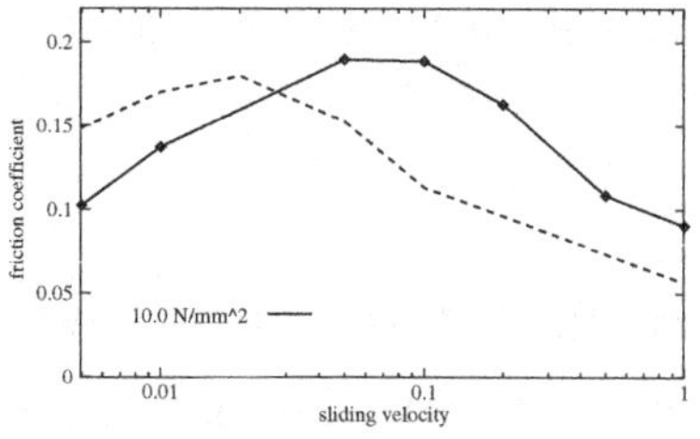

References

1. Klueppel M, Heinrich G (2000) Rubber Friction on Self-affine Road Tracks, Rubber Chemistry and Technology,73,p.578-606
2. Persson BNJ (2001) Theory of Rubber Friction and Contact Mechanics, Journal of Chemical Physics, Vol.115,8
3. Wriggers P (2002) Computational Contact Mechanics, Wiley

Micromechanics of internal friction stress softening and hysteresis of reinforced rubbers

J.F. Meier[1], M. Dämgen[1], and M. Klüppel[1]

Deutsches Institut für Kautschuktechnologie e.V., Hannover, Germany

A recently proposed micro-mechanical model for the stress-strain behavior of filled elastomers has been formulated on the basis of an extended tube model of rubber elasticity that permits a sound molecular-statistic description of bulk polymer networks up to large strains [1]-[4].

For the strain energy density function of the rubber matrix, we follow the approach of the extended tube model [1]-[4]. In the case of filled rubber, additionally the concept of hydrodynamic reinforcement is taken into account. A second contribution to the strain energy density function origins from the filler clusters which deform in an external stress field applied through the surrounding strained elastomer matrix. Therefore, the resulting filler related effects, cluster breakage and reaggregation are considered (fig. 1).

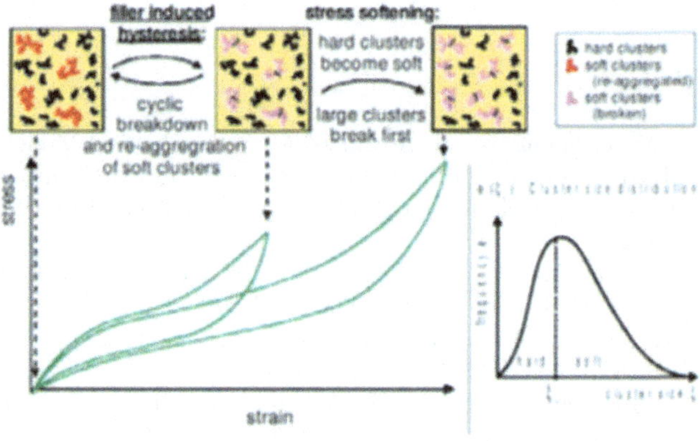

Fig. 1. Stress softening and hysteresis due to cyclic cluster breakage and reaggregation.

Applying this model on experimental stress strain data from uniaxial deformation test for different strain maxima gives reasonable values of the fitted physical parameters. The stress softening is found to occur substantially at strain values higher than 2-5 %, which is in good agreement with the phenomenological behavior of the Payne effect threshold.

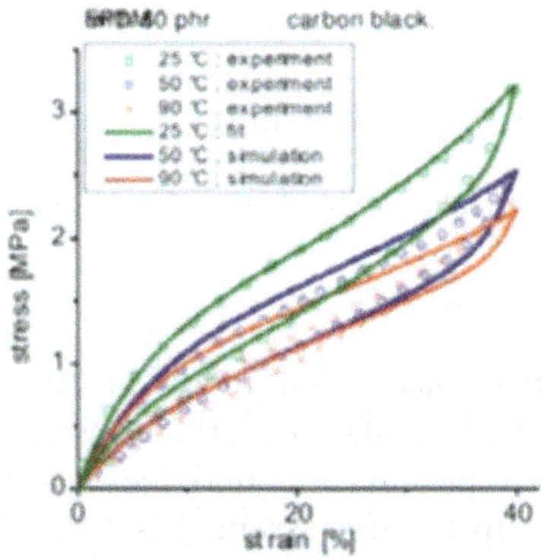

Fig. 2. Prediction of the temperature dependency.

As all the model parameters are of physical meaning, their temperature dependency can be studied separately. Here the activation energies of the tensile strengths of virgin and of damaged filler-filler bonds are acquired from separate dynamic mechanical mastercurves. This activation energies allow for good predictions of the temperature dependency (fig. 2).

References

1. Klüppel, M. & Schramm, J. "An Advanced Micro-Mechanical Model of Hyperelasticity and Stress Softening of Reinforced Rubbers" in Dorfmann, A. & Muhr, A. (eds.), "Constitutive Models for Rubber", A. A. Balkema Publishers, Rotterdam (1999) p. 211
2. M. Klüppel, J. F. Meier "Modeling of Soft Matter Viscoelasticity for FE-Applications"; in D. Besdo, R. H. Schuster, J. Ihlemann (eds.) "Constitutive Models for Rubber II", A. A. Balkema Publishers, Lisse (2001) p. 11
3. M. Klüppel, Adv. Polym. Sci. 164 (2003), p. 1
4. Heinrich, G., Straube, E. & Helmis, G. Adv. Polym. Sci. 85 (1988) p. 33
5. Y. Kantor, I. Webman, Phys. Rev. Lett. 52 (1984) p. 1891

Selected contact problems in human joints after arthroplasty

A. John[1], M. Mazdziarz[2], J. Rojek[2], J.J. Telega [2], and P. Maldyk[3]

[1] Silesian Technical University, Gliwice, Antoni.John@polsl.pl
[2] Institute of Fundamental Technological Research, Warsaw, mmazdz@ippt.gov.pl
[3] Institute of Rheumatology, Warsaw, Poland, kancelaria@ir.ids.pl

Abstract. A general anisotropic model of unilateral contact with adhesion and friction is proposed. The model is applied to the numerical analysis of contact between pelvis and acetabulum after arthroplasty.

1 Introduction

Clinical evidence confirms the role of not only friction but also of adhesion at the bone-implant interface. We have elaborated a general model of unilateral contact with friction and adhesion applicable to the study of such an interface. Numerical implementation has been performed using the finite element method preceded by a regularization of unilateral conditions. Numerical solutions pertain to the development of zone of adhesion loss. For a discussion of phenomena occurring at the bone-implant interface see [3, 4].

2 Numerical implementation & examples

The new contact formulation has been implemented within the dynamic explicit formulation in the finite element program SIMPACT [1]. The numerical algorithm for the calculation of contact interaction forces is based on the regularized contact conditions. Both normal and tangential contact constraints have been regularized using the penalty method. Thus the elastic behaviour has been introduced here for compression in normal contact, and the frictional contact has been reformulated as a problem analogous to that of elastoplasticity. For more details on the numerical algorithm see [5]. The explicit dynamic codes are best suited to carry out simulation of dynamic processes, nevertheless they can be applied to quasi-static simulation as well. We can obtain approximate solutions of linear and non-linear quasi-static problems introducing adequate damping which would damp out. First, we test the elaborated numerical algorithm for contact problem with adhesion and friction. Second, our discretized interface model is applied to the analysis of contact between pelvis and acetabulum after arthroplasty. The geometrical models described in [6] are applied to perform numerical analysis of the contact with friction and adhesion between pelvis and acetabulum after arthroplasty. Loosening is also modeled. Fixation of prosthesis is one of the most important mechanical aspects of the performance of the bone-implant system. Loosening can be caused by variety of mechanical and biological phenomena. We shall focus on the role of contact with adhesion in the fixation of implants.

3 Final remarks

The numerical analysis performed in Sect. 2 pertains to the case of isotropy of bone adhesion but anisotropic model can also be used. Also, our approach applies to any joint after arthroplasty and fixation with or without bone cement. Also, one may envisage applications to dental biomechanics. According to clinical data failure of the hip prosthesis is mainly caused by loosening of the acetabulum and not the stem. Surprisingly, in the relevant biomechanical literature not the acetabulum but stem loosening is most often analyzed. Hence the importance of study like the one performed in this paper. Our next goal is to take into account the influence of wear debris, on prostheses loosening. To this end the model proposed by Shillor et al. [2] will be generalized.

Acknowledgements

The authors were supported by the Ministry of Science and Information Technology through the grant No. 4 T11F 003 25.

References

1. Rojek J., Onate E., Postek E (1998)Application of explicit FE code to simulation of sheet and bulk metal forming processes. J. Mater. Proc. Technology 80-81:620-627
2. Shillor M., Sofonea M., Telega J.J. (2003) Analysis of viscoelastic contact with normal compliance, friction and wear diffusion. C.R. Mecanique 331: 395-400
3. Rojek J., Telega J.J. (1999) Numerical simulation of bone-implant systems using a more realistic model of contact interfaces with adhesion. J. Theor. Appl. Mech. 37:659-686
4. Rojek J., Telega J.J. (2001) Contact problems with friction, adhesion and wear in orthopaedic biomechanics. Part I: General developments. J. Theor. Appl. Mech. 39:655-677
5. Rojek J., Telega J.J., Stupkiewicz S. (2001) Contact problems with friction, adhesion and wear in orthopaedic biomechanics. Part II: Numerical implementation and application to implanted knee joints. J. Theor. Appl. Mech. 39:679-706
6. John A. (2004) Identification and analysis of geometrical and mechanical parameters of human pelvis bone Habilitation thesis, in Polish, Silesian Technical University

Optimal impulsive control of dynamical system in an impact phase

A.A. Galyaev[1], B.M. Miller[2], and E.Y. Rubinovich[3]

[1] Institute of Control Sciences, 65 Profsoyuznaya str., Moscow 117997, Russia, galaev@ipu.rssi.ru
[2] Institute for Information Transmission Problems, 19 B. Karetny, Moscow 127994, Russia, bmiller@iitp.ru
[3] Institute of Control Sciences, 65 Profsoyuznaya str., Moscow 117997, Russia, rubinvch@ipu.rssi.ru

Abstract. There are two approaches to the description of impact phase in impact mechanics, namely, *mathematical and physical* [1]. Both approaches give the same restitution law if external control is absent. However, the situation changes dramatically if one can apply an impulsive control during the phase of contact. Indeed, a real interaction force acts when a contact takes place only. The difference in these two models is as follows. In the case of visco-elastic media the mathematical approach implies an existence of interaction force acting during a whole period of an existence of body-obstacle deformation, which leads to appearance of a drawing force pulling the body into obstacle. In reality (in the physical model) the interaction force vanishes when a contact is lost, even if the deformation still presents. In this paper a control problem based on the physical model (which is more realistic) is considered.

In considering model it is supposed that a rigid body interacts with a visco-elastic Kelvin-Voigt media parameterized by its rigidity where both a control force acting on the body in a contact phase and a total impulse of this force are bounded during the impact period. Two different problems corresponding to maximum and minimum of post-impact velocity of the body are investigated. It is shown that in contrasted to the case of control force absence (in this case a body-obstacle contact deceases when a kinetic energy of the body achieves the first local maximum after beginning of the impact phase) in a controllable case the situation changes and a loss of contact is described by special terminal condition.

In the paper the optimal control is obtained for both types of criteria: maximization and minimization of post-impact velocity of the body. The corresponding restitution laws are derived when the obstacle rigidity tends to infinity.

1 Problem statement and main results

Let a perfect rigid body (a material point of a mass m) with initial velocity v collides with an obstacle arranged at $y \geq 0$. It is assumed that the obstacle possesses visco-elastic properties such as Kelvin-Voigt media. A force acting on the body is caused by both deformation of the obstacle and velocity of this deformation and parameterized by its rigidity

$$F_{el} = -2k\sqrt{\mu}\dot{y}(t) - a\mu y(t).$$

Here k and a are the coefficients characterizing visco and elastic properties respectively, μ is a rigidity, y is a deformation, $\dot{y}$ is a velocity of the obstacle deformation. It is also supposed that a controlled force F_{cnt}, applying to the surface of the obstacle, is parameterized by its rigidity, namely, $F_{cnt} = m\sqrt{\mu}u(t\sqrt{\mu})$, where m is a mass of the body, u is the control effect itself. Let $|u(t\sqrt{\mu})| \leq u_0$, i.e. the control action is restricted. Suppose that a contact begins at a zero moment and continues until the acceleration of the body is less or equal to zero. A motion equation of the body has the form :

$$m\ddot{y}(t) = F_{el} + F_{cnt}. \tag{1}$$

Eqn.(1) takes place if only there is the body-obstacle contact, i.e., when coordinates of the body and the obstacle surface coincide.

Remark 1 *There are two cases when the body-obstacle contact takes place.*

1. The force, acting from the obstacle on the body differs from zero, i.e., $F_{el} + F_{cnt} < 0$. Under this condition the body moves with the acceleration $\ddot{y} < 0$.

2. $F_{el} + F_{cnt} = 0$, but the deformation recovery rate coincides with the body's velocity.

Our aim is to find an admissible control effect that minimizes an absolute value of the body's velocity after the impact. Since this velocity is negative or equal to zero after the impact we need to minimize $-\dot{y}(\tau)$. Here $\tau = \sup\{t \geq 0 : \ddot{y}(t) < 0\}$ is an instant of the body-obstacle separation. Physically, it means that after τ an external action "resource" is not enough to keep the body-obstacle contact. It should be emphasize that after τ the body-obstacle contact desists and the body begins to move with constant and negative (or zero) velocity. To derive an optimal control law the space-time transformation technic [2] and Pontrjagin's maximum principle are used.

If we tend the rigidity to the infinity then the total impulse imparted to the body during the impact will be limited. So, it was ascertained that for kinetic energy minimization of a colliding body with immobile obstacle obtained visco-elastic properties the control force must acts oppositely to the body's velocity up to the moment when the velocity becomes equal to zero. After that moment the controlled force direction must be changed. On the contrary, to maximize the body's kinetic energy a direction of the controlled force must coincides with the body's velocity until the velocity becomes equal to zero and then the direction of the controlled force must be changed as well.

References

1. Kozlov V, Treshev D (1991) Billiards. (Genetic Introduction into Dynamics of Systems with Impacts). Moscow University, Moscow
2. Miller B, Bentsman J (2003) The singularity opening approach to control of mechanical systems with constraints. 2nd IFAC Workshop on Lagrangian and Hamiltonian Methods for Nonlinear Control, LHMNLC'03, Seville
3. Miller B, Rubinovich E (2003) Impulsive Control in Continuous and Discrete-Continuous Systems. Kluver Academic/Plenum Publishers, New York

Tribometric analysis of two tribo-materials with different contact geometries – critical reflection and simulation of the results

F. Grün[1], I. Gódor[2], B. Araujo[1], and W. Eichlseder[1]

[1] CD-Laboratory for Fatigue Analysis, florian.gruen@mu-leoben.at
[2] Department Product Engineering / Chair of Mechanical Engineering, University of Leoben, Franz-Josef-Straße 18, 8700 Leoben, Austria, fatigue@mu-leoben.at

1 Introduction

Tribometers can be used for preselecting tribo-materials in an early development stage. The test set-ups range from simple model tests to complex component tests [1]. All measured tribological variables are not only material-parameters but they also depend on the used tribosystem. Therefore the results of tribometer measurement often cannot be transferred to complex tribological systems or to components. The different test setups can be classified according to the *Stribeck*-Curve. Typical hydrodynamic running components like camshafts operate in the fully lubricated region of the curve. Tribological damage (like seizing) only occurs if the lubricant film fails, for example due to severe overload. These conditions can be reproducibly obtained by choosing the appropriate tribological model, which operates under boundary lubrication conditions. Two different metallic tribo-materials (Al-based and Cu-based) were investigated by using two different configurations (ball-on-disc, ring-on-disc).

The parameters of the ball-on-disc tests were as follows. The used steel balls had a diameter of $5mm$. The speed varied between 10 and $1250\frac{mm}{s}$. The normal load ranged from 0.25 to $5N$. The overlap ratio was $\ll 1$. The test duration was set to $2000rev$. The ring-on-disc set-up enabled an overlap ratio of one at macroscopic level. The diameter of the used specimen was 1". The speed varied between 500 and $2000rpm$. The load was applied stepwise (up to $1000N$).

2 Comparison of the different test geometries

To compare the test geometries the different local loadings in the contact zone were computed with numerical simulation methods and interpreted depending on the physical properties of the materials. Emphasis was put on the used elasto-plastic material model which was based on tensile tests and on a true stress-strain curve. In figure 1 the contact stresses of the ball-on-disc configurations are depicted [2]. With the help of surface analysis techniques the damage symptoms, which are displayed in wear-maps [3], were interpreted qualitatively and quantitatively. A subjective classification of the damage using light microscopy and REM-analyses was performed [4].

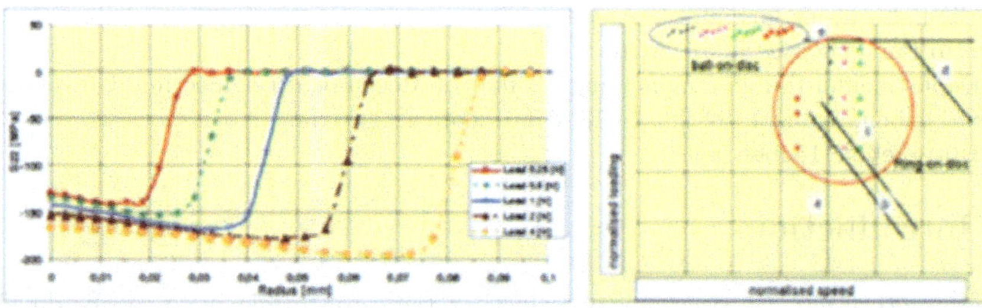

Fig. 1. Contact pressure at ball-on-disc loading, wear-map of the Al-alloy

3 Conclusions

Information on the mode of operation of the tribosystem can be obtained by means of tribometric tests. The contact geometry has an influence on the measured data. To simulate multi-phase materials appropriate models are necessary. The mixed lubrication condition can be investigated by tribometer tests with a higher resolution and reproducibility than component tests allow for. To sum up tribometric tests can be used in an early development phase since they enable a selective design of the materials, but they cannot substitute component tests. Recommendations for further developments are the combination of test engineering and simulation techniques.

References

1. Czichos H, Becker S, Lexow J (2003) Tribologie-Handbuch. Friedr. Vieweg & Sohn Verlag, Wiesbaden
2. Gruen F, Gódor I, Stoschka M, Eichlseder W (2004) Kritische Betrachtung von Tribometeranwendungen. Symposium ÖTG, Seibersdorf
3. Antoniou R, Subramanian C (1988) Wear mechanism map for aluminium alloys. Scripta Metallurg 22:809-814
4. Köberl H (2005) Vergleichende Tribometerversuche an metallischen Tribowerkstoffen. Diplomarbeit, Montanuniversität Leoben

Solvability and limit analysis of masonry bridges

G.E. Stavroulakis[1], G. Drosopoulos[1], M.E. Stavroulaki[2], Ch. Massalas[1], and A. Liolios[3]

[1] University of Ioannina, Greece, {gestavr@cc.uoi.gr, me01122@cc.uoi.gr
[2] Technical University of Crete, Chania, Greece, mstavr@mred.tuc.gr
[3] Democritus University of Thrace, Xanthi, Greece, liolios@civil.duth.gr

1 Unilateral analysis and limit load of masonry

We consider stone bridges with unilateral mortar joints and elastic or rigid stones [1], [2]. Opening or sliding indicates crack initiation and propagation. The ultimate load has been calculated by using a path - following technique. Lack of a solution at a certain level of loading indicates onset of failure, which can be checked by the solvability conditions of the corresponding variational or hemivariational inequality [1]. For validation and comparison the ultimate load is recalculated by the classical collapse mechanism method of Heyman, in a modern implementation with linear programming [3]. The theoretical difficulty of using upper bound style techniques, especially in the presence of friction (in general, nonassociated models), can be followed in [4].

1.1 A static problem. Modelling and comparison

From the application on the model of the Strathmashie bridge, in England, the following conclusions can be drawn. Identical failure mechanisms arise (see Figure 1). The value of the failure load converges to the value of the classical collapse mechanism after a number of 30-40 interfaces equidistangly distributed along the span of the bridge (see comparison in Table 1). The proposed model has advantages when sliding mode failure occurs, usually in low bridges.

1.2 Dynamic analysis

A similar model for the Plaka stone bridge in Epirus, Greece under seismic base excitation has been done. Solvability has to be checked within each time step in the dynamic analysis. Comparison of displacements along the intrados of the arch for various assumptions (Model I, II: interfaces, Model III: with fil, Model IV: reference linear elastic) at a typical time step are shown in Figure 2.

Fig. 1. (a) Four and (b) five hinge collapse mechanisms, for quarter span and middle span loading, respectively.

Table 1. Failure loads of unilateral and mechanism models, for a quarter-span and a middle-span loading.

Loading	Unilateral model	classical mechanism
Quarter span	87.14 kN	88.14 kN
Middle span	168.95 kN	174.83 kN

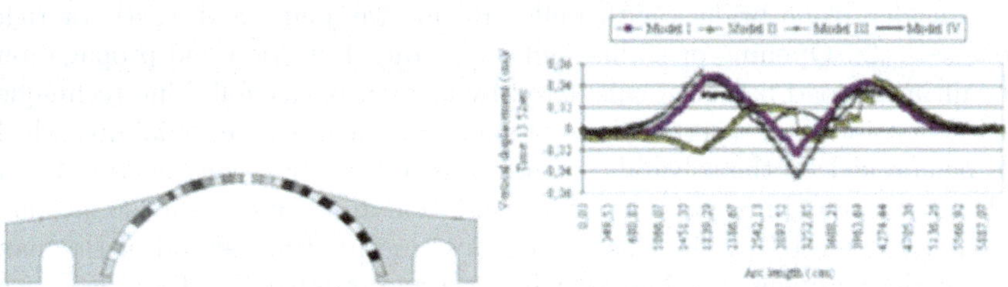

Fig. 2. (a) Plaka unilateral stone bridge model and (b) typical results.

References

1. Panagiotopoulos P. D., *Inequality problems in mechanics and applications. Convex and Nonconvex Energy Functions*, Birkhäuser, Basel 1985
2. Leftheris B. P., Stavroulaki M. E., Sapounaki A. K., Stavroulakis G. E., *Computational methods for heritage structures*, WIT Press, Southampton, U.K, 2005
3. Melbourne C., Gilbert M., *The behaviour of multiring brickwork arch bridges* , The Structural Engineer, 1995, 73, 39-47
4. Ferris M.C., Tin-Loi F., *Limit analysis of frictional block assemblies as a mathematical program with complementarity constraints* , Int. Jnl. Mech. Sci., 2001, 43, 209-224

On the treatment of inelastic material behavior in an ALE-description of rolling contact

M. Ziefle and U. Nackenhorst

Institute of Mechanics and Computational Mechanics, University of Hannover, ziefle@ibnm.uni-hannover.de, nackenhorst@ibnm.uni-hannover.de

Abstract. In FEM analysis of rolling contact problems Arbitrary Lagrangian-Eulerian (ALE) methods are the state of the art. These methods allow mesh refinements concentrated to the contact region and offer a time independent formulation of stationary elastic rolling. In the case of inelastic material behavior history dependent constitutive equations contain convective terms in this description. The treatment of these convective terms is performed by a time-discontinuous Galerkin (TDG) method. For the demonstration of the practicability of the developped algorithms a viscoelastic computation of a typical tire model is discussed.

1 ALE-formulation of rolling contact

Inserting an arbitrary moved reference configuration between the referential and the current configuration leads to a decomposition of motion into the rigid body rotation (Eulerian) and the pure deformation (Lagrangian). Discretizing and linearizing the weak form of the equilibrium equation

$$\operatorname{Div}\mathbf{P} + \rho\,\mathbf{b} - \rho\,\dot{\mathbf{v}} = \mathbf{0} \tag{1}$$

leads to the incremental matrix representation of the ALE-FE-formulation of rolling problems

$$\left[{}^{t}\mathbf{K} - \mathbf{W}\right]\,\left[\Delta\,\mathbf{u}\right] = \left[{}^{t+\Delta\,t}\mathbf{f}_e +^{t}\mathbf{f}_i -^{t}\mathbf{f}_\sigma\right]. \tag{2}$$

Herein, ${}^{t}\mathbf{K}$ is the tangential stiffness matrix, $\mathbf{W}$ the ALE-inertial matrix, ${}^{t+\Delta\,t}\mathbf{f}_e$ are the external forces, ${}^{t}\mathbf{f}_i$ the ALE-inertial forces and ${}^{t}\mathbf{f}_\sigma$ the internal forces. For a detailed description of the used ALE-formulation of rolling contact is referred to [1].

2 Treatment of convective terms

the relative-kinematic description of rolling leads to a relative motion between the fixed finite element mesh and the material points. Therefore the internal variables $\boldsymbol{\alpha}$ have to be transported on the streamlines of the material particles through the mesh by solving the advection equation

$$\frac{\partial \boldsymbol{\alpha}}{\partial t} + \mathbf{v} \cdot \nabla_x \boldsymbol{\alpha} = \mathbf{0}. \tag{3}$$

This hyperbolic partial differential equation is solved by a space-time finite element method based on a time-discontinuous Galerkin (TDG) method [2]. These TDG-methods allow jumps or discontinuities on the time intervall boundaries and present a very good and stable numerical advection solution. For the discretisation in time and in space Lagrangian polynomials are choosen. The order of the time integration scheme depends strongly on the order of the time polynomials. The solution procedure is embedded into the ALE-rolling contact computation by using a so called *fractional step-method.* In a first step the mechanical equilibrium of the nonlinear contact problem is estimated and the convective terms are neglected. In a following step the convective terms are considered and the computed inelastic state has to be advected through the mesh [3].

3 Examples

The rolling tire model in Fig.1a has been analyzed with the assumption of frictionless contact. Using a finite linear viscoelastic constitutive model leads to an unsymmetric contact pressure distribution. This effect is explaned by the relaxation of the particles in the contact zone and results in a frequency dependent rolling resistance torque in Fig.1b.

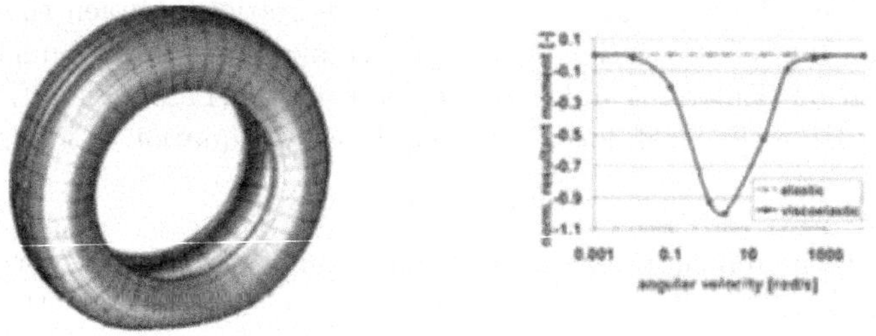

Fig. 1. a) Finite element tire model, b) Rolling resistance torque.

References

1. Nackenhorst, U. (2004) The ALE-formulation of bodies in rolling contact, Comp. Meth. Appl. Mech. Eng. 193
2. Cockburn, B., G.E. Karniadakis, Shu, C.-W. (1999) The development of discontinuous galerkin methods. In: Cockburn, B., G.E. Karniadakis, Shu, C.-W. (eds) Discontinuous Galerkin Methods. Springer
3. Nackenhorst, U., Ziefle, M. (2005) Finite Element Modelling of Rolling Tires, Kautschuk, Gummi, Kunststoffe 6/2005

A formulation to define the contact surface in the 2D mortar finite element method

M. Tur[1], F.J. Fuenmayor[1], and P. Wriggers[2]

[1] Dept. Ingenier´i a Mecánica y Materiales, Univ. Politécnica de Valencia, Camino de Vera s/n 46022, Valencia, Spain, matuva@mcm.upv.es, ffuenmay@mcm.upv.es
[2] Institut für Baumechanik und Numerische Mechanik, Universität Hannover, Appelstrasse 9A 30167, Hannover, Germany, wriggers@ibnm.uni-hannover.de

Abstract. A new implementation of the mortar method is proposed to solve 2D contact problems. The size of the contact zone can be varied continuously by means of the definition of an intermediate contact surface. It has been proved that this is equivalent to using special shape functions for the interpolation of the contact pressure. A formulation is proposed to find the correct size of the contact zone that minimizes the lack of smoothness of the contact pressure.

1 Motivation

Fretting fatigue contact problems are usually modelled using the finite element method. In this kind of contact problems the evolution of the stress distribution near the contact zone has to be obtained with a high accuracy. Since the elastic bodies in contact can be under global sliding condition, in general, the meshes are non-conforming in the contact zone.

The classical node-to-segment approach has not been demonstrated to have an optimal convergence to the exact solution and it is not able to pass the contact patch test.

On the other hand, the mortar finite element method has been successfully applied to solve contact problems using non-conforming discretizations of the bodies in contact [1, 2, 3]. Unlike the classical node-to-segment approach, the mortar method satisfies Babuska-Brezzi stability conditions and it permits to pass the contact patch test for planar surfaces. Also it has been demonstrated to have an optimal convergence rate [4].

2 Definition of the contact surface

The standard mortar formulations may present some problems like oscillations in the contact pressure for discretizations with coarse meshes. This problem appears in incomplete contact problems, where the size of the contact zone depends on the transmitted loads. In the usual implementations of the mortar method, the size of the contact zone is changed when a node in the surface of the body is added to the contact zone (the node is closed) or it is removed from the contact zone (the node is open).

In this work, a new implementation of the mortar method is proposed to solve 2D contact problems that avoids oscillations in the contact pressure. In this implementation the size of the contact zone can be varied continuously by means of the definition of an intermediate contact surface. It has been proved that this is equivalent to use special shape functions for the interpolation of the contact pressure.

Two methods are analyzed to define the size of the contact surface. The first is the minimization of a function that measures the non-smoothness of the contact pressure. The other is the extrapolation (linear or quadratic) from the values of the contact pressure at the end of the contact. Also, an extrapolation method to define the transition between the slip and slip zones is proposed and it is used to solve frictional contact problems.

The proposed implementation uses the Lagrange multiplier approach to impose the contact conditions. Also a Newton-Raphson method is used to solve the non-linear problem together with and active-set strategy. A numerical integration is performed in order to evaluate the contact integrals.

A number of numerical examples with closed analytical solutions are solved using the proposed finite element formulations and the results are compared with th1ose obtained via other formulations like node-to-segment approach or the standard mortar method. The results demonstrate the performance of the proposed formulation.

Ackowledgements

Manuel Tur Valiente would like to acknowledge Universidad Politécnica de Valencia for the financial support received from the Vicerrectorado de Investigación and from Ministerio de Educación y Ciencia by means of the project DPI2004-07782-C02-02.

References

1. McDevitt T W, Laursen T A (2000) A mortar-finite element formulation for frictional contact problems, Int J Numer Meth Engrg 48:1525–1547
2. Yang B, Laursen T A, Meng X (2005) Two dimensional mortar methods for large deformation frictional sliding, Int J Numer Meth Engrg 62:1183–1225
3. Fischer K A, Wriggers P (2005) Frictionless 2D contact formulations for finite deformations based on the mortar method, Comput Mechs 36:226–244
4. Hild P (2000) Numerical implementation of two nonconforming finite element methods for unilateral contact, Comput Meth Appl Mech Engrg 184:99–123

Index

The manufacturer's authorised representative in the EU is Springer Nature Customer Service Centre GmbH, Europaplatz 3, 69115 Heidelberg, Germany. If you have any concerns regarding our products, please contact ProductSafety@springernature.com

Printed and bound by CPI Group (UK) Ltd, Croydon, CR0 4YY
07/01/2026
02030963-0010